Ancient Myths,
Ancient Wisdom

Ancient Myths, Ancient Wisdom:

Recovering humanity's
forgotten inheritance
through
Celestial Mythology

David Warner Mathisen

Published by Beowulf Books, Paso Robles, California

Mathisen, David Warner.

Ancient Myths, Ancient Wisdom / David Warner Mathisen. - 1st ed.

1. Mythology. 2. Astronomy. 3. Spirituality.

ISBN 978-0-9960590-6-0

Dedicated to

the choice of Wisdom
in order to help others
and restore what is right

and to
that Goddess
who always came to the aid of Odysseus
and helped him to defeat the Suitors

and who
also assisted Perseus
in his confrontation with the Gorgons

*May we, individually and collectively,
also choose Wisdom
and
listen to that Goddess,
and realize that with her help
we can overcome the challenges we face today,
though they seem as daunting as the
Suitors or the Gorgons*

SECTIONS OF THIS COLLECTION

Introduction

This book is a collection of selections from essays published on my blog, *The Mathisen Corollary* (also available on my newer website, *Star Myths of the World*), exploring the connection between the world's ancient myths and the heavenly landscape of the constellations and galaxies, through which move the sun, the moon, and the visible planets.

The blog started in April of 2011, and by the fall of 2017 is now nearing its first thousandth post. The essays included here are selected from those first one thousand blog posts, organized by general subject and cross-referenced by page number (so that if one essay refers to something in another essay, the page number will be included, similar to the links within the original posts).

During the earliest history of the blog, many posts were devoted to the subject of geology, and the overwhelming amount of evidence that our planet has suffered a major catastrophe in its past. While this is an extremely important topic, and related to the material that later became the main focus of most discussion (the connections between the stars and the myths), most of the material included here focuses upon the subject of "celestial mythology" -- the evidence that the sacred stories preserved by virtually every culture around the world, on every inhabited continent and island, appear to be built upon a common and very ancient system of celestial metaphor.

This system can already be seen to be in operation and fully mature in the earliest human texts to have survived the ravages of the millennia -- such as the Pyramid Texts of ancient Egypt, or the hardened clay tablets containing the Gilgamesh cycle from the cultures of ancient Mesopotamia. There are also hints (discussed in some of the essays contained within) that this system was in operation in ancient Japan, ancient China, and even perhaps in the extremely ancient and very mysterious Indus-Saraswati Valley civilizations, of which the ancient city-complex of Mohenjo Daro was a part.

Thus this ancient system of celestial metaphor appears (perhaps) to be a legacy of some even more ancient culture or civilization, of which conventional academia knows (or at least admits) little or nothing. It is possible that this system comes from the time of the cultures responsible for even more ancient monuments, of which the buried stone circles of Göbekli Tepe are an example -- cultures so ancient that they predated the earliest dynasties of ancient Egypt by at least as many millennia as those early dynasties of Egypt predate our own time.

This evidence in the world's ancient myths thus points to the possibility that some catastrophe (or catastrophes) destroyed a culture or civilization of great antiquity, of which the later ancient civilizations (and their myths) were but a much later echo -- perhaps even a much later "re-start" after a long period of centuries in hiding, perhaps even physically underground, due to some catastrophe of unimaginable magnitude.

For this reason, the evidence in earth's geology (and in other parts of our solar system) which points to a cataclysm (or cataclysms) in our planet's history is extremely relevant to the entire story of the myths. However, because much of my earlier examination of the geology of our planet was done during a time when I still took the texts of the Bible (and thus the description of the Genesis Flood) literally, and because I have now found abundant evidence to suggest that the Genesis Flood texts can be shown to be built upon celestial metaphor, and thus to describe a *celestial* flood -- whatever the history of cataclysmic flooding on our planet, and there is certainly plenty of evidence for it -- in this volume the blog posts selected will not focus so much on the geological discussions but rather the celestial aspects of the myths, and what they might mean for us today.

The broad subject groupings are used in order to make this volume a little more coherent, but of course all of these topics connect and intertwine, which means that the dividing lines between the various sections are by no means sharp and distinct, and thus in some cases an essay's inclusion in the subject in which

2

you find it could be debated. The general progression of subjects begins with the mechanics of the heavens themselves – what causes the heavenly phenomena we observe from our vantage point on earth, including the motions of the moon and visible planets and of course the constellations – followed by a dive into the connections between these heavenly phenomena and the ancient wisdom preserved in the myths and traditions found in different cultures around the globe, and through the millennia, and concluding with essays exploring the wide-ranging ramifications of all of this evidence.

The posts within each section follow chronological order from their original publication date. They have only been lightly edited (in order to change links into page-number references, for example, or to remove a reference to an embedded video which for obvious reasons cannot be included in a book format). Other than those types of edits, the content of the post is as published on the date indicated – and because there has been some evolution of my thought on certain subjects over the years, some changes may be detected as the reader proceeds through each section.

I feel that some of my best writing is done on the blog – it is a medium and a format which lends itself to essays, and writing them may have a "more relaxed" feel than the act of writing a book. In order to preserve those writings in a format that is not dependent upon electricity or upon access to the world-wide web, selections from the blog have been forced out of the world of datacenters and servers, where those posts have been stored in digital form, and pressed into the pages of the analog volume you now hold in your hands.

The format of a book has certain advantages as well. Beyond the obvious fact that its contents cannot be altered or erased or restricted the way something in digital form can be, it is still much faster to flip through the pages of a physical book in order to peruse hundreds of different pieces of writing. It is true that the "search" function can help one navigate rapidly to any part of the online version of the blog, but searching requires knowing at least something about what you are looking for. Browsing through several hundred past blog posts written over the course of more than 72 months in an online format is much more time consuming and tedious than is flipping through the pages of a physical book.

In addition, there is an index at the back, to help replicate the "search" function which is one of the admitted strengths of the digital format of the blog. When one essay references another in this collection, **a page-number will be indicated using superscript**, which thus functions as a sort of "link" to that other essay, allowing you to turn to that page if you want to see the essay being referenced. If an essay references a post that is not included in this collection, then no superscript number will be indicated, but you can of course look for the title of the post online and find it there.

I hope you will enjoy these selections from the first 1,000 posts, and that their message will in some way be a blessing to you in this incarnate life.

Paso Robles, California
Octber, 2017

Again: as you read through the text of this book, when you see a little number in "superscript" (like the number 331 right here[331]), that number refers to a **page number** to which you may turn for the essay which is being referenced in the text (similar to a "link").

Esotericism and the Ancient System

Know Thyself

2014 March 29

In an important dialogue by Plato known as the *Phaedrus*, the discussion examines the subject of self-knowledge, the meaning of the concept of "knowing oneself," and the role of love in that quest for self-knowledge.

The command to "know thyself" was famously said to have been inscribed upon the temple at Delphi, and Plato has Socrates refer somewhat ironically to this famous dictum early in the *Phaedrus*. As Socrates and Phaedrus are walking along the path of the stream of the Ilissus, Phaedrus asks Socrates whether he was correct in deducing that it was "somewhere about here that they say Boreas seized Orithyia from the river" (referring to a famously beautiful daughter of a legendary king of Athens, who was seized by the god of the north wind, Boreas, and carried away to be his bride, becoming the mother of two of the heroes who sailed on the *Argos* in search of the Golden Fleece -- the incident is described by the later Roman poet Ovid in *Metamorphoses* Book VI, lines 979 through 1038).

Socrates says he believes the abduction took place about a quarter of a mile lower down, and not where the two are currently walking. Phaedrus then asks Socrates whether he believes the story to be true.

Plato has Socrates reply with a wonderful passage in which Socrates says he would be "quite in the fashion" if he disbelieved the tale, and if he came up with some kind of rationalistic explanation for the mythological story, such as if he were to soberly explain that the myth originated when the maiden was blown by a gust of wind over the edge of some steep rocks to her death (quotations from the *Phaedrus* used in this discussion come from the translation of Reginald Hackforth, 1887 - 1957).

Socrates then goes on to say that such theories are "no doubt attractive" but are merely the "invention of clever, industrious people who are not exactly to be envied," (a masterful example of "damning with faint praise") -- in other words, that those pedantic scholars who spend their time trying to reduce mythological stories to literal episodes from some imagined history are completely misguided, and that those who indulge in manufacturing such theories deserve more to be pitied than to be taken seriously.

Plato then has Socrates declare of those who want to reduce every myth to some kind of historical, literal episode:

> If our skeptic, with his somewhat crude science, means to reduce every one of them to the standard of probability, he'll need a deal of time for it. I myself have certainly no time for the business, and I'll tell you why, my friend. I can't as yet 'know myself,' as the inscription at Delphi enjoins, and so long as that ignorance remains it seems ridiculous to inquire into extraneous matters. Consequently I don't bother

about such things, but accept the current beliefs about them, and direct my inquiries, as I have just said, rather to myself, to discover whether I really am a more complex creature and more puffed up with pride than Typhon, or a simpler, gentler being whom heaven has blessed with a quiet, un-Typhonic nature. By the way, isn't this the tree we were making for? 229e - 230b.

It is undoubtedly no accident that Plato has Socrates refer to the inscription from the temple at Delphi at this particular point in the dialogue, nor that Socrates illustrates his ongoing quest to obey that dictum with a reference to a mythological being (Typhon, and the question of whether or not he, Socrates, is "more puffed up with pride than Typhon").

Through Socrates, Plato is here clearly slamming those who completely miss the point of the "ancient treasure" of mythology, and telling us in no uncertain terms that the purpose of the myths is not to preserve some historical, literal event from the past (albeit in slightly exaggerated form, with a girl falling to her death from some rocks transformed into a beautiful maiden being abducted by the god of the bitter north wind), but rather that the purpose of the myths has to do with the Delphic inscription "KNOW THYSELF." To drive the point home, Plato has Socrates illustrate by telling Phaedrus that he himself applies the myth of Typhon to his own examination of himself, and the danger of becoming "puffed up with pride" (like Typhon).

This little passage from the *Phaedrus*, it seems, sheds some extremely helpful light on the famous dictum from Delphi. It reveals that, far from being a mere collection of fanciful tales, or even a compendium of ancient historical events embellished with touches of the fabulous, the sacred myth-traditions of the world were actually an exquisite set of instruments designed to facilitate the quest for self-knowledge, and the removal of the ignorance which Socrates says should be the primary object towards which we devote our time and energy.

But how, exactly, do the sacred mythologies enable us to emerge from our state of ignorance into greater self-knowledge?

As the conversation in the *Phaedrus* moves on from the above passage, it plunges first into a discussion of the nature of love, and then proceeds from there into a discussion of the soul and its incarnation. In 245c - 245e of the dialogue, Socrates determines from his foregoing examination of love that the soul is immortal, that it comes into a body and "*besouls*" the body, and that (at the beginning of section 246), that "it must follow that soul is not born and does not die."

This, in fact, is *precisely* what the ancient mythologies of the world teach us, using an exquisite system of metaphor, according to the penetrating analysis of Alvin Boyd Kuhn, in works such as his *Lost Light* (1940). Through their beautiful allegories, the myths are teaching us just what Plato has Socrates expounding in the *Phaedrus*: that soul is immortal, that we descend into the body only to rise up again into the world of spirit, and descend into the body again, as many times as necessary to obtain the *gnosis* (and overcome the ignorance) that Socrates and the inscription at Delphi are talking about.

In fact, we could tentatively explicate the myth of Orithyia being seized by the wind-god from the river as a metaphorical depiction of the aspect of the soul's journey when it leaves the world of the incarnation (the river, or the body -- the body being composed largely of water and minerals, the lower elements) and returns again to the realm of the spirit (the realms of air and fire, the higher elements or those more illustrative of the spiritual sphere).

The later philosopher (and priest of the oracle at Delphi) Plutarch, in his own dialogue examining the meaning of the inscriptions at Delphi (including the mysterious inscription of the letter "E" at Delphi, which is a subject for another discussion at another time), certainly seems to hint at the same interpretation. In his famous essay *On the 'E' at Delphi*, Plutarch puts these words into the mouth of his own mentor, Ammonius (beginning in section XVII and carrying on into section XVIII and XIX):

> All mortal nature is in a middle state between becoming and perishing, and presents but an appearance, a faint unstable image, of itself. If you strain the intellect, and wish to grasp this, it is as with water; compress it too much and force it violently into one space as it tries to flow through, and you destroy the enveloping substance. [. . .] "It is impossible to go into the same river twice," said Heraclitus; no more can you grasp mortal being twice, so as to hold it. So sharp and so swift its change; it scatters and brings together again, nay not again, no nor afterwards; even while it is being formed it fails, it approaches, and it is gone. Hence becoming never ends in being, for the process never leaves off, or is stayed. [. . .] Yet we fear (how absurdly!) a single death, we who have died so many deaths, and yet are dying. For it is not only that, as Heraclitus would say, "death of fire is birth of air," and "death of air is birth of water"; the thing is much clearer in our own selves. [. . .] What then really is? That which is eternal, was never brought into being, is never destroyed, to which no time ever brings change."

This concept is closely related to the discussion posted over a year ago concerning the myth of Narcissus, a discussion which helps to outline the importance of the concept of love in this whole discussion (the concept of love being the springboard in the *Phaedrus* which Plato uses to launch into his examination of this topic). In a post examining some of the assertions of the neoplatonic philosopher Plotinus entitled "Plotinus and the upward way,"[463] we saw that

> Plotinus seems to teach that love of beauty is an entry-gate to the upward way, but that the "lesson" for the lover of beauty is to learn to disentangle from being enamored with one specific embodied form (whatever form that lover of beauty is enamored with) and to see that specific form of beauty as a pointer to "beauty everywhere" (this being the

very opposite of Narcissus, who could only see beauty in himself), and ultimately to the "One Principle underlying all."

Again, this conclusion has strong resonances with the theme of Plato's *Phaedrus*.

As we begin to wrap up this examination, we might pause on the myth-metaphor of Narcissus, another figure who (like Orithyia) is pictured next to an enchanting body of water. As we saw in that previous examination of Narcissus,[463] certain ancient philosophers appear to have interpreted his myth as symbolic of the descent of the soul into this incarnational world, and his fate as a warning against certain tendencies (perhaps even tendencies related to those which Socrates examined himself for, when he referenced the puffed-up self-pride of Typhon). Socrates would surely laugh at us and imply that we were wasting our time if we were to try to go on a scholarly quest to uncover the "historical Narcissus" and to identify some particularly handsome or vain young prince from history who might have inspired the "legend of Narcissus." Such stories are intended to provide us with a tool for self-reflection and ultimately self-knowledge, knowledge about the human condition and our purpose in this life (or this incarnation, if you believe the interpretation that the ancients and Alvin Boyd Kuhn espouse).

If the famous command from the oracle at Delphi to "Know thyself" was intended to tell us to learn that (in Plato's words) our physical existence is temporary and that in reality, "soul is not born and does not die," and that (in Plutarch's words) "we fear (how absurdly!) a single death, we who have died so many deaths," then it follows that those who -- either mistakenly, or malevolently -- try to reduce the myths to literal or historical interpretations are doing the world a great disservice. They are placing a tremendous obstacle in the path of those who would learn the truth about the human condition, knowledge which is essential in the pursuit of that Delphic command.

11

Unfortunately, such "clever, industrious people who are not exactly to be envied" are perhaps even more prevalent in our day than they seem to have been in the time of Plato, Socrates, and Phaedrus.

Wax on, wax off

2014 May 02

The famous scene from the first *Karate Kid* movie (1984), in which Mr. Miyagi reveals to his student (whom he always refers to as "Daniel-San") the hidden meaning behind all of the hours of chores Mr. Miyagi has been assigning to him is perhaps one of the clearest and most-accessible examples of the concept known as "the esoteric" that can be found in popular culture.

Just about everyone has heard of "wax on, wax off," for the simple reason that it is a profoundly memorable and even moving scene, even thirty years after it was filmed. In this case, waxing the car and all the other tasks (which Mr. Miyagi insisted must be done in a very strict and precise manner) was a way to teach something else: the "hidden" or "esoteric" meaning that lies behind the apparently mundane action of "wax on, wax off."

Although the esoteric is often defined as a "hidden meaning," note that Mr. Miyagi did not select his powerful teaching method in order to *deceive* Danny, or even in order to conceal something from Danny. Mr. Miyagi taught his student that way because he knew that it was the best way to reach Daniel-San, and to convey something on a deep level to Danny's mind -- something that might have been difficult or even impossible to convey in any other way.

Note carefully in the scene that even when Mr. Miyagi finally reveals the esoteric meaning behind the mundane tasks, he does not do so by *explaining* them to Daniel-San: he forces Daniel-San to actually *experience* how it works, so that Danny knows what it feels like when it works. This is a very different approach than trying to appeal to the rational, intellectual, "left brain" part of our thinking.

If Mr. Miyagi, instead of having Danny spend days waxing the cars, sanding the floor, painting the fence, and painting the house, had tried to explain to his student what to do in order to stop a punch or a kick, then Danny would have had lots of questions as his "intellectual mind" tried to make sense of what Mr. Miyagi was telling him -- and Danny might (or might not) have *believed* that what Mr. Miyagi was telling him would work. This is a very different approach, and for some very important things it is not at all the best approach (not even a good approach). *Believing* something with your mind is very different from *knowing* it because you experience it for yourself, the way Danny did when Mr. Miyagi started to throw punches and kicks at him while yelling fiercely!

As it turns out, there are certain concepts which are best conveyed to our mind esoterically -- and that, in fact, are difficult to properly grasp through any other method *besides* the esoteric. The martial arts, in fact, are almost always taught esoterically (usually through forms which contain hidden applications that can be appreciated only after their motions are internalized). For a variety of reasons, the sages responsible for the ancient sacred scriptures and mythology systems of the world also chose to convey the ancient wisdom of mankind esoterically, through allegorical stories which could enable the mind to grasp truths which the "intellectual mind" would choke on.

The ancient myth systems of the world were not designed to create *belief*, any more than Mr. Miyagi wasted any thought asking Daniel-San in the famous scene whether or not he believed in the blocks and hand motions and whether they would be effective. They were designed to create the experience of *knowing* -- which in the ancient Greek was called *gnosis*.

My new book, *The Undying Stars*, demonstrates that the ancient scriptures of the world operate in exactly the same way -- esoterically. They are all a form of "wax on, wax off" which contain an amazing esoteric message that is "hidden" inside. This includes the stories of the gods of ancient Greece, of ancient Egypt, of ancient India, of the Norse, of the Hawaiians and the Maori and the Maya and the Inca and the Native American tribes and nations, and of almost every other culture around the world --

14

and it also includes the stories found in the Old and New Testaments of the Bible.

The masters who created these stories were not trying to deceive us, any more than Mr. Miyagi was trying to deceive Daniel-San by showing him how to wax the car or paint the fence -- and they were not trying to keep these truths "hidden" so that nobody could ever learn them. Quite the contrary: the ancient scriptures and sacred traditions of the world were intended to lead men and women to consciousness, and to awareness of the truth about the human condition -- in fact, to gnosis.

But something most unfortunate for the human race happened along the way: for reasons of their own, a powerful group of families decided to suppress the ancient understanding and teach that they were *not* intended to be understood esoterically. They argued that the ancient scriptures were intended to be understood literally first and foremost: that we were to understand the ancient stories as describing the adventures of literal and historical personages or beings, and not as metaphorical carriers of profound esoteric truths applicable to every single man and woman. This approach can be thought of as the equivalent of telling Danny that his efforts of waxing the car or painting the fence have absolutely nothing to do with learning a martial art -- that there is no connection to learning a martial art, and that he should just focus on waxing the car and painting the fence, because that is all that there is to know about the subject.

To be more direct, they taught people to accept the existence of a literal, historical individual called Christ rather than (as Paul says) "Christ in you" (an esoteric truth conveyed by the allegorical stories). The original gnostic teaching of the stories of the New Testament (and the Old Testament as well) were all designed for the purpose of conveying esoteric truth so that men and women could achieve consciousness, overcome illusion and mind control, and experience gnosis. The evidence suggests that those who wanted to suppress the esoteric truths constructed a literalist

15

religion (built primarily upon *belief* rather than upon *gnosis*) and they married it to the military and economic might of the Roman Empire, creating an extremely powerful system of mental tyranny: using belief and illusion and propaganda and mind control, backed up by the violence.

The Undying Stars examines the evidence that these enemies of human consciousness and gnosis began a campaign to ensure that those teachers and texts imparting an esoteric and gnostic approach and opposing the literalist subversion were marginalized, burned, buried, or otherwise silenced. Then, they expanded that campaign over the next seventeen centuries into the rest of the world, to eradicate the esoteric and shamanic teachings that had survived outside the boundaries of the Roman Empire.

Many readers, especially those with deep attachment to the Biblical scriptures or to the teachings of those who say that these scriptures are first and foremost to be understood literally, will no doubt be profoundly dismayed or even angered at such an assertion, and may decide they want nothing to do with any books that examine the evidence supporting such a view. To them, I would say that I do not wish to insult or offend anyone's sensibilities -- I am not an "authority" and I am only offering the results and insights that I have experienced in my own personal walk. There is a "note of caution to literalist readers" at the front of the book, explaining that the book itself examines evidence and arrives at conclusions which may be extremely damaging to the foundations of literalist belief, and some readers may prefer not to examine that evidence rather than threaten a paradigm in which they have significant personal and psychic investment.

Some readers, however, may decide that if the books arguments and analyses are in error, they can be safely read and then rejected, but if they are correct then it might be preferable to know it. The good news for literalist readers is that they likely are very familiar with the ancient scriptures, and this actually gives them a wonderful advantage: they have been "waxing the car" for

16

years and years, even if no one has ever shown them how it relates to actually doing martial arts!

The fact is that the descendants of those enemies of human consciousness, who conspired to steal the esoteric teaching from humanity and get everyone focused on "waxing the car," were most likely careful to keep the ability to "do karate" for themselves! If so, then those enemies of mankind could very well still be "doing karate" today against the rest of us, and getting away with it, because most of the world only knows how to "wax the car," without understanding the meaning behind the motions.

Like a finger, pointing a way to the moon . . .

2014 May 05

In a famous episode from the film *Enter the Dragon* (1973), Bruce Lee famously explains an aspect of philosophy in action to his young student:

> "It is like a finger, pointing a way to the moon . . .
> Don't concentrate on the finger, or you will miss all that heavenly glory!"

In doing so, the groundbreaking film brought into the popular awareness an ancient principle which was recorded in writing at least as early as the inscription of the text of the Shurangama Sutra, which according to tradition was translated into Chinese in AD 705 from an ancient Indian *sutra* or scripture (sutras are writings, as opposed to other sacred teachings which were not written down but memorized and passed verbally from generation to generation).

As discussed in a presentation by Professor Ron Epstein, published in 1976, there is some controversy over whether or not the Shurangama Sutra is actually a translation of an older sutra or whether it was actually created by the minister Fang Yung, who lived during the period that it was supposedly translated into Chinese. In any event, because records regarding the authenticity of the Shurangama Sutra exist from as early as AD 754, we know that it was in existence by at least that year, and probably before. Further, whether it was originally penned by Fang Yung or was in fact a translation or at least an adaptation of earlier scriptures, its principles resonate with teachings that are much older, and it became a very influential text in Ch'an Buddhism in China (which is philosophically related to Zen Buddhism in Japan -- both the Chinese word Ch'an and the Japanese word Zen are probably linguistically related to the Sanskrit word *dhyana*).

18

You can read a translation of the Shurangama Sutra for yourself at various places on the web, although the version originally cited when this essay was first published is no longer on the web. That version translated the text containing the finger and moon reference as follows:

> This is like a man pointing a finger at the moon to show it to others who should follow the direction of the finger to look at the moon. If they look at the finger and mistake it for the moon, they lose both the moon and the finger. Why? Because the bright moon is actually pointed at; they lose sight of the finger and fail to distinguish between brightness and darkness. Why? Because they mistake the finger for the bright moon and are not clear about brightness and darkness.

Whatever other deep matters this passage is illuminating, the analogy of the finger pointing to the moon provides another powerful illustration of the concept of the esoteric (the inner or the hidden) and the exoteric (the external or the literal), and the danger of losing sight of the esoteric truth by a mistaken focus on the literal or exoteric. This concept was discussed in another post[13] using an example from the 1984 film *Karate Kid* (for a variety of reasons, some aspects of the martial arts have traditionally been taught using esoteric methodologies, as that post mentions).

The finger in this illustration is only an aid, pointing to a higher truth (represented by the moon). To lose sight of the higher truth because one mistakes the finger or the "teaching aid" for the truth itself would be analogous to losing sight of the martial art that the waxing of cars was intended to teach, and to focus exclusively on waxing cars.

Shockingly, there is abundant evidence that this is exactly what has happened through the literalist interpretation of the stories found in the ancient scriptures which became the Old and New Testaments of the Bible -- the literalists have fallen into the exact mistake warned against in the Shurangama Sutra: "they look at the finger and mistake it for the moon" (and in doing so, they lose both the moon and the finger).

For example, in another post entitled "No hell below us . . ." I argue that the scriptures describing hell which are found in the Bible were intended to be read metaphorically, and to refer to that portion of the year in which the sun's daily path (the ecliptic) is below the celestial equator -- and particularly to the winter months at the very "bottom" of the annual cycle, that part of the year on either side of the winter equinox, which is metaphorically speaking the very Pit of hell. In other words, these scriptures are intended to convey an esoteric message, but literalists have interpreted them as describing a literal place called hell where souls are consigned for eternal torment – a mistake of the same magnitude as mistaking the finger for the moon.

Another example would be mistaking the twelve disciples for literal historical figures, when they are almost certainly representative of the twelve signs of the zodiac and the characteristics associated with each. Angrily insisting that they must be studied first and foremost as literal men living in the Roman Empire is akin to reversing Bruce Lee's dictum in the above film clip to say, "Don't focus on the moon -- you must only focus on the finger, such as the disciples in the stories, and must never consider the possibility that they are only a guide to point you towards something else!"

Further evidence that the ancient scriptures of the Bible (and of many other sacred traditions found around the globe) are primarily esoteric in nature rather than literal can be found in my new book, *The Undying Stars*, which also examines some of the history behind the replacement of esoteric truths with a mistaken literalist hermeneutic.

The *Undying Stars* also discusses the profound truths that these esoteric ancient scriptures may have been intended to convey. In other words, it examines the question which one may be thinking upon reading the above discussion, which might be expressed something like this: "OK, if you are saying that the twelve disciples represent zodiac signs, or that the passages about hell represent the lower half of the annual zodiac wheel, then why would anyone write sacred scriptures about that and make such a big deal about those scriptures for so long? What's the point of making a bunch of stories about the stars?"

One important thing to notice in both the segment from *Enter the Dragon* and from the Shurangama Sutra is the fact that in both cases, the moon itself is also being used as a metaphor for something else! In other words, the teachings are not just talking about "the moon," meaning the massive rocky body orbiting our planet at an average distance of 238,857 miles. They are using the moon in a metaphorical sense, just as they are using the overall metaphor of a finger pointing to the moon in a metaphorical sense. The moon in both examples is meant to stand for a higher-mind that is beyond the intellect, a thinking that is beyond or above our ordinary form of thinking (in fact, it is meant to convey a truth which is difficult to express in a sentence, which is why it is best grasped through a metaphor and through the esoteric).

In just such a way, the stars and the motions of the heavens to which the ancient scriptural texts (including those which found their way into the Bible) are themselves an analogy for something else. The ancient scriptures are not just "a bunch of stories about the stars" -- they are esoteric stories related to the motions of the heavens and the heavenly bodies, but they are much more than that. They use the motions of the heavens and the heavenly bodies to express profound truths about the human condition and our purpose in this life, as well as to imply a sophisticated cosmology that appears to anticipate modern quantum physics by many thousands of years.

The sophistication of this ancient cosmology suggests that extremely ancient civilizations may somehow have been possessed of extremely advanced science and even what we can only call advanced technology, and may help to explain some of the ancient accomplishments which are extremely difficult to explain using the conventional historical paradigm. This fact may also help to explain why someone would want to subvert the ancient scriptures which teach it, and to get everyone focused on the finger (and only the finger) . . . and to miss all that heavenly glory!

The undying stars: what does it mean?

2014 May 10

Once it can be demonstrated beyond a reasonable doubt that the ancient scriptures and sacred traditions around the world consist of celestial allegory, with each story describing the motions of the sun, moon, stars or planets, the next question which naturally arises is: *Why?* Why would the cultures of the world ascribe such importance to the circling heavens, so much so that they encoded those motions at the very heart of their most sacred art, literature, and cultural rituals? (For some discussions of the evidence supporting the conclusion that the ancient sacred traditions are all founded upon celestial allegory, see books and videos published since this post was first written, including the *Star Myths of the World* series).

One possibility, much in vogue since the nineteenth century, would be that primitive humans were so over-awed by the glory of the heavens that they worshiped the sun and moon and stars and planets as deities, and that the sacred traditions and ancient scriptures of the world are simply the more-civilized versions of those early religious impulses from the days of the hominids. This theory argues that, beneath the accretions added by later civilizations, one can still detect the outlines of the earliest and most primitive sun-and-star worship, and this fact (in their view) reinforces the storyline of mankind's long and generally linear progress from "early humans" to "primitive hunter-gatherers" to "pastoral herdsmen" to "neolithic farmers" and ultimately to the building of the first true civilizations (and further progress from there).

This explanation, however, either overlooks or deliberately obscures the evidence for the existence of a very sophisticated ancient civilization (or civilizations), with capabilities we still do not fully understand -- capabilities such as the construction of megalithic structures using huge stones, quarried from great distances, set at precise angles, and aligned with great precision to subtle astronomical phenomena. The combination of these subtle alignments and the massive stones transported from great distances implies that the designers of these sites knew exactly what size stones they needed to obtain before they quarried them. Further, these sites often incorporate dimensional measurements and ratios indicating that their designers understood the subtle celestial mechanics of precession, and even the size and shape of planet earth! The existence of such monuments, many of them from great antiquity (predating the earliest known civilizations) tends to upend the conventional storyline (briefly outlined in the previous paragraph) of mankind's progression from primitive humans subsisting as hunter-gatherers and pastoral herdsmen prior to the earliest civilizations.

Further, these incredible ancient monuments (whose construction is still in many cases unexplainable by conventional academics and either un-duplicate-able or extremely difficult to duplicate using today's technologies) are not simply amazing on their own: their relationship to one another forms what can only be described as a "world-wide grid" which indicates that their designers were far more advanced than the conventional timeline can possibly account for.

The profundity of the wisdom preserved in the ancient scriptures and traditions of mankind likewise argues against the theory that they are simply the glorified remnants of the awestruck deification of the sun, moon, stars and planets by "primitive humans." Of these manifestations of ancient wisdom, already evident in some of the earliest writings known to history (such as the Pyramid Texts), Alvin Boyd Kuhn declares: "They were the products, not of early man's groping tentatives to understand life, but of evolved men's sagacious knowledge and matured experience. On no other ground can their perennial durability and universal power be accounted for" (*Lost Light*, 30).

To conclude that just because the ancient myths and sacred traditions (including the scriptures of the Old and New

24

Testaments) are full of allegorical descriptions of the motions of the circling heavens they must ultimately be about the circling heavens themselves is to commit a colossal error. Many traditions of Buddhism contain the metaphor of the "finger pointing to the moon"[18] and advise the listener not to concentrate on the finger or risk missing the heavenly glory of the moon to which the finger is pointing; however, it would be a ridiculous mistake to use this fact to conclude that Buddhism is all about the rocky sphere which orbits our planet!

Likewise, just because it can be conclusively demonstrated that the scriptures of the Old and New Testament describe celestial metaphors, this by no means necessitates the conclusion that those scriptures are all about the heavenly bodies. In fact, just like the moon in the Buddhist saying about the finger and the moon, the motions of the starry heavens are themselves a metaphor for higher truths regarding the human condition and the nature of existence in this material realm.

One aspect of this metaphor discussed in some detail in a previous post entitled "A Hymn to the Setting Sun, and the ultimate mystery of life" (04/19/2014), teaches that each human soul is immaterial and undying, incarnating in successive material bodies and then successively returning to the spirit realm. That post explained:

> One of the metaphors which the ancients used in order to convey this vision of human existence as a cycle of incarnation in matter followed by re-ascent into the spiritual realm was the cycle of the heavenly bodies, including that of the sun, which plunges nightly beneath the horizon (matter) only to rise again each morning clothed in fire into the upper realm of the heavens.

This same illustration of the successive incarnation and re-ascent is also provided by the motion of the individual stars, which plunge into the western horizon to become "enfleshed" in earth and water (the lower elements), only to appear again on the eastern horizon to rise again into the heavens -- the realm of fire and air (the higher elements). In doing so, they perfectly allegorize the incarnation of the soul in a body of earth and water (often described as a body of "clay" in the ancient scriptures) and its subsequent re-ascent into the freedom of air and fire, followed by another descent into matter.

This (apparent) daily turning of the stars across the (apparent) crystal dome is depicted below, showing the course of the stars for an observer located at about 30 degrees north latitude:

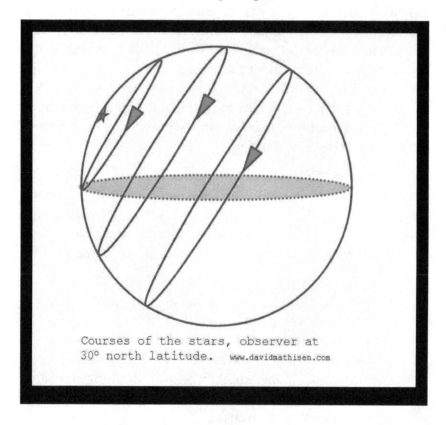

Courses of the stars, observer at
30° north latitude. www.davidmathisen.com

(This diagram has been published in an earlier blog post from August of 2011, entitled "Listening to the greatest navigators our globe has ever seen"[76]). In the diagram, because the observer is located at about 30 degrees north latitude, the pole star about which the heavens appear to turn is located at about 30 degrees

26

elevation above the horizon (the horizon in the above diagram is depicted as the dotted line around the edges of a flat grey disc, out of which the heavenly bodies such as the sun, moon, planets and stars will appear to rise in the east and into which the same heavenly bodies will appear to sink in the west).

Please note that stars which are close enough to the pole star will never actually sink below the horizon (although they will be blotted out by the brilliance of the sun each day when the day-star is above the horizon itself). Those stars which are closer to the pole star in the above diagram (that is, any star located between the smallest circle in the diagram and the pole star itself) will not plunge below the horizon. These stars the ancients (particularly the ancient Egyptians) called the "imperishable stars," or the "undying stars." For an observer at 30 degrees north latitude, these would be the stars which are within about thirty degrees of arc from the north celestial pole.

The diagram below from an old US Air Force manual depicts the stars near the north celestial pole in the modern epoch, with the north celestial pole marked by a small cross and the letters "NCP" in the center of the wheel. Very close to the true north celestial pole is the star Polaris, at the end of the handle of the Little Dipper (Ursa Minor). Pointing to the pole star Polaris are the "pointers" in the bowl of the Big Dipper (an arrow is drawn from the two "pointers" towards Polaris, to help illustrate the concept). Very helpful rings are drawn on the chart, each ten degrees greater from the central point of the north celestial pole (NCP). These are marked with degrees of elevation above the celestial equator: the NCP is at 90 degrees, the first circle from the NCP is marked as 80 degrees, the next circle is marked as 70 degrees, and so on. For an observer at 30 degrees north latitude, all the stars within the first three circles would be considered "the undying stars," and would never dip below the horizon (of course, mountains and trees along the horizon would cause some stars on the edges of the outer circle to dip below the horizon at times).

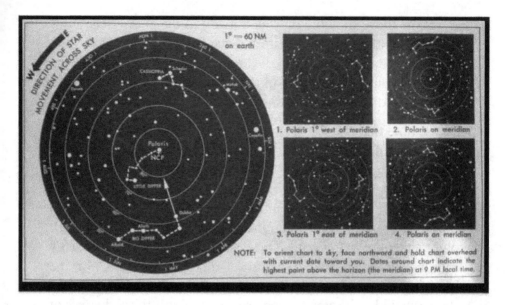

In his 1940 book *Lost Light* (which can be read online at https://archive.org/stream/lostlightanintero29017mbp#page/n5/mode/2up), Alvin Boyd Kuhn explains that one interpretation of the metaphor of the imperishable or undying stars in ancient mythology would be spirits which never have descended into mortal incarnation, but another would be spirits who have, through successive incarnations, surpassed the cycle of reincarnation. He writes:

> Among the ancients the stars that dipped beneath the horizon were emblematic of souls in physical incarnation, in contradistinction to those that never set [. . .] The redeemed souls rejoiced in the Egyptian *Ritual* (Ch. 44) at being lifted up "among the stars that never set." 115.

Later in the same text, Alvin Boyd Kuhn explains that in ancient Egypt,

> The souls having attained the resurrected state in shining raiment were called the Khus or the glorified. [. . .] the Egyptian Khemi, or Akhemu, the dwellers in the northern heaven, as never-setting stars or spirits of the glorified, the Khus or Khuti. 587.

28

This is not to say, as textbooks often authoritatively declare to schoolchildren, that the ancient Egyptians literally believed that they would go dwell in the physical starry heavens at some point, or that when they said "the dwellers in the northern heaven" they believed that this was the *physical* location of the spirits of those who had surpassed the cycle of incarnation. Such an assertion falls into the trap of literalism: of believing that the texts must be read literally and that the authors of those texts intended for them to be read only this way.

These teachings were using metaphor and esotericism to teach profound truths regarding the nature of human existence – not literal directions to the location of disincarnate souls.

As *The Undying Stars* discusses further, these celestial metaphors may also have been preserving and conveying a sophisticated understanding of the nature of our physical universe, one which anticipates modern quantum physics by many thousands of years and which teaches (as does modern quantum physics) that human consciousness actually impacts and in some way *causes to manifest* the apparently physical world we inhabit in this incarnation.

Ultimately, these ancient scriptures can also be said to teach that all stars are really "undying stars" or "imperishable stars," in that they do not cease to exist, even after many successive cycles of incarnation, and in that they represent undying souls who will one day overcome the cycle of incarnation and join the Akhemu, "the dwellers in the northern heaven, as never-setting stars or spirits of the glorified."

Montessori and "thinging"

2014 May 25

The Montessori method, originated by pioneer educator Maria Montessori (1870 - 1952), has a very special place in my heart. I went through Montessori education from preschool to third grade, my children went through Montessori, my sister teaches at a Montessori school, and my mother has run Montessori schools in California for over thirty years. Of course, with so much exposure to Montessori, many of our family's close friends and the fantastic individuals I was around while I was growing up also come from the Montessori community. Some of the most influential teachers I ever had were my early Montessori teachers, and I am tremendously grateful to each one of them to this day.

Not only is Montessori a wonderful approach to education, but it is also centered on respect for the child as an individual and a person, and respect for the child's own initiative and ability to learn by himself or herself. Montessori also inculcates in the child a respect for other children and the ability to work with and help others. All of these wonderful aspects of Montessori are evident in a video available on the web entitled "A Montessori Morning," which shows in about four minutes a series of photographs taken during the course of three hours in the morning of a four-year-old named Jackson, along with his friends at the Dundas Valley Montessori School in Ontario.

In addition to all these outstanding characteristics (and there are many more I have not mentioned), Montessori also provides an excellent example of the *esoteric method* of enabling the human mind to grasp big or profound concepts (previous discussions of the esoteric include "Wax on, wax off"[3] and "Like a finger, pointing a way to the moon . . ."[18]).

Montessori uses ingenious physical materials to represent abstract concepts. In doing so, it echoes the method employed by the sages responsible for the mythologies which make up the world's ancient sacred traditions, according to thinkers such as

Gerald Massey (1828 - 1907) and Alvin Boyd Kuhn (1880 - 1963) --
although most conventional historians and academics erroneously
approach ancient sacred texts and traditions as if they were
intended to be understood literally.

Gerald Massey vigorously refutes the conventional view that the
world's ancient myths were intended or anciently understood to
be literal in the second and third sections of his essay "Luniolatry,
Ancient and Modern," in which he explains:

> They [meaning the conventional historians and professors
> of mythology, several of whom he cites in the essay] have
> misrepresented primitive or archaic man as having been
> idiotically misled from the first by an active but untutored
> imagination into believing all sorts of fallacies, which were
> directly and contradicted by his own daily experience; a fool
> of fancy in the midst of those grim realities that were
> grinding his experience into him, like the grinding icebergs
> making their imprints upon the rocks submerged beneath
> the sea. It remains to be said, and will one day be
> acknowledged, that these accepted teachers have been no
> nearer to the beginnings of mythology and language than
> Burn's poet Willie had been near to Pegasus. My reply is,
> 'Tis but a dream of the metaphysical theorist that
> mythology was a disease of language, or anything else
> except his own brain. The origin and meaning of
> mythology have been missed altogether by these solarites
> and weather-mongers! Mythology was a primitive mode of
> *thinging* the early thought. It was founded on natural facts,
> and is still verifiable in phenomena. [. . .]

> In modern phraseology a statement is sometimes said to be
> mythical in proportion to its being untrue; but the ancient
> mythology was not a system or mode of falsifying in that
> sense. Its fables were the means of conveying facts; they
> were neither forgeries nor fictions. Nor did mythology
> originate in any intentional double-dealing whatever,
> although it did assume an aspect of duality when direct

31

expression in words had succeeded the primitive mode of representation by means of things as signs and symbols.

Alvin Boyd Kuhn picks up on the importance of Massey's concept of "thinging" and says:

> As Gerald Massey says, thinking is in essence a process of "thinging," since thoughts must rest on the nature of things. And things are themselves God's thoughts in material form. *Lost Light*, 42.

This "thinging" that Massey and Kuhn are talking about is perhaps best illustrated by the Montessori materials, some of which can be seen in the beautiful little video above. For example, in the video mentioned above, Jackson works with the Montessori "sensorial material" project known as the trinomial cube beginning at about 2:30 into the video, through about 2:44 in the video (the video moves fast -- you can see the trinomial cube segment following immediately after Jackson and his friends have a snack, at about 2:25 -- immediately following the "window squeegee" scene -- just after Jackson finishes cleaning his dishes from the snack and puts them into a drying rack to air-dry).

The trinomial cube is an example of "thinging" the somewhat abstract algebraic concept of cubing a trinomial (a trinomial is a mathematical expression containing three variables, with the vairables commonly designated as *a*, *b*, and *c*). If we have a trinomial (a + b + c), and we wish to *cube* it, we must multiply the trinomial by *itself* three times (this is the definition of cubing something). In other words, we must multiply (a + b + c) (a + b + c) (a + b + c).

If you remember your algebra, you will remember that the way to tackle this particular operation is to begin with the first term in the first trinomial, and multiply it by each of the terms in the next two instances of the trinomial, and continue this process all the way through the operation.

After multiplying it all the way out, and adding it all together, one finds that the cube of $(a + b + c)$ can be written:

$$a^3 + 3a^2b + 3a^2c + 3ab^2 + 6abc + 3ac^2 + b^3 + 3b^2c + 3bc^2 + c^3$$

The trinomial cube used in Montessori classrooms makes this rather intimidating-looking formula into a thing, into a model which children such as Jackson can manipulate and explore at a very early age (remember that Jackson is four years old, and he can be seen assembling the trinomial correctly in the video).

The way the model cube "things" the expression of the cubed trinomial shown above is ingenious. You can see that in the full formula, the cube of each variable appears one time each, a-cubed, b-cubed, and c-cubed appear at the beginning, the "middle," and the end of the formula, respectively. The variable a is represented by the largest dimension of the blocks in the cube – when a is cubed it is represented by the largest cube in the model, painted red on all surfaces, of length a on each side of the cube. The variable b is the next-largest dimension of the blocks in the cube: when it is cubed it is painted blue on all sides and appears as a cube with sides of length b (a shorter distance than length a). Finally, the variable c is the shortest of the dimensions represented in the cube; when c appears as a cube (which it does one time in the above formula for a cubed trinomial), it is depicted as a cube in which all faces are painted yellow, and the sides are a length c (shorter than b, which in turn was shorter than a).

Note that in the solution formula above, the term following a^3 is $3a^2b$. The term a^2b is "thinged" in the Montessori trinomial cube as a solid with a face that is length a on each side (that is, it is a physical representation of a^2 and it is painted red on the square face), but which is only a depth of b (these sides, b in length, are painted black). Thus, the Montessori trinomial cube represents a^2b as a solid with a height and width of a and a depth of b, and it contains three such solids, to match the a^2b in the solution to the cubed trinomial.

The model of the trinomial has solids to represent each of the terms in the full formula above. It has three that are again a height and width of a but this time only a depth of c, to represent the next term which is $3a^2c$ (and again, the face representing a-squared is painted red). It has three solids which are b in height and width and a in depth, to represent the $3ab^2$ (and this time, of course, the face representing b-squared is painted blue, while the depth representing a is painted black -- colors are only used when a term is either squared or cubed, otherwise the side is black). And it has six solids which have a height of a, a width of b, and a depth of c, which are black on all their sides, and represent the term $6abc$. To help visualize all of this, you can pay a visit to the excellent schematic on the trinomial cube page of Wikisori, which lays it all out visually.

Now, the interesting thing about all of this is that the child learning how to work with the trinomial cube (and its slightly less-complicated cousin, the binomial cube, which represents the binomial a + b multiplied by itself three times) is not taught anything at all about the way that the cube is an ingenious physical representation of a rather advanced and very abstract algebraic concept. That would not really be helpful to a four-year-old child.

However, when the child is old enough, and is being introduced to binomials or trinomials in algebra, then the teacher can explain the connection to the old, beloved, familiar binomial cube and trinomial cube, and show the "esoteric" connection between the physical model and the formula they are learning. What a flash of recognition will go off in the young person's mind! It is exactly akin to the sudden dawning of recognition experienced by Daniel-san when Mr. Miyagi showed him what "wax on, wax off"[13] was really all about!

You can, in fact, see for yourself that the webpage for the binomial cube on the *Montessori World Educational Center* website expressly states: "Do not explain to the child why you are setting the cube out in this order, or talk about the mathematics of the cube." Is this because the Montessori teachers do not want children to know the "esoteric secrets" of the binomial cube? Of course not! The whole point is to eventually help the child to learn about binomials, in a way more profound than the child might ever be able to understand otherwise. But trying to explain it in a "left-brained" way first would just invite confusion and questions as the analytical "left-brain" tries to absorb the abstract and complicated concepts involved, likely causing the brain to "choke" on it (and possibly never feel comfortable around binomials or trinomials ever again). Instead, the webpage advises: "The math is presented to the children when they are older and are ready for it."

This example from Montessori (and there are many others that could be used, such as the bead-chains which you can see Jackson and his friend working with after the trinomial cube segment, beginning at around 2:47 and going to about 3:00) really illustrates Massey's point about the value of "thinging" an abstract concept (a point Alvin Boyd Kuhn also underscores as being of supreme importance). It is easy to see the source of Massey's frustration with conventional academics who insist that the myths were simply a bunch of "fallacies" which ancient men and women believed literally.

Kuhn disagreed with Massey, however, because Kuhn argued that these exquisite mythical metaphors, which so wonderfully "thinged" profound spiritual concepts, could not have originated as a "primitive mode" of early thought. He declares that these incredible metaphors betray the handiwork of sages who already understood completely the deepest spiritual truths, saying:

> Primitive simplicity could not have concocted what the age-long study of an intelligent world could not fathom. Not aboriginal naiveté, but exalted spiritual and intellectual

35

acumen, formulated the myths. Reflection of the realities of a higher world in the phenomena of a lower world could not be detected when only the one world, the lower, was known. You can not see that nature reflects spiritual truth unless you know the form of spiritual truth.
Lost Light, 71-72.

In other words, no one could start with the physical model and come up with the spiritual truths -- the makers of the model had to know the spiritual truths *already*. We can immediately agree that the designer of the trinomial cube had to understand the full formula of

$$a^3 + 3a^2b + 3a^2c + 3ab^2 + 6abc + 3ac^2 + b^3 + 3b^2c + 3bc^2 + c^3$$

before designing the wooden model. By this analogy, it stands to reason that the designers of the exquisite esoteric myths of the world understood the profound spiritual truth they wished to convey before they ever created the myths -- the myths were not the product of "an active but untutored imagination," as Massey thought.

Furthermore, it is also evident that one could learn the trinomial cube as a child (as a four-year-old, for example), and never fathom the connection to the trinomial expression shown above -- even if they later became quite advanced at mathematics and algebra and learned all about trinomials! To make the leap from the model with the solid forms painted red, blue, yellow and black on their various sides, to the formula shown above, is not necessarily intuitive until the connection *is shown* to the student. This concept is expressed in the New Testament book of Acts, in which a man is depicted reading an Old Testament scroll (Isaiah), and is asked if he understands what he reads. He replies: "How can I, unless someone guides me?" (Acts 8:31-32).

What a tragedy it would be if the stories in the ancient scriptures were really intended to act as a sort of "trinomial cube" pointing to profound spiritual truths, but those who were able to teach the

36

connection were prohibited from doing so! It would be as if children were prevented from being shown the true purpose of the Montessori materials, such as how the bead chains teach multiplication and squaring and cubing of the various digits from I to IO.

What a tragedy if all those who knew the esoteric connections were, at some point in ancient history, marginalized and suppressed by people who wanted to teach that these stories should all be understood *literally* first and foremost, and if these literalists did their best to destroy or cast out all the texts which opposed that literal interpretation or said that the scriptures were not really literal but rather esoteric.

Fortunately for the human race, the finely-crafted "Montessori materials" which are the ancient metaphors of the myths of all the world's cultures (including those which were preserved in the scriptures of the Old and New Testaments) are still with us today, and can be turned over in our minds as we might turn over a finely-crafted trinomial cube. The connections to the spiritual concepts that these stories were intended to teach (via the method of "thinging") were not entirely eradicated by the literalists, but survived in various channels over the long centuries, and have been elucidated by various teachers in various texts. The connections can be made again.

The ancient torch that was lighted for our guidance

2014 May 29

At the beginning of his nearly 600-page tour de force *Lost Light*, Alvin Boyd Kuhn declares that his aim is not to attack what he calls at one point "the lovely temple of ancient truth," but rather to restore it.

He proclaims his reverence for "the Bibles of humanity," by which he means the ancient scriptures and sacred traditions of humanity -- and he includes the scriptures of the Old and New Testaments in that group. "In them," Kuhn writes, "were given the ordinances of life, the constitution of the cosmos, the laws governing both nature and mind" (2).

He states that he feels it necessary at the very outset of his work to make this declaration, because he intends to then go on to launch a frontal assault upon the whole "untenable structure" of "ecclesiastical doctrinism" which has grown up to replace the temple of ancient truth found in those Bibles of humanity (1). He minces no words in calling the literalist interpretation of the ancient truth "an unconscionable perversion of its original significance to gross repulsiveness," and saying that "the errors and distortions perpetrated upon it by those of its own household must be ruthlessly dismantled" (1). But he is very clear that he reserves his devastating assaults for the *interpretations* that have been forced upon the ancient texts, and not upon the texts themselves.

This is a critical distinction, and it is one that is all too easily missed, especially by some of those writing today who likewise perceive that the texts are not meant to be taken literally. It is a tremendous error to make the leap from perceiving that the ancient stories describe celestial events and not literal historical earthly events, to the erroneous conclusion that the stories are somehow less profound because of it, or even that they should be criticized or mocked.

It is perhaps understandable that some, in their anger at what Kuhn calls an "endless train of evils, fanaticisms, bigotries, idiosyncrasies, superstitions, wars and persecutions" that followed inevitably from the near-total destruction of the ancient philosophical knowledge in "about the third century of Christianity's development [. . .] ushering in sixteen centuries of the Dark Ages," to wish to lash out at the texts that the perpetrators of those injustices were citing as an authority for their actions. But it is a serious mistake to confuse a perversion of the texts with the value of the texts themselves (both quotations are from page 3). Kuhn throughout his work expresses deep reverence for what he calls "the sage tomes of antiquity" and compares their contents, rightly interpreted, to a light for our guidance – a light which has over the centuries been all but lost (hence the title of his own book).

He writes of "modern man" that, "The ancient torch that was lighted for his guidance he has let burn out. This lamp was the body of Ancient Philosophy" (4). He also refers specifically to the year in which his own book was coming to print, saying: "The present (1940) most frightful of all historical barbarities owes its incidence directly to the decay of ancient philosophical knowledge and the loss of vision and virtue that would have attended its perpetuation" (3).

Again, this reverence for the ancient scriptures (not just those included in the Bible but all the "sage tomes of antiquity" from many different cultures, as well as those ancient scriptures not included in the "canonical" Bible such as the Pistis Sophia and the Hermetic texts, to which he refers in *Lost Light*) sets Kuhn's work apart from many others who have likewise perceived the errors in the literal approach to those scriptures, but who use those literalistic errors to then wrongly denigrate the texts themselves.

We can perhaps detect in this prologue of Alvin Boyd Kuhn's book a caution against falling from one error into an equally harmful opposing error -- either of which can cause us to miss the ancient torch that was lit for our guidance.

I have argued that an important aspect of the core message of the ancient scriptures (whether Norse or Vedic or Biblical or Maya) is a "shamanic - holographic"[167] understanding of the nature of the cosmos and the nature of human existence. If this assertion is correct, and if malevolent forces wanted to suppress this shamanic - holographic understanding (perhaps because it is too empowering to men and women, makes them too difficult to manipulate or enslave or deceive, or for one of many other possible reasons malevolent forces might want to suppress this cosmology), then it would not really matter to them whether the shamanic - holographic was obscured because people took the texts literally and thereby missed their esoteric message,[23] or if they vehemently rejected the literal interpretation and in their vehement rejection went on to denigrate the texts, and thereby missed the esoteric message as well.

There have been some in the past who have written on the celestial allegories which are abundantly evident in the ancient texts -- and some writing today -- who stop there, and do not go on to the "higher message" that these exquisite allegories were intended to convey (this is akin to never teaching the child the connection between the trinomial cube and the trinomial itself,[30] or never teaching Daniel-San[13] the meaning of "wax on, wax off," and leaving him at the mercy of the bullies who continue to physically assault him).

The ancient sacred traditions of all cultures, which share a common system of celestial allegory and which were (I believe) intended to convey a common esoteric knowledge of great power and tremendous benefit to men and women throughout the ages, are indeed an ancient torch that was lighted for our guidance. It is possible that more than one method has been devised to keep this light from being rekindled.

40

One is a literalism which proclaims a high view of ancient scriptures, but which in fact refuses to even entertain the *possibility* that those scriptures are esoteric and not literal (and, in doing so, also denigrates all the other ancient traditions of mankind, traditions which the literalist approach likewise views strictly literally, and thus labels them the "doctrines of demons," rather than seeing them as esoteric also). This literalist approach will obviously reject the shamanic - holographic cosmology which the ancient scriptures were designed to convey.

But the other error can be just as damaging, even as it recognizes that the literalist approach is incorrect. It sees the errors of literalism, and traces these to the texts themselves, which it insults, or mocks, or denigrates. Those falling into this error may even perceive that the texts contain celestial allegories, but they use this fact to further mock the literalists and -- erroneously -- the texts themselves, saying "it's just a bunch of primitive sun-worship" or words to that effect. In doing so, they run the risk of also missing the profound and sophisticated cosmology which these ancient texts were designed to teach.

This is a very important subject, and one which deserves careful consideration.

Scarab, Ankh and Djed

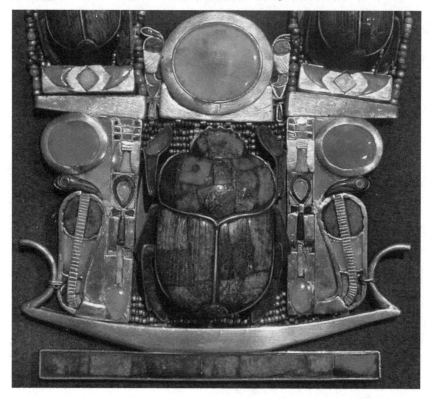

2014 August 23

The importance of the ancient symbol of the Ankh simply cannot be overstated. It is a symbol of eternal life, and as such it is closely associated with two other important ancient symbols, the Scarab and the Djed-column.

Previous posts have explored the abundant evidence which suggests that the Ankh (along with other cross-symbols) represents the two natures which join together in our human existence: the material or animal nature symbolized by the horizontal bar, and the spiritual nature, symbolized by the vertical column, which in the case of the Ankh is surmounted by the circle representative of the infinite or the unending.

The Ankh as a symbol is closely related to the Djed column, which is also depicted as having a horizontal component (when the Djed is cast down, representing our physical incarnation in "animal" matter) and a vertical component (when the Djed is raised up, representing the uplifting of our spiritual nature and representative of spiritual life).

So, the connection between the symbology of the Ankh and the symbology of the Djed is fairly straightforward and easy to understand. But, how are these two symbols connected to the symbol of the Scarab? Let's examine the question more closely -- the answer contains many breathtaking connections and sheds light on the exquisite profundity of the ancient wisdom, bequeathed to us in the mythology and symbology of the human race.

In the image above, an elaborate necklace from the tomb of Tut-Ankh-Amun is depicted, featuring a central figure of a Scarab beetle with uplifted arms, upon what I would interpret as being (based upon evidence presented below) the solar bark. The Scarab is flanked most immediately by two columns which each feature a prominent Ankh symbol (in dark blue) and immediately above each Ankh for good measure is a Djed column (in light blue, with alternating red and blue segments in the "spinal" columns at the top of each Djed).

On either side of the Scarab, just outside of the two Ankh-adorned columns, are two *uraeus* serpents, each with a solar disc above its head, and above the upraised arms of the Scarab itself is another, larger solar disc. On either side of this larger solar disc are yet two more *uraeus* serpents, and suspended from each is another Ankh.

As can be seen from the image above, this elaborate ornament continues on beyond the section in the close-up view shown above: the wide "straps" of the necklace on either side are adorned with another pair of Scarabs, each of which are surmounted by another solar disc (not shown in the image above), this time in gold, and again flanked by two *uraeus* serpents. Above these Scarabs and solar discs can be found yet another pair of *uraeus* serpents (on each "strap"), this time flanking a central Ankh symbol (on each "strap").

I believe that all of these symbols are powerfully depicting variations upon the same theme, which is the raising-up of the immortal, spiritual component in the individual, symbolized by the raising of the Djed column, which is associated with the vertical and immortal portion of the Egyptian cross (the Ankh), and which is *also symbolized by the vertical line or "column" between the two solstices* of the year on the zodiac wheel of the annual solar cycle.

As discussed in other posts, the vertical pillar of the solstices was connected in mythology with *the Djed column raised up*, and also with the constellation of Cancer the Crab, the sign which commences at the point of summer solstice (a fact which is commemorated in the name of the Tropic of Cancer, which is the latitudinal circle designating the furthest north that the direct rays of the sun will reach each year, on the day of the northern hemisphere's summer solstice, at the start of the astrological sign of Cancer). Because the constellation of Cancer itself appears to have outstretched or upraised arms, this "top of the solstice column" is mythologically associated and symbolized by the upraised arms of Cancer the Crab -- and, as we have seen in that same post just linked, by the upraised arms of Moses in the battle against the Amalekites in Exodus 17, as well as the upraised arms of the Egyptian god Shu.

And, as that post also points out, the Ankh symbol (which is closely associated with the symbol of the Djed-column "raised up") itself was often depicted with a pair of human arms raised upwards in just the same way (a famous image from the Papyrus of Ani showing the Ankh with upraised arms, surmounting a Djed column flanked by Isis and Nephthys, has been included in several previous blog posts, such as this one[127]).

44

And with that in mind, we can now understand the symbology of the Scarab, and why it is "of a piece" with the Ankhs and the Djed columns in this necklace!

The understanding that these upraised arms are associated with Cancer the Crab, whose position at the very summit of the year places him at the top of the vertical Djed column that can be envisioned connecting the solstice-points on the zodiac wheel, and whose upraised arms are responsible for the upraised arms that are sometimes depicted on the Ankh-cross, enables us to see that the Scarab itself is another way of recalling Cancer the Crab and the uplifted arms -- symbolic of the vertical, spiritual, eternal force in every man and woman. (Below is an image of the zodiac wheel, with the horizontal and vertical lines depicted: you can see the sign of Cancer with its outstretched arms, looking in this 1618 illustration a bit more like a Lobster than a Crab, at the top of the vertical column and to the "right of the line," just past the point of summer solstice):

DAYS LONGER THAN NIGHTS:
Heaven, Promised Land, Greece, etc.

Horizontal Column:
The Djed cast down

Vertical Column:
The Djed raised up

NIGHTS LONGER THAN DAYS:
Hell, Egypt, Troy, etc.

For this reason, we can safely assert that the Scarab in this necklace, surrounded as it is by Ankhs and Djeds, and depicted as it is with upraised arms, is symbolic of the summer solstice, and that the disc above its head must be a solar disc, and the bark on which it and the *uraeus* serpents are positioned must be a solar bark.

The two serpents, by the way, are also closely associated with the vertical Djed-column -- if we imagine the ancient symbol of the caduceus, we will instantly perceive that these two serpents are positioned on either side of all these central (spinal) column images (the Ankh, the Djed, and the Scarab) in just the same way that the two serpents are positioned at the top of the caduceus column (and intertwine all the way down). The fact that elsewhere upon the same necklace the two serpents are depicted as flanking an Ankh shows that the symbols of the Ankh and the Scarab are closely connected and practically interchangeable here.

As has been explored in numerous previous posts as well, the Djed column is closely associated with the "backbone of Osiris," and hence with the backbone of every incarnated man and woman (Osiris being the deity of the underworld and of our incarnated state, as discussed at greater length in *The Undying Stars*). Most appropriate it is, then, to note the connection pointed out by R. A. Schwaller de Lubicz (and reiterated by John Anthony West in *Serpent in the Sky*) between the imagery of the Egyptian Scarab and the top-down view of the crown of the human skull, discussed on page 96 of that book and accompanied by an illustration similar to the one below:

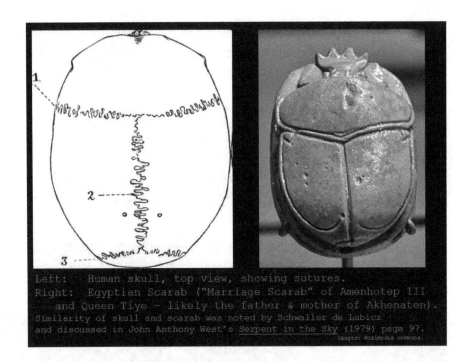

Left: Human skull, top view, showing sutures.
Right: Egyptian Scarab ("Marriage Scarab" of Amenhotep III
    and Queen Tiye – likely the father & mother of Akhenaten).
Similarity of skull and scarab was noted by Schwaller de Lubicz
and discussed in John Anthony West's Serpent in the Sky (1979) page 97.
Images: Wikimedia commons.

If we consider that the top of the skull forms the very pinnacle of the vertical "Djed-column" in each man or woman, corresponding to the very peak or crown of the year at the summer solstice (representative of Heaven itself, and in its very domed shape most representative of the dome of heaven in the microcosm of the human body, which reflects the macrocosm of the infinite dome of the universe), then it is most appropriate that a Scarab symbol (reminiscent of the sign of Cancer the Crab, which is located at the solstice-summit of the year) is found there on the top of each of our heads!

In a future post we will explore further the significance of the name of the Ankh itself, following on the illuminating analysis of Alvin Boyd Kuhn on the subject. Some aspects of this important concept have already been touched upon, in a previous post about the ancient Vedic concept of the Vajra, or Thunderbolt, which we saw in that post to be almost certainly connected to the concept of the raising of the Djed column and the "backbone of Osiris." There, we saw evidence from the work of Alvin Boyd Kuhn that the "N - K" sound of the name of the Ankh is linguistically connected to the name of the practice of Yoga (or *yonga*, in which the "N - G" sound is linguistically related to the "N - K" of the Ankh).

This connection to the practice of Yoga is most revealing, in that Yoga itself is a discipline which concerns the raising of the divine spark up through the chakras of the spine and ultimately up and through the crown chakra, located at a point which is associated with the crown of the head, or the "top of the Scarab" described above!

Below is an image from Wikimedia showing practitioners of Yoga with their arms in a very characteristic and significant position:

Namaste.

The name of the Ankh

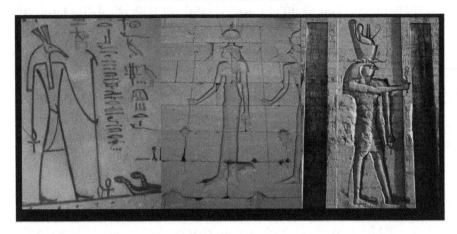

2014 August 25

The previous post[12] explored some of the profound significance of the Ankh and its relation to the symbols of the Djed and the Scarab -- and to the message that we as individual men and women have an unending, spiritual component in addition to the horizontal, animal, and material aspect of our being to which we are currently *joined*.

That post also touched very briefly upon the amazing linguistic analysis Alvin Boyd Kuhn has provided regarding the word *Ankh* itself, and his assertion that the "N-K" sound seen in the word *Ankh* finds its way into an astonishing array of words still in use today, including *Yoga* -- a practice whose central purpose clearly involves the "raising of the Djed-column," so to speak.

Alvin Boyd Kuhn lays out this analysis of the name of the Ankh primarily in his short treatise entitled *The Esoteric Structure of the Alphabet and its Hidden Mystical Language*, a delightful and insightful text which can be read in its entirety online here (among other places), and which can of course also still be obtained in print versions from a variety of sources (click here to go to the Project Gutenberg page for that text, which provides some other digital formats, including pdf).

Kuhn really warms to his theme beginning on page 12 of 88 in that online version linked above (and on page 8 of the facsimile print edition that I use at home, which can be purchased online here or perhaps ordered from your favorite local bookstore), saying:

> Nothing has been more revealing than the list of words, in English, Greek, German, Hebrew, which can be traced to the old Egyptian name of this mighty symbol ["this mighty symbol" meaning the Ankh, that is]. Its central idea, it was noted, is the production of life through the tieing or *union* of spirit and matter. The central clue to the meaning of all these derivatives is the idea of *tieing* two things together. [He then goes on to explain that the root-sound found in the word *Ankh*, the "N-K" sound, sometimes found its way into words in the order "K-N," and sometimes the "N-K" is replaced by "N-G" (note that a "G" is linguistically nearly identical to a "K" except that the "G" is voiced where the "K" is not), and hence it is also indicated by "G-N" as well as by "N-G"]. With these specifications it is possible now to discern a whole new world of meaning in many common words never deemed to have come down from so divine a lineage.
>
> It is seen first in such words as *anchor*, that which *ties* a boat to a fixed place; *knit, knot, link, gnarled, gnaw, gnash* (accounting for the odd spelling); *ankelosis*, a growing together of two bones; *anger, anguish, anxiety*, a tightening up of feelings. But most interestingly it seems to have given name to at least four joints or hinge-points (*hinge* itself seems to be another) in the human body: *ankle, knee, neck* and *knuckles*. *Lung*, as being the place where outside air unites with the inner blood, could perhaps be added. Far away as our English *join* appears to be from a source in A N K H, (N being the only letter common to both), it is certainly directly from it after all. For A N K H was the root of the Latin *jungo*, to join, N K becoming N G through the

50

Greek. From this we get *junction, adjunct, juncture, conjunction,* from the Latin past participle of *jungo,* – *junctus.* But in coming into English through the French, all these words were smoothed down to *join,* joint, and thus carried so far into English as to give us *union,* which is really junction in its primal form. With even the N dropping out we have *yoke,* that which *ties* two oxen together. And in Sanskrit it comes out as *yoga,* which in reality stands for *yonga,* meaning *union.*

He then goes on to argue that the very common prefix *con-* (which means "with" or "together" and which by itself means both of those things in Spanish) comes from the K-N sound and is thus linked to the Ankh. By the same argument, he argues that the extremely versatile English ending *-ing* derives from the same ancient symbol (this time in the form of N-G). From there, he even argues that the word *thing* can trace its lineage to the same source.

But is that all? Far from it -- in fact, he's just getting warmed up!

Next comes one that carries an impressive significance in the study, the common verb to *know,* in Greek *gnosco,* German *kennen,* English *ken.* What constitutes the knowing act? The *joining* together of two *things,* consciousness and an object of consciousness, for there must be something apart from consciousness to be known.

Further arguments bring him to *can, king, angel* (the name for the *messengers* between the heavenly realms and the earthly), *angle, nook,* and of course *Gnosis.* We could perhaps argue that along with *king* could be the corresponding word *queen,* which also contains the K-N sound. As Kuhn explores briefly when discussing the connection between Ankh and king (and we could add, queen), each individual is in some ways a king or queen, "the one who both *thinks* and *knows*" as he says: the ruler and *sovereign* (a word which itself contains the N-G sound, as does *reign*) of his or her own universe, since each individual is a microcosmic reflection of the macrocosm.

Here Kuhn (whose very surname can itself be seen to contain the K-N combination) leaves off the pursuit in this particular text, but he takes it right up again with even more profound effect in *Lost Light* (published in 1940 and available online as well as in print format). There, on page 186 of the version linked in the foregoing sentence, Kuhn provides arguments that the Egyptian tradition of the anointing of Osiris (closely connected to the raising of the Djed-column), and of anointing of the mummy with unguents prior to burial, connects to the A N K H origin as well:

> An item of great importance in this ritual was its performance always previous to the burial. It was a rite preparatory to the interment. Said Jesus himself of Mary: "In that she poured this ointment upon my body, she did it to prepare me for my burial" (*Matt.* 26:12). She was symbolically enacting the Mystery rite of the chrism, and her performance quite definitely matched the previous practices of the Egyptians, from whom it was doubtless derived. But what does such an act denote in the larger interpretation here formulated? If the burial was the descent of the gods into bodily forms, then the anointing must have been enacted immediately antecedent to it or in direct conjunction of it. The etymology of the word sheds much light upon this whole confused matter. The "*oint*" portion of it is of course the French softening of the Latin "*unct*" stem; and this, whether philologists have yet discovered the connection or not, is derived from that mighty symbol of mingled divinity and humanity of ancient Egypt -- the A N K H cross. The word Ankh, meaning *love*, *life* and *tie*, or life as the result of tying together by attraction or love the two nodes of life's polarity, spirit and matter, suggests always and fundamentally the incarnation. For this is the "*ankh-ing*" of the two poles of being everywhere basic to life. The "unction" of the sacrament is really just the "junction"

of the two life energies, with the "j" left off the word. Therefore the "anointing" is the pouring of the "oil of gladness," the spiritual nature, upon the mortal nature of living man. The "unguents" of the mummification were the types of the shining higher infusion, and they prepared the soul for, or were integrally a part of, its burial in the grave of mortality. And the Messiah was then crucified in the flesh.

In other words, Kuhn is here arguing that the scriptures are really teaching that the incarnation of every man and every woman is a form of "crucifixion in the flesh" (that is, the pinning down into a body of a spirit), the *joining* or *ankh-ing* or *yoking* of spirit and matter (or spirit upon a cross of matter). This teaching is depicted in the very form of the Ankh, and in words derived from the N-K sound. The act of *anointing* for burial was a depiction of the teaching that each human life consists of a divine element (represented by the anointing, the unguent, the "oil of gladness" which Kuhn comes right out and defines as "the spiritual nature") poured down upon (and in fact buried within) the body (the mortal, material, and animal part of our earthly existence).

This explanation is central to his argument that the interpretation of the story of *the Christ* is that it is always meant to teach of and point to the "Christ in you" (that is to say, in each and every individual) and not to a literal figure (an argument he makes throughout *Lost Light*, as well as its companion volume *Who is this King of Glory?*). If this argument is correct, then we can see that the "raising of the Djed column," could be seen (according to such a teaching) as central to our human existence in this incarnation: the process of remembering our status as *king* (or *queen*); of *knowing* and achieving *Gnosis*; or even of *anointing* our physical, horizontal, and animal nature with the "vertical component" of the Ankh-cross -- that is to say, our spiritual or even divine aspect -- and in doing so to *raise it up*.

Whether or not one accepts that this teaching is in fact an accurate depiction of our human condition, the linguistic connections that Kuhn finds between words such as *Yoga, unction, angel* and *Gnosis* to the Ankh itself -- and the conceptual connections between these words and the others to the message conveyed by the symbology of the Ankh -- are quite compelling.

To add even more strength to his arguments, we can in fact suggest even more words which appear to have strong linguistic connection to the word *Ankh*, and which are in fact words which connect to the idea of the joining of the material and the physical natures, or to the "raising" of the spiritual consciousness within our human nature that we have seen is central to the "message of the Ankh."

You may have thought of some of these yourself already, as you have been reading along. How about the word *Annunaki*, the beings from the celestial realm who apparently joined themselves to the daughters of men? At this time, I personally believe that this episode was intended to teach the same esoteric concept that has been outlined above (the teaching that we are a mixture of divine spirit and material flesh), and not intended to be understood literally, although some believe that it refers to a literal event. Either way, the name of these beings, *Annunaki*, can most certainly be argued to be connected to the word *Ankh*.

Another one which is almost certainly linguistically related is the name of the amazing complex of *Angkor Wat*, which Graham Hancock has demonstrated to be precisely 72 degrees of longitude east of the Giza pyramids in Egypt, and hence deliberately connected to Egypt (72, of course, being one of the most important precessional numbers). Would it not be too far a stretch to suggest that, given this clear longitudinal connection between the sites, and given the fact that the word *Angkor* begins with an "Ankh," that Angkor Wat was intended to be (like the sacred sites of Egypt) a "place where men and women became gods"?

While we are on the subject of precessional numbers, I have pointed out before (in my first book, and in previous blog posts) that the martial arts of China are replete with precessional numbers. Given the fact just discussed, that Giza in Egypt (source of our knowledge of the Ankh) and Angkor Wat are separated by a significant precessional number, is it not possible that the name by which the Chinese martial arts are widely known, that is to say *Kung Fu* or *Gung Fu*, contains the K-N (and the N-G) sound which Alvin Boyd Kuhn believed to be connected to the Ankh?

Critics may argue that there cannot possibly be any linguistic connection between China and ancient Egypt, and that the name *Kung Fu* is a Cantonese name (Guangdongwa) and that in Mandarin or Poutongwa the art is typically called *WuShu*. However, if we accept the possibility that the word *Yoga* itself is connected to the concept of the Ankh (and the practice of Yoga can certainly be argued to be related to the concept of "raising the spiritual" in conjunction with the physical), then it certainly seems to be a strong possibility that the practice of Kung Fu is also related to the same concept. And, in fact, there are very strong traditions in China itself that Kung Fu anciently came from India and is indeed related in some ways to the practice of Yoga. It should also be pointed out that technically, the terms *Kung Fu* (and *Yoga*) refer to a far broader set of practices and disciplines than they are popularly understood to mean (those terms are traditionally applied to a whole set of other forms of "work" or "discipline" than just to fighting movements or yoga asanas, in other words).

Other names which fit Alvin Boyd Kuhn's thesis include *Angola* in Africa, the name of which country is apparently derived from the title given to the *kings* who ruled in that land, the *ngola*. Along the same lines, it might even be argued that there could be a connection to the name of the *Hmong* people, among whom the surname *Nguyen* is very common.

Another, much more amazing connection might be suggested with the civilization of the *Inca*, whose name can most certainly be argued to have linguistic similarity to the name of the Ankh.

Most revealing is the fact that the Inca themselves did not refer to their empire or their people as "the Inca," but that this name is derived (as with the land of Angola) from the name of the *kings* of that civilization, who were called in their language *the Inka*. This fact fits the arguments of Alvin Boyd Kuhn perfectly, although to my knowledge he never mentioned it. It would seem to provide strong linguistic support to the enormous piles of other evidence pointing to ancient contact across the oceans (as well as the possibility of an ancient common predecessor civilization predating both -- the two possibilities are not mutually exclusive in this case).

There are no doubt many others which I have not thought of yet, but which you have been yelling at the screen as completely obvious: feel free to share them with me and with others through the medium of Facebook or Twitter (or through your own publication and discussion of this subject, if you have your own blog or other outlet).

And, while remaining alert to the manifestations of the incredibly important Ankh around the world, perhaps it is even more important to consider the message that this ancient sign was intended to convey, and to work to raise and *anoint* our individual consciousness and individual *sovereignty*, perhaps through Yoga, or Kung Fu, or some other path . . .

The name of the Ankh, continued: Kundalini around the world

2014 August 31

The foregoing series of posts has been exploring the evidence which suggests that the concept of "raising the Djed" is absolutely central to the ancient wisdom which was apparently given to humanity from some pre-historic source and which manifests itself in the world's sacred scriptures and traditions, from the earliest "historical" civilizations of Egypt and Sumer and Vedic India, ancient China, and around the globe to the lands of the Norse, the Americas, Africa, Australia, the islands of the Pacific, and the vast lands of Asia.

We have seen evidence that the raising of the Djed-column has a celestial component, in the cycles of the heavenly bodies of the sun, moon, stars and planets -- especially in the annual sun-cycle and the "Cross" that is created by the equinoxes (where the Djed is "cast down") and the solstices (where the Djed is "raised up").

And we have seen evidence which suggests that the same raising of the Djed-column has an individual component, in which each and every one of us has the opportunity to recognize (or even remember) the spiritual, celestial, and in fact *divine* nature inside ourselves and to *raise it up* within this material incarnation that we find ourselves in during our earthly sojourn. In doing so, we are connecting with the vertical component of the Cross discussed above, and transforming and transcending the horizontal, material, and animal portion of our human experience, according to the ancient wisdom texts and traditions.

We have seen that this process of "raising the Djed" was also symbolized by the ancient Egyptians using the Scarab and especially the Ankh, and that the name of the Ankh and the linguistic sound of the N-K has found its way into a myriad of words which are associated with the process of raising our consciousness and restoring our cast-down inner divine nature, including the word *Yoga* (a derivation of *yonga*) and the English words *king* and *queen*.

Alvin Boyd Kuhn has demonstrated, in texts referred to in that previous post, that the N-K sound at times shows up as the K-N sound, and sometimes as the N-G sound or the G-N sound.

In light of that fact, it may be instructive to examine still further manifestations of this all-important "sound of the Ankh," and see that they are in almost every case illustrative of the concept of the "raising of the Djed" that the ancient wisdom tells us is so central to our human existence.

Pictured above, for example, is a shrine in a temple in southern India, showing the intertwined and ascending serpents associated with the *kundalini,* the dormant, primordial, and divine life-force-energy in each of us, described as a serpent coiled at the base of the spine (in the Norse Eddas the World-Tree Yggdrasil is always described as having a serpent or serpents at its base) which should and can be elevated through deliberate practice.

58

Obviously, the word itself begins with the K-N sound, which Alvin Boyd Kuhn would argue to be a connection to the name of the Ankh and to the concept of the hidden divine force inside each incarnate man and woman. There is no doubt that the *concept* of kundalini is closely related to the concepts we have been discussing with the Scarab, Ankh, and Djed in previous posts, and it is hard to deny that the *name* of kundalini is closely related as well.

Here is a link to an interesting web page tracing the concept of the serpent-force of the kundalini through various world cultures.

What other manifestations of the "name of the Ankh" can we find around the world which may be similarly instructive to our understanding of this absolutely central ancient teaching? Let's have a look!

The ancient civilizations of South and Central America share a legend of a benevolent civilizing figure who was described using many different names by the Maya, the Inca, the Aztecs, and other cultures of the Americas, but whose characteristics were largely similar across many different legends from many different peoples, as Graham Hancock documented extensively in *Fingerprints of the Gods*, and as Thor Heyerdahl documented in earlier texts including *American Indians in the Pacific*.

Among the names of this divine figure are Conn, Kon-Tiki, Kukulcan, and Quetzalcoatl. The first three clearly contain the "Ankh-sound" in its K-N form. While the fourth does not, its meaning of "Feathered Serpent" (the serpent which can fly, or ascend into the heavens) is clearly related to the concept we have been discussing, and to the upwards motion of the kundalini mentioned previously.

The pyramid of Chichen-Itza (also called the Pyramid of Kukulcan) is well-known for its annual serpent-shadow manifestation, which appears each year on the equinoxes. The equinoxes, of course, create the horizontal line in which the Djed-column is cast-*down*, and so it is appropriate that the serpent in this case is seen coming *down* to earth on those days. Note that linguistically, the word *Chichen* in Chichen-Itza contains the K-N sound in its second syllable (chen), the K-sound in this case being "palatized" to the "ch-sound" (*palatization* is a linguistic term for the softening of the K-sound into a "ch-sound," as happened in English with the word *kirk*, that became *church* in southern parts of the British Isles, when the k-sounds of *kirk* were palatized into the "ch-sounds" of *church*).

Along the same lines, the sound found in the sacred name of the Ankh is also found among the Native peoples of North America in the holy name of the Great Spirit, which among different nations has been spoken as Wakhan Tankh, Wakan Tanka, and Omahank-Numakshi. The names of numerous Native American peoples contain this same sacred sound, among them the name of the Kansa or Kanza tribe, the Mohicans, the Mohawk (whose name for their people is the Kaniankenhaka), and many others too numerous to list within the scope of this short essay but which may be found through study by those interested in the subject.

60

It has already been noted in the previous post about the "name of the Ankh"[49] that the name of the Inca comes from the title given to the kings of that people: he was known as the *Inka*.

The previous point about the "Feathered Serpent" of Kukulcan or Quetzlcoatl being conceptually (as well as linguistically) related to the kundalini serpent should point us to another "winged serpent," and one who is also a "fire serpent" (Alvin Boyd Kuhn has much to say about the "fire serpent" and about the element of fire, to which he devotes an entire chapter in his masterful 1940 text *Lost Light*). The "Feathered Serpent" or "Fire Serpent" I am thinking of here is the *Phoenix,* which traditionally starts out life as a worm or serpent found inside the ashes of the previous Phoenix, and which then grows into the fiery bird that flies upwards and away -- an upwards-rising serpent which is clearly related to the upward-rising motion of the kundalini.

It is certainly possible to argue that the N-K-S sound at the end of the word *Phoenix* is related to the N-K sound of the Ankh, despite being commonly spelled -*nix*. Note also that Chinese legend describes a very important "fire bird" named Feng-huang, also called the "vermillion bird" (more discussion of Phoenix-birds around the world can be found here). That name clearly contains the N-G sound twice.

Along these same lines, we can suspect that the -nx sound at the end of the word *Sphinx* is, like the -nix sound in the name of the Phoenix, associated with the N-K sound of the Ankh.

The Sphinx was also a mythological creature, like the Phoenix, and is found in many myths in addition to being embodied in the famous Giza Sphinx. In some legends, the Sphinx is also depicted as having wings, and in the myth of Oedipus the Sphinx is depicted asking the "Riddle of the Sphinx," which relates to the lifetime of a man, and hence to the incarnation we are discussing in the general topic of the casting down of the Djed-column and the act of raising it up (in the episode of the Riddle of the Sphinx, she gives the answer in terms of the ages of a man, although it could also of course apply to a woman; in any case, it is interesting that like the Phoenix, the Sphinx in mythology is often female, although sometimes male as well -- we might conclude from this that the message was intended to apply equally to all incarnate men and women).

The monument of the Sphinx at Giza faces due east, looking towards the point of the rising sun on the day of the equinox. In *Keeper of Genesis: A Quest for the Hidden Legacy of Mankind*, originally published in 1996, Robert Bauval and Graham Hancock articulate their now-famous thesis that the monuments of the Giza Plateau reflect and model the celestial landmarks, specifically the belt of the constellation Orion and the outline of the constellation Leo (see especially pages 58 through 82).

If so, then they are clearly associated with the "raising up" of the Djed-column (the "Backbone of Osiris"). The Sphinx, who looks towards the rising sun across the north-south watercourse of the River Nile, may also be associated with that "raising up" motion – and there is reason to believe that the Nile itself was esoterically associated with the kundalini-serpent and the human backbone as well (I articulate some of the mythological evidence for this association in pages 137-147 of *The Undying Stars*). And certainly the presence of the N-K sound in the Nile-facing Sphinx upon the Giza Plateau would seem to argue for the validity of this connection.

The connection of the Nile River to the rising "serpent force" is further established by the name of the sacred Nile's counterpart in India -- the sacred River Ganges (Ganga).

The sacred nature of the Ganges to Hindu tradition needs no embellishment here -- it is well attested and continues to play a central role to this day. Clearly, the name *Ganga* can be argued to contain "the name of the Ankh," and the restorative role that the river plays according to sacred tradition would argue that this alleged linguistic connection is not spurious.

It is notable to examine the evidence that there are very profound parallels between the sacred traditions of India and those of ancient Egypt, including the reverence for the Ganges and the Nile but also between the deities Osiris and Vishnu, both of whom are described as being "cast down" (and dismembered) and then subsequently "raised up" (a connection which I explore a post from July 16, 2014 which is available online).

Interestingly enough, there is new evidence that the worship of Vishnu is extremely ancient -- including the significant discovery of Vishnu statues in the region of what is modern-day Vietnam, which Graham Hancock posted as an article on his website.
https://www.indiadivine.org/4000-year-old-vishnu-statue-discovered-in-vietnam/

While that article is noteworthy on several important levels, one point that should not be missed that is very pertinent to the present discussion is the linguistic connection that the article itself makes between the name of the Ganga in India and the name of the mighty Mekong River in Vietnam. The article calls the Mekong *Ma Ganga*, which is also the name given to the Ganges in India in the river's role as "Mother Ganga." There is certainly room to argue a connection between the names of the two sacred rivers. Here is a link to a beautiful post describing some of the points the author visited along the Mekong, and the ancient traditions which have been preserved to this day by those who hold the Mekong sacred (http://www.humanitysvessel.com/2013/07/sacred-mekong.html [unfortunately, the page appears to no longer be active]).

In light of the connections already shown between Vishnu and Osiris, and in light of the newly-discovered ancient Vishnu statuary in Vietnam, it is certainly plausible to argue a possible connection between the N-K or N-G sound of the Ankh and the N-K and N-G sounds of the Ganga and the Mekong. The reverence given to these rivers through the centuries (and the millennia) suggests the clear connection to the human process of "raising the Djed" and "restoring the cast-down" in our individual journeys as well.

Finally, it is perhaps not inappropriate to point out the undeniable linguistic connection to the Sanskrit word for cannabis or hemp, which is of course the word *Ganja.* It is well-known that Ganja is viewed as a sacred plant among Rastafari, and that it is seen as essential to the process of raising consciousness and seeing through illusions.

It can be argued that here again there may be an ancient connection to the mighty Ankh, and to the central task of raising the Djed.

Note that the varied history of the human experience provides clear evidence that it is definitely possible to achieve states of ecstasy (transcendence of the "static" or physical vehicle of the body) without the use of external plant-derived substances, and that many shamanic cultures use a variety of techniques including drumming, chanting, rhythmic breathing, dancing, and other methods to induce ecstasy without the aid of plants. However, it would be ridiculous to deny that the use of plants, including ganja, peyote, ayahuasca, and mushrooms, has also played a central role in many shamanic cultures in shamanic rituals and techniques of inducing ecstasy.

In light of this, and the assertion in the previous post[186] (which is traced out much more extensively in *The Undying Stars*) that all of the world's ancient sacred traditions are or were fundamentally shamanic but that there has been a concerted effort to rob humanity of this shamanic heritage, we must wonder whether the strict prohibitions against the use of these plants is not part of the same ancient campaign.

In any event, there is no doubt that the message of the Ankh and the raising of the Djed is absolutely central to our human experience -- and that tracing out the echoes of the N-K name of this ancient symbol can be greatly instructive.

There are certainly many more places where the name of the Ankh is hidden, waiting for you to discover!

Happy Anniversary to *Moby Dick*!

2014 November 14

November 14: On this day in 1851 appeared in print the US edition of Herman Melville's *Moby Dick*. The work had been previously published in England the month before (October of 1851) as *The Whale*, with a print run of only five hundred copies.

Melville's opus stands as a towering example of the process depicted in the world's ancient myths, by which numinous truth descends from the realm of ideas, the realm of spirit, the realm of form, and clothes itself in the massy, dirty, bloody, ugly, realm of matter . . . for the purpose of rising again back to the world of spirit, dragging the material world (and us with it) along in its train.

In the vast and alien world of whaling, Melville found a canvas broad enough upon which to wrestle with some of the greatest questions of human existence. The book, of course, is not about the endless details of whaling in which readers can sometimes become bogged down or overwhelmed, but rather the deeper questions towards which the physical "teaching aids" of the whaling life are constantly turned throughout the work.

For example, Chapter 60 is entitled "The Line," and it is ostensibly an essay upon the whale-line which connects the harpoon to the whale-boat, about its characteristics, and the various ways in which it is rigged about the boats by the crews of various nationalities, and how it is attached, and the dangers it poses to life and limb as it runs out at lightning speed when a whale is struck by a harpoon and plunges into the deep in a burst of surprise, anger and pain.

But, as with everything else in *Moby Dick*, the intricate detail of the explanation is provided with the intention of suddenly making the leap from the physical, literal details being described, across the chasm to the realm of spirit, the realm of ideas, the realm of philosophy, where the lessons of the whale-line will suddenly be seen to be a metaphor for an aspect of human existence, and where the humble hemp will be shown to crackle with metaphysical meaning.

The final paragraphs of Chapter 60's description of "the line" illustrate this "turning the corner" or "making the leap" quite nicely, although any number of other chapters in *Moby Dick* could be used to illustrate the same move:

> Again: as the profound calm which only apparently precedes and prophesies of the storm, is perhaps more awful than the storm itself; for, indeed, the calm is but the wrapper and envelope of the storm; and contains it in itself, as the seemingly harmless rifle holds the fatal powder, and the ball, and the explosion; so the graceful repose of the line, as it silently serpentines about the oarsmen before being brought into actual play -- this is a thing which carries more of true terror than any other aspect of this dangerous affair. But why say more? All men live enveloped in whale-lines. All are born with halters round their necks; but it is only when caught in the swift, sudden turn of death, that mortals realize the silent, subtle, ever-present perils of life. And if you be a philosopher, though seated in the whale boat, you would not at heart feel one whit more of terror than though seated before your evening fire with a poker, and not a harpoon, by your side.

Clearly, to take *Moby Dick* for a book about whaling, or to complain that its incessant discussion of the grimy, miserable, and often gory details of life aboard a whale-ship often gets in the way of the "adventure story" about the obsessive pursuit of the white whale by Ahab, is to completely miss the point of the book -- to take it "too literally," so to speak.

In fact, to do so is a form of getting "bogged down" in the material world, and of missing the existence of an invisible world which throbs just beneath the surface of this one, always waiting to be called forth. In many ways, it is possible to say that the recognition of this invisible world, this spirit world, in ourselves and in the world around us, is the reason we are here in the material world in the first place.

Happy anniversary to *Moby Dick*, published 163 years ago today! If you feel inclined, read a few chapters . . . and contemplate the ability of just about everything in this material world to serve as a pointer to realms of heavenly glory.

Esoterism, Mystery, and Schwaller de Lubicz: Interview on *Beyond the Veil*, September 02, 2016

2016 September 03

Special thanks to Chris & Sheree Geo for inviting me over to their *Beyond the Veil* podcast at *Truth Frequency Radio* yesterday (September 02, 2016).

The interview runs for approximately an hour, with some commercial interruptions, so we didn't get to quite finish the thought we were discussing at the end -- but it concerns an extremely important subject, which relates to the name of the show, "Beyond the Veil," and the implicit reference to *the Veil of Isis*, referred to by Plutarch in his discussion of the legend of Isis and Osiris, which I highly recommend reading in its entirety.

There, Plutarch makes reference to a famous statue of the goddess Isis at ancient Saïs in lower Egypt, on one of the branches of the Nile's delta -- and to an inscription there which was referenced by other ancient authors and by later philosophers and esoterists down through the ages. According to Plutarch:

In Saïs, the statue of Athena, whom they believe to be Isis, bore the inscription:

"I am all that has been, and is, and shall be, and my robe no mortal has yet uncovered."

The word translated "robe" is also sometimes translated "mantle," and sometimes as "veil."

The profound subject to which these lines are generally understood to be referring involves the mystery of the unseen realm, the invisible realm, the realm of spirit, the sacred realm (set apart from the ordinary realm) . . . the realm of the gods.

Anything we know of that sacred and ineffable realm clearly is never given to us by mortal or earthly agency.

This topic came up beginning around 1:15:00 in the interview, in discussion of the story of Isis and Osiris, and when I mentioned the Veil of Isis, Chris says at about 1:16:06 half jokingly, "Well, you're disclosing the Mysteries here, David!"

This is actually a very serious charge which I believe should be addressed, but because the interview only goes until about 1:17:15, I did not get an opportunity to explain the assertion by R. A. Schwaller de Lubicz that the Mysteries actually *cannot* be "disclosed" or unveiled by mortal means, just as the inscription described by Plutarch from the statue of Isis tells us.

The inscription quite plainly declares that no mortal *can* uncover the Mystery, and never has done so -- and so, whatever is conveyed to us concerning the divine realm *must come from* the divine realm.

In *Esoterism & Symbol*, first published in French in 1960 and first translated into English in 1985, Schwaller de Lubicz declares (in the very opening lines of that work):

Esoterism has no common measure with deliberate concealment of the truth, that is, with secrecy in the conventional sense of the term.

70

He then goes on to say a couple of pages later:

> Esoterism can be neither written nor spoken, and hence cannot be betrayed. One must be prepared to grasp it, to see it, to hear it. This preparation is not a knowing but a being-able, and can ultimately be acquired only through the effort of the individual himself, by a struggle against all obstacles, and a victory over the human-animal nature.

> There is a sacred science, and for thousands of years countless inquisitive people have sought in vain to penetrate its "secrets." It is as if they attempted to dig a hole in the sea with an ax. The tool must be of the same nature as the objective to be worked upon. Spirit is found only with spirit, and esoterism is the spiritual aspect of the world, inaccessible to cerebral intelligence. 3.

In all of this, everything which Schwaller is saying seems to be in accord with the declarations of the famous inscription at Saïs.

If this is true, then, that no mortal power can reveal that which belongs to the invisible realm, the divine realm, then what is the purpose of the ancient myths and scriptures? What is the purpose of esoteric writings and stories in general?

Schwaller addresses that in *Esoterism & Symbol*, as one might expect from the title of the book. It is to facilitate that awakening to the gnosis of the other realm, or (to use a better term) to *evoke* the gnosis that comes from the other realm (from the other realm, and never from this one).

The word "evoke" means to "call out for" or to "call forth" (and has as its root the same sound *voc* which we find in other words such as *vocal, vocabulary*, and *vocation*).

Schwaller says:

> Thus esoteric teaching is strictly *evocation*, and can be nothing other than that. Initiation does not reside in any text whatsoever, but in the cultivation of intelligence-of-the-

heart. Then there is no longer anything occult or secret, because the intention of the enlightened, the prophets, and the "messengers from above" is never to conceal -- quite the contrary. 75.

This important passage shows us what the myths and sacred stories are for: to point us towards the place where the answer can be found -- but which we ourselves must experience for ourselves (no one else can "experience it for us" and then hand their experience over to us to have for our own -- they can only point the way).

To use a metaphor which I believe is very helpful, the esoteric stories function in very much the same way that waxing the car and painting the fence function in the well-known first *Karate Kid* movie from 1984 (discussed here[13]). Mr. Miyagi did not tell Daniel-san to wax the car and paint the fence in order to *conceal* -- quite the contrary.

But neither does Mr. Miyagi *say* very much of anything in the famous scene in which Daniel finally gets his first glimpse into what is going on.

This is because Mr. Miyagi cannot write it down, or speak it -- Daniel-san has to feel it, and once he feels it, then he grasps it. See again the first line quoted from page 3 of Schwaller's work: "Esoterism can be neither written nor spoken, and hence cannot be betrayed. One must be prepared to grasp it, to see it, to hear it." Daniel-san had to be prepared to grasp it. To be prepared to grasp it, he first had to wax the cars and paint the fence.

Thus, I vigorously deny that I am "disclosing the Mysteries here," as mentioned at the end of that interview.

No mortal *can* disclose the Mysteries.

But just as Daniel-san *needed* (desperately) what Mr. Miyagi was pointing him towards, I believe that we also need (desperately) that *towards which* the ancient myths are pointing us.

And I do not believe the myths are trying to *conceal* anything from us.

As Schwaller de Lubicz says: "Quite the contrary."

Celestial Mechanics
and the Heavenly Cycles

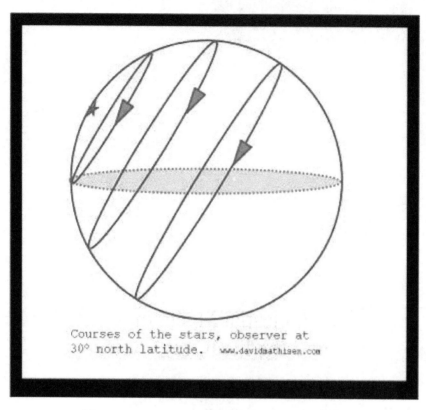

Courses of the stars, observer at
30° north latitude. www.davidmathisen.com

"Listening to the greatest navigators
our globe has ever seen"

2011 August 15

Previous posts (not included in this volume but available online) discuss the awe-inspiring voyages of the Polynesian Voyaging Society's traditional double-hulled ocean-going canoe *Hokule'a* and the nearly-lost navigational techniques they use to travel thousands of miles across the open sea using the stars, the sun and moon, and the subtle directional clues provided by the ocean swells, the colors of the sky and sea at sunrise and sunset, and the activities of marine wildlife and birds.

Polynesian Voyaging Society President and *Hokule'a* wayfinder Nainoa Thompson learned these traditional techniques from master navigator Mau Piailug (1932 - 2010), and he graciously explains some of the outlines of that ancient wisdom on the PVS website at hokulea.com, and passes it along in person to new students of the craft who participate in the ongoing voyages of *Hokule'a*.

On one of their website's pages, entitled "The Celestial Sphere," the PVS explains the celestial mechanics of the circling stars, as well as the celestial mechanics of the paths of the sun and moon throughout the year. It is well worth studying and understanding, and is explained clearly and with excellent diagrams. As explained on that page, an observer at the north pole, looking up at the night sky, would see the entire celestial sphere turning around a point directly overhead (see diagram below).

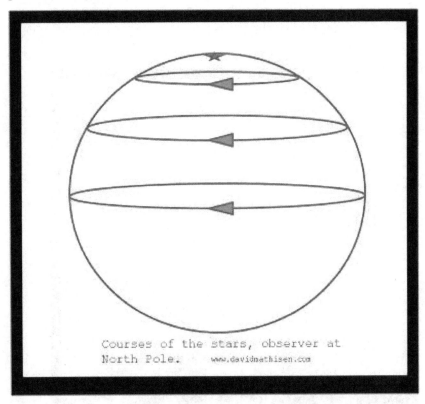

Courses of the stars, observer at North Pole.   www.davidmathisen.com

The north star, marked in the diagram above by a star, would not appear to move (actually, because it is just slightly off of the exact true celestial north pole, it would move in a tiny circle, but for purposes of this discussion it can be understood to mark the celestial north pole and thus to remain stationary while the rest of the sky appears to turn). However, stars that are located some angle away from the north celestial pole would appear to trace out a circle around the celestial north pole as the earth turns. These circles are marked in the diagram above as blue circles with arrows indicating the direction of the star's apparent daily motion (opposite to the direction of the earth's rotation).

The Polynesian Voyaging Society, however, does not typically sail across the north pole, but rather through the Pacific latitudes north and south of the equator. To an observer sailing across the equator, the apparent motion of the stars in their courses would be quite different than to our observer sitting at the north pole. Below is a diagram of the courses of the stars for an observer at the equator.

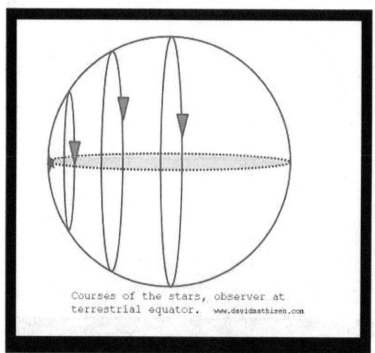

Courses of the stars, observer at terrestrial equator. www.davidmathisen.com

In this image, the surface of the ocean upon which the observer is sailing has been added as a light-grey disc. The north celestial pole and the north star are now located on the horizon at due north (left in this diagram and marked with the north star). The south celestial pole is similarly located on the horizon at due south (not depicted on the diagram above). The courses of the stars will now be perpendicular circles, but half of these circles will take place below the horizon. As the earth turns, the stars will appear to rise out of the eastern horizon in vertical courses, arc overhead, and descend on vertical courses to the western horizon, where they will again disappear.

We can now understand how an expert navigator who knows the stars could set his vessel's course by lining up known sight-marks on the beam of his vessel with a known rising star. If he were exactly at the equator, for example, and wanted to head due north, he could sight to a star known to occupy declination 0° (along the celestial equator) and keep it 90° to his course, thus pointing his prow due north. He could use such a star even after it had risen many degrees in the night sky, because he knows it rises perpendicular to the horizon and thus he can mentally trace its course back down to the horizon and use it for many hours as a guide. If he wanted to take a heading some degrees west of due north, he could sight to a star along the celestial equator but place it those same number of degrees further than 90° to his starboard beam, thus turning his prow west of north by that number of degrees (or, as described on this webpage, within one of 32 headings of 11.25° degrees each, each of which can also be divided for even greater precision).

Likewise, if he knew of a star that was at declination +19° (which is to say, 19° on the north side of the celestial equator) and he did not have a star along the celestial equator to use, he could still orient his vessel due north by lining up that star's rising point 19° forward of due east (or at 81° in his mental compass). If he desired a heading that was west or east of north by some number of degrees, he would simply make the desired adjustment to the alignment that he kept that star.

Between the north pole and the equator, the stars in their courses will not rise perpendicularly out of the ocean as they do at the equator, nor will they make circles in the sky parallel to the horizon as they would at the north pole. Instead, the north celestial pole will be tilted by the same number of degrees that the observer is north of the terrestrial equator. In the diagram below, the observer has proceeded north from the equator to a latitude of about 30° north, and because he is going towards the north star it is rising up out of the ocean as he proceeds north (remember that it was on the horizon at due north when the vessel was at the equator). As it rises up, the stars in their courses which appear to circle the north celestial pole due to the rotation of the earth will still trace out perfect circles, but these circles will now be tilted as indicated in the diagram.

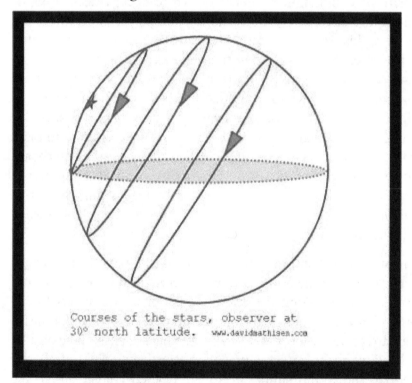

Courses of the stars, observer at 30° north latitude. www.davidmathisen.com

The navigator could still set his course by a star rising along the celestial equator or rising at a known declination on either side of that celestial equator (such as our star tracing a circle at a declination of 19° on the north-star side of the equator), but he must remember that the star no longer rises perpendicularly as it did at the equator, so as it climbs higher in the night sky he must mentally draw it back to the horizon at an angle corresponding to his vessel's northerly latitude.

All of this fascinating detail is described on the Polynesian Voyaging Society website page discussing the Celestial Sphere and also on this page entitled "Holding a Course."

Perhaps the most fascinating detail of the techniques that Nainoa Thompson uses is described on the page entitled "Hawaiian Star Lines and Names for Stars." There, the text explains that, "To help remember the pattern of stars in the sky, Nainoa Thompson has organized the sky into four star lines, each line taking up about one fourth of the celestial sphere." These star lines are quadrants of the celestial sphere like the wedges of an orange cut into four equal wedges. As the earth turns, these wedges or star lines – each containing recognizable constellations such as Orion or Scorpio or the Great Square of Pegasus – rise up out of the eastern horizon and then move overhead, setting later into the western horizon.

What is so fascinating about this mental construct created by Nainoa Thompson is the fact that it sheds some light on the very ancient practice of dividing up the celestial sphere, a practice dating back at least to the very ancient Babylonian mythological records from around 1700 BC (and perhaps even earlier than that, if you believe there was an advanced civilization which bequeathed its knowledge to the ancient Sumerians, Babylonians and dynastic Egyptians, from whence that knowledge was passed on to other successive cultures including the Greeks and the Celts and others).

In Appendix 39 of the indispensable if often-mysterious treatise *Hamlet's Mill: An Essay on Myth and the Frame of Time*, by Giorgio de Santillana and Hertha von Dechend (1969), the authors discuss the division of the celestial sphere found in ancient mythology. There, they explain that in the Babylonian creation epic known as *Enuma Elish*, we learn at the end of the Fourth Tablet and beginning of the Fifth Tablet that Marduk / Jupiter surveyed the heavens and the earth and divided up the world, and specifically that he made "Anu, Enlil, and Ea" to occupy their places, and that he "founded the station of Nebiru to determine their (heavenly bands)" in the translations cited by de Santillana and von Dechend (430 - 431).

Later, the authors explain that these "ways of Anu, Enlil and Ea" were divisions of the celestial sphere, bands running parallel to the celestial equator rather than dividing it up like a quartered orange the way Nainoa Thompson does. Each of these ancient Babylonian celestial bands was approximately thirty degrees wide:

> The "Way of Anu" represents a band, accompanying the equator, reaching from 15 (or 17) degrees north of the equator to 15 (or 17) degrees south of it; the "Way of Enlil" runs parallel to that of Anu in the North, the "Way of Ea" in the South. 434.

Using the spheres in the diagram above, in which the largest circle represents the celestial equator, the reader can easily envision these ancient "ways" dividing up the celestial sphere. Later, the authors discuss this concept even further, and tie it explicitly to the great navigators of Polynesia:

> Mesopotamia is by no means the only province of high culture where the astronomers worked with a tripartition of the sphere -- even apart from the notion allegedly most familiar to us, in reality most unknown – that of the "Ways" of Zeus, Poseidon, and Hades as given by Homer. The Indians have a very similar scheme of dividing the sky into Ways (they even call them "ways"). And so have the Polynesians, who tell us many details about the stars belonging to the three zones (and by which planet they were

"begotten"); but nobody has thought it worth listening to the greatest navigators our globe has ever seen; nor has any ethnologist of our progressive times though it worth mentioning that the Polynesian megalithic "sanctuaries" (maraes) gained their imposing state of "holiness" (taboo) when the "Unu-boards" were present, these carved Unu-boards representing "the Pillars of Rumia," Rumia being comparable to the "Way of Anu," where Antares served as "pillar of entrance" (among the other "pillars": Aldebaran, Spica, Arcturus, Phaethon in Columba). 436-437.

The fact that this was all written and published before the rediscovery of the non-instrument navigation techniques that had been preserved among the people of Satawal and before the first voyage of *Hokule'a* on which Mau Piailug was the navigator is noteworthy -- it indicates that the authors of *Hamlet's Mill* were onto something, although they could not know it.

The fact that Nainoa Thompson has found it useful to divide the sky up into four "star lines," much the way the ancients including the ancient Polynesians divided up the sky into three "ways" is equally significant, and indicates that the ancient mythologies may well have preserved knowledge that was used for open-ocean non-instrument navigation as well.

In fact, it is quite clear that the wisdom and ocean lore preserved by Mau and his ancestors, and passed on to Nainoa Thompson and the other members of the Polynesian Voyaging Society -- where it is still used to great effect on amazing deepwater voyages across the mighty Pacific and beyond -- may be one of the most significant pieces of ancient wisdom that somehow survived through the ages (even though it came very close to dying out).

It can provide a window onto mysteries of mankind's ancient past that we might never have otherwise understood, or might have only been able to guess at without practical testing. Whether or not one believes that there may have been actual ancient contact between people who are traditionally thought to have been isolated by the mighty oceans is actually less important than the fact that such understanding of and division of the celestial sphere is clearly very handy for those who venture out into the great deeps in traditional vessels without modern instruments. Where it appears in other ancient cultures, we might suspect that some form of similar navigational skills might also have accompanied the celestial knowledge that was preserved in those traditions.

In this way, it appears that all of mankind owes the Polynesian Voyaging Society, and Mau Piailug and Nainoa Thompson, a tremendous debt of gratitude and respect.

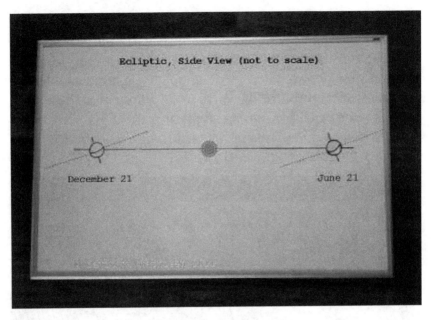

Ecliptic, Side View (not to scale)

December 21   June 21

Winter Solstice, 2011

2011 December 20

Earth is rapidly approaching the point on its orbit at which the north pole points directly away from the sun (while still at its normal angle, which is inclined from the ecliptic plane by ninety degrees less the 23° 26′ of axial tilt of the earth), known as the December solstice. For the northern hemisphere, it is winter solstice.

As various articles available on the web explain, there is one precise moment at which earth will pass through that specific point on the orbit, although it will be a different clock time in all the different parts of the globe when it happens (and the calendar date will be different depending on your location in relation to the arbitrary convention of the international date line).

Here are some diagrams which may be helpful in understanding the celestial mechanics associated with the solstices. Above, the earth is drawn at the two solstices, the December solstice on the left and the June solstice on the right.

The sun is depicted at the center, and the horizontal line represents the plane of the ecliptic. The plane of the ecliptic is the plane of earth's orbit around the sun, which (if seen from the edge-on view depicted here) would look like a horizontal line. If you think of the orbit of the earth as an ellipse around the sun, and turned that ellipse into an elliptical plate of glass, then that glass ellipse would be a plane, the plane of the ecliptic. Seen from the side it would just be an edge of glass.

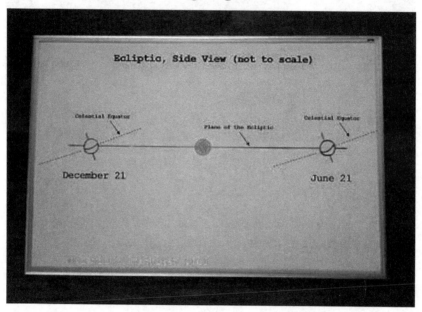

At any point in the northern hemisphere, there is a point in the sky around which the entire sky appears to rotate. This point corresponds to the north pole on earth and is the point in the sky to which the north pole points -- the north star, Polaris, is very close to that point in the sky. Ninety degrees down from this point is the celestial equator, an imaginary line in the sky which is the projection of the earth's equator, a great circle which arcs across the sky at the same angle all year long, depending upon your latitude (you can see more diagrams and discussion of this concept in this blog post, and follow the links to the excellent website of the Polynesian Voyaging Society website, which explains it in even greater detail).

The celestial equator is marked in the above diagram by a dotted line projecting earth's equator into the sky. Again, it forms a great circle, but because that circle is seen "edge-on" in the drawing, it appears as a straight line.

Now, using this diagram, you can envision the reason that the ecliptic reaches its highest point above the celestial equator in the night sky on the winter solstice (the December solstice in the northern hemisphere) and reaches its lowest point below the celestial equator in the night sky on the summer solstice (the June solstice in the northern hemisphere). Looking at the diagram above, you must first consider the fact that observers on the earth-globes in the drawing only see the night sky when they are facing away from the sun (on the side of the globe facing leftward for the left globe at December 21, and on the side of the globe facing rightward for the right globe at June 21). When the earth's daily rotation brings them around to face the sun, it will of course be daytime and we will examine that situation momentarily.

In the night sky, the stars along the celestial equator include the three bright stars of Orion's belt, and the left wing-tip of Aquila the Eagle (Rey 113). The ecliptic passes through the zodiac constellations, and the planets (which orbit the sun along planes that are roughly aligned with earth's orbital plane to create the familiar shape of the solar system) track across the sky along the ecliptic as well.

Obviously, if the ecliptic is below the celestial equator in the summer night sky, and above the celestial equator in the winter night sky, there must be a point in time at which the ecliptic crosses the celestial equator once on the way to being above it and once more on the way to being below it, for a total of two crossings each year. Those crossings are the equinoxes.

On the "day side" of the equation, the ecliptic will mark the path that the sun travels as it arcs across the sky. In the illustration below, an attempt has been made to "shade in" the half of the globe that is not receiving sunlight in the two depictions of earth (the left half in the earth on the left and the right half in the earth on the right). You can see why the north pole (and all the points above the Arctic circle) stays dark all day on deepest days of winter (including of course the December solstice) and why the south pole (and all the points between the Antarctic circle and the south pole) stay light all day on the same December solstice. On the daytime side of each earth-globe in the diagram below, the ecliptic is above the celestial equator in the summer (the June solstice for the northern hemisphere) and below the celestial equator in the winter.

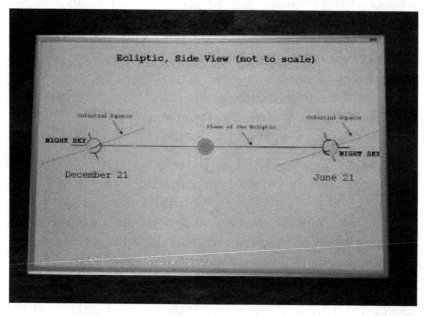

If you pay much attention to the arc that the sun takes through the sky on various days of the year, this will all seem pretty obvious: the sun's arc follows a much higher path through the sky in the summer than it does in the winter. The celestial equator is not marked with visible markers during the daytime the way it is by night with familiar stars such as Orion's belt, but you can always imagine where it is if you know your latitude and the location of the celestial pole and then envision the great circle that runs across the dome of the sky ninety degrees down from that point. The sun's ecliptic path will be above this line in summer

88

(after crossing it on the spring equinox) and below it in the winter (after crossing it again on the fall equinox).

The fact that the steepness of the sun's arc is much more vertical during the summer (most vertical on the summer solstice) and much shallower during the winter (most "laid down" on the winter solstice) can be envisioned by drawing yet another line on our diagram, this time representing the horizon to an observer in the northern hemisphere (who is represented by a light blue rectangle on each earth at December solstice and June solstice).

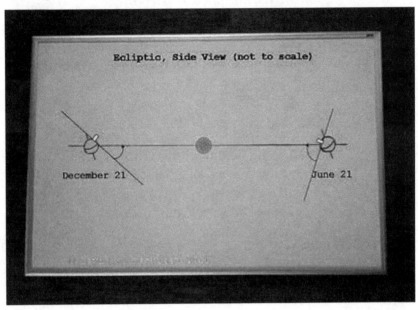

Note that the blue rectangle is at the same latitude in both diagrams: if we imagine the earth rotating as it does each day around its axis, the blue rectangle on the left globe will swing around to being as close to the ecliptic as it is on the right globe, but it will be facing away from the sun (night) on the left globe, while it will be facing towards the sun (day) on the right globe. In the above diagram, the rectangles are both depicted as facing towards the sun (day -- in fact, noon).

In this diagram, the horizon as it appears to the observer at that point is drawn as an additional line. The dotted lines depicting the celestial equators have been removed for greater clarity. However, the line representing the plane of the ecliptic is still drawn, because it indicates the plane the sun takes through the observer's sky during the day. This diagram should make quite obvious the reason that the arc-path of the sun on winter solstice is as low in the sky as it ever gets, and as vertical as it ever gets on the day of summer solstice. The angle of the sun at noon has been indicated with an angle-arc arrow in each case, and the angle can be seen to be much greater (a much higher sun-path) on the June solstice for the northern hemisphere observer.

These diagrams should help explain the mechanics of what is going on during the great annual cycle that gives our year its rhythm. This cycle is every bit as important to the cycles of life on earth as is the daily rythm of night and day – and, of course, the length of each night and each day is directly impacted by this larger cycle.

We are all aware to some degree of the power of the daily cycle upon our minds and our bodies and our energy levels – just try working a "graveyard shift" job for several weeks or months and see if you notice any effect. Certain cultures preserve the knowledge of the fact that different organs of our human bodies have different times during the day at which they are at higher and lower energy levels (traditional Chinese medicine, for example, places great importance on these cycles).

If this is true of the daily cycle of night and day, it stands to reason that it is also true for the annual cycle that is marked by the solstices and equinoxes. For this reason (as well as many others), it is very beneficial to understand and appreciate the annual rhythm of which tomorrow's solstice is such an important station. As you go through the solstice, you can imagine the moving pieces of earth, sun, ecliptic (crossing and then yawning above or below the celestial equator), light and dark.

If you did not understand all those aspects of the solstice before, I hope that this discussion has been helpful.

How the earth-ship metaphor helps explain the sun standing still at the solstices

2011 Dec 23

Many people are aware of the fact that the word "solstice" means "sun stand-still" or "sun stationary," but may not understand exactly *why* the sun seems to come to a standstill at each solstice before turning around and starting towards the opposite solstice.

The word itself is composed of two words from Latin, *sol* (which obviously refers to the sun and gives us the word "solar"), and *sistere* (the Latin verb meaning "to stand," found in words like "resist" and also in less widely-used modern words such as "interstitial," meaning "occupying a small or narrow place in between two other spaces," which is a synonym for the equally obscure but interesting word "liminal," which has to do with things occupying a boundary area between two other areas and comes from the word for "a threshold," and which can metaphorically apply to concepts that occupy a "third space" between two binaries or polar opposites).

Many discussions of the solstices mention the fact that the sun's rising point (and setting point) moves back and forth along the eastern (and western) horizon during the year, and that the solstices mark the furthest north and furthest south points along that annual path. These discussions may also mention that the sun appears to pause at each of these two "turnaround points" and rise (and set) there for a few days in a row before moving back

91

in the other direction. Why does the sun linger at the solstices, when it does not linger at the equinoxes, for example?

To conceptualize why this important "standstill" takes place at each solstice, it is helpful to think of the earth as an old sailing ship from the classic days of sail (such as the full-scale model of the *Columbia Rediviva* pictured above -- the original *Columbia* being the first US ship to circumnavigate the globe, in 1790).

In this "earth-ship" metaphor, the bowsprit (that spar which extends forward from the vessel's prow, right above the figurehead) represents the north pole of the earth. The lantern which hangs from the ship's stern (not visible in this photograph) would represent the south pole of the earth.

Now, the interesting thing about the earth-ship is that, as it makes its annual journey around the sun, its prow stays pointing in the same direction all the time. You might expect that the ship would turn as it goes around the sun, like a racecar would when it goes around the track, but in this case of the earth-ship, the bowsprit of the north pole and the lantern of the south pole always maintain their same orientation as the ship sails around the sun (see diagram below).

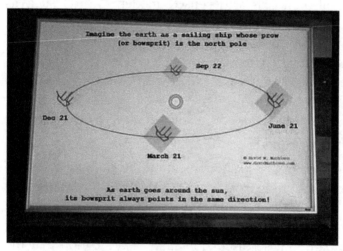

The ship in the above diagram is shown without all the rigging, for simplicity, and it is angled to show the earth's obliquity to the

92

ecliptic plane. Also, if this ship were really the earth, it would be rotating along the axis of the bowsprit once per day, which would make all the sailors on it quite uncomfortable and perhaps even a bit seasick, but for the purposes of this analogy we can imagine that the ship is not rotating like that. The important point to notice is that the bowsprit stays pointing in the same direction all the time as it goes around.

The summer solstice for the northern hemisphere is when that bowsprit is directly pointed at the sun as the ship sweeps by on its rotation. In fact, there is one moment in time at which that bowsprit will point most directly at the sun, and that moment in time will fall somewhere around the calendar date of June 21. The aft or stern of the ship will then be pointing most directly away from the sun, and thus it will be winter solstice to those in the southern hemisphere.

Likewise, there will be one moment on the other side of the orbit at which the lantern in the stern of the ship will point directly at the sun (when the bowsprit is pointing directly away). This moment in time will fall somewhere around the calendar date of December 21 -- we just passed it a couple days ago, in fact. This point will be the winter solstice for those in the northern hemisphere, and the high point of summer for those in the southern hemisphere.

Halfway between these two points, the ship will drift past the sun with the bowsprit and stern lantern aligned with the motion of the ship, and the sun will be "broadsides" to the ship. These two points (when the ship passes by the sun such that it is directly broadsides) will be the equinoxes.

Now, to understand why the sun appears to "pause" or "linger" or even "stand still" at each solstice, while it does not do so at the equinoxes, think of yourself as an observer tied to the bowsprit of the ship (or occupying the position of the figurehead, if you prefer). As the ship approaches the solstices, your view of the sun will be directly in your face as you pass by it at the solstices, and it will not appear to move very much as you stare at it while the ship moves along its orbit in a motion that is almost lateral to the sun.

However, as the ship gets past these points and starts to race down the track towards the next equinox, your view of the sun will start to change rather rapidly. As you get going, your face will be pointing, not at the sun, but towards the direction that earth is moving on its orbit. You will literally "whiz by" the sun at the equinox, and its motion will be quite rapid as you pass it by.

As you approach the next solstice, however, your perspective will again begin to change, and the speed of the sun's passing will seem to slow down again, until it becomes slowest on the days surrounding the solstice itself.

If you have a more mathematical sort of mind and prefer to examine this concept using a graph instead of a metaphor, you can think of the apparent motion of the sun throughout the year as a sinuous wave (like a sine wave) on a graph. We have seen in previous posts that the ecliptic actually traces out a sine wave as it moves up and down across the celestial equator during earth's progress around the sun (this back-and-forth motion of the ecliptic is discussed in this previous post,[85] and a helpful visual display of the sun's noon point on the ecliptic moving up and down across the celestial equator throughout the year can be seen using stellarium, leaving the sun at noon and advancing the days).

An excellent discussion which contains a good graph of the sun's path back and forth across the celestial equator throughout the year can be found at star-www.st-and.ac.uk/%7Efv/sky/standstill.html -- the entire page is actually discussing the concept of lunar standstill and in particular the major standstill of 2006, but the very first graph is a solar wave, and the discussion incorporates the sun's annual motion and is very well done and thorough.

Looking at the sine-wave-like graph of the sun's motion, you can also conceptualize why the sun "stands still" at solstices and not at equinoxes or other days between the solstices. The solstices are at the top and the bottom points of the wave. The vertical change on the graph slows down significantly as it approaches a top (or a bottom), and comes to almost a complete halt (nearly flat) at the actual top and bottom (the solstices and the days on either side of the solstices). However, as the wave begins to dive from the upper solstice down towards the lower one (or to climb from the lower solstice towards the upper one), the wave becomes much steeper. As it screams through the equinoxes, it is changing its vertical position quite rapidly from one day to the next.

You can verify this phenomenon for yourself by marking the sun's rising point on the horizon throughout the year. You can also use the handy calculations provided by the US Naval Observatory's outstanding website here. That calculator enables you to enter latitudes and longitudes and elevations (or cities in the US) and determine the rising times, rising azimuths, and setting times and setting azimuths, for various heavenly bodies (including the sun). The tables below show the rising azimuths for the sun from San Francisco, California during the days surrounding the fall equinox and the days surrounding the winter solstice, for the year 2011.

For San Francisco, CA latitude N37°

Date in 2011	Azimuth in degrees		Azimuth Degrees
September 01	79	December 01	117
September 02	79	December 02	117
September 03	80	December 03	118
September 04	80	December 04	118
September 05	81	December 05	118
September 06	81	December 06	118
September 07	82	December 07	118
September 08	82	December 08	118
September 09	83	December 09	119
September 10	83	December 10	119
September 11	84	December 11	119
September 12	84	December 12	119
September 13	85	December 13	119
September 14	85	December 14	119
September 15	86	December 15	119
September 16	86	December 16	119
September 17	87	December 17	119

September 18	87	December 18	119
September 19	87	December 19	119
September 20	88	December 20	119
September 21	88	December 21	119
September 22	89	December 22	119
September 23	89	December 23	119
September 24	90	December 24	119
September 25	90	December 25	119
September 26	91	December 26	119
September 27	91	December 27	119
September 28	92	December 28	119
September 29	92	December 29	119
September 30	93	December 30	119
		December 31	119

Note that the USNO data only gives whole number azimuths, but even so the difference in the rate of change is quite apparent. Note that during the month of September, as the earth approaches the September equinox point on its orbit, the rate of change is quite dramatic. The sun's rising azimuth changes a full degree every couple of days (note that it is rising north of due east – due east being ninety degrees and true north being zero degrees – as it approaches equinox, and then rises at ninety degrees at equinox, and proceeds south of due east after that).

However, during the month of December (as earth approaches the December solstice), the rate of change has slowed significantly. Now, the rising azimuth of the sun for an observer in San Francisco is much further south, at 117 degrees at the beginning of December (fully 27 degrees south of due east). As the earth progresses, it creeps south still further, to 118 degrees, and then to 119 degrees (which it reached on December 09 for an observer in SF). After that, it continues to rise at 119 degrees for the remainder of the month of December! What a difference from the rapid change of azimuth taking place around the equinox point on the orbit.

This discussion should help you understand the reason why the solstices are the "stationary" places of the sun during the annual rhythm created by earth's orbit and the tilt of earth's axis.

The horizon and the scales of judgment

2014 April 12

In ancient Egyptian texts, such as the Egyptian Book of the Dead, there is much usage of the symbol of the horizon. In the Book of the Dead, for example, the Sun-God Ra is often described as rising on the horizon, and sinking to rest on the horizon.

In the "Hymn of Praise to Ra When He Riseth Upon the Horizon, and When He Setteth in the Land of Life" (which can be found in this online transcription of E. A. Wallis Budge's translation of various Book of the Dead texts contained in the British Museum, a bit more than half-way down the very long web page, in a section entitled "APPENDIX (from the Papyrus of Nu, Sheet 21)" we read:

> Those who have lain down in death rise up to see thee, they breathe the air, and they look upon thy face when the disk riseth on the horizon.

Immediately before that, in "Another Chapter of the Coming Forth of a Man by Day Against His Enemies in Khert-Neter" which can be found on the same web page in the "APPENDIX (from the Papyrus of Nu, Sheet 13)" we read:

> I have divided the heavens. I have cleft the horizon. I have traversed the earth in his footsteps. I have conquered the mighty Spirit-souls because I am equipped for millions of years with words of power.

These are obviously very important and moving passages, but what do they mean? Conventional academia teaches that these passages, along with the rest of the Book of the Dead, express the Egyptian hope for the soul in the afterlife, that the souls of those who "have lain down in death" will somehow rise up to see the sun again and breathe the air again, and that this hope is somehow connected with the knowledge of "words of power."

Alvin Boyd Kuhn, however, gives a very different interpretation, and one which he backs up with hundreds of pages of evidence from these and other ancient sacred texts. He argues that the conventional reading just described falls into the trap of literalism – of reading as literal a piece of literature which was *never meant* to be understood literally, but which instead attempts to convey profound and nearly-ungraspable truths in the form of symbols and metaphors through which those truths *can* in fact be apprehended. The Book of the Dead, he argues, is not talking about those who are *literally* dead, but rather those on earth right now, who are *metaphorically* passing through the underworld, who can be described as "the dead" because they are in fact immortal spirits who have taken on the physical mortal flesh of material existence.

Using this understanding, the metaphor of the horizon becomes, in Alvin Boyd Kuhn's reading, a wonderful symbol of human existence, because it depicts the union of the spiritual realm (the sky) and the material realm (the earth beneath). It symbolizes the human condition, because in our incarnate state we are a union of soul and matter. In *Lost Light*, Kuhn writes:

> And the horizon is half way between heaven and earth, typing, as always, spirit and matter, the two ends of being. The momentous information, then, which is vouchsafed to man in this recondite fashion is that he, as a creature in a stupendous cyclical evolution, stands at the point exactly midway between the beginning and end of the complete area to be traversed. [. . .]
>
> [. . .] this fateful line would at the same time mark the boundary between the two natures in man's constitution, the earthly and the heavenly. 449-450.

Kuhn goes on to explain that once this metaphor is understood, all the imagery in the ancient Egyptian texts referring to a great battle which is fought "on the horizon" can be understood to refer to the great metaphorical battle which each individual must wage during his or her time in this material realm in which the spirit is enmeshed in earthly matter. This world in which we find ourselves, composed of "the earthly and the heavenly," is the place where we must learn to reconcile the two natures. Kuhn declares:

Straight and clear is Egypt's proclamation of this sterling truth: "He cultivates the Two Lands; he pacifies the Two Lands; he unites the Two Lands." Man is "the god of the two mysterious horizons," and the glowing pronouncement of his final evolutionary triumph is given in the words: "Thou illuminest the Two Lands like the Disk at daybreak." 451.

Closely related to the symbology of the horizon is the symbology of the Judgment Hall in the Egyptian scriptures. Here again, the conventional interpretation of the famous Judgment Hall scenes in the Book of the Dead is that the text refers to a judgment which takes place over the soul in the afterlife, but Kuhn demonstrates that, like the metaphor of the horizon, the Hall of Judgment symbolizes the soul's journey in this life, during which our daily actions and experiences, our "living activity and expression," is measured and recorded as if in a book (451). Below is an image from the famous Hall of Judgment vignette in the version of the Book of the Dead found in the Papyrus of Ani.

In fact, as Alvin Boyd Kuhn demonstrates in *Lost Light*, our judgment in the horizon of this life and in the scales of the Hall of Judgment are closely related. For one thing, the Judgment Hall in the Book of the Dead is also referred to as the "Hall of Two Truths," and we have already seen that the symbology of the horizon depicts the "two truths" of mankind's human condition -- that we occupy the boundary between the realms of the earthly and the heavenly, and that we embody and encompass both aspects in our human nature (*Lost Light*, 483). Also, the scales of judgment are often described as being located upon the horizon, or upon a hill or mount, which is a terrain feature which suggests the horizon.

Intriguingly as well, the zodiac sign of the Scales of Judgment -- Libra -- is located upon the "horizon" of the zodiac wheel, for Libra begins just at the border formed by the September equinox, where the ecliptic path of the sun crosses below the celestial equator again after the summer months in which the sun's daily ecliptic path has been above that line (see diagram below). As Alvin Boyd Kuhn writes about this symbology:

> And this at once opens the way for the introduction of the whole range of symbolic values connected with the sign of Libra, the Scales of the Balance, and the Scales of Judgment. And precisely at the horizon's western terminus stands the Libra sign! The Judgment is a corollary aspect of the horizon typism and will be treated in a following chapter. 451.

We can now see that there are plenty of cogent reasons for interpreting the passages in the Book of the Dead, including those cited at the beginning of this essay, as describing the condition of the individual soul *in this life*, and not in a life hereafter. But what are the implications of the Book of the Dead's teachings which declare our earthly incarnation to be a "battle at the horizon," or a "weighing of the heart" in the scales of truth? And what does the Book of the Dead mean to tell us when it depicts the triumphant soul declaring that it is "equipped for millions of years with words of power"?

We have already seen in a previous meditation upon the Hall of Judgment scenes in the Book of the Dead that the texts surrounding the scene put a great emphasis upon the soul "telling the truth" or "not going about with deceitful speech while upon the Earth." And, we have seen in our examination of the famous command inscribed in stone at the Oracle at Delphi that "Know thyself" appears to have been understood by Plato and by other philosophers including Plutarch as a command to realize that we are immortal souls enfleshed in a body during successive incarnations which each have an important purpose. If this is the case, then the Book of the Dead's emphasis upon "declaring the truth" and on "not going about with deceitful speech" while we are here between or upon the horizons -- while we are being weighed and measured in the balance-scales of the Hall of Two Truths -- may also be telling us that an important part of our activity in this life includes first learning (or, more precisely, *remembering*) the truth of our true human condition, appreciating all the ramifications of this truth, and then speaking and acting in a way that acknowledges this truth about who we are.

Acting deceitfully (or "going about with deceitful speech") would then include efforts to deny or obscure the truth that, as Kuhn eloquently describes it:

this life is the period of its [the soul's] trial and testing. The soul is drawn here to exercise her undeveloped powers, as Plotinus has so well told us. Without such a testing she would remain forever ignorant of her own latent capacity, or would never bring it to expression. Here is where she is thrown into the scales of balance, in Libra on the horizon, and here is where she is being weighed. 485-486.

One way this truth is obscured is by those who insist that ancient scriptures such as the Book of the Dead were anciently understood literally and woodenly, describing a fantastic judgment scene in an afterlife-world: an interpretation which obscures all the teachings we discover when we see that these texts employed exquisite metaphors to convey profound truths about our human condition in *this life* itself. It is this knowledge, Alvin Boyd Kuhn tells us, that the ancient Egyptians believed to be the touchstone for the soul during the daunting passage through the material life in each successive incarnation. He writes:

> The *Ritual* [that is to say, the Book of the Dead] speaks of the secret knowledge of the periodicities and cycles of incarnation as requisite to render safe the passage through all the trial scenes in the Judgment Hall [that is to say, the trials of this life here on earth]. The salvation of the deceased depended on his having the facts treasured up in his memory. As the soul walked through the valley of the shadow of death, his security depended upon his knowledge that he was a divinity threading his way through the dark underground labyrinth of matter. His memory of his intrinsically deific nature would be his safeguard; and this memory was his book of life and character, for it was his own self, come hither to purify itself of dross. 489-490.

These, then, constitute the "words of power" which equip the soul "for millions of years." That is to say, it is the soul's memory of its eternal, divine nature that equip it for the long journey through successive incarnations over vast stretches of time, and

102

which safeguards the soul from being swallowed up in the animal nature of the physical body, forgetting where it came from.

We should all be grateful to Alvin Boyd Kuhn for illuminating the esoteric truths contained within the Egyptian Book of the Dead, and for demonstrating the importance of approaching those texts esoterically, rather than literally. And, we should be grateful as well to the ancient sages who composed those incredible texts, and passed on their understanding of the human condition to generations who would live thousands of years after the kingdom of Egypt had been buried beneath the sands of time.

"Thou journeyest on the road whereon the gods journeyed"

2015 June 25

The earth has now passed the point of the June solstice, and the sun is now beginning its long decline towards the bottom of the year (for those in the northern hemisphere: for the southern hemisphere, the sun's path will now be climbing in the sky).

This long descent of the sun each year furnishes the ancient mythology of the world with a powerful metaphor for the descent of the god out of the fiery realm of the heavens and into the miry realm of earth and water: the incarnation.

In *Lost Light*, Alvin Boyd Kuhn argues that all the ancient myths of the god or goddess descending to earth -- there to be cut up and scattered abroad, or to search for the cut up body of the god, or to dwell among mankind, or to be hidden in a secret place --

were intended to teach that individual men and women (and babies and children too) each contain a divine and immortal spark, temporarily buried inside material flesh.

In other words, the gods and goddesses of the myths are *in humanity*.

Examples cited by Kuhn in chapter VII of *Lost Light* include the tearing to pieces of Dionysus or Bacchus, the cutting in pieces of Osiris, the descent to earth to search for the pieces of Osiris by Isis, the descent of Persephone into the underworld for six months of each year, the bringing of fire to mankind by Prometheus (here Kuhn quotes Thomas Taylor who says that Prometheus concealing the fire in the Thyrsus reed was representative of "leading the soul into the body"), the descent of Eros the god to wed the human bride Psyche, the descent of Orpheus into the underworld to pursue Eurydice, the descent of Hercules into Hades to bring up the three-headed dog Cerberus, and many more.

In support of his argument, Kuhn cites a remarkable passage from the Pyramid Texts of the Egyptian king Pepi I (reigned circa 2289 BC to 2255 BC). On page 121 of *Lost Light*, Kuhn writes:

> In Egyptian scriptures we encounter the promise that "if Pepi falleth on to the earth, Keb [Seb] will lift him up." Pepi here stands for the divinity in man, the god come to earth. To him in another place it is said: "Thou plowest the earth . . . Thou journeyest on the road whereon the gods journeyed." Here is identification of the earth as the place to which the gods were sent to travel the road of evolution.

Note what Kuhn is asserting in the above lines: that the scriptures in the Pyramid Texts are identifying the "road whereon the gods journeyed" as this human world of the incarnation.

In other words, the gods and goddesses of the myths are *in humanity*.

Here is a more modern translation of the same passage from the Pyramid Texts of Pepi I, this time from the 2005 publication *Ancient Egyptian Pyramid Texts,* translated by James P. Allen and edited by Peter der Manuelian, available to read in its entirety online here (but worth owning in hard copy as well, of course). On page 107, we find the passage that Kuhn quotes in his argument: it is from Recitation 37, *Address to the Spirit as Osiris in the Duat:*

> The earth has been hacked for you and a presented offering laid down for you before you, and you will go on yonder path on which the gods go.

The implications of this passage are profound. The ancient text is here confirming the interpretation presented by Kuhn, and elaborated in *The Undying Stars,* that the celestial motions of the sun, moon, stars, and planets which are allegorized in all the ancient myth-systems of the world (including those in the Old and New Testaments of the Bible) were intended to convey an esoteric teaching: that the gods descend and walk the path of the underworld, and that this path is the path we ourselves are walking ("thou journeyest on the road whereon the gods journeyed").

The gods, of course, are the sun, moon, planets, and stars -- after all, are the planets not named Mars, Mercury, Jupiter, Venus and Saturn? And yet, as they journey along their "road" through the heavens, each of these eventually plunges down into the western horizon, there to toil through the underworld until they rise again on the eastern horizon. This allegorizes the human condition in incarnation, divine sparks plunged into bodies made of clay (water and earth).

And the assertion that the gods and goddesses of the myths are to be found inside each one of us is reinforced in this passage by the text's calling attention to the "road" or "path" on which the gods go. The sun, moon, and visible planets travel through the sky along the ecliptic band: this is the road whereon they journey, and they cannot deviate from it. That ecliptic band contains the stars which make up the zodiac constellations – and we have already seen that the ancients taught that there is a direct zodiac correspondence in the human body ("as above, so below"). In other words, the zodiac not only stretches across the infinite heavens in the macrocosm, but is also contained in the microcosm of the individual body of each man or woman – from head to toe (Santos Bonacci has many outstanding videos in which he expounds upon this concept).

So, if the planets (that is to say, "the gods") travel along the zodiac path through the heavens, but that zodiac path is also reflected within the human body, we can see once again the clear teaching that . . . the gods and goddesses of the myths are *in humanity.* And, as discussed in a previous post (10/03/13), there are ancient wisdom traditions which teach a direct connection between each major organ in our bodies and a specific planet in the heavens.

While literalist theology as it has been taught for the past seventeen centuries would generally tend to reject the assertion that "the incarnation" refers to the spark of divinity coming down and incarnating in every single human being, there is some clear support in the scriptures of both the Old and New Testaments for the position that this is exactly what they always intended to convey. In the Old Testament, for example, Kuhn argues in that same chapter VII of *Lost Light* that myths such as Abram being sent out of Ur of the Chaldees into a land to the west is yet another clear celestial allegory of the descending heavenly spark, representative of incarnation (he argues on page 113 that the word *Ur* itself means "fire," as in the realm of the fiery stars, and notes that the Duat of the Egyptians was located *in the west,* where the stars set due to the turning of our planet on its axis). He also points to the "descent" of Abram into Egypt (allegorical for the lower half of the zodiac wheel, the underworld) for a time as telling the same incarnational story, and again that the descent of the twelve sons of Jacob later in Genesis tells the same (113).

Another well-known Old Testament verse which Kuhn does not directly point to in *Lost Light*, but which seems to have a clear parallel to the metaphor of Prometheus bringing down divine fire concealed in a smoking reed or handful of reeds is found in Isaiah 42 (quoted again in the New Testament book of Matthew): "A bruised reed shall he not break, and the smoking flax shall he not quench: he shall bring forth judgment unto truth" (Isaiah 42:3; Matthew 12:20 modifies the last part somewhat by saying "till he send forth judgment unto victory," which could be esoterically understood as saying that our incarnations here will continue until the cycle ends in judgment unto victory).

In the New Testament, Paul teaches in many places "the mystery of Christ in you," for example in 2 Corinthians 13:5, as well as in his own description of his first receiving the revelation of this mystery, in Galatians 1:16.

And, Kuhn argues that this understanding of the New Testament metaphors (that they teach the exact same esoteric truth as that which was conveyed by the descent of the gods to earth in the myths of Dionysus or Osiris) helps shed light on the otherwise difficult-to-explain passage in Luke 14:26, in which Jesus orders his followers to hate their father, mother, brother, sister. In Kuhn's interpretation, this discussion of "leaving one's family" is metaphorically describing the leaving of the heavenly realms to plunge into exile in matter (the incarnation). Seen in this light, Kuhn argues, it is not talking about our human parents and siblings at all, and it certainly is not teaching "hatred" of those human relations (*Lost Light*, 126).

On the contrary, all these ancient teachings should impart to us a new appreciation of everyone we meet, as well as ourselves. In this view, each person contains an infinite universe, because each is a "microcosm" of the infinite cosmos. Each, too, embodies within himself or herself "the gods." That includes the baby screaming in the seat behind us on an airplane (or the 4-year old kicking the back of our seat)! This should engender in us a profound sense of wonder for everyone we meet, and a new appreciation of who we are as well. It should also point us towards the natural-law principles of nonviolence and respect for the lives and bodily safety of those with whom we come in contact (and our own as well).

September Equinox, 2014

2014 September 22

Earth will speed past the point of equinox at 0229 GMT on 23 September, which will be 10:29 pm Eastern time on 22 September for those along the eastern edge of North America, or 7:29 pm Pacific time on 22 September for those on its western edge.

As it does so, the northern hemisphere will pass from the "upper half" of the zodiac wheel into the "lower half," as nights begin to be longer than days (a phenomenon which takes place a few days after the fall equinox, owing to the fact that the size of the sun disc plus the bend of light from the earth's atmosphere act to lengthen the days ever so slightly at each sunrise and sunset, delaying the actual day upon which night takes over as being longer than daytime: see the work of Professor Gordon Freeman on this subject).

For those in the southern hemisphere, of course, this equinox marks the complete opposite phenomenon, being the gateway to the return to the "upper half" of the year and the climb towards the point of southern-hemisphere summer solstice.

But for the northern hemisphere, this equinox marks the point of descent, an annual crossing point of very great metaphorical import in nearly all the sacred scriptures and traditions the world over. The passage down to the lower half of the year, enroute to the very Pit of the year at the low-point of winter solstice, symbolized the incarnation: the plunge of the soul from the fiery realms of spirit into the miry realms of matter.

It is perhaps for this reason that the Chichen Itza pyramid was designed to manifest the "serpent of light" going *down* from heaven to earth upon the day of the equinox (image above).

This descent into the body was figured in ancient myth and legend in countless metaphors: as the death of Osiris and the corresponding "casting down" of the Djed-column or backbone of Osiris, as the plunging of Narcissus into the pool of water, to the myths of the god or goddess who was held captive in the underworld for half the year, such as Balder in northern Europe and Persephone in the Mediterranean – the latter being central to the rites and rituals of the Eleusinian Mysteries,[173] which included symbolically rich metaphors such as the taking of *piglets down to the waters* of the ocean by the initiates, and of their being mocked and humiliated as they *crossed over a bridge* on the way to Eleusis, and finally of searching *all through the night* for the *missing or hidden* goddess Kore (Persephone), symbolizing the hidden and nearly lost state of our spiritual or immortal part within our animal nature during the incarnation.

The annual Eleusinian Mysteries, by the way, commenced at the end of September each year and ran through early October – until they were finally shut down by the Roman emperor Theodosius[693] in AD 392.

The ancient writings of the philosophers were filled with references to the fact that this material world through which we pass in our incarnate state is not the real world, but is rather the shadowy projection of that real but invisible world of spirit -- a worldview we have seen to be common to shamanic cultures worldwide, and which thus supports the argument that this worldview is part of the shared heritage of humanity (which the forces of which Theodosius was a part have been trying to stamp out and deny to the people of the world for at least seventeen centuries).

In his masterful 1940 text, *Lost Light*, which traces this theory regarding the ancient teaching of the soul's plunge into the body and its manifestation in metaphorical myth and in the discourse of ancient philosophers, Alvin Boyd Kuhn writes on this subject:

> In the *Phaedrus* Plato, in the beautiful allegory of the Chariot and the Winged Steeds, portrays the soul as being dragged down by the lower elements in man's nature and subjected to a slavery incident to corporeal embodiment. Out of these conditions he traces the rise of numerous evils that disorder the mind and becloud the reason. [. . .] The rational element, formerly in full function, now falls asleep. Life is thereupon more generally swayed by the inclinations of the sensual part. Man becomes the slave of sense, the sport of phantoms and illusions. This is the realm in which Plato's *noesis*, or godlike intellect, ceases to operate for our guidance and we are dominated by *doxa*, or "opinion." This state of mental dimness is the true "subterranean cave" of the Platonic myth, in which we see only shadows, mistaking them for reality. 145.

All this, Alvin Boyd Kuhn argues, was associated with the fall into the "lower half" of the year, when darkness begins to dominate again – the half of the year which represents our earthly sojourn in this body which commences metaphorically at the fall equinox and ends with the triumphant return of the spirit to the heavenly realms at the "opposite horizon" of the spring equinox.

Discussing this realm of darkness further, he writes:

112

"The dark night of the soul," no less than the *Gotterdammerung*, was, in the ancient mind, just the condition of the soul's embodiment in physical forms. [. . .] All this is the dialectic statement of the main theme of ancient theology -- the incarnation of the godlike intellect and divine soul in the darksome conditions of animal bodies.

The modern student must adjust his mind to the olden conception -- renewed again by Spinoza -- of all life as subsisting in one or another modification of one primordial essence, called by the Hindus *Malaprakriti*. This basic substance was held to make a transit from its most rarefied form to the grossest state of material objectivity and back again, in ceaseless round. Darkness was the only fit symbol to give to the mind any suggestive realization of the condition of living intellectual energy when reduced in potential under the inertia of matter. 146-147.

From all of the foregoing, we can readily see why the points of equinox were metaphorically portrayed in myth-systems across the planet as points of *sacrifice*. For the September equinox which falls between Virgo and Libra in the astrological system that informs much of ancient myth, this often involves myths about the sacrifice of a virgin, as with the sacrifice of Iphigenia (discussed in the first three chapters of *The Undying Stars*; the Iphigenia discussion begins on page 34 of the book pagination) or the sacrifice of Jephthah's daughter.[297]

These myths, and some of the philosophical passages cited above, depict the point of incarnation in negative and even horrifying tones, but the counter-balancing component present in all of these same myth-systems is the restoration of the spiritual component, the raising up of the Djed-column that has been cast down, the remembering and recognition and then elevation of the divine spark within the individual, which the ancient myths portray as one of our central tasks here in this "lower realm of darkness."

If the point of incarnation depicts the temporary burying of spirit within matter, and the temporary increase of darkness over light, the act of calling forth the spirit again and letting it shine through and ultimately elevate the material world which is "covering it" is an act of great importance, and one which was discussed in the previous post about "blessing" as a daily requirement (09/19/2014), and one with an individual component that "faces inwards" (so to speak) but also one that "faces outwards" towards others and towards the natural world. Such is the opposite of the burying and hiding of the spiritual, in that this concept of blessing is one of evoking the spiritual and enlivening the material world which after all contains and is infused with the hidden realm of spirit, and which in fact ultimately springs from the world of spirit, as even the passage from Plato discussed above asserts.

These are uplifting thoughts to consider at the significant point in the year at which the sun's path again crosses the celestial equator. The human condition may be deeply confusing, and our "mixed" state of "spiritual-animal" may cause us to feel like the participants in the Eleusinian mysteries, stumbling about in the darkness and tripping over everything, but the human condition is after all a wonderful mystery as well, and one in which we can experience the increase and eventual triumph of the light again, the restoration of the Djed, and the *shining through* of the world of spirit which permeates and sleeps within all of the natural universe around us, as well as in ourselves

The profound lessons of the Lantern Festival

2015 March 16

While on the subject of important festivals and holidays taking place this month, let's briefly discuss some of the significance of the traditional Chinese celebration of the Lantern Festival, which takes place each year on the night of the fifteenth day after the date of the lunar New Year.

Because the traditional lunar New Year takes place on the first New Moon after winter solstice, a festival timed to take place fifteen days after a New Moon is a festival which will always correspond to the Full Moon (for a discussion of the lunar New Year and moon cycles, see the post from 01/31/14, for example).

Thus we can immediately perceive that a celebration in which participants (and especially children) carry around lanterns through the streets and pathways and gardens underneath the first Full Moon of the year is declaring, establishing, and reinforcing a connection between the events taking place in the heavenly sphere and the events taking place here on earth, and between the motions of the heavenly bodies and our lives in these physical bodies: a proclamation of *"as above, so below."*

The image above, most likely painted in the year 1485, is a detail of a long series of painted panels showing the Ming Emperor Xianzong enjoying the Lantern Festival. In the image, children can be seen carrying their lanterns suspended from long red poles. Some of the lanterns are in the shapes of animals or people, while others are more traditional globes with tassels. Vendors can be seen providing the lanterns to the children, from portable tables or booths, each with their own colorful canopies.

The progress of the moon, from New Moon to Full Moon, celebrated in the lunar New Year and the Lantern Festival, was understood by the ancient cultures around the world as conveying deep truths about our condition in this human life. Along with the motions of the sun, the planets, and the glorious backdrop of stars in the celestial sphere, the moon's monthly progress was understood as depicting for us the dual spiritual-physical nature of this universe we inhabit, and the dual spiritual-physical nature of our own human condition in this incarnate life.

In a short lecture on some of the spiritual truths that the ancients saw depicted in the motions of sun and moon, entitled *Spiritual Symbolism of the Sun and Moon*, Alvin Boyd Kuhn (1880 - 1963) said of the moon:

> It is a reflector of the sun's light, and this reflection is made at night, when the sun is out of sight of man! The moon is our sun-by-night, and it is well to delve more deeply into the implications of this datum. The moon conveys to us the sun's light in our darkness. What does this mean on the wider scale of values? It means this: as the night typifies our time of incarnation, the diminished solar light reflected on the lunar surface is an index of the fact that by no means the full power and radiance of the sun (our divine light or spirit) can fall upon us or shine for us while in the life in body. As the moon stands for the body, the reflected light of the sun upon it and from it to us betokens that we can have access only to as much of the spiritual glory as our bodies can give passage to, or give expression to, or become susceptible to. In incarnation we are in spiritual darkness, or have access only to that spiritual force and radiance that can get down to us through the intervening medium of the physical

116

mechanism. [. . .] When in incarnation, we are deprived of the full glow of our inner light. Our god is then as the hidden sun, and we must get its rays through a reflecting-medium, the body. [. . .] It thus also symbolizes the human body, which is surely not the light of spirit, yet in its structure and in its outward countenance, it reflects and bears witness to the divine spirit that animates it, the god hidden within. A man's spirit shines out on his face as the sun shines on the moon! 7 - 8.

Note that, as with other authors from previous generations, Kuhn is accustomed to using the term "man" to refer to "humanity," but that he is quite explicit, both in this lecture and in his other work, to indicate that he is referring to both men and women equally when discussing these great spiritual truths.

It is also important to note that Kuhn is not in any way indicating that he believes the moon's importance is somehow "lesser" as a heavenly teacher than the importance of the sun or any other heavenly light: he is simply saying that by nature of the moon's unique properties, the ancients saw it as perfect for conveying certain truths about our human condition that related to the fact that in these bodies we inhabit, our divine invisible spirit nature is *hidden*, and not seen directly or in *full force*. The contrast between the moon and the sun was seen as perfect for conveying that message.

Later in the same portion of his lecture, Kuhn goes on to explain that the moon was in no way inferior as a symbol of our human condition and as a teacher of our purpose in this life of remembering the spiritual that is within and behind everything physical that we see, in other beings and in ourselves, and calling forth the spirit, raising it up, and in doing so *blessing* them and creation:

But the most sublime element in the spiritual symbolism we are trying to depict comes next in the development of our theme. This is the eternal meaning connected with the sun's light on the moon that we are desirous of impressing in unforgettable vividness upon the imagination. This is the great fact which we would have you call to mind whenever you gaze upon the silvery orb from night to night. As the young crescent fills with light and rounds out its luminous circle, it is writing our spiritual history! It is preaching to us uncomprehending mortals the gospel truth about our own divination. The growing expanse of light on the moon, we repeat, is the sign, symbol and seal of our own transfiguration into godhood! The spark of divinity implanted in our organisms must, to use one Biblical figure, gradually leaven the whole lump; to use another, must illuminate the whole bodily house. [. . .] As we gaze upon the lunar crescent and see it go on toward the full, the vision should fortify us with the profoundest and sublimest truth about this mortal existence of ours, viz.: that we are in process of filling our very bodies with the mantling glow of an interior hidden light, which will steadily transform our whole nature with the beauty of its gleaming. 9.

The existence of Lantern Festivals such as that celebrated in China and other nearby countries and cultures on the fifteenth night after the lunar New Year "brings down" the message of the heavens to us here below, so that we can be reminded of our true condition as beings dwelling indeed in physical bodies, but possessed of an invisible, interior, submerged and hidden divine spark. The fact that children are the ones given the lanterns to carry about seems to emphasize this message – their carrying about their "little moons" serves as a way of teaching it to them (through symbolism) and of reminding those of all ages who see this drama enacted each year of this message which the world always seems to be conspiring to make us forget.

Lest any skeptical readers doubt that the celestial connotations of this festival have long been part of the culture that observes this annual festival of lanterns, they should consider this beautiful poem by Xin Qiji 辛棄疾(AD 1140 - AD 1207) which evokes the

Lantern Festival and is "set to the Tune of the Jade Table" -- a very well-known poem in China:

东风夜放花千树，
更吹落，星如雨。
宝马雕车香满路，
凤箫声动，
玉壶光转，
一夜鱼龙舞。

蛾儿雪柳黄金缕，
笑语盈盈暗香去。
众里寻她千百度，
蓦然回首，
那人却在，
灯火阑珊处。

Personally, I am not currently capable of reading and translating all of that, but it is included for those who can, as translations inevitably must sacrifice some aspect of the. The translation from *An Introduction to Chinese Literature* by Liu Wu-chi (page 122), reads as follows:

> At night the east wind blows open the blooms on a thousand trees,
> And it blows down the stars that shower like rain.
> Noble steeds and carved carriages -- the sweet flower scent covers the road;
> The sound of the phoenix flute wafts gently;
> The light of the jade vase revolves;
> All night the fish and dragons dance.
>
> Decked in moths, snowy willows, and yellow-gold threads,
> She laughs and talks, then disappears like a hidden fragrance.
> Among the crowds I have sought her a thousand times;
> Suddenly as I turn my head around,
> There she is, where the lantern light dimly flickers.

The possibility that the opening two lines of verse (which are rendered rather nicely in this translation as "One night's east wind adorns a thousand trees with flowers / and blows down stars in showers") refers to not only the lanterns of the festival but also the people themselves, who are themselves stars "blown down in showers" to earth and incarnation by the night's east wind (that is, a wind which *originates* in the east, but proceeds *towards the west* and thus towards the place of incarnation, where stars plunge beneath the western horizon, out of the heavens and into the realm of earth and water), would seem to be very defendable.

In other words, the poem evokes the Lantern Festival, and the poignant search for the woman the speaker has seen talking and laughing, but who disappears among the crowds . . . but it also evokes our condition in *this incarnate life*, in which we are like stars blown down from another world, searching for something of great beauty which we have glimpsed but lost among the jostling crowds of this physical world, but which we suddenly encounter again when we are least expecting it (perhaps when we are not even trying: in the peace of utter stillness).

Further support for the assertion that this poem is working on such multiple levels (even as the Lantern Festival itself is working on these multiple levels) can be found in the imagery of the "fish and dragons" in the final line of the first stanza, referring to the lanterns in the shapes of fish and dragons, carried by the revelers through the streets, but also evoking the tradition in China that carp can eventually transform into dragons. Both creatures, of course, have long barbels at their mouths, which suggests that dragons may once have been carp, or that carp may one day become dragons: but carp are creatures of water and hence represent our lower, human condition in these human bodies, seven-eighths of which are water, while dragons are creatures of air and fire, and represent the unbound power of the spirit-nature, which we all contain.

The poet's choice of creatures, fish and dragons, in the lanterns described as dancing through the night at the festival thus carries the idea of our dual nature, just as the lanterns themselves with their glowing inner candle shining within the delicate outer paper wrapper also depict an aspect of our condition in this incarnate passage through night, when we carry an inner divine spark dimly glowing through our fragile "reflecting-medium" of the body (as Alvin Boyd Kuhn describes it in the passages cited above).

The fact that this poem is "To the tune of the Jade Table" might also be a hint, in that the "Jade Table" may poetically refer not just to a specific tune about a piece of furniture but to this green table of the earth upon which we have been blown by the east wind to blossom like spring flowers on the trees, or upon which we fall as a rainstorm, even though we are (spiritually speaking) stars who come to earth from the realm of spirit.

And then there is this legend associated with the origin of the Lantern Festival itself, which says that in ancient times, a beautiful bird beloved of the heavenly Jade Emperor flew down to earth, and the people in their ignorance chased the bird and killed it. Enraged, the Jade Emperor vowed to destroy the people by raining fire down upon them, and what is more to do it upon the night of the fifteenth lunar day of the year.

The Jade Emperor's beautiful daughter heard of this terrible plan, and sent warning to the villagers of earth. One of the people, an old man who was very wise, came up with an idea to avert the storm of fire. He instructed all the people to hang lanterns on that night, so that the Jade Emperor and his heavenly army would look down and see what looked like a river of fire already blanketing the towns and villages. The old man also instructed them to set off firecrackers all night, so that the heavenly army would think that all the people were already perishing. In this way, they would no longer feel the need to rain fire down upon the villagers.

And so, the lanterns and fireworks averted the vengeance of the Jade Emperor for the loss of his favorite bird of heaven, and the event is still observed all these thousands of years later on the same night (and perhaps to ensure that the Jade Emperor continues to hold off the rain of fire -- who knows?).

The story can also be found in numerous children's books describing the origin and meaning of the Lantern Festival.

Now, this story is a clear giveaway that the Lantern Festival has celestial origins and foundations, because a descending heavenly bird and fire from heaven should by now ring bells for readers who have followed some of the Star Myth explanations detailed in previous posts and publications such as books and videos which show the celestial foundations of myths, scriptures, and sacred traditions from around the world.

Below is an image of the portion of the sky which I believe contains the constellations that form the foundations of this myth regarding the origin of the Lantern Festival.

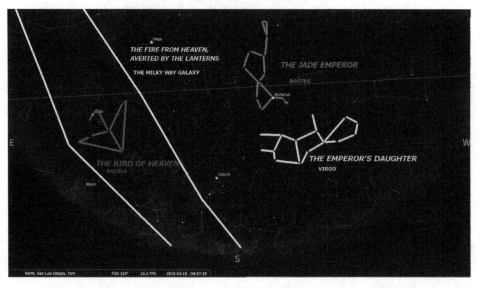

The parallel white lines indicate the smoky column of the Milky Way galaxy, rising up from a point just to the left of the large red "S" that indicates the center of the southern horizon (this screen view simulates an observer in the northern hemisphere, looking towards the south).

I believe the Milky Way represents the threatened "rain of fire" from heaven, but also the celestial counterpart to the Lantern Festival that is taking place down on earth. When the Jade Emperor and his heavenly captains see the glowing river of lights from the lanterns of the people, they stay their hand and do not release the storm of fire. The Milky Way represents the lanterns of the people, which the Emperor and his commanders see, causing them to avert the threatened vengeance.

The Bird of Heaven who flies too close to earth and is pursued and killed by the people is probably the great Eagle, Aquila. The Eagle and the Swan face one another in the Milky Way galaxy itself, and they figure prominently in many myths around the globe. The Jade Emperor is most likely the constellation Boötes, the Herdsman. The reason I believe he is played by Boötes is the fact that Boötes is seated, as if on a throne, and the Jade Emperor is usually presented (not surprisingly) the same way. In some images of the Jade Emperor, one leg is even tucked-in at a sort of "semi-cross-legged" angle, which may be reflective of the short, bent leg of this constellation.

Even more importantly, Boötes is a northern constellation, and one who is very close to the pivot of the heavens, the central point around which all of the celestial attendants must turn: the north celestial pole. His "pipe" reaches nearly to the handle of the Big Dipper, and the Dipper of course points to the current Pole Star, Polaris. This fact would make Boötes a strong contender for the Jade Emperor, who holds court at the very center of the heavenly dome.

Another good reason to believe that the Jade Emperor is played by the celestial giant Boötes is the fact that Boötes is very close to Virgo and appears as her father (or grandfather) in many, many Star Myths around the world.

We obviously have a "daughter" appearing in this story: the daughter of the Jade Emperor, who takes pity on the people of earth and warns them of her father's terrible plan. The daughter is almost certainly played by Virgo in this Star Myth about the origins of the Lantern Festival.

Connections like this in the foundational myth of the Lantern Festival indicate that its celestial significance was well understood by at least some segment of the culture down through the millennia that this annual celebration has been going on.

And so, the festival of lanterns enacted each year at the Full Moon of the first month after the lunar New Year serves an important purpose: it connects heaven and earth, and reminds us that, though we move about on this terrestrial surface (this "Jade Table") in these human bodies, we are also truly spiritual beings, from another realm not mingled with earth or water – the realm of invisible spirit, the realm of the gods.

The lanterns and the firecrackers themselves serve to remind us and re-awaken us to that fact -- and they seem in fact to be symbols most excellently selected to do so. Below is another segment from the Ming Dynasty painting shown earlier -- in this portion, several people are setting off fireworks, and the Emperor himself can be seen enjoying the festivities from a porch above, under a yellow awning or silken tent.

And so, let us conclude with a final quotation from the lecture by Alvin Boyd Kuhn mentioned above, in which the spiritual significance of the moon in its phases, and the message the ancients believed that this glorious and nearest of heavenly bodies conveys to us, progressing as it does each month not only from New Moon to Full Moon and back again, but also from the western horizon (where the New Moon is first seen, trailing the sun which has "lapped it" again at each New Moon and then proceeds to get further and further "ahead" of the moon in its course across the sky) to the eastern horizon. He says of the moon:

> As just seen, the new moon is conceived in the west, the region of all beginning or entry upon incarnate life, the place of descent into the underworld. It has its birth and begins its career of growth in the west, moving night by night further toward the east. Man, the soul, enters his journey toward divinity in the west, and life by life moves further toward the east, the place of fulfillment and glorious resurrection. What more fitting, then, that the rising of the moon in its full glory, when it typifies the completed and perfected human-divine, the man become god, should take place in the east, the gate of the resurrection! The young new moon appears and mounts in the west; the full moon in the east! 13 - 14.

With this added insight from Alvin Boyd Kuhn, we can now see that the "Lantern Festival Night" poem by Xin Qiji incorporates *both* the motion of the incarnation or the "descent into the material realm" (with its reference to the wind that moves *towards the west*, and which sends stars to fall down to earth like rain, or causes flowers to bloom on the trees, representative of our incarnation in these fragile physical forms which are akin to flowers or to paper lanterns) *and* the motion of the movement back towards the east – the motion of greater spiritual awareness and awakening -- in the evocation of the motion of the moon from New Moon to the Full Moon, a motion during which the moon's position at the same time each night will be seen to be further and further *east* in the sky.

The moon is currently voyaging back towards the point of New Moon, having passed through the night of Full Moon and the night of the Lantern Festival, but it is still the "first moon" of the lunar year, and we can get up early in the morning and see it in the sky before dawn (rising less and less distance ahead of the sun each day), and think about the spiritual truths that the moon conveys to us.

In fact, we can do so throughout the year, and as we do so give thanks for the moon's constant reminder that we and the universe are composed of a physical aspect but that within each of us there is a spiritual component which we should be bringing forth in every way that we can.

> One night's east wind adorns a thousand trees with flowers
> And blows down stars in showers
> Fine steeds and carved cabs spread fragrance en route
> Music vibrates from the flute
> The moon sheds its full light
> While fish and dragon lanterns dance all night
>
> In gold-thread dress, with moth or willow ornaments
> Giggling, she melts into the throng with trails of scents
> But in the crowd once and again
> I look for her in vain
> When all at once I turn my head
> I find her there where lantern light is dimly shed

DAYS LONGER THAN NIGHTS:
Heaven, Promised Land, Greece, etc.

Isis and Nephthys: March Equinox, 2015

2015 March 20

As the earth speeds past the point of March equinox, days in the northern hemisphere begin to grow noticeably longer than nights, "crossing over" from the half of the year in which night rules over day, and ascending into the "upper half" of the year in which day, light, and warmth become more and more dominant.

The spinning earth will hurtle past the exact point of the equinox at 2245 UTC on 20 March this year, which corresponds to 1845 or 6:45 pm for observers in North America on the east coast, and 1545 or 3:45 pm for observers in North America on the west coast (and, if you are not in one of those two slices of the planet, you should be able to calculate the local time for your particular spot on the spinning globe based on the difference in longitude to your location from the line of UTC).

As previous posts have explained, the equinox marks the crossing of the plane of the ecliptic (the arc along which the sun appears to travel each day, due to the fact that we are standing on a spinning earth) and the celestial equator (that imaginary line in the sky that is 90 degrees down from the north celestial pole, or 90 degrees up from the south celestial pole, which are best visualized using the diagrams in this previous post,[76] in which the celestial equator is the third and largest of the circles in each diagram), such that the sun's daily arc is north of the celestial equator during the day, which means it will be "above" it and closer to the apex-point of the sky for viewers in the northern hemisphere, and "below" it or closer to the horizon for viewers in the southern hemisphere.

Other significant actions will be taking place this year on the same day -- even as the "ship" of the earth passes "broadsides" to the sun (in the "earth-ship metaphor" in which the prow or bowsprit of the ship equates to the north pole, and the lantern in the stern of the ship equates to the south pole, as described in this previous post from winter solstice of 2011,[85] the solstices being the two points at which either the bowsprit or the stern are pointing at the sun), the moon which is orbiting in circles around the earth is also passing directly between the earth and the sun itself, creating the monthly phenomenon of New Moon. The moon will pass that point of New Moon at 0938 UTC on 20 March.

Additionally, because the moon will simultaneously be crossing the ecliptic plane as it passes through the point of New Moon this month, it will create an eclipse of the sun which will be visible to many observers in northern portions of the eastern hemisphere.

Because the moon does not orbit the earth on the plane of the ecliptic, but along a plane that is inclined to the plane of the ecliptic, it does not create an eclipse every time it passes between the earth and the sun. However, during its monthly orbit, its tilted orbital plane crosses once from "above to below," and once more from "below to above," the plane of the ecliptic. These lunar "crossing points" are referred to as the two "lunar nodes," also known as the "draconitic points." When the moon crosses through a lunar node at the point of New Moon, it blocks the sun for observers on some parts of our globe, and creates a solar eclipse (and when it passes a lunar node at the point of Full Moon, the earth blocks the sun and creates a lunar eclipse). These lunar nodes or draconitic points are discussed in a previous post (07/07/11), along with some diagrams and helpful links.

So, as the earth hurtles past one of the two annual "crossing points" that cause the ecliptic line to cross the celestial equator, the moon is also passing through one of the two monthly "crossing points" at which its path crosses the ecliptic, and it is doing so at the point of New Moon, to create a solar eclipse (the moon is "crossing down" on this particular New Moon).

Truly a stunning array of crossings! And one which cannot fail to produce feelings among us here on earth that these motions must somehow have significance in our own lives and persons.

In fact, the March equinox, when we leave the "lower half" of the year (for those in the northern hemisphere), was invested with profound significance in the ancient systems of sacred metaphor.

It was encoded in myth and sacred scriptures and legends as the point of *crossing up out of the land of bondage,* or out of the underworld, such as in the Old Testament story of the escape from Egypt by means of the *crossing of the Red Sea* (see discussion in the "Myths" section of www.starmythworld.com).

It was also seen as a place of sacrifice, as was the downward crossing point on the other side of the year, at the September equinox.

As discussed in *The Undying Stars* and several online posts, the story of Abraham taking Isaac up the sacred mountain in order (he thinks) to sacrifice him contains numerous clues which show that it encodes the point of the year of the March equinox, when the sun and its ecliptic path are "climbing up" towards the summit of the year (for observers in the northern hemisphere). Because the upward crossing point at the March equinox took place each year when the sun was rising in the sign of Aries the Ram (during the precessional Age of Aries, discussed here and explained further in the video here), that story of sacrifice involving a trip *up* the mountain concludes with the substitutionary sacrifice of a *Ram* rather than of Isaac himself.

Similarly, the sacrifice of the Passover lambs took place immediately prior to the escape from enslavement in the land of Egypt (the "house of bondage," an esoteric and allegorical portrayal in myth of the *lower half of the year*), and the crossing of the Red Sea discussed earlier. The lamb is another indicator of the sign of Aries the Ram, the sign in which the sun would rise on the spring equinox during the Age of Aries. This is why the date of the Passover is tied to the March equinox, as is the Easter celebration of the sacrifice and subsequent *rising* of the Christ (described in Revelation 13:8 as "the Lamb who was slain from the foundation of the world").

These sacred writings are describing the annual *crossing upwards*, from death (the lower half of the year) to life.

But it is very easy to make the mistake of concluding that this discovery means these ancient sacred stories are "only" about the natural phenomena of the annual cycle, the return of spring, the rebirth of flowers and growing things, etc.

In other words, some who have noticed the undeniable connection between the scriptures and the motions of the heavens have concluded that this somehow *diminishes* their message, and makes it less about spiritual things and merely an ancient attempt to explain and perhaps to honor our amazing physical world.

This conclusion would be mistaken.

It is in fact self-evident that the ancient scriptures and sacred traditions of humanity are concerned with imparting profoundly spiritual truths, *in addition* to the fact that they demonstrate an incredibly sophisticated understanding of the glories of our universe and the motions of the earth in relation to the sun, moon, planets and stars, and the cycles of the seasons and the heavenly bodies and even the cycles of the subtle motion of precession.

130

The originators of the myths and sacred stories and ancient scriptures of the world do not go to the trouble of encoding all these motions in their stories about men and women, gods and goddesses, giants and demons and spirits, just because they were trying to explain where lightning came from or why eclipses occurred: they were using these cycles of heavenly bodies *that we can see* in order to impart knowledge of what we cannot see – the invisible realm of spirit which lies within and behind everything in this physical universe, and which may actually be the true source from which the material universe is projected.

And, by the very act of turning the motions of these heavenly bodies into stories involving men and women, they are showing that we are intimately connected to the entire universe, and that thus we ourselves have a physical and a spiritual nature, a *dual nature* just like the dual nature of the infinite universe which we ourselves embody and reflect and even contain.

This dual nature – of both the macrocosmic universe and the individual man or woman who contains and represents the universe as its microcosmic reflection – is conveyed in the world's myths through the dual nature of the zodiac wheel (divided into the upper half and lower half, the land of spirit and the land of incarnation), the dual symbols of the equinoxes (one marking the crossing point down, into material incarnation and hence into the underworld of incarnation and death, and one marking the crossing point up, into the realm of spirit and spiritual life), and the dual nature of the moon (passing from New Moon to Full Moon, and crossing through the lunar nodes upwards and downwards just as the ecliptic crosses upwards and downwards during the solar cycle).

As has been discussed in countless previous posts, this dual nature of our incarnate existence is seen in the ancient symbols of the Cross and the Ankh, each with its horizontal beam (representative of being *cast down* into matter, and hence representative of our physical, animal nature while incarnate in these physical bodies) and its vertical pillar (representative of the spirit *rising up*, the invisible aspect of our nature which overcomes and transcends the physical body and which, unlike the body, does not experience physical death).

It was also conveyed in the ancient myths involving the Djed column of ancient Egypt, which was cast down when Osiris entered the realm of death and was layed out horizontally, and which is then raised up to represent the same "vertical component" of spirit and triumph over death and incarnation.

The equinox points, then, represent the "crossing" points of the spirit, down into incarnation (depicted by the lower half of the year, and by the underworlds of myth, including the "house of bondage" of Egypt in the Old Testament, or of Tartarus and Hades and many others around the world) and up into greater spiritual awakening and transcendence (and ultimately into spiritual life beyond the body).

Alvin Boyd Kuhn has offered the tremendous insight that, in order to convey this important understanding of our dual physical-spiritual nature, these two crossing points were sometimes depicted in mythology around the world as the two mothers of the god, sometimes seen as two goddesses stationed at those two important crossing points of the year, the fall and spring equinoxes. There, they would *give birth* to the hero, or the god, one being a birth down into physical incarnation, and the other being a second birth or spiritual birth, upwards into greater consciousness and spiritual life.

He notes that these two goddesses, or two mothers, are depicted on either side of the Ankh cross of ancient Egypt in some important representations -- such as the Ankh with upraised arms shown above, from the Papyrus of Ani, in which the goddesses Isis and Nephthys are shown on either side of the Ankh itself, in the positions corresponding to the two equinoxes (the upraised arms represent summer solstice and the constellation of Cancer the Crab at the top of the year during the Age of Aries, as discussed in previous posts such as the post from 06/21/2014).

132

Elaborating on the importance of these two goddesses and the two births, Alvin Boyd Kuhn writes in *Lost Light* (1940):

> Man is, then, a natural man and a god, in combination. Our natural body gives the soul of man its baptism by water; our nascent spiritual body is to give us the later baptism by fire! We are born first as the natural man; then as the spiritual. Or we are born first by water and then by fire. Of vital significance at this point are two statements by St. Paul: "That was not first which is spiritual, but that which is natural"; and, "First that which is natural, then that which is spiritual." Again he says: "For the natural man comprehendeth not the things of the spirit of God, neither can he." [. . .]
>
> Using astrological bases for portraying cosmic truths, the ancients localized the birth of the natural man in the zodiacal house of Virgo and that of the spiritual man in the opposite house of Pisces. These then were the houses of the two mothers of life. The first was the Virgin Mother (Virgo), the primeval symbol of the Virgin Mary thousands of years BC. Virgo gave man his natural birth by water and became known as the Water-Mother; Pisces (*the Fishes* by name) gave him his birth by the Fish and was denominated the Fish-Mother. The virgin mothers are all identified with water as symbol and their various names, such as Meri, Mary, Venus (born of the sea-foam), Tiamat, Typhon and Thallath (Greek for "sea") are designations for water. On the other side there are the Fish Avatars of Vishnu, such as the Babylonian Ioannes, or Dagon, and the Assyrian goddess Atergatis was called "the Fish-Mother." Virgo stood as the mother of birth by water, or the birth of man the first, of the earth, earthy; Pisces stood as the mother of birth by spirit or fire, or the birth of man the second, described by St. Paul as "the Lord from heaven." Virgo was the water-mother of the natural man, Pisces the fish-mother of the spiritual man.

[. . .]

The ancient books always grouped the two mothers in pairs. They were called "the two mothers" or sometimes the "two divine sisters." Or they were the wife and sister of the God, under the names of Juno, Venus, Isis, Ishtar, Cybele or Mylitta. In old Egypt they were first Apt and Neith; and later Isis and Nephthys. Massey relates Neith to "net," i.e., fish-net! Clues to their functions were picked up in the great Book of the Dead: "Isis conceived him; Nephthys gave him birth." Or: "Isis bore him; Nephthys suckled him," or reared him. 6 - 8.

Just to make the positioning of the two goddesses or two mothers discussed in this passage by Kuhn more visually clear to the reader, note in the diagram below that the positions of Virgo and Pisces are each directly before the crossing points of the year, and are outlined on the zodiac wheel in blue (for the birth by water, into physical incarnation) and red (for the birth by fire, into spiritual life) on the modified picture here:

One of the passages of the Book of the Dead that Kuhn quotes above is, interestingly enough, section 134, which can be found on page 39 of this online pdf version of the old translation by Budge (1895). That section is known as the *Hymn to Ra on the Day of the Month (the Day of the New Moon) When the Boat of Ra Saileth.*

The title of that Hymn to Ra seems most appropriate for this particular day of the year, which is the Day of the New Moon as well as the Day of the Equinox, and of the equinox in which the sun in the Solar Bark is leaping upwards, across the horizontal line of the equinoxes, into the upper half of the year, out of the lower house of bondage and into the heavens!

In that hymn (as translated by Budge) we read:

> The heart of the Osiris Ani, whose word is truth, shall live. His mother Isis giveth birth to him, and Nephthys nurses him, just as Isis gave birth to Horus, and Nephthys nursed him. (In Budge translation of Book of the Dead, 1.71 - 1.72)

This is very significant, as Ani was a human priest of ancient Egypt, and yet he is here identified explicitly (by name) with "the Osiris" and he is told that he too shall have two births, by two goddess mothers, one of whom gave birth to him and one of whom nurtured him towards his spiritual birth.

In other words, this description is not a myth: it is teaching a truth that applies to actual human beings in this incarnate existence -- it teaches us of our dual physical-spiritual nature (in a universe which itself has a dual physical-spiritual nature), and that we are to be aware of and growing in awareness of our spiritual life even as we inhabit a physical body in a world that tries to tell us that the physical is all that matters.

Note also that just because Kuhn uses the masculine pronoun, and just because the ancient symbology uses two mothers for a male god (such as Osiris and Horus in ancient Egypt), does not mean that this truth is not meant to apply equally to every man or woman who has ever lived. Elsewhere we cited the extremely memorable and helpful quotation from Alvin Boyd Kuhn in a different work, in which he said that "the actors [in the sacred myths of humanity] are not old kings, priests and warriors; the one actor in every portrayal, in every scene, is the human soul."

And so, as we ponder this important day of equinox (and of New Moon!), we realize that it can point us towards deep spiritual truths concerning our own human condition. It can remind us of the invisible spirit-world around us, and within us -- and as we become more and more aware of the truth of this dual nature, it may cause us to be resolved to live a life that is more and more about blessing[550] (lifting up the spirit of others and of creation, and our own as well), and not cursing (casting down the spirit, degrading, or brutalizing). *This* is what the ancients intended, when they encoded these awesome and majestic celestial motions into the actions of the sacred stories and scriptures of the human race.

And, as we participate in celebrations (such as Easter, or Passover, or Equinox observances, or others) which tie the motions of the heavens to our own motions and actions here on the ground, we declare the truth "as above, so below" -- that we ourselves contain an invisible and infinite spiritual side, the very truth that the heavens proclaim.

Mid-Autumn Festival 中秋節
and the
Total Lunar Eclipse ("Blood Moon") of September 2015

2015 September 25

This Sunday, September 27, marks the beginning of the traditional celebration of mid-Autumn festival in China and Vietnam. It is a very ancient holiday, its observance stretching back to as early as 3600 years ago, and perhaps even earlier, and it is one of the most important holidays in Chinese culture. Great effort is usually made to travel and be with family on this day, much like Thanksgiving in the US, and for several days around the holiday many businesses and markets are closed as people make their way back to the places where they grew up, in order to celebrate with their extended families.

The Chinese characters for this holiday are 中秋節 which is pronounced *Zhong Qiu Jie* in Mandarin and *Jung Chau Jit* in Cantonese, and which translates literally into "Mid-Autumn-Day" or "Middle-Fall-Holiday" (or even more literally the "Mid-Autumn-Node").

Jung Chau Jit is celebrated on the fifteenth day of the eighth lunar month, the fifteenth day corresponding in general to the full moon in a lunar month (because a lunar month commences with a new moon, and the moon waxes for fourteen days to become full, which happens on the fifteenth day, and then wanes for fourteen more days to the point of another new moon), and so this festival always falls very close to or directly upon the day of a full moon, as it does this year.

Thus, the Mid-Autumn Holiday is also a Moon Festival, and is in fact often called the Moon Festival, and an important tradition during the days (weeks!) leading up to this holiday and on the day of the holiday itself is the giving of round "mooncakes," light gold in color and filled with a variety of different kinds of heavy, sweet fillings, and sometimes with a candied egg yolk:

These are traditionally served by being cut carefully into four equal quarters (a little combination cutting-and-serving implement, something like a small version of a cake trowel, is often included in commercially-sold mooncake boxes or packages), with each person present being given one section. The cakes themselves often have "blessing" words baked into the top of them.

Being a Moon Festival, the holiday is also closely associated with the Moon Goddess, pictured at top, whose name is 嫦娥 which is pronounced *Chang Er* in Mandarin and *Seung Ngo* in Cantonese and translates rather directly into "Chang the Beautiful" or "Seung the Beautiful."

There is a legend about Seung Ngo and her husband, 后羿, being banished from the heavenly realms by the Jade Emperor (whom we met in the earlier discussion of the Lantern Festival, which takes place in the first lunar month) and having to live down upon the earth as mortals (his name is pronounced Hou Yi in Mandarin and Hau Ngai in Cantonese, and it means something like "King Archer").

In the legend, he is distraught at the idea that his beautiful wife, having been banished from the celestial realms, is now faced with mortality, and so he seeks and eventually obtains an elixir of immortality which will restore their immortality to them. However, as so often happens in such myths, the plan goes awry, when she is forced to drink it all herself (either to keep it from a marauding robber who breaks in to steal it from her while her husband is away, or because she is overcome with curiosity while he is asleep, and drinks the whole elixir without knowing the consequences).

As soon as she does, she feels herself floating up into the heavens, without her unfortunate husband, who is left behind as a mortal. The two are thus separated forever, but Seung Ngo settles on the Moon, where she can look down upon Hau Ngai, and he can gaze up to her new home and think of her.

Having examined some of the most prominent aspects of this important ancient holy day, we are now in a position to benefit from the deep knowledge contained within its symbols and forms.

Because this poignant myth, and all the other symbols of the Mid-Autumn Festival, are powerful symbols which speak to truths about our incarnate existence, this existence in which we find ourselves crossing the "underworld" of the material realm in a physical body -- which is closely associated with the figure of the moon in the ancient system of celestial metaphor -- but doing so with the dimly-remembered awareness that we are separated from our true home (and disconnected from our higher "divine twin") during this earthly sojourn, and that we are in fact actually spiritual beings as much or more than we are physical beings.

The festival, positioned in the time of year next to fall equinox, contains the same symbols of a goddess and the fall from the celestial realm into the mortal incarnate life associated with the point of autumn equinox worldwide in the ancient myths.

Among them:

- The presence of a *goddess-figure* (in this case, the goddess Seung Ngo, or Chang Er), goddess figures being shown in the post from 09/23/15 to be associated in ancient myth *the world over* with the point of fall equinox and the plunge into incarnation.
- A myth in which there is a prominent theme of expulsion from the heavenly realm and banishment to the earthly realm (the plunge into this lower realm), featuring a duo in which one of the pair is mortal and one divine: just as we, in this incarnate life, find ourselves "crossed" with a physical body and an internal divine spark.
- The incorporation of moon-themes to go along with the incarnation theme of the fall equinox (dominated by the presence of a goddess at the point of incarnation). As Alvin Boyd Kuhn demonstrates in extended discussions found in *Lost Light*, published in 1940, the ancient myths and sacred traditions very often used the moon to symbolize our incarnate form, and the sun our divine spirit, which lights up and animates our physical body in

the same way that the sun gives its light to the moon (see pages 115 and following, for example, or 520 and following, or 139 and following, or 521 and following). In that exploration of ancient myth, Kuhn says quite explicitly: "The sun types soul, always, the moon, body" (479), and elsewhere: "The moon being the parent of the mortal body, lunar symbolism was prominently introduced into the portrayal "(140).

- The connection of the *moon* (our incarnate side) with the idea of water, seas, oceans, and incarnation (through the tides, and also through the internal tides of our body), which also connects with the goddess-ocean connection discussed in the post from 09/23/15 (with examples which demonstrated the "mother-ocean" connection inherent in the names of Mary, Tiamat, and Aphrodite, as well as in the Chinese ideograms for *mother* and *ocean*).

- The tradition of gathering together with family at the Moon Festival, representative of the idea that we align the cycle of our personal lives and our physical motions (often traveling great distances) with the cycles of the earth, sun, moon and stars: reinforcing the profound connection between "microcosm" and "macrocosm" discussed in the preceding post (and many others), a connection which the ancient myths and sacred traditions of the human race the world over all seek to convey.

- The tradition of gathering together with family at the Moon Festival, which also commemorates our physical, material entry into this incarnate life, which is celebrated when we honor our family and especially our parents.

- Traditions in this holiday (especially as celebrated in Vietnam) which focus on children and proclaim it to be a holiday which honors young children, who are just embarking upon their journey through the incarnate human life.

The tradition that mooncakes are cut up into four *quarters*, which is clearly connected with the lunar symbology, but also with the concept of "crossing" or the crucifixion of this incarnate life (see numerous previous posts which demonstrate that the Great Cross of the year was associated in ancient myth with the twin components of incarnate human existence: the horizontal component representing the physical, "dead," "animal" nature of our body, and the vertical component representing the spiritual, divine, celestial component of the invisible and infinite realm which the ancient myths tell us is actually all around us and also within us and within every other being with whom we come into contact).

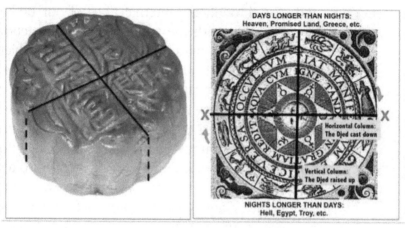

Clearly, then, the Mid-Autumn Festival preserves a great many symbols which carry a profound spiritual message, using the symbology of the moon (associated with incarnation), the casting down from the spiritual realms into incarnate existence (in the story of Hou Yi and Chang Er, or Hau Ngai and Seung Ngo), the myth regarding a married couple who are extremely close but who find themselves in the condition of one divine and one mortal (the "divine twin" pattern found around the world, including in the myth of Castor and Pollux but also of Jesus and Thomas and many others), the traditions of gathering with family and ordering our lives in accordance with the cycles of earth, sun, moon and stars, and the traditions of honoring our physical family and our parents, who brought us into this incarnate body in the first place.

142

It is also worth pointing out, in passing (although it could become a full-length examination and discussion) that a great many Chinese characters which use the symbol for "the moon" actually refer to our physical human body. The Chinese ideogram for "moon" is:

月

Other words whose ideograms use this as a "radical" in their Chinese character, and which relate to the physical human body, include:

The liver:

肝

The ribs or chest:

肋

The armpit, or arms:

肐

The elbow:

肘

Pelvis, groin or thighs:

胯

The diaphragm:

膊

Internal organs, guts, viscera:

脏

A gland:

胞

Fat, plump, or obese:

服

And there are many others.

Some scholars or those familiar with Chinese radicals may argue that none of the above characters are actually connected with "the moon," even though the radical looks just like the Chinese symbol for the moon, because the actual radical for "meat" -- which looks like this --

肉

-- ends up looking *like* the symbol for "the moon" when it functions as a radical in a compound character.

That is a valid argument, but we must ask ourselves why that "meat" symbol turns into a "moon" symbol in all of these ideograms? The answer, of course, could very well be the fact that the ancient wisdom of the human race universally acknowledged an esoteric connection between *the moon* and the physical, corporeal, carnal ("meat") body.

And so it becomes clear that all of the symbology of the culturally significant and anciently-established Mid-Autumn-Festival can be shown to be connected to other mythological symbols used in other myths around the world -- all of them designed to impart to us profound gnosis regarding our human condition here in this incarnate life, including the fact that we are not merely physical beings but that our human nature consists of *both* a physical and a spiritual component, that our physical "moon" form (associated with water) is illuminated by our spiritual "solar" and divine nature (associated with fire and with air -- or *spirit*).

Now, very briefly, let us also note the fact that because the Mid-Autumn-Festival always falls on or very near a full moon, it will also periodically happen that this anciently-ordained observation will coincide with a lunar eclipse. Previous posts on the actual celestial mechanics of the moon phases have explained why lunar eclipses must always coincide with a full moon, and why solar eclipses must always correspond to a new moon (not every full moon is a lunar eclipse, of course, nor every new moon a solar -- but *every* lunar eclipse occurs at full moon, and *every* solar eclipse occurs at new moon).

This September 27 full moon also happens to take place when the moon is passing through a "lunar node" (a "crossing point" with the plane of the ecliptic of the earth) and will therefore result in a total lunar eclipse visible for most of the Americas, Africa and Europe.

Not only is this a total lunar eclipse, but it is also a total lunar eclipse which corresponds to the moon's closest approach on its orbit around the earth, when it is physically closer to us and thus appears physically larger in the sky -- all of which add up to the promise of a spectacular heavenly event this weekend.

This particular moon (all month long) is in fact known universally as the Harvest Moon (in China also), which is traditionally understood to be the brightest moon of the year.

As the moon enters the shadow of the earth, it will take on a dusky red hue -- which (only recently) has begun to be designated as a "blood moon" by some in the popular media and in certain evangelical circles (largely based upon a literalistic interpretation of certain Old and New Testament scriptures which I believe can be definitively shown to be *esoteric* in nature and not *literalistic* in nature). Scriptures in the Old and New Testament which describe the moon as turning to blood or being bathed in blood include the following texts:

Joel 2:31 "The sun shall be turned into darkness, and the moon into blood, before the great and terrible day of the LORD come."

Acts 2:20 "The sun shall be turned into darkness, and the moon into blood, before that great and notable day of the Lord come:"

Revelation 6:12 "And I beheld when he had opened the sixth seal, and, lo, there was a great earthquake; and the sun became black as sackcloth of hair, and the moon became as blood;"

Alvin Boyd Kuhn actually addresses many of these Biblical passages directly, and argues (with extensive textual evidence) that the description of the "moon becoming as blood" only emphasizes even more dramatically the esoteric symbolical connection between the moon and our physical body in this human life.

Discussing the passage cited above from Revelation 6, he explains among the metaphors given:

> along with the darkness over the earth, the veiled sun, the blood-stained moon, is that "the stars from the heavens fell." In the same place we read that "when the message of the third angel was sounded forth, a great star went down from heaven and it fell upon the earth." Another star fell at the sounding of the trumpet of the fifth angel. The various legends, then, of falling stars become invested with unexpected significance as being disguised allusions to the descent of the angelic myriads to our shores, -- to become our souls. 116.

In other words, Kuhn here argues that the metaphors in Revelation 6 (and indeed throughout the Bible) *all* have to do with our incarnate condition, consisting of a "crossing" between spirit (symbolized by the sun) and matter (our material bodies, symbolized by the moon).

146

This interpretation (according to Alvin Boyd Kuhn) would include the metaphor of the earth being enshrouded by darkness -- because we plunge down to incarnation in the *lower* half of the zodiac wheel, as described in numberless previous posts. The lower half of the wheel is the half in which *night* triumphs over daylight (initiated by the fall equinox, when the hours of darkness begin to be longer than the hours of light, in each 24-hour period):

DAYS LONGER THAN NIGHTS:
Heaven, Promised Land, Greece, etc.

NIGHTS LONGER THAN DAYS:
Hell, Egypt, Troy, etc.

It would include (according to Kuhn) the moon being bathed in blood -- because the moon represents our incarnate condition, our sojourn in a body composed of water and blood and clay, our crossing of the "Red Sea" (which can be metaphorically seen to be the crossing which each and every human being undertakes, going through life in a human body through which courses the "red sea" of the blood in our veins and arteries).

It would include (according to Kuhn) the stars being cast out of heaven and forced to "fall upon the earth" – for this is the very condition in which we find ourselves, as human souls who dimly realize that we come from a spiritual home, but who have been exiled (just like Hau Ngai and Seung Ngo) upon this material plane.

In other words, the passages in Revelation (and all the other Biblical scriptures) are describing *our own human experience,* our experience as divine beings who have been "crossed with" physical, material, animal bodies during this incarnate life.

And this is just what all the other Star Myths of the world are trying to convey to us as well! (Note that it can be conclusively demonstrated that the passages of the book of Revelation involving the opening of the seven seals are absolutely based upon metaphorical descriptions of *the constellations in our night sky,* as I demonstrate briefly in a post from July 6, 2012 regarding Revelation Chapter 9: they are all allegorical celestial metaphors which use the awe-inspiring motions of the heavenly cycles to convey truths to us about the invisible realm).

Indeed, all of these metaphors and sacred scriptures are designed to convey to us the very same truths conveyed through the ancient metaphors connected to the Mid-Autumn-Festival celebrated in China and Vietnam and some other surrounding cultures from time immemorial.

As the day of the first full moon after fall equinox approaches, it is a time for contemplation and reflection upon our human condition in this incarnate life – our "plunge into matter" which in ancient myth was associated with the point of the fall equinox, with the goddess at the edge of the ocean (or the goddess of the Moon), and with the "crossing" of our divine nature with a physical body.

148

And yet, even as we are plunged into this physical human form, we are given forms and symbols and myths and stories and scriptures to remind us that this material world that is visible and perceptible to our senses is *not* all that there is, and that this physical "animal" human body we inhabit is *not* all that we are.

Just as the moon is illuminated by the fire of the sun's life-giving rays, so our material nature is illuminated and animated by a higher spiritual self that exists "above and beyond" our merely physical carcass.

Just like the mooncake in the Jung Chau Jit celebration, which is divided and quartered into four equal sections, we ourselves are made up of a "cross," a "crucifix," a "quartered whole" consisting of both a horizontal line (between the equinoxes, and associated with matter) and a vertical line (between the solstices, reaching towards infinity, and associated with all that is spiritual, and with raising the spiritual aspect within ourselves and with calling it forth in those we meet and indeed in all of creation around us).

I sincerely wish you a very blessed Mid-Autumn Festival, and harmony between the microcosm and macrocosm. May all beings be freed from suffering and filled with peace and joy, love and light.

"Here it has reached the turning-point" -- the celestial map of the soul's spiritual trek

We are now drawing so near the point of solstice that some of the great monuments of deep antiquity are already beginning to hint at their stunning alignments to the sunrise of that significant day.

For instance, if you have the opportunity to visit Stonehenge this week, and if the weather conditions permit a view, you can station yourself near Aubrey Hole 40 and look back across the stone circle between Sarsens 22 (on the north or left as you look east) and 21 (to the south or right as you look east), on through the gap between West Trilithon Sarsens 58 and 57 (likewise left and right, respectively) and past Bluestone 69 just beyond them (framing the right edge of the gap, along with the edge of Sarsen 57), you will be able to see the sunrise sight line for the winter solstice as it would have appeared 4,000 years ago, according to the analysis of Professor Gordon Freeman in his landmark text *Hidden Stonehenge* (beginning at page 93 and illustrated with full-color photos in image 4-13, 4-14, and 4-15, captured in 1997).

150

Even if you do not have the opportunity to visit Stonehenge in person, you can still contemplate the silent, massive stones, patiently marking out the great pivots of the year, as they have been doing year after year, as centuries draw on into millennia, and the earth wings its course around the fiery ball of the sun.

You can consider the vision and the skill of those incredible and now-unknowable minds who conceived of this incredible monument, who placed the stones and engineered those alignments which still remain in effect to this day, through all that has come and gone in between -- alignments (some of them) which remained hidden and all but forgotten, until new souls such as Gordon Freeman came and unlocked them to share with humanity once again.

And, considering this almost inconceivable concept, you can also cast your mind around the planet to other ancient places, where temples face the sunrise on winter solstice or mark our sun's rising point with similarly precise alignments: the great Temple of Karnak in Egypt, the mysterious observatory known as the Caracol in what is today called Belize, the incredible passage mound of Newgrange in Ireland, the more recently-discovered Goseck Circle in northern Europe which is thought to date to as early as 4,900 BC (an amazing 6,900 years ago), and many more.

All testify to the ancient and enduring importance of the great turning points of the annual cycle -- turning points which, according to the inspired analysis of Alvin Boyd Kuhn, were marked primarily for their *spiritual* significance as a metaphorical "map" of the circling path of the soul itself.

When we understand the ancient allegorical outline of the year as a representation of our own soul's journey down into the physical body (the first birth) and a state of spiritual amnesia (a form of spiritual sleep or even, metaphorically speaking, spiritual "death"), followed by the awakening of the realization of our true spiritual nature and the spiritual nature of the seemingly-material

151

universe around in which we find ourselves, followed by our increased communication and communion with our inner divine nature and eventual return to the invisible realm – then the importance of the great cycle of the year as a beautiful, visible, ever-present reminder to us becomes clear.

Alvin Boyd Kuhn sketches the outline of this ancient spiritual map most clearly and succinctly, perhaps, in his essay entitled *Easter: The Birthday of the Gods*, available to read online in various places.

The description is so important that it is worth citing at some length:

> Using solar symbolism and analogues in depicting the divine soul's peregrinations round the cycles of existence, the little sun of radiant spirit in man being the perfect parallel of the sun in the heavens, and exactly copying its movements, the ancient Sages marked the four cardinal "turns" of its progress round the zodiacal year as epochal stages in soul evolution. As all life starts with conception in mind, later to be extruded into physical manifestation, so the soul that is to be the god of a human being is conceived in the divine mind at the station in the zodiac marking the date of June 21. This is at the "top" of the celestial arc, where mind is most completely detached from matter, meditating in all its "purity."

> Then the swing of the movement begins to draw it "downward" to give it the satisfaction of its inherent yearning for the Maya of experience which alone can bring its latent capabilities for the evolution of consciousness to manifestation. Descending then from June it reaches on September 21 the point where its direction becomes straight downward and it there crosses the line of separation between spirit and matter, the great Egyptian symbolic line of the "horizon," and becomes incarnated in material body. Conceived in the aura of Infinite Mind in June, it enters the

152

realm of mortal flesh in September. It is born then as the soul of a human; but at first and for a long period it lies like a seed in the ground before germination, inert, unawakened, dormant, in the relative sense of the word, "dead." This is the young god lying in the manger, asleep in his cradle of the body, or as in the Jonah-fish allegory and the story of Jesus in the boat in the storm on the lake, asleep in the "hold" of the "ship" of life, with the tempest of the body's elemental passions raging all about him. He must be awakened, arise, exert himself and use his divine powers to still the storm, for the elements in the end will obey his mighty will.

Once in the body, the soul power is weighed in the scales of the balance, for the line of the border of the sign of Libra, the Scales, runs across the September equinoctial station. For soul is now equilibrated with body and out of this balance come all the manifestations of the powers and faculties of consciousness. It is soul's immersion in body and its equilibration with it that brings consciousness to function.

Then on past September, like any seed sown in the soil, the soul entity sinks its roots deeper and deeper into matter, for at its later stages of growth it must be able to utilize the energy of matter's atomic force to effectuate its ends for its own spiritual aggrandizement. It is itself to be lifted up to heights of cosmic consciousness, but no more than an oak can exalt its majestic form to highest reaches without the dynamic energization received from the earth at its feet can soul rise up above body without drawing forth the strength of body's dynamo of power. Down, down it descends then through the October, November and December path of the sun, until it stands at the nadir of its descent on December 21.

Here it has reached the turning-point, at which the energies that were stored potentially in it in seed form will feel the

153

first touch of quickening power and will begin to stir into activity. At the winter solstice of the cycle the process of involution of spirit into matter comes to a *stand-still* -- just what the solstice means in relation to the sun -- and while apparently stationary in its deep lodgment in matter, like moving water locked up in winter's ice, it is slowly making the turn as on a pivot from outward and downward direction to movement first tangential, then more directly upward to its high point in spirit home.

So the winter solstice signalizes the end of "death" and the rebirth of life in a new generation. It therefore was inevitably named as the time of the "birth of the Divine Sun" in man; the Christ-mas, the birthday of the Messianic child of spirit. The incipient resurgence of the new growth, now based on and fructified by roots struck deep in matter, begins at this "turn of the year," as the Old Testament phrases it, and goes on with increasing vigor as, like the lengthening days of late winter, the sun-power of the spiritual light bestirs into activity the latent capabilities of life and consciousness, and the hidden beauty of the spirit breaks through the confining soil of body and stands out in the fulness of its divine expression on the morn of March 21. [. . .]. (pages 8 - 11 as paginated in this version).

A couple quick things to point out might include, first, the fact that Kuhn does not intend to exclude women when he uses the masculine pronoun (as was the custom in grammatical usage when he was writing) -- he specifically makes this clear at numerous different points in his voluminous writings.

Second, the ancient sacred myths, scriptures, and traditions of humanity from all over the globe can now be *conclusively demonstrated* to use the above heavenly cycle (and many other heavenly cycles, including those of the moon and planets but most especially the motions of the stars and constellations) as part of their inspired method for conveying to us the most profound and necessary spiritual knowledge to aid us in this incarnate life.

154

Alvin Boyd Kuhn wrote the above incredible explication of the annual solstice and equinox points as spiritual analog for the "soul's peregrinations round the cycles of existence," but while he correctly tied the esoteric stories of the Bible to these stations on our spiritual pilgrimage, the level of celestial correspondence to the specific constellations that can be demonstrated in the stories of the Old and New Testament, and in the other myths of the different cultures all around the globe (probed more thoroughly later by Professors Hertha von Dechend and Giorgio de Santillana in *Hamlet's Mill*, published in 1969, but without perceiving their spiritual depth in the same way that Kuhn had decades earlier) had not yet been fully appreciated.

The more we begin to understand the specific celestial correspondences of the various gods and goddesses and spiritual beings who are found in the different myths of the world, the more we can begin to see where they might fit and what roles they might play in the great spiritual cycle elucidated by Alvin Boyd Kuhn in the passage above, and better understand their significance and meaning for our own spiritual growth.

This is actually a matter of absolutely the utmost importance, I am convinced, and opens tremendous new avenues of communication with the incredible ancient wisdom given to humanity in the "high and far off times" (as Giorgio de Santillana called it) -- perhaps in the same millennia that the great stone henges and circles and towers and temples were being erected, or even millennia before that.

I hope that as we approach the powerful and significant turning point of December solstice this year, you will have the opportunity to contemplate these matters at some length and that meditating upon this ancient wisdom will be a blessing to you in your life and in the years to come.

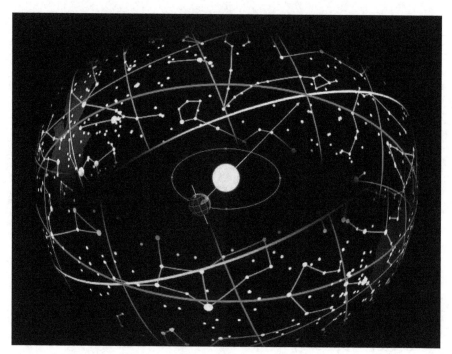

Gaining fluency in the language of the myths

2016 August 14

I have said many times, in interviews and in published materials (written and video), that the concept of "speaking in different languages" provides a very helpful metaphor regarding the interpretation of ancient myths, scriptures, and sacred stories.

If the myths arrive in our village speaking a language we don't understand (perhaps a language our ancestors understood, but a language we have long forgotten), and someone offers to interpret what they say for us -- but they don't actually speak the language either -- they could offer all kinds of misinterpretations which could lead to all kinds of confusion and chaos.

There have probably been scenes in movies or plays in which just such a scene is acted out for humorous effect -- perhaps with the visitor saying (in his or her own language), "This food is delicious -- I am honored that you are sharing it with me," and the deceptive translator telling the anxious listeners, "He said your food is disgusting, and he is extremely insulted that you have offered him such disgraceful refuse."

The point of this metaphor is my assertion (which has been made by others in the past -- stretching back at least to Plato[6]) that the myths, scriptures, and sacred stories of humanity are not speaking to us in a *literal* language, but rather in the language of *celestial metaphor*. Therefore, if we try to interpret them as if they are speaking a literal language, we are almost certain to introduce mis-interpretations, mis-translations, and mis-understanding (with potentially disastrous results).

The fact that the myths of humanity, from virtually every culture in virtually every inhabited corner of the globe, can be shown to be built upon a common system of celestial metaphor is astonishing, and has incredible implications for our ancient history. Setting aside the question of how this situation came about (because there are numerous possible hypotheses which could be offered, each of which deserves careful investigation -- the question is by no means settled and the solution is by no means obvious or self-evident), the fact that this same system can be seen to underlie the myths and scriptures and sacred stories of peoples around the world, from Africa to Scandinavia to Australia to the Americas and all points in between, argues that someone in the distant past believed their message to be incredibly valuable and worth safeguarding and preserving with the utmost reverence.

If we wish to understand this ancient message, preserved in the sacred myths which are the precious inheritance of the human race, then it would behoove us to try to listen to them *in the language that they are speaking*, and not in some other language that we *wish* that they were speaking, or that someone else *tells us* that they are speaking.

(That last category includes me, myself: don't simply take my word for it when I assert that they are speaking in the language of celestial metaphor -- examine the evidence for yourself; I offer hundreds of examples from around the world which I believe point to the conclusion that the myths and scriptures, including those in what has for centuries been called "the Bible," are built upon a foundation of celestial metaphor).

If one examines the evidence and concludes that the myths are indeed speaking a celestial metaphor, then the next question might be to ask *how we understand this language*. It is a very good question.

Some conventional scholars, detecting some of the solar and celestial and "seasonal" components in the myths (by "seasonal" I mean having to do with summer or winter or spring or fall, usually by the incorporation of metaphors or observations centered on the equinoxes and solstices, obvious examples being the celebrations of Christmas and Easter / Passover around the time of the December solstice and March equinox, respectively) assert that these elements in the myths and sacred traditions of the world are simply remnants of a time when humanity was working out the civilizational tools of agriculture and keeping a calendar, both of which do in fact involve complex sets of knowledge which would be helpful to preserve and pass on, once people have figured them out.

However, although that knowledge may certainly be *part* of the story (and while I certainly do not deny that the skills of keeping a calendar and successfully growing crops in order to keep a civilization alive are critically important skills, whose knowledge ancient peoples would want to preserve), I believe that there is more than sufficient evidence to argue that the myths are speaking a celestial language which can actually be described as a *spiritual* language (and even a *shamanic* language, as I have argued extensively in previous books and previous posts on this blog).

So, how does understanding their celestial foundation help us to hear the ancient myths in the language that they are in fact speaking? By understanding the way that the myths of the world use the heavenly realms, and the cycles of the heavenly bodies which move through those realms, as a way of describing a worldview or cosmology which includes both the Visible Realm and the Invisible Realm -- the second being the realm of spirit, the realm of infinite potentiality, the realm of the gods.

This Other World is very real, but because it is invisible to us (most of the time), the myths and scriptures use figures and metaphors in order to convey teaching about it to our understanding. And the figures and metaphors they use to convey that knowledge are drawn from the heavens above -- the heavens *which are in fact* infinite in their own right, and which thus form a perfect vehicle for conveying to us truths about the infinite realm.

Alvin Boyd Kuhn, who was in my opinion one of the most articulate and insightful teachers regarding the way that the myths use the heavenly cycles as a language for conveying spiritual truths, gives an outstanding (and essential) translation of the spiritual meaning of the annual cycle which forms the Great Wheel of the year, divided into four different parts by the great stations of the two equinoxes and two solstices. He published thousands of pages of explication of the myths and their spiritual language, and the clearest and most concise explanation of the spiritual "code" in the annual cycle is perhaps found in a little volume entitled *Easter: The Birthday of the Gods* (based on a lecture he delivered in 1936).

It is my belief, after some years of study on this subject, that in addition to the references to the solstices and equinoxes found in the myths, we can also perceive references to *specific* constellations in the myths and sacred stories themselves -- constellations which point us towards specific places within that larger framework, and which thus act as pointers to help us understand what specific themes and teachings that particular myth may be expounding for our understanding and our benefit.

This fact may be the key to understanding the language which the myths and sacred stories of humanity are speaking to us -- and the more we understand this "language," and the more time we spend "conversing" with the myths using that language (which is a spiritual language, and even a shamanic language, as most traditional peoples on earth already understand, but as the proponents of a literal approach to the Biblical texts wish to deny), the more we will gain "fluency" in hearing their message.

The Invisible Realm and the Shamanic

The centrality of ecstasy, according to ancient wisdom

2014 May 14

In an important text entitled "Man in search of his soul during fifty thousand years, and how he found it!" poet and Egyptologist Gerald Massey (1828 - 1907) declares:

> The ancient wisdom (unlike the modern) included a knowledge of trance-conditions, from which was derived the Egyptian doctrine of spiritual transformation. [See section 17 of Massey's text].

This statement is worth careful examination. First, judging by the second half of the sentence ("from which was derived . . ."), it is evident that Massey considers the ancient wisdom of which he speaks to have *predated* ancient Egypt, and to have been the even more ancient source from which the Egyptian doctrines were derived. In other words, Massey is referring to a wisdom which predates one of the most ancient civilizations known to history.

Second, we see that the heart of his sentence is the declaration that this ancient wisdom concerned "knowledge of trance-

162

conditions," and that this knowledge was somehow intimately involved with the concept of spiritual transformation.

By "trance-conditions," the rest of the essay makes clear, Massey is referring to the process of "entering the spirit world as a spirit" and then returning to the material world of everyday life. Even further, he explains that in trance one actually "enters the eternal state" (both quotations from section 17). From this contact with the eternal state, one can gain the knowledge that comes only by personal experience, which the ancients called "gnosis."

Note that in the previous post entitled "Wax on, wax off,"[3] describing the famous scene from *Karate Kid* (1984), it was pointed out that in the video clip that Mr. Miyagi never actually "explains" anything to Daniel-San: Daniel-San *experiences* it for himself (no matter how many times you have seen it in the past, or how well you think you know that scene, it is worth watching it again for its beauty and power). Neither does Mr. Miyagi ask Daniel if he "believes" the moves will work: Daniel-San experiences *that* for himself too.

Massey makes this very point in part 20 of the same essay where he asks:

> What do you think is the use of telling the adept, whether the Hindu Buddhist, the African Seer, or the Finnic Magician, who experiences his "Tulla-intoon," or supra-human ecstasy, that he must live by faith, or be saved by belief? He will reply that he lives by knowledge, and walks by the open sight; and that another life is thus demonstrated to him in this. As for death, the practical Gnostic will tell you, he sees through it, and death itself is no more for him! Such have no doubt, because they know.

In the first quote cited above, Massey declared that the ancient wisdom (in contrast to the modern) included this very knowledge: the knowledge of ecstasy, of leaving the "static" of physical realm ("ecstasy" comes from "ex-stasis," or outside of the static or solid). The ancient wisdom was centered around the experiential knowledge -- the *gnosis* -- of the non-material realm: this is what distinguishes the ancient wisdom from the modern, which Massey plainly says is completely impoverished in comparison to the ancient gnostic wisdom.

The Undying Stars explores the evidence that all the ancient vessels containing the ancient wisdom (wisdom which predates even ancient Egypt) have ecstasy and gnosis as their central concern. In other words, the ancient wisdom can in an important sense be described as being in some sense shamanic. The ancient Egyptians appear to have possessed a shamanic tradition of out-of-body contact with "the spiritual realm" or "the eternal," as shown by Lucy Wyatt and other researchers who are cited in her book *Approaching Chaos* (2010). The ancient Greek civilization had oracles who entered into a trance-state, including the most famous of the ancient oracles, the Pythia at Delphi. There were also the ancient *mysteria*, found in many ancient cultures, which involved profound experiences and possibly altered states of consciousness, all designed to impart gnosis to the participants.

There is substantial evidence of a similar centrality of "ecstasy" or "trance-conditions" in many other sacred traditions around the globe, including those whose shamanic wisdom survived right up through the nineteenth and even the twentieth centuries (and, in some more remote locations, survives to this day).

The Undying Stars further makes the argument that the esoteric texts which were incorporated into the Old and the New Testaments were also originally intended as vessels to convey this very same ancient wisdom, just like all the others. It was the literalists who destroyed this understanding and who insist that their texts have no kinship with all the other closely-related sacred traditions of the world. It was the literalists who denied the clear fact that these texts (those in the Old and New Testaments, just as much as those of ancient Egypt or of ancient Greece) are meant to be understood gnostically -- that is to say, esoterically rather than literally, perceiving that they are designed to convey the same *knowledge* born of the experience of the eternal. And it

164

was the literalists who labeled all the other sacred traditions of the world as "pagan" or "heathen" and who commenced a centuries-long campaign to destroy their ancient knowledge, convert their people at the point of the sword to the literalist religion, or kill them if they would not convert (see, for example, the history of the brutal campaigns of Charlemagne).

In doing so, those behind the campaign against the shamanic and ecstatic and gnostic have robbed humanity of its ancient heritage, inflicting untold death and misery in the process. Further, if the experience of the ecstatic is essential to the kind of *knowing* described above by Massey, they have robbed a large number of individual men and women of the opportunity to develop a vital gnosis of their own.

But there is much more to the loss even than that, because the evidence also suggests that the ability to cross over into the other realm and return with the knowledge available there benefited the entire community or civilization, and may well be the source of technical and medical knowledge which could not be obtained otherwise, which shamanic cultures demonstrate (including that of the ancient Egyptians), and which conventional historical paradigms absolutely *cannot explain*. If so, then the deliberate eradication of the shamanic, first in the areas conquered by the Roman Empire and then – systematically – in other cultures in other parts of the globe, may have stunted the civilizations that have developed ever since. Who knows what shape cultures would take today, and what capabilities and peaceful arts we might enjoy, had the esoteric vessels that carried the ancient wisdom not been deliberately fouled by the literalists.

There is one more aspect of this discussion which is worth considering, and that is the logical possibility that those who deliberately suppressed the esoteric, gnostic, shamanic understanding of myths and sacred scriptures like those found in the Old and New Testament (and then destroyed the gnostic and esoteric and shamanic understanding of other cultures as well, forcing them at the point of a sword to accept the literalist religion) kept the ancient knowledge for themselves, while denying it to everyone else. Given the evidence that the ability to cross the boundaries between the realms may have been connected with the inexplicable medical and architectural achievements of ancient civilizations (and some of the knowledge of modern-day shamanic cultures found in remote regions, such as isolated parts of the Amazon rain forest), there are some rather thought-provoking ramifications if that is indeed the case.

It may be that the first step towards regaining some portion of what was lost begins with the appreciation for the esoteric system of interpreting sacred traditions, scriptures, and myths, coupled with an awareness of how this esoteric approach points to the outlines of an ancient cosmology which understood the nature of the visible *and invisible* worlds (and our place in them).

166

Shamanic - Holographic

2014 May 21

Previous posts have made the assertion that virtually all of the world's ancient sacred traditions incorporate celestial allegory, and that this celestial allegory is central to their esoteric message.

The system of celestial allegory remains the same across traditions which are superficially very different, but which share at their core a common system in which the sun, moon, and the visible planets, along with the silent and majestic backdrop of the fixed stars (and especially those groupings of stars along the ecliptic path, the twelve constellations of the zodiac) have identifiable characteristics which can be used to identify them in the widely varying mythologies from very different cultures and even from different millennia. As some of the posts linked above assert (and as *The Undying Stars* argues at some length), the scriptures which were included in the Old and New Testament

167

can be shown to be built upon this same esoteric system of celestial allegory.

This aspect of the Biblical scriptures has been noted in the past, although different authors have advanced widely different arguments regarding the reason that the world's sacred stories would be so fixated upon the motions of the heavenly actors. One of the more cogent and well-supported arguments comes from the writings of Alvin Boyd Kuhn, who explored the subject at great length in two major works, *Lost Light* (1940) and *Who is this King of Glory?* (1944), as well as in many shorter studies and essays. His thesis involves the argument that the sacred traditions of the world incorporated examples from nature in order to more clearly convey profound truths regarding spiritual matters -- to use the visible in order to teach the invisible -- and that they fastened upon the motions of the heavens as being among the most majestic natural spectacles in all of physical creation, as well as being ideally suited for allegorizing the aspects of the incarnation of the soul in matter and its eternal survival through endless cycles. Aspects of Kuhn's argument have been discussed in previous posts such as "A Hymn to the Setting Sun, and the ultimate mystery of life," "The horizon and the scales of judgement," and "The undying stars: what does it mean?"[23] as well as a post exploring the question of whether the Bible in fact teaches reincarnation (from 05/19/2014).

It is possible to agree with Kuhn's arguments, while simultaneously arguing that these celestial metaphors at the heart of the world's ancient traditions go even further -- and indeed, this is the approach taken in *The Undying Stars*. While Kuhn was extremely insightful in his analysis, and uncovers profound esoteric teachings while plumbing the depths of the ancient metaphors, there were some scientific developments which had not yet come to light at the time he was doing most of his writing. One of these was the development of holograms visible to the naked eye, which require the use of a light source producing extremely coherent light such as a laser (discussed in this short YouTube video), and the other was the discoveries by theoretical physicists that our universe actually resembles or behaves like a hologram in many important ways.

The Undying Stars demonstrates that, in many incredible ways, the system of allegory found in the world's ancient sacred traditions (including the Bible) can be seen to anticipate the "holographic universe" theory which modern physicists only began to discuss in the 1960s (after the invention of lasers and holograms produced using coherent light).

In creating a hologram, a beam of light is first split and one part of the beam is sent to the object being captured on the holographic film, and the other part of the beam is routed is such a manner as to reach the holographic film from a different angle and create an interference pattern with the first beam. By capturing the interference pattern in the holographic medium, a light source can be used later to create a holographic image of the original object, even though the object is not actually there. This holographic projection is called a *hologram.*

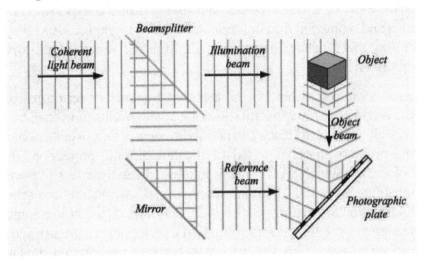

As explained for example in Michael Talbot's *Holographic Universe* (1992), many theoretical physicists have found the metaphor of the hologram to be extremely helpful for describing a model of the universe which fits the often counter-intuitive

169

discoveries of quantum physics, which began to come to light in the first part of the twentieth century and (after much argument and head-scratching) have become widely accepted among the scientific community and are generally regarded as proven.

In this model, physicists posit that in some ways, the *information* which makes up the universe which we inhabit (or which we think we inhabit) can be said to be stored in a form that is very much like the information stored in the holographic film or holographic medium described above. According to this model, the universe we inhabit (or believe we inhabit) is in many ways very much like the holographic projection or the *hologram.* There are many recorded lectures available on the web in which professional physicists discuss this model, and the evidence which led physicists to begin to propose it (most of them take about an hour -- the explanation in this paragraph is a gross over-simplification).

The important point for this discussion is the fact that all of the ancient scriptures of the world can be shown to be teaching a cosmology, or a vision of the universe and of the experience of men and women in that universe, which is holographic. And they convey that teaching esoterically, through the system of allegory described above.

Even more astonishing is the fact that the ancient scriptures of the world not only appear to describe a holographic universe, but they also appear to teach the possibility of crossing over between the ordinary material world of the holographic projection (or hologram), which we normally inhabit (or at least feel like we inhabit) and the hidden, compressed world where the information for that projection is actually stored (the holographic medium or the holographic film, in the metaphor we are using to explain it here). This ability to cross between these two realms is often described today using the term "shamanic," because this ancient ability survived into modern times primarily among shamanic cultures, but there is strong evidence that *all* the ancient sacred traditions of the world could be described as shamanic, in that they taught both the possibility *and* the supreme importance of crossing between those two aspects of our holographic universe.

The Undying Stars attempts to articulate the connection between the ancient system of celestial allegory and this ancient emphasis on the shamanic.

For more on this subject, see the discussions in "The centrality of ecstasy, according to ancient wisdom,"[162] among others. Authors who have demonstrated that the ancient cultures of the world were essentially shamanic include Jeremy Naydler and Lucy Wyatt.

Gerald Massey, who declared that all the sacred repositories of ancient wisdom "included a knowledge of trance-conditions," argued in various places (including the essay "Man in search of his soul during fifty thousand years, and how he found it!" which is the source of that "trance-conditions" quotation), believed that the ascent to the other realm was symbolized by the raising of the *tat cross* (the ancient Egyptian symbol sometimes depicted as the backbone of Osiris, and usually rendered into our alphabet as the *Djed* column by more modern scholars), which symbolized the ascension of the spirit in a man or a woman, just as the same column lying on its side or horizontally represented the animal nature or the material prison of the body (emblematic of the coffin or the corpse, in which the spirit is imprisoned during the incarnation in this realm, or of the animals and four-footed beasts who go about in a horizontal fashion).

Massey briefly mentions the *tat* cross in conjunction with other symbols of spiritual vivification in section 40 of that essay, and dwells in more detail upon its importance as the Tree of Knowledge, a symbol found across many cultures and traditions which is related to the same *tat* cross or *Djed* column, often in conjunction with the symbol of the serpent, and often in close vicinity to a body of water or a sacred pool (see for instance Massey's discussion in sections 8 and 9 and 34 through 36). The same symbology is evident in the photograph above of shamanic poles raised in modern times near Lake Baikal.

We may not be accustomed to thinking of the scriptures of the Old and New Testaments as being "shamanic" in nature (in fact, such thinking has been strictly prohibited for many or most of the past seventeen centuries at least), but it can be demonstrated that the same symbology is present and in fact extremely prominent and important in those scriptures, from the first chapters of Genesis right through to the Revelation.

The fact that the ancient sacred traditions of the world can be shown in some sense to be both "holographic" and "shamanic" is extremely significant, especially if (as authors such as Lucy Wyatt and Jeremy Naydler have demonstrated) the techniques of shamanic experience were seen as critical for the health and well-being of the cultures who knew and used them, and as a source for beneficial knowledge and technologies which even today we cannot fully understand.

Eleusis Transcendent

2014 July 12

The late afternoon sun bathed the city in a warm gold reminiscent of summer, but now a gentle breeze blew in from the sea, the cooler salt air invading the hillsides where the warmth of the sun had left a relative vacuum, bringing with it the hint of the approaching autumn. From the hills of Athens, the sun in its arc descended towards the west, in the direction of Eleusis and the Gulf of Eleusis, and the island of Salamis rising up out of the Aegean and separated by only a short stretch of water from the site of sacred Eleusis.

Salamis, of course, brought to mind Athenian naval prowess, but at this time of year, all the city-states of Greece observed a holy truce: each year, messengers from Athens traveled throughout the land to announce the celebration of the Eleusinian Mysteries, and to guarantee safe passage for an entire month leading up to the great festival, and for most of the month afterwards, to anyone who wished to attend in person. All were invited, whether man or woman, free citizen or slave -- as long as they could speak the Greek language, and as long as they had never committed the crime of murder.

The truce would last from the time of the full moon in the month of Metageitneon (the middle of the second month after the annual cycle of the Attic calendar began with the first new moon following summer solstice, which would put the fifteenth day of Metageitneon towards the beginning of our September), when the runners went out, through to the tenth day of the fourth month after summer solstice, Pyanepsion, which would be towards the end of October. But the runners had been sent out nearly a month ago now, and it was now the fourteenth day of Boedromion (and the eve of a full moon), that third and final "summer month" after the solstice, towards the end of what moderns would someday call September. Athens was now packed with those who had streamed in from all over the civilized world to observe the sacred *mysteria*.

Earlier in the day, the procession of the priests and priestesses of the mysteries had left Eleusis and walked from there to Athens, the priestesses bearing the sacred and secret objects in closed containers, out of public view.

A thrill of anticipation ran like a steady electrical current through all those who would participate for the first time in the rites which would begin the next day. They did not know, nor did any who had not already experienced the Eleusinian mysteries themselves, the exact details of the secret rites in which they would participate on the fifth day after the sacred festival began, for the penalty for divulging the secrets of those rites was death, and the ancients honored the prohibition so steadfastly that none can say for sure to this day exactly what went on within the sacred sanctuary of the Telesterion at Eleusis.

But, the public portion of the mysteries was widely known, for it began in the public places of Athens on the first day of the festival (the fifteenth day of Boedromion), on which day the formal invitation was proclaimed in the streets. The following day, the command of *Halade Mystai!* would go up ("Initiates, to the Sea!"), and all the people desiring to be initiated (the *mystai*) had to go down to the waters of the Aegean below Athens to bathe in a ritual act of purification, taking with them a piglet.

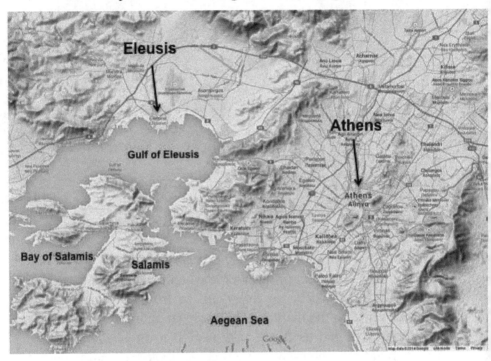

The following days there would be sacrifices and celebrations, and then at last would come the long-anticipated processions of the initiates, led by the priests and priestesses with the sacred objects still in their containers. Significantly, this ritual procession would include a stop at a cemetery, a crossing of a bridge over the river Cephisus, and a ritual mocking of the initiates by onlookers along the way. There was also a point where the initiates had to shout obscenities, in recollection of the time that Demeter, whose rites these Eleusinian mysteria were, had been made to smile by the off-color words and antics of the old nurse Iambe, who had cheered the mourning goddess during her search for her missing

175

daughter Persephone. Both Demeter and Persephone were the goddesses of the Eleusinian rites, but Persephone was always referred to as Kore (meaning "the Maiden") in connection with the Eleusinian Mysteries.

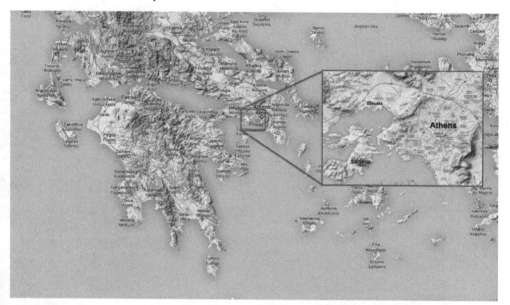

Imagine the excitement of the *mystai* when they finally arrived in Eleusis! They would be nearly exhausted from the day's march and the events on the march itself, but what an impression it had left upon them and as they finally rested in the shadow of the sacred buildings themselves, they could look back and reflect upon all of them, as well as the emotion-laden days in Athens on the previous days. Perhaps many would then continue further back, and reflect upon how long they had anticipated this day, when they would finally walk the path to Eleusis itself -- perhaps a lifetime goal for many who were there now for the first time.

For the poets, and no doubt those who were already initiated, made no secret of the knowledge that those who witnessed and experienced the nights that were to come, here among the pillars of the holy buildings dedicated to Demeter, would be forever changed, and would be freed of the dread of death, and would be promised a better condition in the invisible realm of the dead.

But, the *mystai* no doubt had also heard that the experience itself -- the secret rites that would commence the following night -- would move them profoundly, and would contain moments of terror that would fill them with dread (how exactly this would take place, we can today only speculate).

And so it would be with a mixture of high emotions and awful anticipation that those undergoing the initiation would have watched the sun sink once again into the west. They felt that events were now in motion that they could not stop, even if they wanted to back out now, drawing them onward almost in spite of themselves.

They might reflect, too, on the knowledge that these rites had been taking place since time immemorial, right there on those slopes of Eleusis facing the water, each and every year at this particular season, when the sun was crossing the great line of the celestial equator and the nights were balanced with the days, with the festival really getting underway after that line had been crossed and the kingdom of the night was expanding further and further, the days steadily retreating in their length on the way to the lower half of the year.

No one knows exactly when the Eleusinian mysteries had been founded, but they had been taking place at least since what we would call the eighth century BC, and possibly for many centuries before that based upon the archaeological evidence. We know that the so-called Homeric Hymns, some of them probably composed in the seventh century BC, describe the search of Demeter for Persephone after Zeus permitted Hades to kidnap the maiden and take her to the underworld to be the bride of the god of death, and that Eleusis receives prominent mention in the Homeric hymn to Demeter. The hymn to Demeter is the second in the Homeric hymn series, and it is rich in mythical detail that help us to understand the profound themes which the Eleusinian Mysteries esoterically convey.

What experiences awaited those anxious *mystai,* as well as the *epoptai* (those who had already been initiated, and were returning to Eleusis to participate again as initiated "seeing-ones," possibly to have revealed to them additional parts of the ritual which took place while the *mystai* were still blindfolded or in the dark)? They could only wonder and imagine that first night, as the cold night winds accompanied the sun's plunge into the western seas and rustled the leaves on the trees upon the darkened hills.

We, too, can only wonder what awaited them, for the Eleusinian Mysteries, those most celebrated in the ancient world, enacted for two thousand years in a row (a few centuries longer than the period which separates us today from the last participants), were shut down along with many other sacred conveyances of the ancient knowledge, at the dawn of the era of literalist Christianity, by the decree of the Roman Emperor Theodosius in AD 392. Flavius Theodosius began the third dynasty of Christian emperors, following Gratian (who had outlawed the Vestal Virgins in Rome) and the Valentinians, who themselves followed the Constantinians which began with Constantine himself, who had declared Christianity the religion of the empire in AD 324, and who had died in AD 337.

No longer would the secret experiences of Eleusis be available to all who chose to participate.

However, from the imagery of some of the artwork surrounding the Eleusinian Mysteries -- including the tablet shown below dedicated by a participant named Ninnion (her name can still be seen on the dedicatory inscription, at the lower left corner of the tablet), scholars have formulated guesses regarding what would have taken place on the night of the second day after the arrival of the initiates (behind the procession of the priests and priestesses carrying the hidden sacred objects).

We also have veiled hints of what took place which have survived in some ancient accounts, including those of the ever-helpful Plutarch. From these accounts, and from the scene above, we can surmise that the transformative events that awaited the *mystai* on the night following that first night after their arrival at Eleusis would involve a harrowing search by torchlight, during which they would no doubt be enacting the desperate search of the bereaved Demeter, and during which some have speculated that they would be startled by sudden visions of the two goddesses themselves, Demeter and Kore, illuminated by bursts of light with pyrotechnic effects and perhaps thunderous music or percussion, techniques perfected for and used in ancient theater as well.

Ancient descriptions suggest that they would have wandered in circles, calling out loudly, perhaps blindfolded during the initial windings and searchings, and that they would have also been given glimpses of sacred and symbolic objects representative of the cult of the two goddesses. One ancient source also declares that at the end of the search there would have been a triumphant revelation and a celebration, and much flinging about of torches also!

The most sacred aspects of the central rites probably took place in the cave-like settings inside the inner sanctuary buildings, and included a ritual drink and perhaps even a handling of the sacred objects themselves, as well as the revelation perhaps of the goddess or goddesses themselves. Some have speculated that the drink given to the initiates possessed hallucinogenic properties, or that hallucinogenic substances were otherwise involved, accounting for the powerful visions and the feeling of dread described.

But we simply do not know. The contents of the mysteries have been sealed off from our view, and may (as far as we can tell at this point in history) retain their secrets for eternity.

What we can say for sure is that, for those who have become familiar with the ancient system of celestial metaphor which underlies all of the most ancient sacred rites and scriptures of the planet, there are abundant clues in the imagery that has survived, as well as in the ancient Hymn to Demeter itself, to allow us to gather just a taste of the significance of Eleusis.

First of all, it is clear from the specific station of the year when these rites would be celebrated -- just after the fall equinox, and after the full moon began to wane -- that the Mysteries of Eleusis involve the descent of the great heavenly cycle into the lower half of the year, the dark half, the portion signifying the underworld. From the extensive analysis and evidence offered by Alvin Boyd Kuhn (so extensive that it is really beyond dispute, in my opinion), we can also declare that this descent was seen by the originators of the ancient myths to be representative of the descent of the fiery soul into the material realm: the incarnation in the body.

DAYS LONGER THAN NIGHTS:
Heaven, Promised Land, Greece, etc.

NIGHTS LONGER THAN DAYS:
Hell, Egypt, Troy, etc.

Note all the imagery which confirms this interpretation surrounding the rites of Eleusis: there is the descent from Athens into the sea by the participants, during their ritual cleansing at the start of the festival; there is the crossing of the bridge, during which crossing the initiates are ritually mocked (note the *crossing* points of the year marked on the zodiac wheel, above – the downward crossing is at the sign of the Virgin); there is the story of Persephone, seized and taken down into the underworld; there is the fact that Persephone is always referred to as Kore, the Maiden (the word "maiden" means "virgin") in the rites of Eleusis, just as the sign of the Virgin presides over this downward crossing on the annual cycle; there is the association of Demeter

181

with grain and the harvesting of wheat, just as Virgo traditionally carries a sheaf of wheat, associated with her brightest star, Spica; and there is the imagery of the torches, which we have already seen were anciently associated with the equinoxes and the crossing of the blazing ecliptic path of the sun downward or upward across the unchanging line of the celestial equator (this specific connection is also discussed in *Hamlet's Mill*).

The goddess Hecate is often shown in imagery surrounding Eleusis and the search for and return of Persephone/Kore, and when she is shown, she is often depicted carrying two torches.

Further, the imagery of the goddesses Demeter and Kore, seated upon their thrones, are often configured to resemble the distinctive features of the constellation Virgo. Note in the Ninnion Pinax shown above the posture of the two goddesses, who are seen on right side of the plaque (facing to the left). Demeter is above, and Kore is beneath. Both are seated on thrones, and both have the extended hand that is very distinctive of Virgo. Below is the same ancient scene, with the outline of the stars of Virgo superimposed:

182

Further, we have already discussed at length the thesis put forward by Alvin Boyd Kuhn that the search for the hidden god or goddess over all the lands inhabited by men and women was meant by the ancients to convey the sacred truth that the divine spark is hidden within all men and women, but that it is seemingly "lost" and must be pursued, reawakened, and found again.

We note that the *mystai* stopped at a graveyard on their way down to Eleusis: suggestive of the ancient allegorization of the plunge of the soul from the world of spirit as a "death," and this life as a passage through the "underworld" itself (which, relative to the world of spirit symbolized by the ethereal spheres that circle above our heads, this world in fact is -- again see Kuhn's extensive analysis and support of this assertion). We also note the symbolism that on the day in which they were *ordered* down into the sea (symbolic of the command to incarnate), they had to carry with them a live piglet: symbolic of the fact that when the spirit incarnates, it must take on the "animal" body, and become a cross of the spiritual and the carnal (how many ancient allegorical myths warn against becoming too comfortable with the carnal side of this combination, and of thus "turning into swine").

And, note carefully the incorporation into these rites of the obscenities shouted by the participants, in commemoration of the actions of Iambe in bringing a smile to the lips of the grieving goddess: this is a component of goddess-myths the world over, found in Norse myths (Loki and Skadi), in Japanese myths (Uzume and Amaterasu), and in the Old Testament story of Sarai/Sarah and her secret smile (all of these are discussed in a previous post from 06/07/2014, and involve the faint stars forming "smile" on the face of the constellation Virgo: see page 303).

It is very interesting to note that sexually explicit antics are often involved in the myths involving the smile – this is true of the antics of Loki, of Uzume, and (in many versions of the tale) of the old Iambe (sometimes named Baubo in some versions of the Greek or Latin myth of Demeter) as well. This is highly significant, as the stars themselves can hardly be said to suggest such a constant theme, and it is difficult to argue that this very specific and distinctive aspect of the myth would arise independently in so many different cultures. And yet Japan is very far from Scandinavia, and both are far from Greece as well – how did this myth-detail arise in such far-flung mythologies? There are a few possible answers, but I believe the most likely is the fact that all these vessels of ancient wisdom are descended from a common, even more ancient source (the "lost civilization" -- and one that other evidence reveals to have been very advanced and very sophisticated).

Esoterically, however, we can speculate that the connection to the more physical and generally private aspects of the human body (specifically, the sexual functions) at which the goddess or maiden is made to smile in these myths connects to the greater theme of incarnation as well: this specific part of the solar cycle (the autumn equinox and the plunge into the lower half) and of the lunar cycle (after the full moon is over, and the moon begins to wane towards "death") and of the elemental cycle (plunging from the air and fire of the heavenly journey into the earth and water that all the circling heavenly bodies encounter when they sink down into the western horizon) figures the plunge of the disembodied soul into the incarnated physical vessel of the human form.

Perhaps these myths are meant to hint at the idea that the soul finds this condition somewhat uncomfortable at times, especially the most carnal aspects of its incarnation. Certainly the myths involving the smiling goddess seem to poke fun at this aspect of our incarnate state. The mocking of the initiates, and their explicit participation in the obscenities of Iambe, might certainly have been intended to convey to the *mystai* this esoteric message.

We do not know what exactly took place on those powerful nights of mystery which the initiates could look forward to as they marveled at the experiences which they had undergone thus far. But we can guess that the truth that the ancient Eleusinian

184

Mysteries meant to have them experience -- and hence to *know*, deep in their bones, not just by faith but by *gnosis* -- was that this mortal material life is not all that there is, and that the physical world and all that seems so solid and imprisoning can actually be transcended. If they actually underwent an out-of-body experience during the rites themselves, as some believe they did, then they would have experienced that aspect of transcending apparently un-crossable boundaries right then and there!

This is exactly what the evidence tells me that the ancient universal myths bequeathed to humanity were *all* meant to convey. They all intended to teach a shamanic, holographic vision of the universe we inhabit, and that we can grasp hold of the shamanic and holographic within this very life, to overcome and to create and to transcend.

Unfortunately -- tragically -- this ancient message was somehow subverted, and an active and deliberate campaign to stamp it out was initiated, even as the Eleusinian Mysteries were still going on. Their shutting down by Theodosius was a decisive event in that anti-shamanic campaign (discussed at greater length, and using other clear examples from history, in my book *The Undying Stars*). The sanctuary and the Telesterion at Eleusis would be sacked in AD 396 and never rebuilt.

But the message of the Eleusinian Mysteries does not depend upon the physical stones of the sanctuary, nor even in the secret objects or rituals (whatever they were) enacted on those final nights. The message is still available today, and can be read in the book of the starry heavens and the books of the sun and the moon in their cycles, open for all to see every day and every night.

The shamanic foundation of the world's ancient wisdom

2014 August 29

The previous post on "The sacrifice of Odin"[291] presented abundant evidence that the important Norse god Odin is a shamanic figure, frequently depicted as undertaking journeys in search of hidden knowledge, and knowledge which specifically can only be obtained through shamanic methods.

The most central and most shamanic of all of these vision-quest journeys undertaken by Odin is undoubtedly his ascent to hang himself upon Yggdrasil, sacrificing in his own words "myself to myself," wounded with "the spear" which we can assume would likely mean deliberately and with his own spear Gungnir, and through a nine-night-long ordeal eventually obtaining a breakthrough into another reality in which he sees with non-ordinary vision the secret of the runes.

We saw that the power of the runes is far more than "just writing" (as if the power to write, which most of us take for granted, is not incredible enough in and of itself): the ability to see and know and use the runes implies the ability to create worlds through the power of words, sounds, language, speech, and mind. In a very real sense (as Shakespeare, George Orwell, and a host of other thoughtful writers have perceived) we are composed of our thoughts and thought-patterns and narratives, and those thoughts and thought-patterns and narratives are ultimately composed of words and of language, that is to say of symbols -- and we could say *of runes.*

Students of Old English will know that the very word "spell" which in modern English means a formula to alter reality was the Old English word *spel* that meant generally "word" or "message" (and hence the English word *gospel* is derived from the combination of the Old English words *god* pronounced "gode" and meaning "good" and *spel* meaning "word"). This fact reflects and illustrates the reality-altering power of words, language, and runes.

Interestingly enough, in light of the tremendous reality-altering power of words (and runes) is the fact that in order to obtain the knowledge of the runes, Odin had to undertake a journey that is clearly shamanic in its elements, including the ascent up a pole or tree: examples abound of the use of a pole or "tree" in the ritual shamanic journeys described in Mircea Eliade's compendium of shamanic observations from around the world entitled *Shamanism: Archaic techniques of ecstasy* (originally published in French in 1951). It is quite clear from the details of many of these shamanic poles that they represent the celestial pole, which is in fact the World-Tree, and thus they correspond directly to the "pole" upon which Odin had to ascend during his own ordeal to transcend ordinary reality and obtain the power of runic reality-creation and reality-manipulation.

Eliade offers numerous examples of shamanic rituals which involve, "as an essential rite, climbing a tree or some other more or less symbolic means of ascending to the sky" (123) including the "South American consecration, that of the *machi*, the Araucanian shamaness," who undergoes an initiation ceremony centered upon "the ritual climbing of a tree or rather of a tree trunk stripped of bark, called *rewe*. The *rewe* is also the particular symbol of the shamanic profession, and every *machi* keeps it in front of her hut indefinitely" (123). Eliade informs us that the *rewe* is always nine-feet tall in this particular culture, and that the multi-day ceremony involves drumming, drum circles, dancing, stripping naked, the sacrifice of lambs, falling into trance or the state of ecstasy, and the ritual cutting of the fingers and lips of both the shamaness candidate and the initiating shamaness, using a white quartz knife (123-124). Eliade then goes on to describe a shamanic initiation rite among the Pomo of North America involving "the climbing of a tree-pole from twenty to thirty feet long and six inches in diameter," and similar (and sometimes even more dangerous) symbolic ascents among shamanic cultures from the regions of Hungary, Iran, Australian aborigines, the Sarawak of Malaysia, and the Carib shamans of Dutch Guiana (125-131).

If the reader is not thoroughly convinced that this most central vision quest undertaken by Odin indicates his shamanic nature -- and is thus additional powerful evidence that all the ancient sacred mythologies are in fact shamanic in their core message -- there is the additional evidence that he is known for riding through the heavens upon his eight-legged horse, Sleipnir (shown in the upper section of the carved runestone above).

While other Norse gods and goddesses of course had horses too, Odin's was the horse most well-known, most unique, and most associated with his wild journeys through the heavens in the company of the wild band of the Valkyries (in this he resembles Dionysus, who was often accompanied by Maenads – and whose rites in the hills and wilderness were described in terms indicating that they involved ecstasy). As the authors of *Hamlet's Mill* point out, the shaman's drum was described as the "horse" that serves to carry him or her into the state of ecstasy and to enable the shaman's soul to ascend to the sky (*Hamlet's Mill*, 122).

Odin's horse, Sleipnir, was notable for having eight legs -- four in the front and four in the back -- making him twice as fast as any other horse. Celestially, since Odin embodies the characteristics of the planet Mercury (who was also a transcendent god associated with breaking through barriers and with language), the fact that his swift steed Sleipnir had eight legs may be a mythological embodiment of the fact that Mercury is the swiftest of the planets (by virtue of its being so close to the sun). In fact, as you can easily confirm for yourself, the orbital period of Mercury is . . . 88 earth days! So, of course, Odin's steed would be expected to have eight legs -- what other number would have been appropriate?

But, if we see that Odin is clearly a shamanic figure, and that the shaman's horse is his or her drum, then the rhythmic drumming that would be produced by the hoofbeat of an eight-legged steed would be quite rapid, and quite apropos of the very rapid drumbeat used to produce a state of ecstasy in shamanic cultures around the world. So, the eight-legged nature of Odin's steed works to convey esoteric knowledge to us on many levels.

The previous post[291] also demonstrated that the shamanic nature of Odin's sacrifice upon the Tree has direct parallels to the sacrifice of Christ upon the Cross. In *The Undying Stars*, I explore the ways in which the realization that all the myths of the world (including those found in the Old and New Testaments) unites the world's ancient wisdom, and leads to the possible conclusion that they were all at their very core conveying a message that is essentially and profoundly *shamanic* (that is, in fact, what I call shamanic-holographic[167]).

This assertion is bolstered by the evidence that the celestial Tree which Odin must ascend (and which the shamans ascend in the ceremonies cited by Eliade) corresponds to the Djed-column of Osiris which must be "raised up" and to the Ankh or Cross of Life[42] of ancient Egypt which has a horizontal component representing the "cast down" nature of our material existence (in which we must go about in an "animal" body), but which also has a vertical component representing our spiritual nature which comes down from above and which is immortal (a fact emphasized on the Ankh itself by the unending loop at the top of the cross), and which represents both the motion of our rise and return to the spiritual realms *after* each incarnation and also the motion of the raising of the inner spiritual component or fire which we can perform during *this* life as an essential part of our mission in this earthly existence.

We have also seen evidence that this "divine spark" in each individual man or woman is associated with the fire brought down from heaven by Prometheus in the ancient Greek mythos, and with the Thunderbolt or Vajra found in the ancient Vedic texts, and that the mission of recognizing this inner divine element and of raising it up is central to our overcoming our cast-down state.

And -- although "orthodox" (a word that means "straight-teaching" or by implication "right-teaching") and literalist Christianity would strongly object to such an assertion -- this mission of recognizing and the of raising up the divine inner spark can clearly be seen to be a possible interpretation of the message taught by Paul in some of his early letters urging his listeners to recognize *the Christ within* (Galatians 1:16, Colossians 1:27, 2 Corinthians 13:5) and to realize that they themselves undergo the process of being crucified and raised by virtue of this mystical identification with *the Christ within* (Galatians 2:20).

This connection advances the strong possibility that the patterns found in the ancient scriptures preserved in the Bible were actually the very same patterns found in the myth-system of ancient Egypt and the Djed-column and Ankh-Cross imagery associated with the Osiris, and the very same patterns found in the myth-system of the Norsemen and the World-Tree sacrifice associated with the shamanic questing of Odin.

It also supports the conclusion that -- like those other world-myths -- the symbology and esoteric message of the Bible scriptures is in fact deeply shamanic, and pointing towards the same individual ascent and breaking free of the bonds of the material body and the material world undertaken by shamans in the rituals recorded by Eliade and other researchers in the early twentieth century and in the centuries immediately preceding.

Powerful evidence, perhaps even conclusive evidence, to support this conclusion -- the conclusion that the imagery employed by Paul and the other *early pre-literalist* teachers was actually composed of exquisite metaphors designed to teach a message closely aligned with the message embodied in the Osirian imagery of "the Djed-column cast down" and "the Djed-column raised up," the same message found in the sacrifice of Odin and the Thunderbolt of Indra (the Vajra) and in the ascent to the heavens by the shaman along the celestial tree -- can be seen in the fact that the traditional symbology surrounding the Crucifixion of

190

Christ quite clearly reflects the imagery surrounding the Osirian imagery of the Djed cast down and the Djed raised up.

Below is an image from the temple of Seti I at Abydos which comes from a series of images depicting scenes from the myth-cycle of Osiris, Isis, Set and Horus. Specifically, the image shown below depicts Isis retrieving the casket containing the slain body of Osiris from the King of Byblos.

Significantly (as I discuss in some detail in my first book), the casket containing the body of Osiris had lodged in a tamarisk bush and then been concealed when the tamarisk grew into a tree around it, which the King of Byblos then cut down to use as a pillar in his palace, thus connecting the body of Osiris to the World-Tree which is cut down in many myths around the world (including to Yggdrasil, which ultimately cracks apart and falls at Ragnarokk) and thus to the unhinging of the world-axis and to the precession of the equinoxes.

191

This aspect of the story links the Djed-column (also called the "Backbone of Osiris") even more strongly to Yggdrasil and the sacrifice of Odin as alleged in the previous post -- and we can see that, sure enough, in the image above the column that the King of Byblos is handing over to Isis has the horizontal "vertebrae" lines that indicate it is a Djed-column and the Backbone of Osiris.

Although you may see or hear some people describe the image above from the temple of Seti I at Abydos as depicting the "raising of the Djed-column," it actually is not showing the *raising* of the Djed. In fact, it is showing the "bringing down" of the Djed and the corpse of Osiris, preparatory to his being laid in the tomb (in later scenes). Only later will Osiris be "raised up."

This fact is very important, because it is my assertion that the above scene is analogous to the *taking down of the body of Christ from the Cross* (sometimes called "the Descent from the Cross")!

If all the foregoing discussion and analysis is correct, and the myths from around the world (including those found in the Bible) are actually closely connected, and that they teach a shamanic message, and that they often use the absolutely central symbol of the Djed-column/Cruciform Cross/Ankh Cross/World Tree/Shamanic Pole to embody that message (a message of the "divine spark within" or the "Christ in you," as Paul phrases it), then the symbology of the "casting down" of the Christ into the tomb prior to his subsequent "raising up" is another manifestation of the same pattern, and the taking down of Christ from the Cross would parallel the taking down and giving to Isis of the Djed-column containing the corpse of the now-dead Osiris.

The imagery surrounding the Descent from the Cross supports this connection in absolutely breathtaking fashion. See, for example, this collection of images taken from art through the centuries of this event.

Even more striking, however, is the Christian art in the category known as the Pietà and depicting the Virgin Mary holding the body of Christ after the Crucifixion.

192

Below is perhaps the most famous such Pietà, that by Michelangelo situated in the Vatican:

If we remember from previous posts that the "Djed cast down" corresponds to the horizontal line between the equinoxes, then the imagery of Isis and of Mary receiving the "cast down" or "nearly-horizontal" body of Osiris (ancient Egypt) and of Christ (New Testament) makes perfect sense: the sign of Virgo is positioned at the point of the fall equinox -- when the sun is *declining down* towards the grave, but *just before* the exact horizontal point of the equinox!

DAYS LONGER THAN NIGHTS:
Heaven, Promised Land, Greece, etc.

Horizontal Column:
The Djed cast down

Vertical Column:
The Djed raised up

NIGHTS LONGER THAN DAYS:
Hell, Egypt, Troy, etc.

The "Virgo imagery" in both the above images (of Isis from the temple of Seti I, who died in 1279 BC and of the Virgin Mary from the work of Michelangelo who died in AD 1564) should be quite clear by now to anyone who has read *The Undying Stars* or looked at some of the images provided in previous posts about the constellation Virgo in the world's mythology (see for instance page 182).

Specifically, look at the "outstretched arm" -- which is one of the most characteristic aspects of the Virgo constellation and which is embodied in ancient myth (and ancient art depicting Virgo-connected figures) over and over and over again. It is most evident in the image of Isis receiving the tilted (descending towards the horizontal) Djed-column from the King of Byblos, but the exact same outstretched hand is also present in Michelangelo's masterpiece:

194

Now that it is pointed out, you can see that the outstretched arm in the Isis image is over-elongated -- as if to ensure that you do not fail to notice it.

For those who may not be as familiar with the constellation Virgo and the way this constellation overlays on ancient sacred art, take a look at the image below from ancient Greece, circa 440 BC, depicting the Pythia: a priestess whose very role was to go into a trance or state of ecstasy in order to obtain knowledge from the other realm which could not be obtained in "ordinary reality." The outline of Virgo (with distinctive outstretched arm) is superimposed:

What does all this mean?

I would submit that it proves the connection of the world's ancient myths -- from ancient Egypt, to ancient Sumer and Babylon (who also had a central story of a "World-Tree" in the mighty cedar whose top reached to the heavens), to ancient India, to ancient Greece, to the myths found in the Old and New Testaments of the Bible, to the myths of the Norse, and the list goes on and on (to Africa, North and South America, the islands of the Pacific, Australia, east Asia . . .).

It also proves the connection and close kinship of all these myths, their central symbology, and most importantly *their esoteric message* with each other and with the world's surviving shamanic cultures and traditions.

This connection suggests an even more radical and even more transformative ramification for what we have discovered above, because the esoteric and shamanic nature of the world's ancient wisdom-texts and traditions indicates that these teachings are meant to be put into practice by each man and woman who is incarnated in a body: by each man and woman who, these ancient scriptures teach, embodies a divine spark, a divine Thunderbolt, a divine "Christ within."

This evidence above suggests that it is part of our purpose here in this incarnation (perhaps even our central purpose) to recognize and to raise that inner spark of divinity, that "vertical portion of the Ankh," that Djed-column which we each share with Osiris, along that central axis that inside the human microcosm reflects the celestial axis of the World-Tree found in the macrocosm.

Perhaps this can be done through the practice of Yoga (whose name itself we have seen to be connected to the Ankh and hence to the Djed).

Perhaps this can be done through the practice of Kung Fu (whose name may also be related to the "name of the Ankh,"[49] and which is most definitely related to the precession of the equinoxes and the other celestial cycles which allegorize our divine spark cycling back upwards after first plunging downwards).

Perhaps this can be done through art and the creative force (as eloquently argued by Jon Rappoport, who connects that activity to the smashing of artificial realities embodied by trickster gods including Hermes, and by John Anthony West, who demonstrates that the ancient Egyptians appear to have had strong ideas about the transformative and consciousness-raising power of the artistic process of creating itself).

Perhaps this can be done through meditation, which science has shown can send the brain into a altered state -- perhaps even akin to a shamanic state – when performed by those who have spent long hours practicing the discipline.

Perhaps this can be done through rhythmic chanting, which appears to have been a central component in the ancient wisdom and which amazingly seems to share a fairly similar form or pattern across many cultures and languages around the world.

Perhaps this can be done through the use of special plants and organisms such as mushrooms, which can be ingested or brewed into teas (please note the strong words of warning regarding the dangers of mistakenly consuming the wrong mushrooms posted on the website of mushroom expert Paul Stamets).

And certainly this can be done through the practice of what we commonly label as shamanic techniques (deliberately inducing states of ecstasy or the experience of *non-ordinary reality*, through a variety of methods available to humanity, including shamanic drumming): as we have seen, there is strong evidence to believe that *all* of the world's ancient wisdom was at one time shamanic, a fact which suggests that part of the world has been deliberately robbed of its shamanic heritage.

In other words, the ancient myths were not intended to teach that Osiris "raised the Djed-column" *so that we don't have to.* The ancient myths were not intended to teach that Christ "raised the

Djed-column" *so that we don't have to.* The ancient myths were not intended to teach that Odin "raised the Djed-column" *so that we don't have to.*

They contained those stories, and showed that pattern so many times, because it is what we are here to do.

Humanity's shared shamanic heritage

2014 September 01

When considering the subject of shamanism and shamanic experience, many "Westerners" (that is to say, those who have grown up in the parts of the world that were actually ruled by the Roman Empire, specifically the western empire, as well as those parts of the world that the later European states descended from the western empire influenced heavily, and in particular those areas which were deeply committed to literalist forms of Christianity for many centuries in a row) may find the subject to be uncomfortable or even threatening.

This discomfort may be due to a variety of factors.

Some of it may be due to the complete unfamiliarity of the entire landscape of the shamanic, and a feeling that the concept is so alien as to be almost completely inaccessible to those coming from any of the cultures encompassed by "Western culture" as broadly described above.

Some of it may be due to the heavy stigma which the literalist forms of religion that have dominated Western culture for at least seventeen centuries have placed on forms of human experience involving contact with spirits or the spirit-world -- a stigma which retains some of its force even among descendants of Western culture who no longer accept the literalist interpretation of ancient scriptures or one of the various forms of literalist Christianity which opposed such experience based on specific and overt doctrines or teachings.

And some of it may be due to the idea that there is a deep and nearly impassible divide between different cultures, and especially between "Western" cultures and those retaining some aspect of the shamanic worldview, to the point that it is seen (perhaps by descendants of cultures on *both* sides of the divide) as "inauthentic" or "invasive" or in some other way "wrong" for anyone from a primarily Western background to wish to explore and especially to practice aspects of shamanic experience.

These barriers to the investigation of shamanic teaching and experience are unfortunate – especially if it turns out that the outlines of the shamanic world-view are in fact accurate in their description of our universe! That is to say, such rejection of the relevance of the shamanic to *everyone* in the world (including those from a primarily "Western" background) would be unfortunate indeed if it turns out that there is *in fact* a realm corresponding to that realm described in various shamanic cultures as the Other World, the Spirit World, the Invisible World, the Dreamtime, or the Realm of the Gods, and if that other realm actually connects to and "interpenetrates" the more familiar or ordinary or material realm in which we spend most of our waking hours, such that changes in one realm can have real and lasting impact on what takes place in the other.

As I explore in *The Undying Stars*, there is indeed evidence that this situation is in fact the case: that is to say, that the universe we normally experience and think of as "reality" is in fact interpenetrated by an invisible world, or that the "explicate" world we inhabit somehow "unfolds from" an invisible "implicate" realm of pure potentiality. Modern theoretical physicists have been forced by the outcomes of various experiments conducted since the end of the nineteenth and especially during the twentieth century to radically reshape their models of the universe, and frameworks have been proposed including the "holographic universe" model which closely resemble the shamanic worldview in nearly every detail, other than the labels given to the two different realms of existence (or the two different modes of the "expression of information" or data).

The fact that shamanism anticipated this modern "discovery" by many thousands of years, and that, in addition to understanding it, shamanism has a rich tradition of techniques for actually moving between these different realms of existence in order to gain knowledge and make changes which cannot be effected in any other way, should alone be enough to recommend a careful reconsideration of the profound value and pertinence of shamanic thought and practice for all humanity.

But even more important, perhaps, to breaking down the unfortunate potential sources of "Western discomfort" with the shamanic that I listed above would be the understanding that in fact the shamanic worldview appears to have been deeply ingrained in ancient sacred tradition in all the places we think of today as "the West" and that this knowledge was deliberately stamped out only as recently as the fifth century AD within the Roman Empire -- and even later in other parts of Europe and the West.

In other words, what today we label as "the shamanic" is part of the heritage of all peoples -- but there has for centuries been a deliberate and very effective campaign to steal this heritage from humanity!

The very fact that we label the worldview broadly described as *shamanic* with an adjective derived from the Tungusian word *shaman* reinforces the extent to which this worldview was hunted to near extinction within the regions conquered or heavily influenced by the western Roman Empire and its successor western European states, and only survived in areas outside of that influence (such as eastern Europe, very far northern Europe, most of Africa and Asia -- including the region of modern-day Siberia where the Tungus peoples live -- as well as the continents of North and South America and the islands of the Pacific Ocean including the island-continent of Australia).

In fact, as I labor to demonstrate in *The Undying Stars* (and many of the posts here on this blog), it can be shown that nearly all the world's sacred myths, teachings and scriptures share a common underlying celestial foundation which actually *unites them rather than divides them*, and that the purpose of the esoteric celestial allegory employed in all these cases was to convey a vision of the universe and of human experience that is essentially what we today would call *shamanic*. For more on that possibility, see for example a post from 08/16/14 containing an index of links to posts detailing the celestial aspects of over fifty different myths from different world cultures (including myths in the Old and New Testaments), and also some of the previous posts which have discussed the possible shamanic purpose of these "star myths" such as "Shamanic-Holographic,"[167] "The shamanic foundation of the world's ancient wisdom,"[186] and "The ancient torch that was lighted for our guidance,"[38] among others.

Although this shamanic worldview took on different forms in cultures such as ancient Egypt, ancient Greece, or the cultures of the Druids, Celts, Norse, and others, it was nevertheless characterized in all of these different cultures by features that are essentially and specifically shamanic: the awareness of "the other realm" or "world of the gods" in addition to the world of ordinary reality, and the practice of techniques for actually traveling between the realm of ordinary reality and the realm of the gods in order to obtain knowledge or effect change not possible to obtain or effect through *any other method.*

Previous posts have described, for example, the important work of Dr. Jeremy Naydler in demonstrating that what we would call shamanic travel or shamanic journeying was an integral part of ancient Egyptian civilization, and that the pharaoh appears to have regularly and deliberately undertaken out-of-body travel to the realm of the gods or *neters* in order to provide benefits for the entire society by doing so: see for instance this post.[222] During the 1960s, the authors of *Hamlet's Mill* also outlined a strong case for a shamanic connection in the myth and practice of ancient Egypt, most explicitly in chapter 8.

Other posts have demonstrated evidence for similarly shamanic worldviews in operation in ancient Greece, including at the oracle at Delphi, and in the long-standing Mysteries of Eleusis,[173] both of which appear to have reinforced the necessity of acknowledging the invisible world and of crossing to the other side even during this life in order to obtain knowledge or make change which could not be accomplished any other way.

The shamanic aspects of the Norse myths are clear and compelling, and are especially evident in the central sacrifice of Odin, in which knowledge is shown being obtained through a ritual that is essentially shamanic -- knowledge that can be obtained by no other method. And, while Odin's ascent on the World-Tree is perhaps the most obviously shamanic episode in Norse mythology, there are many other Norse myths which can be convincingly shown to contain clear shamanic elements, including the myth of Odin and Gunnlod and the mead of poetry (as well as the many stories of Freya and her ability to transform into a falcon, and of the Valkyries who are also able to ascend to the heavens and who wear garments of feathers in some cases, a characteristic of the shaman's costume the world over).

It can be demonstrated that this shamanic worldview, and the practice of crossing over to the other realm in order to obtain knowledge or effect change, continued in what would become the "Western world" right up through the fifth century AD within the Roman Empire, during the period in which the hierarchy of literalist Christianity was actively suppressing esoteric -- we might even say shamanic -- interpretations of the texts that became the Biblical canon.

In fact, this previous post entitled "The centrality of ecstasy, according to ancient wisdom,"[162] cites the analysis of Gerald Massey (1828 - 1907) who concluded that the distinguishing hallmark of all ancient wisdom was "knowledge of trance conditions." This knowledge was found in all the ancient scriptures and cultures in "the West" prior to the advent of literalist Christianity, and this knowledge survived into the modern period in those cultures that were outside the areas that literalist Christianity stamped it out during the ancient period and up through the middle ages.

Elsewhere in his work, Massey puts forward the important theory (which he backs up with compelling evidence), that the author of the earliest Pauline letters in the New Testament was clearly teaching a worldview we would today call shamanic, including an emphasis on individual experience and direct revelation, and even out-of-the-body travel. Massey believes that this original intent was subverted by later literalist teachers and by the creation of letters (such as the "Pastoral letters") not written by the original author, and which taught an opposite worldview.

This information turns conventional understanding on its head, and should go a long way towards overcoming the three main areas of discomfort or objection cited at the beginning of this post.

204

It suggests that what we think of today as the shamanic is actually the heritage of all humanity – but that this heritage was deliberately stolen from a large segment of mankind many centuries ago, and that the campaigns against other shamanic cultures that took place in more recent centuries may in fact be part of the same "stamping out" that took place in the West long before.

It also suggests that the hunger for the exploration of the shamanic among people descended from Western cultures may represent a longing for something that was once part of their own heritage, but from which they are now separated by long centuries of isolation from such experience.

It further suggests that shamanic practice can and does take on many different cultural forms, even as it retains some central features which characterize the shamanic worldview. The external trappings of that worldview looked very different in ancient Egypt, for example, than it did in Eleusis in ancient Greece, or in northern Europe among the Norse and Germanic peoples – and these external differences are real and undeniable. However, the core understanding that there is a spirit world or realm of the gods (or realm of the "implicate order," in terms of modern holographic universe theory) and that deliberate travel to that realm is both possible and at times very necessary and potentially very beneficial, is common to all of the pre-Christian "Western" cultures just mentioned, just as it is common to the many different cultures where the shamanic worldview survived to the present or closer to the present day.

There are real and undeniable differences between more recent shamanic cultures, for instance between those found in the Amazon and those found in Siberia, but there are core similarities as well – especially the core belief in the possibility of such shamanic travel, its potential benefits, and its necessity in some circumstances.

We might also conclude that, given the number of centuries that have intervened between the time that this worldview was stamped out in "the West" and the present day, anyone coming from a primarily Western culture who wants to investigate traditions where this kind of knowledge has survived is of necessity forced to do so among the knowledge that survived in non-Western cultures. This does not mean that someone who does so is trying to "appropriate" or "steal" from the culture where that knowledge has been preserved, or trying to turn into someone that they are not: it is more a case of someone from a culture where long generations have now elapsed since this ancient light was put out going to someone where that flame was preserved right up to recent memory.

We might say it is also like someone from a boat or a ship that has been blown to pieces, and who is now floating in the ocean, paddling over to boats or vessels that may on the outside look very different from the one that they were originally from, but that were also designed with the same primary purpose in mind. Recreating the old boat is pretty much out of the question at this point: Eleusis went silent so many centuries ago that there are now none living who can say with any certainty what techniques were used in their mysteries.

When these refugees learn the techniques that have been preserved in other cultures and other places and then head out to try to navigate the waters of this life using what they have learned, their "boats" and methods of sailing may and probably will have a different look and feel. That fact should not lead to their being criticized or rejected as somehow being inauthentic or counterfeit. The evidence presented above shows that the broadly "shamanic worldview" is the heritage of all humanity, even though it is and probably indeed must be expressed differently by different cultures living in different parts of the world or different centuries and using different technologies. The fact that it will be expressed differently by practitioners in our modern day who have their own different cultural backgrounds and baggage should not be cause for division or criticism or rejection of the desire to follow this ancient path in the circumstances of today's world.

In fact, given some of the evidence touched on above, it could be argued that we do not have the luxury of declaring the pursuit of shamanic experience to be "off limits" to any group or family of humanity. It is clear from evidence stretching all the way back to the Pyramid Texts (and perhaps much further back even than that) that the knowledge of and the ability to enter into altered states of consciousness and in doing so to travel into non-ordinary reality (the other realm, the implicate order) in order to gain information or to effect change that cannot be accomplished any other way is absolutely *essential* to individual health and to the health of society at large.

If "Western society" and the world at large is dangerously off-track or imbalanced, then this fact itself would argue that the *recovery* and *active practice* of that shamanic worldview which was lost (and, it could be argued, deliberately stolen) must be given the highest priority.

How many ways are there to contact the hidden realm?

2014 September 03

Recent posts have advanced the argument that the world's ancient scriptures and traditions share a common, unifying, and shamanic worldview.

Together, they provide evidence that cultures around the world and across the millennia, from ancient Egypt to the steppes of Mongolia, and from the far northern boundaries of Scandinavia to the southern continent of Australia, at one time shared a worldview characterized by the understanding that our familiar, material, "ordinary" reality exists in conjunction with and is interpenetrated by another reality: the seed realm, the hidden realm, the realm of the spirits, the realm of the gods.

This shared shamanic worldview was characterized not only by an awareness of this other realm, but by the understanding that it was possible in this life to deliberately undertake journeys to the spirit world in order to obtain knowledge or effect change that could not be accomplished in ordinary reality.

There is also abundant evidence that this worldview has been deliberately stamped out over the centuries and that practice of shamanic techniques of ecstasy (or transcending the boundaries of the static, physical, ordinary reality) has been discouraged, stigmatized, and even prohibited by law in some places right up to the present day or very recent decades, and that the tools used to cross the boundary to the other realm -- the shamanic drum in particular -- have been outlawed, seized, and deliberately destroyed.

The extent of this persecution of the shamanic worldview across both geographic space and historical time leads to the possibility that those responsible for the campaign are not persecuting this worldview because they believe that it is *false*, but rather because they know that it is *true*, and that there actually is knowledge which can only be obtained and change which can only be effected through shamanic techniques.

Mircea Eliade's *Shamanism: Archaic techniques of ecstasy*, first published in 1951, was the first text to attempt to attempt to map the outlines of the entire broad landscape of the phenomenon of shamanism, and to attempt simultaneously to situate the shamanic worldview within the history of human religion. As such, it contains many first-hand accounts describing shamanic technique from parts of the world where the old traditions were still relatively undisturbed.

Let's examine the various methods recorded in Eliade's work by which men and women from traditional shamanic cultures were able to journey to the world of the spirits and to return.

Eliade himself does not actually provide a single succinct list in his book, although he describes and comments upon a wide variety of methods from many different shamanic cultures. Here is a non-exhaustive list of some of the techniques he covers from page 220 of *The Undying Stars*, showing the wide range of methods employed by different people in different places and times -- below we will examine a few of them using quotations from Eliade's work:

> In *Shamanism: Archaic techniques of ecstasy,* Mircea Eliade catalogues many of the rituals and practices used by shamans around the world to enter altered states of consciousness, including ecstatic dance, whirling, rhythmic drumming, chanting, songs, music involving various instruments and especially flutes, fasting, the use of entheogenic substances derived from plants, the use of difficult exercises or postures similar to or including Yoga, the undertaking of deliberate spirit quests, the use of constricted and enclosed spaces, the use of very crowded spaces, the imposition of long periods of solitude, rubbing the body with rock crystals, rubbing together two stones for days or weeks on end, elaborate initiatory processes involving experienced guides, and many others, as well as many variations and combinations of the techniques listed here.

The same passage also notes that Eliade records evidence that some shamans gained the ability to cross into the spirit world as a result of accidentally being hit by lightning, bitten by a poisonous snake, or experiencing a traumatic accident or illness. As part of an examination of the possibility that the life-changing experience of those who participated in ancient *mysteria* such as those at Eleusis and the life-changing experience of those who have reported out-of-body experiences in modern settings are also related to shamanic travel (in other words, the possibility that they are all going to the same hidden plane of existence, the same unseen realm), I conclude:

> The point to be made is that the techniques of inducing ecstasy in the human consciousness are profuse and multifarious -- suggesting that the human consciousness is perhaps designed to be naturally capable of achieving this state -- and that therefore the techniques that were used by

the mystery cults may have included almost any combination of those listed, as well as many others. 220.

Below is a partial list of techniques of ecstasy, chronicled by Eliade, with quotations from his landmark study of the subject:

- *Use of drums and rhythmic drumming:* "The drum has a role of the first importance in shamanic ceremonies. Its symbolism is complex, its magical functions many and various. It is indispensible in conducting the shamanic seance, whether it carries the shaman to the "Center of the World," or enables him to fly through the air, or summons and 'imprisons' the spirits, or, finally, if the drumming enables the shaman to concentrate and regain contact with the spiritual world through which he is preparing to travel" (168). The importance of the drum is indisputable, and it is used to accomplish the shamanic journey in cultures around the world.

- *Use of other musical instruments, including rattles:* "In North America, as in most other regions, the shaman uses a drum or a rattle. Where the ceremonial drum is missing, it is replaced by the gong or the shell (especially in Ceylon, South Asia, China, etc.). But there is always some instrument that, in one way or another, is able to establish contact with the 'world of the spirits.' This last expression must be taken in its broadest sense, embracing not only gods, spirits, and demons, but also the souls of ancestors, the dead, and mythical animals" (179).

- *Use of chanting:* "He sways, chanting, his eyes half closed. First it is a humming in a plaintive tone, as if the shaman wanted to sing despite some inward pain. The chanting becomes louder, takes the form of a real melody, but still hummed. [. . .] The song is repeated ten, twenty, thirty times in succession, uninterruptedly, the last note being immediately followed by the first of the beginning, with no musical rest. [. . .] The shaman sings only a few measures by himself. At first he is alone, then there are a few voices, then everyone. Then he stops singing, leaving the task of attracting the *damagomi* to the audience. [. . .] As for the shaman, he meditates deeply. He closes his eyes, listens. Soon he feels his damagomi arriving, approaching, fluttering through the night air, in the bush, underground, everywhere, even in his own abdomen . . ." (305-306).

- *Use of dancing:* "From the earliest times, the classic method of achieving trance was dancing. As everywhere else, ecstasy made possible both the shaman's 'magical flight' and the descent of a 'spirit'"(451). "The Kirgiz *baqça* does not use the drum to prepare the trance, but the *kobuz*, which is a stringed instrument. And the trance, as among Siberian shamans, is induced by dancing to the magical melody of the *kobuz*. The dance, as we shall see more fully later, reproduces the shaman's ecstatic journey to the sky" (175).

- *Use of masks or related face-coverings:* "In some places the mask is believed to aid concentration. We have seen that the kerchief covering the shaman's eyes or even his whole face plays a similar role in certain instances. Sometimes, too, even if there is no mention of a mask properly speaking, an object of such nature is present – for example, the furs and kerchiefs that, among the Goldi and the Soyot, almost cover the shaman's head" (167).

- *Use of the tobacco plant:* "The apprentice shaman of the Conibo of the Ucayali receives his medical knowledge from a spirit. To enter into relations with the spirit the shaman drinks a decoction of tobacco and smokes as much as possible in a hermetically closed hut" (83). "Throughout the instruction period fasting is almost absolute; the apprentices constantly smoke cigarettes, chew tobacco leaves, and drink tobacco juice. After the exhausting night dances, with fasting and intoxication superadded, the apprentices are ready for their ecstatic journey" (129).

- *Use of the cannabis plant:* "The *kapnobatai* would seem to be Getic dancers and sorcerers who used hemp smoke for their ecstatic trances" (390). "Herodotus has left us a good description of the funerary customs of the Scythians. The funeral was followed by purifications. Hemp was thrown on heated stones and all inhaled the smoke; 'the Scythians howl in joy for the vapour-bath'" (394). "One fact, at least, is certain: shamanism and ecstatic intoxication produced by hemp smoke were known to the Scythians. As we shall see, the use of hemp for ecstatic purposes is also attested among the Iranians, and it is the Iranian word for hep that is employed to designate mystical intoxication in Central and North Asia" (395).

- *Use of mushrooms:* "After fasting all day, at nightfall he takes a bath, eats three or seven mushrooms, and goes to sleep. Some hours later he suddenly wakes and, trembling all over, communicates what the spirits, through their 'messenger,' have revealed to him: the spirit to which sacrifice must be made, the man who made the hunt fail, and so on. The shaman then relapses into deep sleep and on the following day the specified sacrifices are offered" (221). "In a number of Ugrian languages the Iranian word for hemp, *bangha*, has come to designate both the pre-eminently shamanic mushroom *Agaricus muscarius* (which is used as a means of intoxication before or during the seance) and intoxication; compare, for example, the Vogul *pânkh*, "mushroom" (*Agaricus muscarius*), Mordvinian *panga, pango*, and Cheremis *pongo*, "mushroom." In northern Vogul, *pânkh* also means "intoxication, drunkenness." The hymns to the divinities refer to ecstasy induced by intoxication by mushrooms" (401). Note that this linguistic analysis provides yet further support for the arguments put forth in previous posts about the N-K sound, in "The name of the Ankh"[49] and "The name of the Ankh, continued: Kundalini around the world."[57]

- *Use of ascetic disciplines:* "The power of flight can, as we have seen, be obtained in many ways (shamanic trance, mystical ecstasy, magical techniques), but also by a severe psychological discipline, such as the Yoga of Patañjali, by vigorous ascetism, as in Buddhism, or by alchemical practices" (411). "Finally, we will briefly point out some other shamanic elements in Yoga and Indo-Tibetan tantrism. 'Mystical heat,' which is already documented in Vedic texts, holds a considerable place in Yogic-tantric techniques. This 'heat' is induced by holding the breath and especially by the 'transmutation' of sexual energy, a Yogic-tantric practice which, though quite obscure, is based on pranayama and various 'visualizations.' Some Indo-Tibetan initiatory ordeals consist precisely in testing a candidate's degree of preparation by his ability, during a winter night snowstorm, to dry a large number of soaked sheets directly on his naked body" (437).

Even this dizzying list of widely varying techniques is by no means exhaustive: Eliade discusses and documents many others in his study. Further, those catalogued by Eliade are themselves by no means exhaustive: it seems that the methods for inducing ecstasy or trance are as widely varied as the human experience itself.

What can we conclude from the above examination of the techniques of shamanic travel found around the world?

I believe we can conclude for certain that there is no single "right" way to initiate contact with the unseen realm. While some shamanic cultures utilize psychotropic or narcotic plants, these are by no means the only methodology used. While drumming appears to be one of the most widespread techniques of initiating shamanic journeys, Eliade notes that even drumming is not universally practiced even in some deeply shamanic cultures. It appears that there are an almost infinite variety of methods which can be used to make contact with the spirit world – almost as if someone wanted to make sure that men and women would

214

always have *some* method available to them, no matter where on the planet they might find themselves!

The vastness of the range of techniques by which men and women in shamanic cultures have accessed the hidden realm also suggests the probability that human beings, by their very makeup, are inherently "wired for ecstasy." We can access the world beyond the five physical senses by so many pathways that it is no exaggeration to suggest that we seem to possess a sort of "innate shamanic sense" or "sensitivity."

This possibility is attested to by modern shamanic practitioners and teachers, who have guided hundreds of modern people from all backgrounds in their first experiences of contact with non-ordinary reality. In *Shamanic Journeying: A Beginner's Guide*, Sandra Ingerman states:

> I have never met a person who could not journey. However, I have met many people who tried journeying many times before they felt that something was happening. I suggest that you keep up the practice -- relax, keep breathing into your heart, open all of your senses beyond just your visual awareness, set an intention, and in time, you will be journeying. 42.

She explains those concepts in her books -- you can find the books and other information at Sandra's website at www.sandraingerman.com.

In *The Shamanic Drum: A Guide to Sacred Drumming*, Michael Drake (discussing the work of Mircea Eliade, and coupling it with his own experience of guiding and teaching shamanic drum and journeying for many years) declares (p. 14):

> All people, therefore, are capable of flights of rapture. Ecstasy is a frequency within each of us. Like tuning a radio to the desired frequency, the drum attunes one to ecstasy.

Michael Drake's website can be found at http://shamanicdrumming.com/index.html. On one of the pages of his site, he reiterates his belief that virtually every man or woman is capable of such travel: "Researchers have found that if a drum beat frequency of around four-beats-per-second is sustained for at least fifteen minutes, most people can journey successfully, even on their first attempt."

The evidence from history -- and from the personal experience of longtime practitioners and teachers of shamanic journeying -- appears to be overwhelming: we are designed to be able to access the hidden realm of non-ordinary reality. This fact seems to fit well (fit perfectly) with the possibility that, as previous posts have explored, the testimony of the ancient wisdom inherited by virtually every culture on our planet, appears to declare a complementary message: that the ability to access the hidden realm is absolutely essential to human existence.

The heron of forgetfulness

2014 September 06

In the previous post entitled "How many ways are there to contact the hidden realm?"[208] we saw the breadth and variety of the techniques which human beings have used around the world, across different cultures, and across the centuries to achieve a condition of ecstasy: of freeing their consciousness temporarily from the material anchor of the body and its normal senses.

While many of the techniques employed do indeed make use of substances including cannabis, tobacco, mushrooms and others, it is also notable that many of them do not. While many of the consciousness-altering substances used in traditional shamanic cultures to induce ecstasy (including mushrooms, peyote, cannabis, and ayahuasca) have been outlawed over the years, it is similarly noteworthy that possession of "consciousness-altering drums" has also been widely and vigorously persecuted for centuries -- probably even more widely and for longer periods of time than have the consciousness-altering plants and mushrooms.

Interestingly enough, in the Poetic Edda, where the account of the shamanic self-sacrifice of the Norse god Odin is described, the same section containing the account of Odin's vision-quest to obtain the knowledge of the runes (the *Havamal*) also contains a warning against intoxication (in this case, by drinking too much mead or beer). Beginning in stanza 13, we read:

> 13. Over beer the bird of forgetfulness broods,
> And steals the minds of men;
> With the heron's feathers fettered I lay
> And in Gunnloth's house was drunk.

> 14. Drunk I was, I was dead-drunk,
> When with Fjalar wise I was;
> 'Tis the best of drinking if back one brings
> His wisdom with him home.
> [. . .]

> 19. Shun not the mead, but drink in measure;
> Speak to the point or be still;
> For rudeness none shall rightly blame thee
> If soon thy bed thou seekest.

The above lines can be found by scrolling down to the page marked [31] in the online version of the Elder Edda (Poetic Edda) linked above. That online version is not the easiest to navigate, but by looking for the bracketed "page-numbers" the above verses can be found.

These verses, coming as they do in the same section of the Edda in which the shamanic ordeal is mentioned, seem to establish a fairly sharp contrast between the idea of becoming intoxicated (discussed in primarily negative terms, and as a condition to be generally avoided) and traveling out of the body (the result of which, in Odin's case, is presented as positive, and the methodology of which is presented as necessary).

218

The image of the "bird of forgetfulness" brooding over the beer, repeated a couple of lines later as a heron which traps the intoxicated with the "fetters" of its feathers (perhaps we might call these the "fettering feathers of forgetfulness") is pretty unforgettable. It's powerful imagery, coupled with delightful alliteration (Norse poetry, as also Anglo-Saxon poetry, made much use of alliteration), and causes a pang or two of regret among those of us who have met that heron a few too many times.

Notably, however, there is within the warning lines (which generally present drunkenness in an entirely negative light) a reference to the mead of Gunnlod, when the speaker switches to first-person in stanza 14, indicating that it is now Odin who is speaking, and that the quest to obtain the mead of poetry (which has many aspects of a shamanic journey, especially the transformation into an eagle but also the descent into a hole in the mountain and the retrieval of hidden knowledge that could not be obtained in any other way) did involve intoxication as part of the process.

This tension between the consciousness-lifting process of crossing over to the other realm, and the generally consciousness-deadening condition of becoming "drunk, dead drunk" and imprisoned by the feathers of the bird of forgetfulness (forgetfulness being pretty much the polar opposite of the usual purpose of the shamanic journey, which is to *obtain* knowledge rather than forget it) is found in other cultures as well -- to the point that it is worth exploring further.

In Mircea Eliade's encyclopedic 1951 study of shamanic culture and practice (*Shamanism: Archaic techniqes of ecstasy*, also mentioned in the post cataloguing shamanic technique), some representatives of shamanic cultures seem to indicate that the need to use of mind-altering substances to induce ecstasy is seen in a somewhat negative light, at least in some cases and in some cultures. Eliade points to the existence of opinions and attitudes that: "Narcotics are only a vulgar substitute for 'pure' trance [. . .] the use of intoxicants (alcohol, tobacco, etc.) is a recent innovation and points to a decadence in shamanic technique. Narcotic intoxication is called on to provide an imitation of a state that the shaman is no longer capable of attaining otherwise" (401).

This attitude (part of a pattern Eliade finds stretching across numerous shamanic cultures of a belief in the "decadence of shamans," in other words, a belief or tradition in the shamanic cultures themselves that shamanic technique and capability had decreased over time) is extremely interesting: there appears to be some recognition that, while the use of intoxicating substances may be a path to the shamanic state of ecstasy, their use can also be a crutch -- and even worse, can lead to an *imitation* of shamanic ecstasy and not the real thing.

This tradition, from some of the shamanic cultures that Eliade and his other sources examined, would seem to be an important warning to those of us coming from "Western cultures" where knowledge of the shamanic was largely hunted down and violently suppressed for centuries. The danger posed by those offering "imitation" shamanic journeys, based more in intoxication than in the travel to the actual realm of the spirits, is one that we may want to keep in mind.

It is also notable that the great Sioux leader Crazy Horse, whose own personal vision was recounted by the Lakota holy man Black Elk, was known for his refusal to drink alcohol (as Stephen F. Ambrose points out on page 220 of his 1975 book about Crazy Horse).

And it is perhaps also worthy of noting that in the Ghost Dance movement, which used dancing and drumming late into the night on multiple nights in a row in order to enable participants to achieve a state of ecstasy, alcohol was similarly discouraged. It might even be worth pointing out that in Rastafari practice, while ganja is revered, alcohol was also traditionally frowned upon.

All of this is not to suggest that one method of achieving contact with the hidden realm is "good" while others are "bad," or to "privilege" one method over another -- far from it. In fact, the whole point of examining the incredibly multifarious array of methodologies utilized around the world and across the centuries was really to point out that men and women appear to be inherently designed to be able to make contact with the other realm by methods that will be available no matter what type of climate or environment or culture they happen to find themselves in.

However, the fairly widespread evidence of a clear tension between paths that lead to "intoxication" or "forgetfulness," and paths that lead to what Eliade called "pure" trance (putting "pure" in quotation marks himself) suggests that we should carefully ponder the warning that these voices from traditional shamanic cultures are giving us, to be careful of shortcuts, "imitation ecstasy," and the feathers of that dreaded heron of forgetfulness.

The shamanic journey described in the Pyramid Texts

2014 September 12

The previous post (from September 11, 2014; not included in this collection), entitled "The Cobra Kai sucker-punch (and why we keep falling for it, over and over and over)," presented the argument that the ancient wisdom encoded in all the world's sacred mythologies taught that this universe contains both the apparently material and ordinary reality with which we are all familiar, and an unseen realm which is actually the "seed realm" from which the apparently material and ordinary reality *is projected*, much like a hologram is projected from the holographic film.

Using an analogy from the original *Karate Kid* film (1984), that post argued that the ability to tap into that unseen realm, and even to journey there in order to bring back information or to effect changes which impact the ordinary realm, is analogous to a form of "karate" or "kung fu" that was secretly taught within the esoteric myths of the world's sacred traditions (just like the karate that Daniel-San learns in the movie was "hidden" inside of ordinary-looking motions such as "wax the car" or "paint the fence").

The post then argued that this ancient knowledge of travel to the other realm was deliberately stamped out in "the West" by a group that wanted to keep the system of "karate" to themselves (although it did survive in other parts of the world for many centuries, including in the shamanic cultures found in parts of the globe that the Roman Empire and the later Western European nations descended from the Western Empire did not conquer).

One extremely important body of evidence which supports the above interpretation of world history can be found in the Pyramid Texts of ancient Egypt. The Pyramid Texts are found inscribed within pyramids left by the kings and queens at the end of the Fifth Dynasty and into the Sixth Dynasty (in the pyramid of the final Fifth Dynasty King, Unas, who reigned from 2375 BC to 2345 BC) and those of his successors in the Sixth Dynasty (Teti, Pepi I, Merenre, and Pepi II and Pepi II's wives Neith, Ipwet, and Oudjebten), probably ending in 2184 BC.

These texts are in fact the oldest known examples of extended writing (longer than brief prayers or simple tomb inscriptions) from the human race.

As discussed in previous posts and at length in *The Undying Stars* book, Egyptian scholar Dr. Jeremy Naydler has presented thorough arguments which demonstrate that these pyramids -- and the texts that they contain -- were *not* primarily funerary structures (as they are almost universally taught to be) but rather sacred enclosures in which the king as part of his Sed Festival rites would lie inside a stone sarcophagus and undergo a shamanic out-of-body journey to the other realm on behalf of the entire kingdom. His book is entitled *Shamanic Wisdom in the Pyramid Texts: The Mystical Tradition of Ancient Egypt.*

Below is an image from within the sarcophagus chamber of the Pyramid of Unas:

This thesis, if correct, has profound significance for humanity. First, it would demonstrate that the very oldest extended scriptures that we currently possess are shamanic in nature. Second, it would demonstrate that "Western civilization" possesses a shamanic heritage, despite later efforts to suppress the shamanic worldview in "the West." And third, it raises the possibility that the incredible accomplishments of ancient Egypt were aided in some way by deliberate, regularly-scheduled trips to the unseen realm. There are many other implications as well, of course, but these three are some very significant ramifications of the assertion that the Pyramid Texts contain clear evidence of shamanic travel.

The Pyramid Texts as they are arranged in the Unas Pyramid can be viewed in their entirety online at the excellent website, *The Pyramid Texts Online*. They can be quite daunting at first, but if examined in conjunction with the helpful "guide" of Jeremy Naydler's book, the process of becoming more familiar with their contents can be greatly facilitated.

Below are just a few selected "utterances" from the Pyramid Texts which provide very clear support to the assertion that these texts were used in conjunction with deliberate shamanic travel by the king. All selections can be found in the *Pyramid Texts Online* site. At the end I will provide a brief commentary, but really these texts speak for themselves.

Utterance 253
275: *To say the words:*
"He is purified, who has purified himself in the Fields of Rushes.
Re has purified himself in the Fields of Rushes.
He is purified, who has purified himself in the Fields of Rushes.
This Unas has purified himself in the Fields of Rushes.
The hand of Unas is in the hand of Re.
Nut, take his hand!
Shu, lift him up! Shu, lift him up!"

Utterance 302

458: *To say the words:*

"Serene is the sky, Soped lives, for it is Unas indeed who is the living [star], the son of Sothis.

The Two Enneads have purified [themselves] for him as Meskhetiu [Great Bear], the Imperishable Stars.

The House of Unas which is in the sky will not perish,

the throne of Unas which is on earth will not be destroyed.

459: The humans hide themselves, the gods fly up,

Sothis has let Unas fly towards Heaven amongst his brothers the gods.

Nut the Great has uncovered her arms for Unas. [...]

461: He [Unas] ascends towards heaven near you, Re,

while his face [is like that of] hawks, his wings [are like those] of apd-geese,

his talons the fangs of *He-of-the-Dju-ef-nome.*

463: Upuaut has let Unas fly to heaven amongst his brothers, the gods.

Unas has moved his arms like a smn-goose, he has beaten his wings like a kite.

He flies up, he who flies up, O men!

Unas flies up away from you!"

Utterance 304

468: *To say the words:*

"Hail to you, daughter of Anubis, she who stands at the windows of the sky,

you friend of Thoth, she who stands at the two side rails of the ladder!

Open the way for Unas that he may pass!

Utterance 305

472: To say the words:

"The ladder is tied together by Re before Osiris.

The ladder is tied together by Horus before his father Osiris, when he goes to his soul.

One of them is on this side,

one of them is on that side,

while Unas is between them.

473: Are you then a god whose places are pure?
[I] come from a pure place!
Stand [here] Unas, says Horus.
Sit [here] Unas, says Seth.
Take this arm, says Re.

474: The spirit belongs to heaven, the body to earth. [. . .]

Utterance 247
259: Unas there! O Unas, see!
Unas there! O Unas, look!
Unas there! O Unas, hear!
Unas there! O Unas, be there!
Unas there! O Unas, arise on your side!
Do as I order, [you] who hate sleep, you who are tired!
Get up, you who are in Nedit!
Your fine bread is made in Buto.
Receive your power in Heliopolis!

The texts were selected to indicate aspects of the shamanic journey -- aspects which are common to the methods of embarking upon shamanic journeying taught today by shamanic teachers and practitioners, many of whom learned their techniques from shamanic teachers from traditional shamanic cultures in places as widely dispersed as Mongolia, Siberia, North America, South America, or Australia.

The selections above begin with Utterance 253, in which an actual spoken formula is recited. One can almost hear the "singsong tone" that might have been used, based upon the repetition of certain phrases, and based upon the common "chant-tone" which has survived in many widely-separated cultures on our planet to this very day, and which was remarked upon and illustrated in videos that can be found within this previous post.[541]

Note that in the final lines of the utterance, the speaker is describing Unas as taking the hand of the god Re (or Ra) and the goddess Nut, and of being "raised up" by the god Shu (whose upraised arms are described in this previous post[542]). It is difficult to deny that this sounds very much like the start of a shamanic

227

journey (those familiar with modern techniques and teachings of shamanic travel may have personal experience to draw upon to support this connection, and Dr. Naydler provides plenty of more academic evidence which also supports the same conclusion about this and other texts).

In the next utterances cited, Unas is described as flying up to the region of the sky in which dwell the Imperishable Stars (also known as the Undying Stars). He is also described as transforming into a bird, or taking on the features of a bird, which is a common shamanic motif (one of the most common and widespread, in fact -- see discussions here, here and here). Elsewhere in the Pyramid Texts, Unas is described on his journey as transforming into a falcon, a hawk, a powerful bull, a crocodile, a serpent, and other animals, and as taking on aspects of various gods and goddesses. These are all elements which can be present in shamanic travel.

In Utterance 304, the Pyramid Texts describe the daughter of Anubis, who stands "at the windows of the sky." Dr. Naydler points out:

> In the text, the word for "opening" or "window" is *peter*, which as a verb means "to see," implying some kind of aperture in the sky that opens onto the spirit realm. Shamanic accounts of passing through such apertures relate that they open for just an instant, and only the initiate is able to go through them into the Otherworld. 272.

Travel through such an "opening" or "window" is a very common feature of shamanic accounts, as well as of the accounts given by those who have undergone Near-Death Experiences (some of which are discussed in previous posts such as several which were published in June and August of 2012).

It is also intriguing to wonder if the name of the New Testament apostle Peter, who is given the keys to the kingdom of heaven, might not be related to this Egyptian concept of the "window" through which one can see the Otherworld, as Dr. Naydler says.

Of course, we are told in the New Testament that the name comes from the word for "rock" (Latin *petrus*), and this certainly seems likely, but the connection to the Egyptian word, in the name of the apostle who is traditionally described as "guarding the gates," is most remarkable (the celestial or zodiacal reason that Peter is the one who guards the gates to the kingdom of heaven is discussed in a post from May 13, 2014).

In any case, it is difficult to deny that this "window of the sky" has strong analogs in shamanic travel.

Next comes Utterance 305, in which Unas is described as climbing what Naydler describes as "the celestial ladder," a concept he relates to the Djed pillar -- one of the most important symbols of ancient Egyptian myth, and one that we have seen has undeniable connections to the shamanic worldview (see for instance this previous post[86]).

Previous posts including that one (and others linked in that post) demonstrate that the Djed symbology embodies both the "earthly" situation of the individual in his or her incarnate state (the "Djed cast down") and also the "heavenly" component in each man or woman, which he or she must remember, recognize, and seek to elevate during this earthly sojourn (the "Djed raised up"). In light of that fact, it is most significant that this same utterance that talks about the "heavenly ladder" also includes the line at 474 which declares "the spirit belongs to heaven, the body to earth."

Dr. Naydler presents analysis to argue that both Utterance 305 and the following Utterance 306 teach "a linking of heaven and earth" (274), a concept which we recently saw to be embodied in the ancient symbol of the vesica piscis (as well as in the widespread symbol of the Cross).

The final selection cited above is Utterance 247 (note that these numbers are not necessarily in chronological order: the order that the walls of each pyramid should be read is debated among scholars, and the texts themselves do not always appear in the same order in different pyramids or in later writings or inscriptions). There, we see what could be interpreted as a very obvious "call to return" from the shamanic state of ecstasy.

This is indeed how Dr. Naydler interprets these lines, and the repetition of commands for Unas to come back to his body, to see, to hear, to "be there," to turn onto his side -- all suggest that the priests who have been in attendance, and who probably assisted in the king's achieving the state of shamanic trance (using whichever of the many known techniques for achieving travel to the unseen realm -- or perhaps a technique that is now unknown), are calling Unas back from his voyage.

These descriptions of what appears to be a deliberate journey to the seed realm, the realm of the gods, are incredible, and their importance to our understanding of human history cannot be over-emphasized.

There is abundant evidence from around the world that such journeys to the unseen realm can and do create a real impact on the ordinary realm. If so, then we can begin to understand why certain people would want to take away this knowledge from the rest of humanity, and keep it all to themselves.

But the evidence of what was once known cannot be suppressed forever. It is there, in plain sight, in the oldest scriptures left anywhere on the planet. It is also preserved in the other sacred traditions of the world's cultures, including the scriptures of the Old and New Testaments, as well as the myths of the ancient Greeks and the Norse, and nearly everywhere else around the globe. These texts are waiting there patiently for us to recognize them for what they are.

If you are in the habit of contacting the other realm, perhaps you will find ways to employ some of these ancient aspects of what may well be the oldest extended description of shamanic travel known to humanity.

Direct unmediated access to the sacred realm

2014 December 27

The previous two posts have examined the assertion of Alvin Boyd Kuhn that the sacred scriptures and traditions of humanity cannot be fully grasped without the understanding that they do *not* describe the "experiences of people not ourselves" but rather that they are meant to convey "that which is our living experience at all times." They do not describe "incidents of a remote epoch" or time of legends, but rather they describe "the reality of the living present in the life of every soul on earth" (see here[331]).

This understanding leads directly to the conclusion that, if the sacred stories are meant to describe "our living experience at all times," then we as individuals actually have access to the reality of the super-material world at all times, and we have access to it *immediately*: that is to say, without the mediation of any other human being.

Note that this conclusion is quite the opposite conclusion of that taught by the literalist approach taken by the west for the past seventeen hundred years, which teaches that the stories are meant to be understood as literally describing the experience of someone else living in some other time. If the myths are about someone else, then it stands to reason that we might require a go-between to intercede between us and them. If the stories are actually about us, about our experience of taking on flesh to experience this material realm without losing our inherent nativity in that realm of spirit from which we came and to which we will return, then we have as much right to direct access to our native realm as any alleged mediator can claim.

There is abundant evidence that, prior to the dawn of literalist Christianity, the fundamental importance of the individual's capacity for direct communication with the realm of the gods was well understood. For instance, in the Mysteries of Eleusis, where men and women experienced direct contact with the realm of the gods and of which we have numerous ancient accounts by participants who reported that it changed their life, there is no evidence that any mediator tried to "explain" the meaning of what the participant experienced directly, and every evidence that whatever happened during the life-altering experience was between the gods and each individual man or woman who went through the ritual.

There is further abundant evidence that, in the lands where literalist Christianity did not stamp out the traditions of direct contact with the realm of the gods, the idea that each and every individual has the capacity for direct and unmediated access to the other world was almost universal.

In his extended examination of the subject in his landmark book *Shamanism: Archaic Techniques of Ecstasy* (1951), Mircea Eliade explains that, "A shaman is a man who has immediate, concrete experiences with gods and spirits; he sees them face to face, he talks with them, prays to them, implores them [. . .]" (88). Note that this translation may sound to us today as though Eliade was only talking about "men," but this is almost certainly a function of the way this passage was translated from the original French: it is quite clear from Eliade's text that he would include both men and women in this description, and that shamanism has been and continues to be practiced around the world by both men and women.

Further, it is quite clear that, while Eliade would certainly assert that the evidence overwhelmingly indicates that in shamanic cultures there are specific individuals who are marked out and distinguished as shamans by calling, in shamanic cultures there is a universal understanding that direct contact with the spirit world is available to everyone. He cites extensive evidence to support the conclusion that "nowhere does the shaman monopolize" the access to such direct contact (297). "Every individual" can seek contact with "certain tutelary or helping 'spirits'" (297). In other words, each and every man and woman has access to teaching (or tutelage) which comes, not from other human beings, but directly from the realm of the spirits or gods. Elsewhere, he writes that: "Besides the shamans, any Eskimo can consult the spirits" (296).

This conclusion is borne out by other testimony, such as the extremely important record of the wisdom of the Lakota holy man Black Elk, who describes the power of vision which Crazy Horse received from the spirit realm, a vision which gave him power throughout his life at important times, even though by all accounts Crazy Horse himself was not technically a "shaman" as his primary calling.

Those shamanic practitioners today who have decades of personal experience communicating with the spirit world often express the importance of the direct and unmediated contact with the spirit world available to each and every individual.

233

In *Awakening to the Spirit World: The Shamanic Path of Direct Revelation* (2010), shamanic teacher Sandra Ingerman says:

> [. . .] first and foremost, shamanism has always been a practice in which each practitioner gets unique directions and guidance from their helping spirits -- those same transpersonal beings that are often referred to as spirit guides and angels. [. . .]

> There is [. . .] a pervasive tendency for people to give their power away to others. Such seekers often desire to find a teacher who will act as an intermediary between themselves and their helping spirits -- a trait that is more characteristic of our organized religions in which bureaucratized priesthoods stand between us and the sacred realms. This is not typical of the path of shamanism and it is not a path of direct revelation. x - xi.

In the same book, shamanic teacher Hank Wesselman relates a similar emphasis on direct revelation from his own decades of experience:

> Perhaps the most fundamental shamanic principle from which everyone may benefit is that in the shaman's practice, there is no hierarchy or set of dogmas handed down to supplicants from some higher religious authority complex. Shamanism is the path of immediate and direct personal contact with Spirit, deeply intuitive, and not subject to definition, censorship, or judgment by others. On this path, each seeker has access to this transcendent connection and all that this provides. xix.

And again, shamanic teacher Michael Drake writes in the beginning of his book *The Shamanic Drum: A Guide to Sacred Drumming* (2009):

> No intermediary such as the church or priesthood is needed to access personal revelation and spiritual experience. All

234

dimensions of reality and the mystical knowledge and powers they contain are available to one who practices shamanism. Every shamanic practitioner becomes his or her own teacher, priest, and prophet. Shamanic practice brings one ultimate power over one's own life and the power to help others do the same. 9.

From the above discussion, it should be evident that this tradition of direct revelation is directly empowering to each individual man and woman -- and that this empowerment is completely in line with the assertion that the sacred traditions of the human race are in fact meant to describe the living experience of each individual soul. It would not be too great a stretch to assert that this understanding of the availability of direct and unmediated access to the transcendent is profoundly antithetical to the concept of "mind control" -- while the opposite teaching that we must be dependent upon others who will act as our intermediaries tends to lend itself to mind control and the "giving away of our power to others."

In fact, the previous post entitled "Crazy Horse against mind control" discussed the high regard for human dignity and freedom exemplified by individuals such as Crazy Horse and Sitting Bull, who each clearly had a strong personal understanding of direct access to the spirit world (03/07/2014).

Over a hundred years ago, self-taught scholar Gerald Massey (1828 - 1907) articulated the contrast between direct revelation and mediated revelation, and the threat that direct revelation posed to those who wish to proclaim their own monopoly on revelation, and to those who wish to use that self-proclaimed monopoly to inflict violence upon other men and women. In an essay entitled "Man in search of his soul during fifty thousand years, and how he found it!" Massey writes that in ancient times, before the literalist doctrine took over the west, the immortality of the soul was not believed as an article of faith, but rather it was known from the actual experience of personal contact with and travel to the spirit world -- what Eliade carefully defined as the distinguishing feature of the shamanic. Massey writes:

> So Nirvana becomes a present possession to the Esoteric Buddhist, because in trance he can enter the eternal state.

> This Gnosis included that mystery of transformation which was the change spoken of by Paul, when he exclaimed -- "Behold, I tell you a mystery," "We shall not entirely sleep, we shall be transformed!" according to the mystery that was revealed to him in the state of trance. This was the transformation which finally established the existence of a spiritual entity that could be detached, more or less, from the bodily conditions for the time being in life, and, as was finally held, for evermore in death. [. . .]

> What do you think is the use of telling the adept, whether the Hindu Buddhist, the African Seer, or the Finnic Magician, who experiences his "Tulla-intoon," or supra-human ecstasy, that he must live by faith, or be saved by belief? He will reply that he lives by knowledge, and walks by the open sight; and that another life is thus demonstrated to him in this. As for death, the practical Gnostic will tell you, he sees through it, and death itself is no more for him! Such have no doubt, because they know. The Mosaic and other sacred writings contain no annunciation of a mere doctrine of immortality, and the fact has excited constant wonder amongst the uninstructed. But the subject was not told of old, as matter of written precepts, but as matter of fact; it was a natural reality, not a manufactured idealism. It was not the promise of immortality that was set forth, or needed, when a demonstration was considered attainable in the mysteries of the abnormal human conditions, which were once common enough to be considered a known part of nature!

Massey makes it quite clear that this direct access to the spirit world, and to direct personal knowledge (as opposed to "faith" or "belief") came from what today we typically indicate by the practice of the shamanic: "by those who could enter the abnormal conditions, and be as spirits among spirits."

He then makes clear that the teachings of literalist Christianity, which he asserts to be built upon a misinterpretation of ancient Egyptian teachings, stand in direct opposition to this universal possession of the pre-Christian understanding:

> What has the Christian Church done with the human soul, which was an assured possession of the pre-Christian religions? It was handed over to their keeping and they have lost it! They have acted exactly like the dog in Aesop's fable -- who, seeing the likeness of the shoulder of mutton reflected in the water, dropped the substance which he held in his mouth, and plunged in to try and seize its shadow! They substituted a phantom of faith for the knowledge of phenomena! Hence their deadly enmity against the Gnostics, the men who knew. [. . .] They parted company with nature, and cut themselves adrift from the ground of phenomenal fact. They became the murderous enemies of the ancient spiritism which had demonstrated the existence and continuity of the soul and [which had] offered evidence of another life on the sole ground of fact to be found in nature. And ever since they have waged a ceaseless warfare against the phenomena and the agents -- which are as live and active to-day as they were in any time past.

But note that Massey in the passage quoted above clearly argues that the ancient scriptures -- including those he calls "the Mosaic and other sacred writings" -- were all originally shamanic in nature: they all actually teach the direct access of the soul to the spirit-world, the direct unmediated experience by the individual *in this life* to the transcendent (rather than the description of the transcendent and the requirement to accept it on faith). And note that he asserts at the end of the passages quoted that the phenomena of direct contact, and the practice of such direct

237

contact by shamanic personages or "agents," is as alive today as ever in the past.

In fact, I believe that it can be convincingly demonstrated that *all* the world's ancient sacred scriptures and traditions, from dynastic Egypt to the Lakota or the Inuit, and from the Norse myths to the "Mosaic scriptures," can be seen as being shamanic in nature, and that direct access to the divine is taught by all of them.

I also believe that this conclusion directly flows from an understanding that the world's ancient myths and sacred stories are telling the story of each and every human soul, and were not originally intended to be understood in a primarily literalistic way. I believe that as we again begin to understand them in this light, we will become more aware of the birthright of each and every individual to direct and unmediated access to the transcendent, and that this in turn cannot but help to be a powerful antidote to mind control, violence, and an artificial and disastrous disconnection from nature.

Bundesarchiv, Bild 135-S-13-18-38
Foto: Schäfer, Ernst | 1938/1939

The Bible is essentially shamanic

2015 February 07

> "*I was in a deep sleep on my face toward the ground . . .*"
>
> Daniel 8: 18

Previous posts (and now the entire 766-page volume *Star Myths of the Bible*, published in 2016) explore the evidence that many of the events in the Biblical scriptures describe the motions of the celestial realms, in metaphorical language.

I believe that all of this evidence strongly suggests that the Biblical texts, in common with other sacred texts and stories from around the world, are profoundly shamanic in nature, using celestial imagery and the heavenly realms in general as a metaphor for *the unseen realm* or *spirit world*, which shamanic cultures the world over can be broadly shown to understand as intertwining and interpenetrating this physical or material universe, and in fact to be the true source from which our more familiar visible world is actually generated or projected.

The shamanic aspects of the Bible (in common with the other mythologies of the world's cultures) are explored and discussed at length in *The Undying Stars*, as well as in many previous blog posts such as here[186] and here.[199] This post will examine a few more aspects of this thesis.

First, it is very noteworthy that the visions which in the above-linked discussions can be shown to depict the motions of the celestial realms, as well as other similar visions described in the Biblical texts, are often described in conjunction with the seer of the vision falling into a deep sleep, sometimes with the additional detail that they are lying with their face to the ground at the onset of the vision.

In each of the visions of Ezekiel, for example, the text (which describes the visions in the first-person perspective) states "I fell on my face," or "I fell down upon my face" as the divine glory which marks the beginning of a vision appears (see Ezekiel 1:28, Ezekiel 3:23, Ezekiel 9:8, and Ezekiel 11:13).

In the book of the prophet Daniel, Daniel is twice described as being in a deep sleep during which he meets an angelic being and has a transcendent vision: in Daniel 8:18 he says, "I was in a deep sleep on my face toward the ground," and in Daniel 10:9 he says, "then was I in a deep sleep on my face, and my face toward the ground."

And, in the extremely important vision of Abram in Genesis 15, we read in verse 12: "And when the sun was going down, a deep sleep fell upon Abram; and, lo, an horror of great darkness fell upon him."

These descriptions of falling into a deep sleep and then obtaining a vision of the spirit world are extremely characteristic of shamanic experience the world over. Similar descriptions can be read again and again in the encylopedic catalog of shamanic technique collected in Mircea Eliade's *Shamanism: Archaic techniques of ecstasy* (1951), many of them reported by first-hand observers in previous centuries visiting cultures where shamanic traditions had remained largely undisturbed by modern incursions.

Images from observers of the shamanic culture of the Sami people of the far northern regions of Scandinavia, show individuals stretched out face-down upon the ground with arms forward also along the ground, very similar to the image at the top of this post showing a similar posture being used by an individual in Tibet in the 1930s.

The degree to which these images correspond to Biblical verses such as the verse in Daniel which says, "then was I in a deep sleep on my face, and my face toward the ground," is extremely noteworthy.

In the New Testament as well, there is a passage in 2 Corinthians which appears to refer to an ecstatic experience, and an ecstatic experience by the apostle Paul (although he relates the event in the third person, while including hints that he is describing his own experience). At the beginning of the twelfth chapter, we read:

> 1 It is not expedient for me doubtless to glory. I will come to visions and revelations of the Lord.
> 2 I knew a man in Christ above fourteen years ago, (whether in the body, I cannot tell; or whether out of the body, I cannot tell: God knoweth;) such an one caught up to the third heaven.
> 3 And I knew such a man (whether in the body, or out of the body, I cannot tell: God knoweth;)
> 4 How that he was caught up into paradise, and heard unspeakable words, which it is not lawful [margin note offers as an alternative translation "possible" rather than "lawful"] for a man to utter.

Based upon this text, as well as a host of other evidence which he discusses in more than one of his essays, poet and esoteric scholar Gerald Massey (1828 - 1907) argued that Paul was actually teaching a doctrine which can broadly be described as Gnostic, meaning that Paul taught an allegorical understanding of the scriptures and that Paul experienced a personal vision of the spirit world on at least one occasion, but that forces during the first four centuries AD supplanted the original Gnostic teachings of Paul and others with a completely different system based upon a literalistic interpretation of the scriptures rather than a Gnostic one.

In an essay entitled "Paul the Gnostic Opponent of Peter, not an Apostle of Historic Christianity" (Massey uses the term "Historic" to describe those teaching that the scriptures of the Old and New Testaments were intended to describe *historic* events that took place in literal history), Massey says:

> Paul, on his own testimony, was an abnormal Seer, subject to the conditions of trance. He could not remember if certain experiences occurred to him in the body or out of it! This trance condition was the origin and source of his revelations, the heart of his mystery, his infirmity in which he gloried -- in short, his "*thorn in the flesh.*" He shows the Corinthians that his abnormal condition, ecstasy, illness, madness (or what not), was a phase of spiritual intercourse in which he was divinely insane – insane on behalf of God -- but that he was rational enough in his relationship to them. [. . .] Paul's Christ, the Lord, is *the* spirit; his gospel is that of spiritual revelation, the chief mode of manifestation being abnormal, as it was, and had been, in the Gnostic mysteries.

> The Gnostic Christ was the Immortal Spirit in man, which first demonstrated its existence by means of abnormal or spiritualistic phenomena. It did not and could not depend on any single manifestation in one historic personality. And when Paul says, "I knew a man in Christ," we see that to be in Christ is to be in the condition of trance, in the spirit, as

they phrased it, in the state that is common to what is now termed mediumship.

Being in the trance condition, or in Christ, as he calls it, he was caught up to the third heaven, and could not determine whether he was in the body or out of the body. Paragraphs 24 - 26.

Along with these admissions that he is prone to being "caught up" into the state of ecstatic trance, Paul also declares plainly that the stories in the Hebrew Scriptures are intended to be understood as an allegory.

Writing in the fourth chapter of his epistle to the Galatians, Paul presents an argument in which he explains that the story and circumstances regarding the birth of the two children of Abraham (namely Isaac and Ishmael) are given as an allegory in order to convey spiritual truths, saying:

22 For it is written, that Abraham had two sons, the one by a bondmaid, the other by a freewoman.
23 But he who was of the bondwoman was born after the flesh; but he of the freewoman was by promise.
24 Which things are an allegory: for these are the two covenants; the one from the mount Sinai, which gendereth to bondage, which is Agar.
25 For this Agar is mount Sinai in Arabia, and answereth to Jerusalem which now is, and is in bondage with her children.
26 But Jerusalem which is above is free, which is the mother of us all.

This passage is clearly of tremendous importance. First of all, Paul directly and bluntly states that these events "are an allegory" (Galatians 4:24). Then, he goes on to explain what he believes this allegory was intended to convey to us.

He says that the two women by which Abraham is said to have had the two sons are actually two "covenants." The first woman,

243

Agar (more commonly spelled and referred to as Hagar in most later English translations), actually "is mount Sinai in Arabia," Paul says, "and answereth to Jerusalem which now is, and is in bondage with her children." Then, Paul declares that in contrast to the "Jerusalem which now is, and is in bondage with her children," there is a "Jerusalem which is above" and which is free -- and that this Jerusalem "above" is in fact "the mother of us all."

What could he mean?

In light of the discussion above, and in the posts linked at the beginning of this discussion, and all of the evidence pertaining to the worldwide symbolism of the zodiac wheel and the great cross of the year which depicts a horizontal component representing the spirit or the Djed column "cast down" into material incarnation and a vertical component representing the spiritual nature being called forth and "raised up" again, I believe that this passage from Galatians 4 should be interpreted as follows:

> The two sons of Abraham, one by a bondmaid and the other by a freewoman, are an allegory. They represent two *covenants*, a word which literally means "coming together" [the prefix *co-* meaning "with" and related to or shortened from the prefix *con-*, which is still seen in the Spanish language meaning literally "with," and the word *venire* in Latin meaning "to come" and seen in other English words such as *intervene* meaning "come between" and *invent* meaning to find "come upon," as well as in the Spanish descendant word *venir* meaning "to come"]. These two "coming togethers" or covenants are, in the zodiac wheel, found at the two points of equinox, where the ecliptic path crosses the celestial equator two times during the year, once at the fall equinox when the sun is on its way *down* to the winter solstice, and once at the spring equinox when the sun is on its way back *up* to the summer solstice, the very pinnacle of the year. These two points can be allegorically seen as representing two different births: one of them the birth from the bondswoman, after the flesh, and the other of

244

them the birth from the free woman. If you need me to spell this out to you, the one from the bondswoman which is the birth "after the flesh" is the birth into the material realm, when we take on a physical body, and are born into this human life. On the great cross of the year, and in the allegories of the sacred stories, this takes place at the point of fall equinox, when the heavenly sun passes down into the lower half of the year, representing the experience of each one of us: we are each a heavenly spirit from the unseen realm, sojourning in this material realm for a time. This is what I mean when I say that this birth gives us "the Jerusalem which now is, and is in bondage with her children" -- this is all of us, trapped here in this material realm below, imprisoned within a physical body in order to learn and do and accomplish certain things which could only be accomplished by entering into this realm. But everything here is based upon a heavenly pattern, a spiritual pattern -- everything here contains and reflects and is patterned after and even projected from the other realm, the heavenly realm, the realm of spirit. That is what I mean by the "Jerusalem which is above, and is free." The covenant that marks the beginning of that "upper realm" is of course the spring equinox. A major part of what we are supposed to be doing in while we are toiling through this "lower realm" is to be remembering and recognizing the fact that we and everyone around us each comes from the spirit realm, from the "Jerusalem above which is the mother of us all," and we are to be elevating and uplifting the spirit in ourselves and in others, and in fact in the entire creation, all of which contains and is interpenetrated with and projected from that unseen world to which we travel when we go "out of the body." Got it?

In other words, it is most significant that the same Paul who tells us that he experiences the ecstatic condition of being "caught up into paradise" also tells us that the scriptural characters and events are actually "an allegory," and that in fact the allegory has to do with the lower realm into which we are born as if into a prison, and the upper realm to which we actually all belong, and which in fact is "the mother of us all."

The two go together: knowledge of the absolutely central importance of the ecstatic or trance condition (which can also be contacted through dreams and many other methods which are as varied as are the myriad different cultural expressions and experiences of the human race), and knowledge that the sacred stories are allegorical in nature and intended to convey to us the understanding of the importance of the spiritual realm from whence we come and to which we journey when we go into non-ordinary reality.

If that upper realm is the mother of us all, that means that we are all actually "native" to the spirit realm. The scriptures, with their celestial allegories, are meant to tell us that and remind us of that fact. So are the teachings of spiritual seers such as Paul.

The scriptures themselves plainly proclaim that they, along with the other sacred traditions the world over, are in fact shamanic in nature. The broadly shamanic (or broadly Gnostic) understanding of the scriptures was very widespread during the early centuries of Christianity, during the period that the advocates of the historical interpretation were hard at work establishing a literalistic and hierarchical form of Christianity to supplant the Gnostic understanding, and during which the literalist Christian authorities published numerous texts which had as their express central purpose the demonization of the Gnostics and everything that they taught and did.

Massey also refers to these opponents of Gnosticism, who became the "fathers" of the literalist Christian faith, as "the literalisers" and the "de-Spiritualizers."

We can see that these two labels place them squarely at odds with Paul himself, who taught an allegorical rather than a literal understanding of the texts, and who taught that the purpose of the allegory was to convey an understanding that is Spiritualizing or broadly shamanic in nature: an understanding of this material universe as being interpenetrated by and indeed generated from an invisible spirit realm, and an understanding of our human nature as being essentially native to the realm of spirit but temporarily plunged into physical incarnation, which is akin to a state of bondage.

This knowledge is extremely uplifting and empowering. It is plainly and abundantly evident throughout the scriptures of the Old and the New Testaments. The only real question we should ask ourselves is why someone would want to suppress it?

Whether in the body, or out of the body, I cannot tell: Paul, the Gnostic Opponent of Literalism

2015 March 13

No one can accuse the author of the New Testament letters attributed to the man using the name "Paul" of displaying anything less than an ardent, burning desire to convey what he believed to be an absolutely urgent message.

However, when it comes to what that message actually was, there are some researchers who have built a very strong case which suggests that it may not mean "what you think it means."

248

They argue that Paul might actually have been trying to urgently and insistently convey an esoteric, anti-literal, broadly gnostic, and even shamanic message: one he identified as "the mystery, which was kept secret since the world began" (Romans 16:25), a mystery which he calls "the hidden wisdom" (1 Corinthians 2:7), and a mystery which was actually the *very opposite* of the literalistic interpretation of the scriptures that Paul is normally believed to be championing.

It was a mystery which even Paul himself indicated would be opposed by clever and deceitful men, opponents he believed would not be above altering his letters to make them seem to say something different from what he was really trying to say, or even forging letters purporting to be from him, in order to keep the truth that he was urgently trying to teach from getting out.

In a previous post not included in this collection, we explored some of the penetrating insights and connections offered by Peter Kingsley in his 1999 text, *In the Dark Places of Wisdom*, providing evidence that ancient philosophy actually relied upon inspiration obtained by entering into a state of incubation or trance, contrary to the understanding of ancient philosophy commonly advanced by most conventional academics today.

It was the suppression of such techniques of ecstasy, of the ability to journey to the invisible realm *with which we are all in fact inwardly connected at all times*, the realm of non-ordinary reality, or non-local reality, along with the suppression of the knowledge of the indispensability of such travel, which ultimately led – according to Dr. Kingsley's thesis – to the incessant pursuit of satisfaction through *external* sources (whether materialistic or spiritual) that became the defining feature of "western civilization."

That preceding post offered a series of what I believe to be some of the most striking and revealing quotations from *In the Dark Places of Wisdom*, including some of the words with which ancient accounts tried to describe and convey the characteristics of this vitally-important state of coming into contact with the other realm (a concept often described today using the word "shamanic," which Peter Kingsley's book also uses).

As Peter Kingsley (who is a professional scholar in this field) describes it, the ancient inscriptions and texts tell us that those ancients who participated in these journeys reported that:

> they'd enter a state described as neither sleep nor waking – and eventually they'd have a vision. Sometimes the vision or the dream would bring them face to face with the god or the goddess or hero, and that was how the healing came about. People were healed like this all the time. What's important is that you would do absolutely nothing. The point came when you wouldn't struggle or make an effort. You'd just have to surrender to your condition. 80.

He cites a passage by the ancient philosopher Strabo (c. 64 BC to AD 24) which describes just such a process of entering into deep trance, like an animal in hibernation or in a state that is neither ordinary sleep nor ordinary awareness.

Note the important insight that this deep state was achieved not by struggling after something external, but that it was actually within the individual already, if they would just "surrender." In one of the other quotations cited in the preceding post, Dr. Kingsley explains (on page 67) that the ancient inscriptions seem to indicate that "We already have everything we need to know, in the darkness inside ourselves. *The longing is what turns us inside out until we find the sun and the moon and stars inside*" (emphasis added: italics not present in the original).

In another significant discussion of this state, Dr. Kingsley says:

> If you look at the old accounts of incubation you can still read the amazement as people discovered that the state they'd entered continued regardless of whether they were asleep or awake, whether they opened their eyes or shut

them. Often you find the mention of a state that's like being awake but different from being awake, that's like sleep but not sleep: that's neither sleep nor waking. It's not the waking state, it's not an ordinary dream and it's not dreamless sleep. It's something else, something in between. IIO-III.

In light of the clear evidence of the importance of this non-ordinary state, and in light of Peter Kingsley's explanation that this deep state was achieved not by struggling after something external, but that it was actually within the individual already, it is most intriguing to examine the passage in one of the letters attributed to Paul, in the New Testament book of 2 Corinthians chapter 12, in which he says:

1 It is not expedient for me doubtless to glory. I will come to visions and revelations of the Lord.

2 I knew a man in Christ above fourteen years ago, (whether in the body, I cannot tell; or whether out of the body, I cannot tell: God knoweth;) such an one caught up to the third heaven.

3 And I knew such a man, (whether in the body, or out of the body, I cannot tell: God knoweth;)

4 How that he was caught up into paradise, and heard unspeakable words, which it is not lawful [margin note says "Or, *possible*"] for a man to utter.

Note the striking similarity here to the ancient descriptions of the trance-state from the tradition Peter Kingsley examines (a tradition of which the profound pre-Socratic philosopher Parmenides or *Parmeneides* was part) and the descriptions given by Paul of his own personal experience of something which certainly appears to be the same type of encounter: one is described as being a state that is "like being awake but different from being awake, that's like sleep but not sleep: that's neither sleep nor waking," and the other is described as a state in which it is impossible to determine whether one is "in the body or out of the body." In both cases, the participant has clearly entered a state which is completely different from ordinary experience, in which normal terms do not seem to apply and normal descriptions fail to capture the condition in which one finds himself or herself.

Note also that a sense of "amazement" appears to be present in the accounts cited by Peter Kingsley and the account as recorded in 2 Corinthians by Paul. Paul specifically says he knows it is "not expedient" for him to glory in this experience, but it was clearly an experience that was most incredible, the wonder of which Paul finds he can barely conceal or contain.

Finally, we can see in verse 2 that in this experience, Paul describes himself (speaking in the third person) as being "caught up" -- it is not something that he caused to happen, but something that just "took him," in much the same way that the participants were described as "surrendering" to the experience (rather than pursuing it or bringing it about by their own efforts) in the ancient texts and inscriptions Peter Kingsley cites.

In an important lecture given by Gerald Massey (1828 - 1907) entitled *Paul, the Gnostic Opponent of Peter, not an Apostle of Historic Christianity,* Massey seeks to establish that the author of most of what we find in the Pauline epistles of the New Testament was not teaching anything like literalistic Christianity, but rather that he vehemently opposed literalism and in fact taught something quite different.

The full structure of Massey's argument is beyond the scope of this particular post (the interested reader is advised to read the lecture linked above -- preferably several times -- which can also be obtained in published print format here), but Massey ascribes great importance to Paul's clear testimony that he experienced ecstatic trance, preserved for us in 2 Corinthians 12.

Of this ecstatic experience, Massey says in paragraph 24 (as numbered in an online version available at http://www.gerald-massey.org.uk/massey/dpr_02_paul_as_a_gnostic.htm):

But, we have not yet completely mastered the entire Mystery of Paul for modern use; and it is not possible for any one but the phenomenal Spiritualist, who knows that the conditions of trance and clairvoyance are facts in nature; only those who have evidence that the other world can open and lighten with revelations, and prove its palpable presence, visibly and audibly; only those who accept the teaching that the human consciousness continues in death, and emerges in a personality that persists beyond the grave; only such, I say, are qualified to comprehend the mystery, or receive the message, once truly delivered to men by the Spiritualist Paul, but which was throughly perverted by the *Sarkolators*, the founders of the fleshly faith. [. . .] Paul, on his own testimony, was an abnormal Seer, subject to the conditions of trance. He could not remember if certain experiences occurred to him in the body or out of it! This trance condition was the origin and source of his revelations, the heart of his mystery, his infirmity in which he gloried -- in short, his "thorn in the flesh." He shows the Corinthians that his abnormal condition, ecstasy, illness, madness (or what not), was a phase of spiritual intercourse in which he was divinely insane -- insane on behalf of God -- but that he was rational enough in his relationship to them.

Note that in the above passage as it is found in almost every online transcription of the original lecture, the word "accept" in the long first sentence is erroneously written as "except" -- the printed text from 1922 linked above shows that this is someone's mistake (which has been duplicated several times on the web) and that the original printed text reads "accept" at that point, not "except."

Note also that although Massey here uses the term "men" to refer to "humanity in general," he was absolutely *not* referring to "men as opposed to women." The reader will find that in this same lecture (particularly in paragraph 20) he explicitly acknowledges that what he is saying applies to men *and* women, and even goes so far as to explain that:

> this manifestor of the re-birth might be feminine as well as masculine. In fact, the female announcer was first, and there are mystical reasons for this in nature. [. . .] Some of the Gnostic sects assigned the soul to the female nature, and made their Charis not only anterior, but superior, to the Christ. In the Book of Wisdom it is Sophia herself who is the pre-Christian *Saviour* of mankind. [. . .] This complete reversal of the Christian belief is to be found in the Hidden Wisdom!

Massey then goes on to make the extraordinary assertion, based upon the phrasing in verse 2, that this condition of ecstasy is what Paul describes as being "in Christ"! Massey writes:

> And when Paul says, "I knew a man in Christ," we see that to be in Christ is to be in the condition of trance, *in the spirit,* as they phrased it, in the state that is common to what is now termed mediumship.

Massey also argues that those who find themselves in "this same spirit" will manifest it in "various spirit manifestations," and that the list Paul gives in 1 Corinthians 12:7-12 is a list of the different manifestations of entering into contact with the realm of spirit (paragraph 26).

Those who are shocked by Massey's arguments, and the basis he finds for his arguments in the New Testament letters themselves, will doubtless point to passages in those same letters, or in other New Testament letters attributed to Paul, which appear to teach the doctrines of literalist or "historic" Christianity (namely, the coming of a literal individual Christ in history, rather than the teaching Massey is asserting that Paul maintained, of a *Christ within* that was not external or historical).

But, as Massey points out, *Paul himself* cautions the readers of his letters that there are those who seek to subvert what Paul is

254

teaching, and who are not above forging letters in Paul's own name in order to suppress Paul's gnostic and spiritual teaching! The second verse of 2 Thessalonians warns the recipients not to be "shaken in mind" or "troubled" even if they hear deceptive reports or even if they receive such reports in a "letter as from us" - - indicating an awareness that adversaries would not be above inserting different teachings into letters actually penned by Paul, or into letters falsely claiming to be from the hand of Paul.

He also makes the insightful point that, since we can clearly find passages today in the New Testament's Pauline epistles which are completely gnostic and anti-literalist, and other passages which are clearly literalist and anti-gnostic, it is far more likely that the literalist doctrines were inserted by literalists than that the gnostic passages were snuck in by gnostics (paragraph 17).

For one thing, the faction that clearly won out was the literalists, who crushed out virtually all of the gnostic teachers and outlawed their writings (on pain of death for any caught in possession of them) during the second through fourth centuries, and so it is probably they who introduced alterations as they saw necessary to letters which had originally been anti-literalist.

For another, he points out that many of the "founders of the Fleshly Faith" did not really understand the gnostic teachings, and thus it is far more likely that the literalists added passages and even entire letters supporting literalism, while failing to perceive the full extent of the gnostic teaching that remained, than that the opposite scenario took place.

The striking thing to notice is that we here find very powerful evidence that the same kind of "ancient wisdom" which Peter Kingsley finds in operation during the time of Parmenides (and his long line of predecessors and successors, before such knowledge was somehow subverted and stamped out). And we find that, just as Dr. Kingsley asserts took place with the ancient knowledge that inspired Parmenides and other ancient lovers of wisdom, the teachings of Paul were co-opted and subverted and made to appear to be part of a very different tradition -- one that basically teaches the very opposite of what Paul actually taught!

And there is still more.

Because, as Peter Kingsley makes clear, the ancient tradition of which Parmeneides was a part, and the practice of going deep into an altered state to make contact with the divine spiritual force, was closely associated with the god Apollo, whom we know of as a solar god but who was also associated with the underworld (where the sun appears to spend half its time) and with crossing the boundary into the non-ordinary realm in order to obtain prophetic messages in the trance state, as at Apollo's temple at Delphi. For the discussion from *In the Dark Places of Wisdom* of the mysterious god Apollo, who has also been "rationalized" and depicted today as something much less than what he was in antiquity, see especially pages 77 - 82.

In fact, in three important inscriptions which were discovered during the twentieth century at the side of ancient Elea (or Velia) and which feature prominently in the mystery-story that Peter Kingsley explores in his text, the three persons commemorated take for their own name one of the names of Apollo himself: Oulis.

Peter Kingsley writes:

> Oulis was the name of someone dedicated to the god Apollo -- to Apollo Oulios as he was sometimes called.
>
> Apollo Oulis had his own special areas of worship, mainly in the western coastal regions of Anatolia. And as for the title Oulios, it contains a delightful ambiguity. Originally it meant 'deadly,' 'destructive,' 'cruel': every god has his destructive side. But the Greeks explained it another way, as meaning 'he who makes whole.' That, in a word, is Apollo -- the destroyer who heals, the healer who destroys.
>
> If it was just a matter of a single person called Oulis you couldn't draw too many conclusions. But a string of three

inscriptions all starting with the same name, this name, isn't a coincidence; and the way each of the men is referred to as Oulis makes one thing very plain. As the first people who published the texts already saw, these were men connected with Apollo not on a casual basis but systematically -- from generation to generation. 57.

Now what is absolutely stunning about all of this, in light of the connections we have just seen from the writings of the New Testament epistles of Paul, is the fact that *Paul himself carries a name which refers to the god Apollo*!

This is something that the mighty explicator of the celestial foundations of the Bible, Robert Taylor (1784 - 1844) makes quite clear in some of his lectures (which were also published posthumously). The tradition is that this author we call "Paul" was originally known as . . . "Saul"! And of course, as Robert Taylor points out, that word is pronounced just as the word "*sol*" is pronounced -- the very word that means "Sun" in Latin but which also can be found in the Hebrew Old Testament of course, not only in the character named *Saul* but also *Solomon* -- while his new name "Paul" is cognate with the "*pol*" that is found in both "Apollo" and "Pollux."

And so here we have a rather textbook example if ever there was one which fits the description from the passage just quoted out of Kingsley's *In the Dark Places of Wisdom*: "these were men connected with Apollo not on a casual basis but systematically -- from generation to generation."

Paul appears to have been part of that systematic, generational line preserving the ancient knowledge.

Which means that it didn't die out with the arrival of Plato or the Platonic school.

But it certainly appears to have come under heavy fire during the period from AD 100 through AD 500, when the literalists crushed out the gnostics and their teachings, so that their work had to go underground for the next seventeen or more centuries.

And note as well that, just as Dr. Kingsley's thesis argues in regards to the suppression of the ancient wisdom known to Parmeneides but later forgotten, those who suppressed this wisdom point to solutions that must be *pursued* externally to the individual, as opposed to the teaching that we are actually already in deep mystical connection with the divine, if we only learn how to "surrender" to it or be "caught up" by it.

There is much more that could be drawn out from further study of *In the Dark Places of Wisdom* in conjunction with Massey's enlightening lecture from nearly a hundred years before. But perhaps one of the most important is an almost-offhand remark which Massey includes at the end of paragraph 26 in his lecture.

There, he says that this destruction of the true teaching which Paul was so zealous to try to pass on to his followers, and its replacement with "historical Christianity," has been "the greatest of all obstacles to mental development and the unity of the human race."

Several previous posts have discussed the reasons that a literalistic interpretation tends to divide men and women from one another -- and to divide them from nature, and even to divide them from themselves and from their true source of "stillness" and satisfaction, so that they must run after it elsewhere, futilely, in an endless cycle of frustration.

The horrible results can still be clearly seen today.

For this reason, it is absolutely essential that we take these matters to heart.

Please share this information with as many people as you believe might find it to be beneficial.

Peace and blessings.

258

Star Myths and Astrotheology

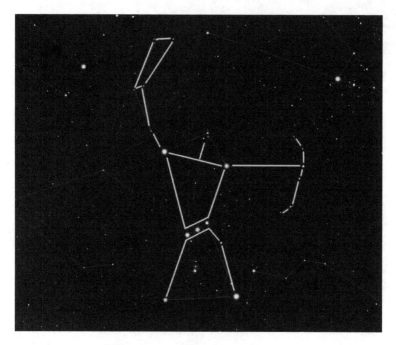

The importance of Orion

2011 May 14

Orion is one of the most distinctive and important constellations in the sky. Even if you cannot recognize any other constellations, you are probably familiar with Orion, with his spectacular belt of three stars and his dominant position in the sky during the winter months.

His famous belt is located very close to the line of the celestial equator, which means that if we think of ourselves standing upon a miniaturized earth orbiting a miniaturized sun inside of a familiar room in our house, we are looking "outward" towards Orion rather than "upward" from earth when we look at Orion, which means that his constellation is not among the undying stars but is obscured by the sun during part of the year. If you think about the earth's orbit around the sun, and imagine it takes place in a large dining room, the stars in the center of the ceiling would be visible all year around from an observer in the northern hemisphere, but the stars on the walls would only be visible on certain parts of the earth's circuit. The stars on the wall across from the sun would be obscured by the sun until the earth made its way around to the other side of the room, at which time those stars would be visible to observers on the side of the earth that was turned away from the sun (which happens every night as the earth spins).

As the earth makes its way around the sky, Orion rises and sets four minutes earlier each day, until he is rising during the day. Currently, from a latitude of 35° north, he is rising around 9:00 am, reaching his highest point around 3:30 in the afternoon, and setting at around ten minutes before 10:00 in the evening. As the earth continues around the sun and these rising points get earlier and earlier, he will rise and set during the day, until his rising becomes early enough to be seen low in the sky prior to the sunrise: an important date of return and a phenomenon known as heliacal rising (a term derived from the name of the ancient sun god Helios).

The rising and setting times of every star should be the same on any given day of the year: if the earth is back at precisely the same point on its journey, the background of stars on the "walls" and "ceiling" of our imaginary room should look precisely the way they did the last time the earth was at that exact spot. In general, they do -- except for the fact that there is a very slow shifting going on due to the phenomenon of precession. This wobble in earth's axis moves the sky by a mere 1° every 71.6 years (we can round it to 72 years for convenience) -- barely enough to be noticed in one human lifetime (especially since most people aren't observing the stars very precisely under the ages of eight or nine years old).

The motion of precession delays the time of the heliacal rising by about four minutes every 72 years, barely enough to make much difference in one lifetime, but enough that over 2,160 years the date of the heliacal rising will be an entire month later. Another way to think of this phenomenon is that the preceding constellation will be rising on the expected day, while the expected constellation is "delayed." This shift to the preceding constellation is the reason this phenomenon is called precession. The entire process is explained with numerous diagrams of the celestial spheres and earth's annual path in the *Mathisen Corollary*.

It is quite obvious that very ancient man understood this phenomenon long before conventional history teaches. In *Hamlet's Mill*, the authors make a compelling argument that the legend of the murder of Osiris by his brother Set is directly related to the failure of Orion to rise on the expected day due to the ages-long delaying action of precession. In *Death of Gods in Ancient Egypt*, author Jane B. Sellers elaborates on their argument with great clarity and additional insight.

The authors of *Hamlet's Mill* trace out the echoes of this same legend throughout many cultures over many centuries. Part of the reason for their title is that the Hamlet legend clearly parallels the legend of Osiris: a wicked uncle has killed his brother (Osiris in the Egyptian myth and Hamlet's father in the story of Hamlet), and he must be avenged by the son (Hamlet, and in the Egyptian legend the god Horus son of Osiris).

One of the many fascinating aspects of this particular connection is the name of Hamlet's father. In Shakespeare he is mainly known as Old King Hamlet, but in some of the earlier manifestations that probably served directly or indirectly as the general source for Shakespeare's version, he is known as Horvandillus, Horwendil, Orendel, Erentel, Erendel, Oervandill, and Aurvadil. You can read an English translation of the *Gesta Danorum* by Saxo Grammaticus (probably composed in the early 13th century AD) online: the story of Horwendil and his son Amlethus, as well as the murder of Horwendil by his brother, can be found in Book Three of Saxo's text.

The authors of *Hamlet's Mill* cite Frederick York Powell's (1850 - 1904) introduction to Oliver Elton's translation of Saxo, in which Powell states: "The story of Orwandel (the analogue of Orion the

Hunter) must be gathered chiefly from the prose Edda." In other words, Powell noted the linguistic similarity of the name Orwandel (or Orendel) with Orion. This connection supports the theory that the death of Osiris (which parallels that of Hamlet's father) is related to the failure of the constellation to appear on time after many centuries.

Another fascinating aspect of the name of Orion and Orendel is the connection to the work of J.R.R. Tolkien, who was an accomplished Old English scholar. As early as 1913, he wrote that he was struck by the great beauty of the Old English lines in Cynewulf's *Christ* which begin:

> *éala éarendel engla beorhtast*
> *ofer middangeard monnum sended*

> Hail, Earendel, brightest of angels thou,
> sent unto men upon this middle-earth!

[from *Hamlet's Mill* 355 – part of an extensive discussion of Orendel in Appendix 2]

Tolkien incorporated the beautiful name Earendil in his *Lord of the Rings*, as an elven king who carries the morning star on his brow and is the father of Elrond. The light of Earendil's star is in the Phial of Galadriel given to Frodo. In Shelob's lair at the end of the book *The Two Towers*, Frodo spontaneously shouts an elven phrase containing Earendil's name when he draws out the elven-glass of Galadriel.

Most fans of the *Lord of the Rings* may not be aware of the connection between Earendil and Orion. Now you know.

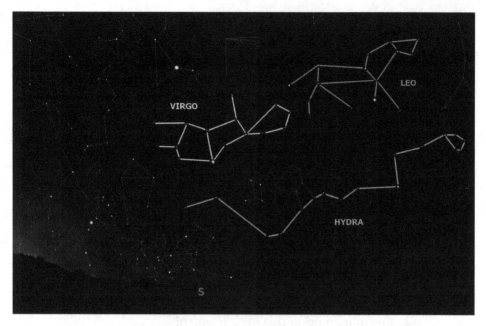

Ever wonder why Ishtar, Cybele, Rhea, and other aspects of the Great Goddess ride lions?

2012 January 14

Have you ever wondered why so many goddesses in mythology are described as riding on a lion, riding in a chariot that is pulled by a lion, or sitting on a throne flanked by lions? Take a look at the sky chart above, and see that the constellation Virgo follows the constellation Leo, and then ask yourself the same question again.

These goddesses were extremely important in ancient mythology, and usually are identified as different aspects of the same goddess (given different names by different civilizations or languages). This goddess in her many civilizations and under her many different names was often given the title of the Great Goddess, or the Queen of Heaven.

In ancient Greece, the goddess (or Titaness) Rhea was often shown seated on a throne flanked by lions. The same goddess was also known as Cybele (often called the Earth Mother) who was also associated with lions and closely identified with Rhea by scholars. Cybele is usually described as originating in Anatolia or Phrygia. The Babylonian goddess Ishtar was also closely associated with lions, her symbol, and the Ishtar Gate of course features lions. The Sumerian goddess Inanna is often identified with Ishtar, as is the Ugaritic Ashtoreth or Asherah.

In the Stele of Qadesh from ancient Egypt, shown below, the Goddess not only stands upon a Lion but also holds a Serpent in one hand and a Fan or Papyrus in the other: likely representative of the constellations Hydra and Coma Berenices.

In India, the goddess Durga is often depicted riding on a lion to slay her enemies.

Here is a website [*Queen of Heaven*: https://thequeenofheaven.wordpress.com/2010/11/16/asherah-part-iii-the-lion-lady/] discussing various manifestations of the Great Goddess and their association with lions. It contains numerous excellent examples of images of this goddess from various cultures.

In spite of the great volume of literature written about this extremely important goddess in the ancient world, very few historians appear to make the connection that the fact that Virgo follows Leo probably accounts for the fact that this goddess either rides in a chariot pulled by a lion or rides on a lion herself.

However, if you go out and watch the constellations of the night sky over the next several weeks, you will be able to see it for yourself.

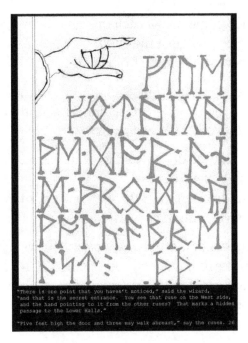

"There is one point that you haven't noticed," said the wizard, "and that is the secret entrance. You see that rune on the West side, and the hand pointing to it from the other runes? That marks a hidden passage to the Lower Halls."

"Five feet high the door and three may walk abreast," say the runes. 26

"Five feet high the door and three may walk abreast"

2012 August 08

In *Hamlet's Mill*, the seminal 1969 examination of the transmission of ancient wisdom through myth, written by Giorgio de Santillana and Hertha von Dechend, the authors cite the *Grimnismal*, the "sayings of Grimnir," in the Old Norse *Poetic Edda*, as an example of esoteric myth containing a surprising hidden precessional number.

They write:

> It is known that in the final battle of the gods, the massed legions on the side of "order" are the dead warriors, the "Einherier" who once fell in combat on earth and who have been transferred by the Valkyries to reside with Odin in Valhalla -- a theme much rehearsed in heroic poetry. On the last day, they issue forth to battle in martial array. Says the Grimnismal (23): "Five hundred gates and forty more -- are in the mighty building of Walhalla -- eight hundred

'Einherier' come out of each one gate -- on the time they go out in defence against the Wolf." That makes 432,000 in all, a number of significance from of old. 162.

You can read the passage yourself online in the 1936 translation by Henry Bellows here. His translation is slightly different from that used in *Hamlet's Mill*, and reads:

23. Five hundred doors | and forty there are,
I ween, in Valhall's walls;
Eight hundred fighters | through one door fare
When to war with the wolf they go.

Interestingly enough, that translation has always reminded me of a significant passage in the beloved tale by J.R.R. Tolkien (1882 - 1973) which first came to the attention of a publisher in the same year (1936), none other than *The Hobbit: or There and Back Again.*

As everyone knows (or almost everyone, and the rest will soon enough, with the release of a much-anticipated movie version by Peter Jackson), the plot of that adventure involves regaining the mighty halls of Thror (King under the Mountain) from the dragon Smaug. Vital to the plans of the thirteen dwarves and the hobbit Bilbo Baggins is the knowledge of a secret door, disclosed on a map made by Thror which is revealed to Thror's grandson Thorin by Gandalf inside the parlour of Bilbo's hobbit-hole.

Take a look at the runes above and listen to Gandalf's translation and see if they do not remind you somewhat of the cadence of the description of Valhalla's doors in *Grimnismal* 23:

"Five feet high the door and three may walk abreast," say the runes, but Smaug could not creep into a hole that size, not even when he was a young dragon, certainly not after devouring so many of the dwarves and men of Dale. 26.

While it may not seem like a direct parallel, it is a fact that in addition to authoring *The Hobbit* and the *Lord of the Rings* trilogy, Tolkien was a respected scholar of Old English and an

268

Oxford Professor, and he acknowledged the influence of Old Norse and Old English sources upon his own fiction, including the influence of the *Poetic Edda.*

I believe it is highly likely that this important passage from the *Grimnismal* lent its cadence to the description of the size of the secret door in *The Hobbit.*

The encoding of the important precessional number 432 in the ancient Eddas is discussed in context in the video "Precession = The Key" if you want a fuller discussion of the connection between the doors of Valhalla and the celestial mechanics of the circling skies above us [this refers to a video I made and placed on YouTube in 2012].

As it turns out, Professor Tolkien used another clear reference to a very important "precessional" figure in his Middle Earth books, and that is his reference to Earendil, who -- it turns out -- is none other than Orion, who is Osiris (for more on those connections, see also "Leo, the Lion King, Hamlet and Osiris"). The fact that Tolkien clearly borrowed the name Earendil lends some support to the assertion that his runic description of the secret door in the Lonely Mountain may also have had influences from early poetic texts, texts which he clearly loved and devoted much study towards.

Incidentally, it is also clear that he created the story of *The Hobbit* prior to the publication of the 1936 translation of the *Grimnismal* quoted above, so it is almost certain that he had his own personal translation of the Eddas, perhaps even including the formulation of "eight hundred can walk abreast."

In any event, it is an interesting connection for anyone who loves *The Hobbit,* and it makes one wonder how much else J.R.R. Tolkien knew about precession and the esoteric encoding of ancient knowledge in the stories of "myth."

Odin and Gunnlod

2014 May 27

In *The Undying Stars*, I make the argument that all the world's ancient sacred traditions are built upon a common esoteric system[18] of celestial allegory – and that the message that these esoteric myths intended to convey includes a shamanic-holographic[167] cosmology of tremendous sophistication and profound import.

Some readers may initially find the claim that essentially all of the world's sacred traditions – from the Vedas of ancient India to the myths of Osiris and Isis and Horus in Egypt, and from the legends of the North American native peoples to the stories of the Old and New Testaments – share a common system of celestial allegory to be just too much to swallow.

However, once one understands the ancient system – which is expounded in the book as clearly as possible, with accompanying illustrations – the connections between the ancient sacred traditions are undeniable.

Although Norse myths are not addressed in great detail in *The Undying Stars*, they are very special and personal to me, having grown up listening to the stories in D'Aulaires' *Norse Gods and Giants* from before I could read them myself, and reading them over and over once I could read (it is simply one of the best books you can give to a child, and is mentioned many previous posts). Also, of course, my father's father came to America from Norway, and so I always looked upon the Norse myths as the heritage of my ancestors (or, as the Old English word puts it, my "old-fathers").

Currently, the recumbent form of the constellation Virgo is high in the southern sky (for viewers in the northern hemisphere) during the hours of darkness prior to midnight – it is one of the best times of the year for observing Virgo. Behind her, rising up out of the eastern horizon, one can now see the dazzling sinuous form of Scorpio, and further north along the line of the eastern horizon are now rotating into view the twin forms of the two majestic birds of the Milky Way: Aquila the Eagle and Cygnus the Swan.

These constellations are incredibly important in the sacred mythologies of the world. Together, they participate in one of the most important stories in the Norse myth-cycle: the stealing of the mead of poetry from Gunnlod, the beautiful daughter of the jotun Suttungr (or Suttung, as his name is rendered in D'Aulaires' version – the name means "Old Giant," according to a note in the Henry Adams Bellows translation of 1923 of the Elder Edda or Poetic Edda, available at Project Gutenberg [link: http://archive.org/stream/poeticeddaoobelluoft/poeticeddaoobelluoft_djvu.txt] see note at stanza 104 in the Havamal, on page 49 of the original pagination, which is indicated by numbers inside square brackets, thus [49]).

The story of the stealing of Gunnlod's mead is told in the Elder Edda in the words of Odin himself, in stanzas 104 through 110. In the Bellows translation linked above, the verses read as follows (Bellows spells Odin as Othin, and Gunnlod as Gunnloth):

> I found the old giant, now back have I fared.
> Small gain from silence I got;
> Full many a word, my will to get,
> I spoke in Suttung's hall.

> The mouth of Rati made room for my passage,
> And space in the stone he gnawed;
> Above and below the giants' paths lay,
> So rashly I risked my head.

> Gunnloth gave on a golden stool
> A drink of the marvelous mead;
> A harsh reward did I let her have
> For her heroic heart,
> And her spirit troubled sore.

The well-earned beauty well I enjoyed,
Little the wise man lacks;
So Othrorir now has up been brought
To the midst of the men of earth.

Hardly, methinks, would I home have come,
And left the giants' land,
Had not Gunnloth helped me, the maiden good,
Whose arms about me had been.

The day that followed, the frost-giants came.
Some word of Hor to win,
Of Bolverk they asked, were he back midst the gods,
Or had Suttung slain him there?

On his ring swore Othin the oath, methinks;
Who now his troth shall trust?
Suttung's betrayal he sought with drink,
And Gunnloth to grief he left.

The verses may seem mysterious to one not familiar with the story
(again a reason to own the D'Aulaires' book!), but some help is
found in the so-called "Younger Edda" or Prose Edda of Snorri
Sturleson (AD 1178 - AD 1241), which can be found online here:
http://www.gutenberg.org/ebooks/18947?msg=welcome_stranger
In that translation from Rasmus Anderson, first published in
1879, we find the story of the marvelous mead in Part IV, "The
Origin of Poetry."

The interested reader may wish to follow that link to take in the
full account from Snorri, but the short version is that the mead
itself traces its origin to a pact of peace between the Aesir gods
and the Vanir gods after their terrible battle and subsequent
truce, during which truce the Aesir and the Vanir both spit into a
jar, out of which mixed spittle they fashioned an entity known as
Kvaser, who was so wise that none could ask him any question he
could not answer, and who traveled the earth to teach wisdom to
human beings. However, two dwarfs treacherously invited him
into their house and slew him, letting his blood run into a kettle

known as *Odrarer* (in the anglicization selected by Rasmus Anderson in the Younger Edda translation linked above, or as *Othrorir* as rendered by Bellows in the translation of the Elder Edda linked above), and into two smaller jars called Bodn and Son. The dwarfs mixed honey with the blood and produced a mead with the magical property of giving to whomever drinks it the gift of becoming a skald and a sage. These three containers of the magic mead eventually came into the possession of the jotun Suttung (who came to avenge the deaths of his mother and father, also killed by the same dwarfs, and who put them on a rock at sea where the tide would rise and finish them; they begged for Suttung to spare their lives and he did so in exchange for the marvelous mead). Here is the rest of the story from the Younger Edda, as translated by Anderson:

> Suttung brought the mead home with him, and hid it in a place called Hnitbjorg. He set his daughter Gunlad to guard it. For these reasons we call songship Kvaser's blood; the drink of the dwarfs; the dwarfs' fill; some kind of liquor of Odrarer, or Bodn, or Son; the ship of the dwarfs (because this mead ransomed their lives from the rocky isle); the mead of Suttung, or the liquor of Hnitbjorg. [. . .]

> Odin called himself Bolverk. He offered to undertake the work of the nine men for Bauge [one of the jotun brothers of Suttung; the nine men were Bauge's field laborers, whom Odin caused to slay one another with a scythe -- probably related to the stars of the Big Dipper], but asked in payment therefor a drink of Suttung's mead. Bauge answered that he had no control over the mead, saying that Suttung was bound to keep that for himself alone. But he agreed to go with Bolverk and try whether they could get the mead. During the summer Bolverk did the work of the nine men for Bauge, but when winter came he asked for his pay. Then they both went to Suttung. Bauge explained to Suttung his bargain with Bolverk, but Suttung stoutly refused to give even a drop of the mead. Bolverk then proposed to Bauge

that they should try whether they could not get at the mead by the aid of some trick, and Bauge agreed to this. Then Bolverk drew forth the auger which is called Rate, and requested Bauge to bore a hole through the rock, if the auger was sharp enough. He did so. [. . .] Now Bolverk changed himself into the likeness of a serpent and crept into the auger-hole. Bauge thrust after him with the auger, but missed him. Bolverk went to where Gunlad was, and shared her couch for three nights. She then promised to give him three draughts from the mead. With the first draught he emptied Odrarer, in the second Bodn, and in the third Son, and thus he had all the mead. Then he took on the guise of an eagle, and flew off as fast as he could. When Suttung saw the flight of the eagle, he also took on the shape of an eagle and flew after him. When the asas [that is, the Aesir] saw Odin coming, they set their jars out in the yard. When Odin reached Asgard, he spewed the mead up into the jars. He was, however, so near being caught by Suttung, that he sent some of the mead after him backward, and as no care was taken of this, anybody that wished might have it. This we call the share of poetasters. But Suttung's mead Odin gave to the asas and to those men who are able to make verses. Hence we call songship Odin's prey, Odin's find, Odin's drink, Odin's gift, and the drink of the asas.

Now, this incident is of tremendous importance, and I would submit that it is also clearly celestial in its major outline. The maiden Gunnlod, placed within the mountain Hnitbjorg by Suttung to guard the precious mead, and whom Odin treacherously deceives into giving him three drinks of her mead (after swearing a troth to her, as indicated in the Poetic Edda, upon his ring), is described in the Elder Edda as sitting upon a golden stool: this detail alone should alert readers of this blog to the possibility that the maiden (or virgin, which the word signifies) is a manifestation of the sign of Virgo the Virgin. See for example the discussion in "The Pythia" from May 18, 2014.

The outline of Virgo in the sky (which you can see this very evening, if you have good weather) clearly resembles a woman seated upon a throne or a golden stool, and goddesses who are related to this constellation are often depicted upon a throne in sacred traditions around the world (see image on page 195, for example, of an ancient depiction of the Pythia or priestess at Delphi, seated upon her tripod -- which clearly corresponds to the outline of Virgo in the heavens).

Further support for the identification of Gunnlod with Virgo comes from the fact that she is described as dwelling within the mountain or rock called the Hnitbjorg, which Maria Kvilhaug (following the analysis of Svava Jacobsdottir) translates as the "Collision Cliffs" or "cliffs which crash together" and identifies them with the Symplegades of Greek mythology, in her important examination of the maiden and mead theme in Norse mythology entitled *The Maiden with the Mead* (published in 2004; see page 49 for the discussion of Hnitbjorg and the Symplegades).

The Symplegades or "clashing rocks" are almost undoubtedly a myth-metaphor for the equinox, as discussed in some detail in a post on Columba the Dove from 01/26/2014, and in more detail in *Hamlet's Mill* (1969; see page 318), by Giorgio de Santillana and Hertha von Dechend (as well as in my previous book, *The Mathisen Corollary* -- see page 85). The constellation Virgo, of course, is located at the very gate of one of the equinoxes: in fact, at the fall or autumnal equinox, where the sun is plunging down from the bright world of the summer half of the year, in which days are longer than nights, to the cold world of the winter half of the year, in which nights are longer than days. In other words, Virgo is located at the gates of the metaphorical "underworld," and so (as Maria Kvilhaug convincingly argues in the text linked above) is Gunnlod.

The image below, discussed in previous posts (which explain how the lower half of the year is allegorized as Hell or the

Underworld in various myth-traditions, including those in the Bible), shows Virgo at the edge of the fall equinox and the gateway to the underworld (she is drawn as a queen in the zodiac wheel shown in the image; the two equinoxes are each marked with a red X):

DAYS LONGER THAN NIGHTS:
Heaven, Promised Land, Greece, etc.

NIGHTS LONGER THAN DAYS:
Hell, Egypt, Troy, etc.

Note also the point included in Snorri's prose version of the event in which we find that Odin (under the name of Bolverk), worked for Bauge all summer, but when winter came he asked for his pay. A clearer indication that we are discussing the transition point between the upper and lower halves of the wheel of the year could not be asked for.

277

But the identification with Virgo is supported by much more celestial evidence than even this (in case any readers remain skeptical at this point). It is a clear fact that Virgo is situated directly above a constellation known as Hydra, the serpent (the longest constellation in our skies, according to H.A. Rey), upon whose back sits a constellation known as Crater the Cup. This cup features in many ancient myths from around the world, and it is almost certainly the inspiration for the containers of precious mead in the myth of Odin and Gunnlod.

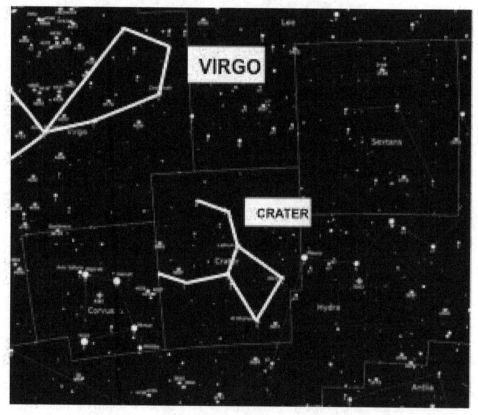

But that's not all, because in the myth of Odin and Gunnlod, we see Odin transform himself into a *serpent* in order to bore his way into the cave of Gunnlod, and then into an *eagle*, in order to fly away with the stolen mead after he betrays the beautiful maiden who gave him her trust and shared with him her bed. It hardly needs to be stated at this point that we find both of these constellations in close proximity to Virgo.

The serpent of Hydra, of course, has already been stated – although the slithery form of Scorpio (also nearby) was anciently depicted as a serpent in some myths and sacred stories as well.

The constellation of Aquila the Eagle rises directly above Scorpio as one traces northward from Scorpio along the shining path of the Milky Way. Clearly, then, the Eagle is very close to the Virgin in the sky, flying as it does above the Scorpion, who follows the Virgin in the zodiac (behind the faint scales of Libra, which are between Virgo and Scorpio and can currently be easily located due to the fact that the planet Saturn is located in Libra in the night sky).

The forms of Aquila the Eagle and Cygnus the Swan, flying in close proximity in the belt of the Milky Way, are very important in ancient myth and legend, and they are breathtaking in the night sky (currently they are rather low in the east until after midnight, but later in the summer they will be high in the sky and together with the bright star Vega in Lyra they form the famous Summer Triangle with each of their respective brightest stars). With this image in mind, the next phase of the myth of Odin's acquisition of the mead, in which Suttung also turns into an eagle and chases after the fleeing thief, becomes quite evident.

If these celestial connections are not enough to convince the skeptical reader, there is yet one more that can be touched upon, and that is the conclusion to the chase, in which Odin spews out the mead that he has carried inside of him, most of which is collected by the Aesir in pots that they set out on the heavenly fields as they see the pair approaching, but some of which falls to earth for the benefit of anyone who finds it. Here is an illustration from an Icelandic manuscript from the 1700s of the pursuit of Odin by Suttung, and the spewing-out of the marvelous mead:

Knowing that the two mighty birds of the celestial realm are both found flying in the midst of the stream of the shining Milky Way, is it not possible that the story of the spewing forth of the mead in this particular myth is connected to the band of the Galaxy, which can be seen descending to earth in a misty ribbon like a silvery waterfall during the time of the year that the constellations Aquila and Cygnus are aloft?

Now, some readers may object that showing the connections from this profound Norse myth, which is full of tremendous drama and pathos and insights into the human condition, to the constellations of the night sky will somehow rob it of its magic (just as Odin himself robbed Gunnlod of her poetic mead). But I

280

would argue that the opposite is true! For, as Maria Kvilhaug herself has powerfully demonstrated in *The Maiden with the Mead*, the import of this myth touches upon deep matters of initiation, shamanic transformation, and ecstatic travel across the boundary of this world and into the "other world" (the ecstatic or mystical or shamanic journey). And, as I labor to demonstrate in *The Undying Stars*, which I wrote and published before I even became aware of Maria's work, this is exactly one of the esoteric teachings which this universal system of celestial allegory was intended to convey. The constellations of Eagle and Swan can also be shown to be quite important in shamanic cultures worldwide, as argued in a post from 08/08/2013.

Not wanting to know the esoteric connections hidden in these exquisite and moving ancient myths is like never wanting to be shown how the trinomial cube (itself a beautiful piece of material artwork) relates to the higher concept of the trinomial equation which it represents and which it was intended to teach (see this previous post, entitled "Montessori and 'thinging'"[30]).

Finally, it is important to be able to show these connections between the events in a mysterious and austere Norse myth found in the Elder Edda, and the stories found in other sacred traditions -- including the stories of the Old and New Testament (in which Virgo furnishes the original for many Biblical characters, including Eve, Sarah, and Mary, among others). Some of these stories, the reader may note, also involve a deceiving serpent. He just happens to be named "Odin" in the Eddas.

This fact demonstrates that the ancient sacred traditions, as they were originally intended to be understood, were all close kin. It was only when the literalist tradition arose, and the esoteric understanding of the Biblical stories was rejected in favor of a literalist approach which insisted the stories be read primarily as historical accounts of literal persons who walked on earth, that followers of that literalist path declared their faith to be totally unrelated to all the other sacred traditions of the world.

Brisingamen, the necklace of Freya

2014 June 20

Now is a particularly good time of year to go out after dark and enjoy the spectacular constellations visible along the zodiac band in the hours after nightfall and leading up to midnight (and the next week and a half will be particularly excellent, as the moon is waning and rising later and later in the night -- that is, closer and closer to sunrise -- as the sun prepares to "overtake" the moon and give us another new moon on June 28th, after which there will still be a few nights of good star-gazing while the moon sets relatively soon behind the sun).

High in the center of the sky you can locate the constellation Virgo, one of the most important constellations in the sky, and one who plays numerous roles in the ancient mythologies of the world, as demonstrated in the previous series of posts which presented some of the connections between the mythical stories of cultures as widely diffuse as those of ancient Japan, of the Indians of North America, and of the Norsemen of Scandinavia.

Virgo is easy to spot if you can locate her brightest star Spica, which Corvus the Crow is constantly staring at as described in Rey's book on the stars (page 50), and her directly outstretched arm which is a very prominent and distinctive aspect of Virgo.

Above her (all relative positions described from the perspective of a viewer in the northern hemisphere; please adjust for your own latitude on our planet) you can now easily make out the stars of Boötes the Herdsman, featuring the brilliant red-orange star Arcturus. A commonly-cited memory aid for locating both Arcturus and Spica tells us to follow the sweep of the handle of the Big Dipper and draw an "arc" to Arcturus, and then to continue along the same general direction and draw a "spike" to Spica. This method works quite well, especially as the well-known constellation of the Big Dipper is now high in the sky for observers in the northern hemisphere.

Another aid to locating Boötes which always helps me to find him is the fact that his long pipe reaches very nearly to the tip of the handle of the Dipper. Once you locate the Dipper's handle, you can trace the Herdsman's pipe back to his large (but faint) head, and then continue to trace out the rest of the constellation. Both Virgo and Boötes are depicted in the star chart below, which is now familiar to readers of the three articles linked above which make the case that all the myths discussed utilize a common system of celestial metaphor, one that connects virtually all the world's ancient sacred traditions no matter how widely dispersed across the globe (including those in the Old and New Testaments).

If the night is dark and Boötes is high in the sky, you can clearly make out the beautiful semicircle of stars known as the Corona Borealis, or Northern Crown, which can be seen in the diagram above just to the left of the Herdsman's head, and which is labeled "Crown." It really is very close to the outline of the head of

Boötes, and the best way to locate it is to look right at his head and it will be seen to be almost touching him. The diagram below shows both Boötes and the Crown, and labels the brightest star in the Crown, which is known as Gemma or Alphekka or Gnosia.

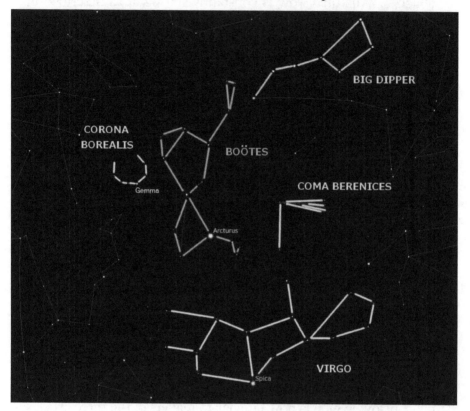

We have seen from the examination of the mythology of ancient Japan that the Northern Crown was described in the Kojiki as an "augustly complete string of jewels eight feet long," which should give us a clue that another marvelous string of jewels belonging to a beautiful goddess may also be connected to this semicircle of stars next to Boötes and above Virgo: the dazzling fire-gold necklace of Freya, the Norse love-goddess.

Freya's necklace is called the Brisingamen, and it is featured in the Norse poetic Edda in the section known as the Thrymskvitha (or "Lay of Thrym"), which can be found beginning on page 173 of this online version of the poetic Edda.

The Brisingamen is also featured in the prose Edda of Snorri Sturluson, particularly in the section known as the Skaldskaparmal, in which the theft of the necklace by Loki is alluded to although not described at length.

Other early Norse poetic compositions outside of the Eddas describe the theft of the Brisingamen by Loki with more detail, saying that Loki (who is a master of shape-shifting) turned himself into a fly in order to steal into Freya's bedchamber while she slept, buzzing around her face until she batted at him with the hand which even in sleep rested upon the clasp of her precious necklace. The moment she did this, Loki transformed in a flash into his normal self and stole the necklace. Snorri also refers to the theft of the necklace, citing passages from other poets which refer to the theft as well, showing that the episode was well-known by his day and probably much earlier.

If any doubt remained about the identification of Loki with Boötes in the episode of Skadi's laughter described in that post on the mythology of ancient Japan, the fact that Loki is described as the thief of the Brisingamen from the sleeping goddess and the fact that the starry necklace in the celestial realm is located not on Virgo where it belongs but next to Boötes who hovers over Virgo's recumbent form should "put those doubts to bed" once and for all (so to speak).

Additional evidence comes from the fact that Snorri mentions that it was Heimdal who challenged Loki over the suspected theft, and who fought Loki for the necklace and eventually beat Loki and returned the necklace to Freya. The authors of *Hamlet's Mill* present extensive evidence that Heimdal, the "son of nine mothers" and the one who, Snorri tells us, is also referred to by the name Vindler (which the authors of *Hamlet's Mill* tell us is associated with turning, or a turning handle). The authors of *Hamlet's Mill* argue that these clues tell us that Heimdal is associated with the "handle" that turns the entire night sky around the central pole -- and that he is in fact associated with the entire "equinoctial colure," which stretches from Ares to Virgo through the north celestial pole (and around from Virgo to Ares through the south celestial pole as well, although this half of the colure is less appropriate to this discussion about the northern constellation Boötes and the northern myth of Loki).

Based upon their arguments, if Heimdal is associated with the handle of the Dipper and the north celestial pole, we can surmise that it is only natural that Norse myth might describe him as the arch-rival of Loki, if Loki is associated with the nearby constellation of Boötes the Herdsman, who appears to be tied to the handle of the Big Dipper [*later note*: it is quite likely that Heimdal is also associated with Boötes, because Heimdal holds the famous horn, the Gjallarhorn, which he sounds in order to warn Asgard of impending danger, and Boötes of course features a distinctive "pipe" which could be envisioned as a trumpet or a horn -- if so, then it is even more understandable that Loki and Heimdal should be depicted as rivals in the Norse myths].

The fact that Virgo's arm is raised as if in the act of "swatting away" the thief of her necklace should be even further proof that this set of constellations furnished the material for this particular episode from Norse myth.

Any doubts which still remain regarding the identification of Loki the thief who steals from the goddess her most precious possession with the constellation Boötes above Virgo can be laid to rest by noting another nearby asterism seen in the star chart above, located just above the head of Virgo and to the right of the figure of Boötes the Herdsman, the constellation known as Coma Berenices or Berenice's Hair (and marked as such on the diagram). In his outstanding book *The Stars: A New Way to See Them*, author H.A. Rey says of Berenice's Hair:

> Small and very faint. Contains a group of dim stars, visible only on clear, moonless nights when the constellation is high up [like, right now]. 36.

He goes on to explain that:

> The constellation owes its name to a theft: Berenice was an Egyptian queen (3rd century BC) who sacrificed her hair to thank Venus for a victory her husband had won in a war. The hair was stolen from the temple but the priests in charge convinced the disconsolate queen that Zeus himself had taken the locks and put them in the sky as a constellation.

This story as related by H.A. Rey almost certainly has it backwards: the story of the queen who sacrificed her hair to the goddess Venus is most likely a legend inspired by the constellations Virgo and Coma Berenices (and not an original event that happened on earth and which later inspired the naming of constellations in the sky).

Those familiar with the Norse myths will immediately be reminded of yet another theft by Loki of the treasured possession of a beautiful goddess: this time, the theft of the golden-red hair of Sif, the wife of the thunder-god Thor. The myth of the theft of Sif's hair by Loki is clearly a dramatization of these three constellations: the disembodied hair of Coma Berenices, floating above Virgo and just next to Boötes. In all of this, it can be seen that our identification of Loki with Boötes has ample reinforcement.

This analysis provides further support for my assertion regarding the identity of Loki and Skadi in the episode of the laughter of Skadi (Loki is again Boötes, and the beautiful Skadi is Virgo, who takes on the female role in a great many of the world's myths). It also further supports the connections we saw, discussed in the post entitled "The celestial shamanic connection: Ancient Japan," between the Norse myth related in the Eddas in which the gods must make the beautiful jotun maiden Skadi laugh, and the Japanese myth related in the Kojiki that brings laughter to the assembled gods when the goddess Amaterasu hides herself in a cave. In the Japanese myth, it is the sexually explicit dance of the goddess Uzume which brings the laughter, and in the Norse myth it is the equally graphic antics of Loki which finally bring laughter to Skadi.

In the discussion, I make the argument that both of these myths clearly involve the constellation Virgo the Virgin and and the surrounding constellations in that region of the sky, and that in the Norse myth Skadi plays the role of Virgo and that Loki is Boötes the Herdsman -- a correlation I have not seen explicitly put forward anywhere else before (although the authors of *Hamlet's Mill* were clearly aware of some relation between the myth of Skadi and the myth of Uzume and Amaterasu, they never tell us directly that the connection specifically involves Virgo and Boötes, or trace out the connections between these myths and those stars).

The details which indicate that Loki's role in the tale come from the location of Boötes are conclusive, in my opinion, particularly the fact that Loki eventually precipitates himself into Skadi's lap in order to finally bring a smile to her lips -- a detail which can be readily understood from the relative location of Boötes and Virgo shown in the star chart. But the further evidence we have seen for Loki as Boötes in the myth of the theft of the necklace of Freya and in the myth of the theft of Sif's hair should put the matter beyond any doubt.

And so, if we have established that Loki is Boötes in numerous episodes from Norse myths, this serves to reinforce the assertion that the episode in which Loki makes Skadi laugh and the episode in which Uzume makes the assembled gods laugh and the goddess Amaterasu come out of her cave share a clear celestial connection, in that both the Norse and the Japanese myth use many of the exact same celestial components.

Further, the fact that we have now established the Northern Crown as the celestial counterpart of the mythical Brisingamen, the gorgeous necklace of Freya, reinforces yet another connection between the Norse and the Japanese myth-systems, in that the oldest surviving Japanese text containing these myths, the Kojiki, describes a jeweled necklace in conjunction with the episode in which Uzume dances for the assembled kami. Both systems are clearly employing many common elements in their myths involving the constellations surrounding Virgo in this particular part of the night sky.

Be sure to note also the fact that both myth systems, from Japan and from Scandinavia, are doing so in texts which can be shown to date from long before the conventional paradigm would allow for contact between cultures situated so far from one another on the globe. The Kojiki was composed no later than AD 711 or AD 712 (and probably contains myths that are centuries older than that). The age of the Poetic Edda is debated among scholars, but its original composition probably predates Snorri's Prose Edda of about AD 1220, and it may contain material that had been passed down for centuries before it was ever written down. In any case, contact between the cultures of Japan and Scandinavia prior to AD 711 is not consistent with the dominant conventional narrative of history, so what can explain the existence of a common system of celestial metaphor in the mythologies of such widely-separated peoples?

There are many possibilities, but almost all of them set the conventional historical paradigm on its ear. One possibility is that there was ongoing transoceanic contact between these cultures during the centuries that these works were composed, or at some time prior. Another possibility is that both cultures (and the many others around the world whose mythologies share the same universal allegorical system) are descended from some even earlier common predecessor civilization, perhaps one which left this ancient esoteric system as a precious inheritance for all humanity.

In any case, if it is at all possible for you to do so, now is an excellent time to head outside in the hours after nightfall, and to identify the constellations discussed, such as Virgo, Boötes, the Northern Crown, and even (if the night is dark enough and the sky clear enough) Berenice's Hair. As you do so, you can think of the legends of the beauty of Freya, and of her dazzling necklace, the Brisingamen. And as you contemplate the theft of the heavenly necklace by Loki (and his theft of another heavenly treasure, that of Sif's hair), you can reflect on the possibility that this once-universal system of celestial metaphor, which Aritsotle himself referred to as the "ancient treasure" and which may represent the legacy of some far older and possibly far more advanced predecessor civilization, has effectively been stolen from humanity, and knowledge of it suppressed, for at least the past seventeen centuries.

The sacrifice of Odin

2014 August 28

For the third edition of the "Ankh trilogy" of posts (which began with "Scarab, Ankh, and Djed"[42] and continued with "The name of the Ankh"[49]), let us continue our investigation of this most central theme by looking at the connections to another manifestation of the Cross of Life (which the Ankh and the Djed represent, as does the Scarab with its upraised arms): Yggdrasil, the Tree of Life found in Norse and Germanic mythology.

The World-Tree Yggdrasil is described in the Elder Edda and the Younger Edda as a mighty ash-tree whose roots penetrate to the deepest underworlds and whose branches reach to the highest heavenly realms. At its base is the holy fountain of Urd, associated with the Norns who tend to the Tree and who, the Younger Edda tells us in Chapter VII, "shape the lives of men."

The waters at the foot of the tree are also associated with Mimir's Well. In the Younger Edda, as part of the question-and-answer session between Odin in the guise of Ganglere (might we not read the same "root sound" of "the name of the Ankh" here as well?) and three divinities who are simply named Har ("High"), Jafnhar ("Equally High") and Thride ("Third"), we read:

> Then said Ganglere: Where is the chief or most holy place of the gods? Har answered: That is by the ash Ygdrasil. There the gods meet in council every day. Said Ganglere: What is said about this place? Answered Jafnhar: This ash is the best and greatest of all trees; its branches spread over all the world, and reach up above heaven. Three roots sustain the tree and stand wide apart; one root is with the asas and another with the frost-giants, where Ginungagap formerly was; the third reaches into Niflheim; under it is Hvergelmer, where Nidhug gnaws the root from below. But under the second root, which extends to the frost-giants, is the well of Mimer, wherein knowledge and wisdom are concealed. The owner of the well hight Mimer. He is full of wisdom, for he drinks from the well with the Gjallar-horn. Alfather once came there and asked for a drink from the well, but he did not get it before he left one of his eyes as a pledge. Younger Edda, Chapter VII.

This famous incident, of course, is responsible for Odin's having only one remaining eye. But there is another episode in Norse myth in which Odin had to undergo tremendous sacrifice in order to gain wisdom, an episode also closely associated with the holy ash Yggdrasil, and an episode which clearly connects the World-Tree with the concepts and symbology that has been discussed in the previous two posts surrounding the Ankh or Cross of Life,[49] and the Djed-column[42] or Backbone of Osiris: the famous sacrifice of Odin in which he hangs himself upon the tree, described in a somewhat fleeting passage found in the Elder

Edda, in the portion known as the Havamal or Hovamol, beginning in stanza 139 (in the online edition of the Elder Edda linked above, it begins on page 59 -- that online text is a little difficult to navigate: the best way is probably to look for the "page numbers" contained within brackets, scrolling down until you reach [59]):

> I ween that I hung on the windy tree,
> Hung there for nights full nine;
> With the spear I was wounded, and offered I was
> To Othin, myself to myself,
> On the tree that none may ever know
> What root beneath it runs.
> None made me happy with loaf or horn,
> And there below I looked;
> I took up the runes, shrieking I took them,
> And forthwith back I fell.

This passage describes Odin "raised up" upon the Tree, hanging upon it in a sacrifice or crucifixion, Odin sacrificed to Odin, and through this ordeal after nine full nights he obtains a new vision which he did not have previously -- the vision to see the runes, and to take them up. It is in many ways analogous to the ordeal he had to go through in order to obtain the wisdom of Mimir from the well, and also to the adventure he had to undertake in order to obtain the mead of poetry from Gunnlod[270], and yet this incident is at once more primordial and defining of the Alfather Odin than any of the others.

It is through this sacrifice that Odin obtains the gift of the runes, the gift of encoding information in symbolic form, the gift of the manipulation of language. We can begin to realize the depth of power that this gift truly contains when we recognize the ordeal Odin had to undergo in order to obtain it.

Previous posts have examined the concept that it is in many ways *through language* that reality is created and that worlds are shaped. In Genesis, of course, it is through the word of God that all Creation is spoken into existence. Modern science tells us that it is through the combination of the four "letters" (dare we call them "runes"?) in the strands of DNA that all our body's characteristics are spun-out in the cells of our being (perhaps these are the strands that the Norns are spinning?).

And Odin wins the ability to see the runes by his hanging upon the Tree.

Many other researchers have observed that the double-helix shape of the DNA strand recalls quite strikingly the two serpents of the caduceus staff (carried, of course, in Greek and Roman myth by Hermes or Mercury, who is in many ways associated with Odin, interestingly enough). But we have seen in the previous two examinations of the Ankh and Scarab and Djed[42] that the caduceus staff is clearly a "Djed-column" type of symbol, representative of the Backbone of Osiris "raised up," and of the vertical column of the year which reaches from the very lowest pit at the winter solstice to the very summit of the year (highest heaven) at the summer solstice.

Clearly, Odin's hanging upon the Tree relates to this same concept.

This profound episode also relates to the concept of "the shamanic," in that Odin by his ascent to hang on the World-Tree penetrates beyond the realm of the ordinary to bring back knowledge that can be obtained by no other means. This is one of the defining characteristics of the shamanic techniques of ecstasy described in the work of Mircea Eliade, and in fact it can be easily demonstrated that shamans around the world often use a vertical pole or "climbing the tree" as part of their shamanic travel. There is clearly a powerful stream of connectivity which flows between the ancient wisdom preserved and conveyed in the myths of Osiris and the Djed, the myths of Odin and the World-Tree, and the shamanic practices of the world.

Finally, we must notice the clear connections between the sacrifice of Odin described above and the sacrifice of Christ on the Cross described in the New Testament. Most obviously, both involve a crucifixion upon a Tree (and the Cross is literally referred to as "the tree" in Biblical verses such as Acts 13:29 and 1 Peter 2:24).

Additionally, in Odin's description of his own sacrifice, he declares that "with the spear I was wounded," which is obviously an element that is present in the sacrifice described in the New Testament as well. Critics might argue, because our records of the Norse myths were written down after Christianity was already known and was spreading throughout Europe, that this element was "imported" into Norse mythology from Christianity, but there is absolutely no evidence that this is the case, and there is no need to assume such an importation. Odin is very closely associated with his powerful weapon the Gungnir, the mighty spear which never misses its target and which Odin used to indicate which force would be victorious when two contending sides met on the field of battle. That he would be wounded by his own spear when he sacrificed himself to himself is clearly not inconsistent with the tenor of what is taking place.

There is also the shout or shriek which Odin utters at the end of his ordeal, when he has finally won the victory and obtained what he sought.

By this episode, and by other arguments I present in *The Undying Stars*, I would argue that the verses preserved in the Old and New Testaments of the Bible are actually shamanic in nature, and were originally intended to be regarded as such. By the aggressive literalizing that has taken place in history, this shamanic vision (and their connection to the myths of Osiris and Odin) has been covered-over and obscured.

And yet, like the hidden runes which Odin found, which have the ability to carry world-changing information to faraway places and even to distant times (to those not yet born, even), the words and letters preserved in the Bible itself continue to patiently carry their message down through the centuries. Their kinship with the myths of the world, from Egypt to Greece to the lands of the Norse, and to the shamanic practices found across so many cultures, from North and South America to Siberia and Mongolia and Australia and Africa, is undeniable.

Jephthah's Daughter

2014 September 18

In the Old Testament scriptures, in the Book of Judges, we encounter the horrifying story of Jephthah and his daughter. If ever there were a Biblical passage which renders an absolutely hideous message when taken literally, while yielding a completely satisfactory conclusion when understood astronomically, this story is it.

In chapter II of Judges, after a description of the elders of Gildead requesting that Jephthah be made head and captain over the children of Israel, and a description of a series of battles between the children of Israel and the children of Ammon, we arrive at verse 29, where we read:

> Then the Spirit of the LORD came upon Jephthah, and he passed over Gildead, and Manasseh, and passed over Mizpeh of Gilead, and from Mizpeh of Gilead he passed over unto the children of Ammon.
>
> And Jephthah vowed a vow unto the LORD, and said, If thou shalt without fail deliver the children of Ammon into mine hands,
>
> Then it shall be, that whatsoever comth forth of the doors of my house to meet me, when I return in peace from the children of Ammon, shall surely be the LORD's, and I will offer it up for a burnt offering.
>
> So Jephthah passed over unto the children of Ammon to fight against them; and the LORD delivered them into his hands.
>
> And he smote them from Aroer, even till thou come to Minnith, even twenty cities, and unto the plain of vineyards, with a very great slaughter. Thus the children of Ammon were subdued before the children of Israel.
>
> And Jephthah came to Mizpeh unto his house, and, behold, his daughter came out to meet him with timbrels and with dances: and she was his only child; beside her had neither son nor daughter.
>
> And it came to pass, when he saw her, that he rent his clohtes, and said, Alas, my daughter! thou hast brought me very low, and thou art one of them that trouble me: for I have opened my mouth unto the LORD, and I cannot go back.
>
> And she said unto him, My father, if thou has opened thy mouth unto the LORD, do to me according to that which hath proceeded out of thy mouth; forasmuch as the LORD hath taken vengeance for thee of thine enemies, even of the

children of Ammon.

And she said unto her father, Let this thing be done for me: let me alone two months, that I may go up and down upon the mountains, and bewail my virginity, I and my fellows.

And he said, Go. And he sent her away for two months: and she went with her companions, and bewailed her virginity upon the mountains.

And it came to pass at the end of two months, that she returned unto her father, who did with her according to his vow which he had vowed: and she knew no man. And it was a custom in Israel,

That the daughters of Israel went yearly to lament the daughter of Jephthah the Gileadite four days in a year.

And the men of Ephraim gathered themselves together, and went northward, and said unto Jephthah, Wherefore passedst thou over to fight against the children of Ammon, and didst not call us to go with thee? we will burn thine house upon thee with fire.

If this passage is understood to be describing the literal and historical actions of a literal and historical judge and war-chief of the ancient children of Israel, who swears a vow to sacrifice the first thing he sees upon returning home from battle and burn it as a burnt offering to the Almighty, it would surely seem to be a horrible episode and one that probably does not feature too often in sermons.

Certainly it could be used as a stern warning against swearing to rash vows (the episode is often referred to generally as "Jephthah's rash vow"), but even if it is used as an example of the dire consequences of swearing too rashly, that still leaves the gaping question of whether such a vow must then be fulfilled, even to the extent of killing another person -- even to the extent of sacrificing one's only daughter. Can this scripture possibly be implying that once such a vow is sworn, to break the vow is considered impossible, and worse than actually taking someone else's life -- let alone the life of one's own beloved daughter?

The passage itself gives us no help in this regard: it simply records that Jephthah groans with pain but clearly does not consider it possible to break the vow, and Jephthah's daughter understands and says that he must do it, especially since he was given victory in the battle over the Ammonites after swearing the vow.

The verses which follow the human sacrifice likewise do not give any hint of whether the community thought Jephthah had acted rightly or wrongly: immediately after the verse about the daughters of Israel mourning Jephthah's daughter once a year, at the end of chapter 11, the following verse (at the beginning of chapter 12) has the men of Ephraim gathering and preparing to burn down Jephthah's house down with him inside, but not because he has killed his daughter and burned her, but because he did not take them along when he went to fight the Ammonites.

Again, a literalist encountering these verses is left with a sickening scenario in which all values seem to be inverted and violence and darkness reign supreme. I am aware that the verses could potentially be used as a "type" or foreshadowing of the sacrifice of Christ by the Father, but even that hermeneutical move would have to be used with extreme caution, as the circumstances surrounding Jephthah's sacrifice are simply so shocking and so horrifying that there seems to be nothing uplifting in the passages whatsoever (to which the reply would be that this is the contrast between the sinfulness of humanity and the perfection of the redemption -- which seems to be the only way these verses could possibly be made to serve a positive purpose in a sermon). Even such a rhetorical move would still leave the question of whether Jephthah then was right in proceeding with the killing and burning of his daughter.

Clearly, a strictly literalistic interpretation of these verses leads into an absolute swamp filled with pitfalls from which it becomes more and more difficult to extract one's self, the further down into it one charges.

300

However, like the horrifying verses about the prophet Elisha calling two she-bears out of the woods to tear apart the forty-two youths in the Old Testament book of 2 Kings (an episode almost as sickening as the Jephthah incident, and yet even that one is less hideous than this one in many respects, except for the obviously higher body-count), this passage contains clear elements of celestial metaphor which indicate that it was never intended to be understood as an account of something that took place on the earth involving human beings: it takes place in the sky.

Very briefly, because the elements of the system of celestial metaphor have now been explained in some detail in numerous previous posts, this incident has all the markers indicating a sacrifice at the point of equinox: those two points on the annual cycle where the two great "hoops" of the ecliptic and the celestial equator "cross" one another, allegorized in countless different forms as a sacrifice (previous posts as well as the first three chapters of *The Undying Stars* which are available for free perusal online have demonstrated that the equinox crossings form the foundation for the myths of the sacrifices of Iphigenia, of Isaac by Abraham, of St. Peter in early church tradition, and even of Christ -- which explains the "echoes" that typologists can find between some of these Old Testament star-myths and the sacrifice on the Cross in the New Testament).

Which of the two equinoctial points we are dealing with here should be fairly obvious from the clues that have been included in the scriptural passage: the sacrifice is of a young virginal daughter (her virginity is emphasized several times in the episode), and so we should be fairly confident in identifying the September equinox. For those in the northern hemisphere, this would be the autumnal or fall equinox, when the days cross over from being longer than nights to the half of the year in which they are shorter (each of the two equinox-points are marked on the zodiac wheel below with a red "X" and the one on the right as we look at the wheel is the one at Virgo, where the sun which moves in the direction of the arrows as the year progresses is declining towards the lower half of the year and the winter solstice at the bottom of the circle). This fact supports the possibility that the "children of Ammon" that Jephthah was battling before he returned to his house to encounter his daughter (returning to the house of the sign of Virgo, that is) represent the upper, sunny, summery half of the year.

DAYS LONGER THAN NIGHTS:
Heaven, Promised Land, Greece, etc.

NIGHTS LONGER THAN DAYS:
Hell, Egypt, Troy, etc.

The exact correspondences of the different opponents Jephthah battles with are much more difficult to identify with exactitude, although we can be fairly confident that this is a celestial battle describing the circle of the zodiac (along with many, many other examples from both the New Testament and the Old Testament, as well as from other mythologies such as the Greek myths about the Trojan War). The identity of the daughter with the stars of Virgo, however, is very certain.

We have already seen that the passage itself takes care to emphasize her status as a virgin. The other distinguishing feature of the daughter of Jephthah, however, is her timbrel -- which is to say, her tambourine. It just so happens that the constellation Virgo has some distinguishing features which are often included in ancient art: one of these is an outstretched arm, and one is the fainter circle of stars that are in front of her face and above this outstretched arm. In the ancient Greek art depicting the Pythia of Delphi, for example, the outstretched arm represents the arm with the sacred laurel-branch, while the circle of stars corresponded to her circular dish or platter holding the holy water, as described in ancient in a post from May 18, 2014. In the case of Jephthah's daughter, this circle becomes a timbrel.

Note also in the painting above, by Giovanni Antonio Pellegrini (1675 - 1741), the artist has incorporated the outstretched arm and the timbrel of the girl, in a way that is most suggestive of the possibility that he understood her connection to the Virgin of Virgo (either that or, in his formal art schooling, these elements were passed down to him without his understanding). This indicates that the esoteric aspects of these myths was known by some and passed down among some circles, without being taught to those seekers in the churches, who were being taught that these stories all represent events that took place in history among literal people.

Below is close-up of the stars of Virgo: you can see the faint "circle" of stars that becomes the dish in depictions of the Pythia, the circular hoop in images of Rhea seated on a throne, and the timbrel of the daughter of Jephthah:

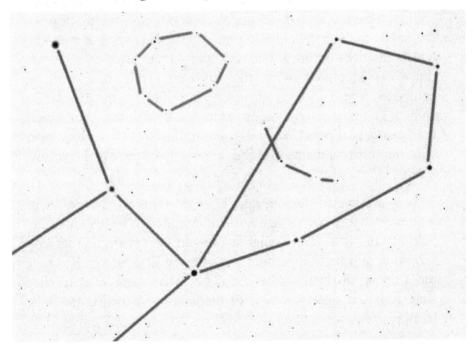

The final clincher for me that we are dealing with a star-myth of Virgo and the equinox (point of many sacrifices) is the presence of fire -- almost always present in equinox-myths, since the equinoxes are the two points where the fiery path of the sun (the ecliptic) crosses through the celestial equator. Previous posts detailing this include "The Old Man and his Daughter" [not included in this collection but discussed in *Star Myths of the World, Volume One*] and "Common symbology between Mithraic temples and the Knights Templar.[653]" In the myth of Jephthah, we find that he must make his daughter (Virgo) into a *burnt* offering, and that immediately thereafter some men come and threaten to *burn* his house down.

And so, we see that what would be an absolutely execrable story if interpreted "historically" (the way most of those with positions of authority inside the various Christian churches have generally approached the scriptures for the past seventeen hundred years or more) is actually just another star myth built upon a structure that is found in the Greek myths and in Native American traditions all the way across the globe, intended (I believe) to convey a positive and uplifting message to all people, embodied in the motions of the stars, the sun, the moon, and the planets, and a message which describes our incarnate condition as the "Djed-column cast down" -- identified with the horizontal line between the equinoxes, the line of sacrifice, the line of being made "like an animal" and plunged down into a similitude of death -- but which teaches the real possibility and even inevitability of reconnection with the spiritual realm, and the "raising up of the Djed column" again.

This is a powerful lesson, and I would think a very powerful argument that these passages are not intended to be read literally. Additionally, the above analysis should demonstrate that the scriptures and traditions the world over are all close kin, and that the artificial distinctions between "Christian" and "pagan" that were imposed upon the rise of literalist Christianity are both harmful and false.

The Festival of Durga

2014 October 02

It is currently the festival of Durga Puja, the celebration of the worship of the goddess Durga, also known by her many other names Bhavani, Kanaka Durgammathalli, and even as Adi-Parashakti the Great Mother Goddess.

The festival of Durga Puja (also known by its other names Durgotsava, Sharadotsav, Akalbodhan, Navaratri Puja, and many more) also specifically celebrates the goddess's triumph over the buffalo asura or demon Mahishasura. In fact, in the depiction of the goddess shown above, the bloodied head of the buffalo can be seen at her feet, directly underneath the lower end of her long scepter (this long scepter-shaped weapon strongly resembles the Vajra).

There are many aspects of the symbology of Durga which indicate that she is a celestial deity, and in fact that she is associated with the constellation Virgo, one of the most important of zodiac signs and one who takes on the form of many, many goddesses and other important female figures in the sacred scriptures and mythologies of the world. This blog has previously discussed the almost certain correspondence of the goddess Durga with the sign and stars of Virgo in posts such this one.[264]

The clear symbolic indications that Durga corresponds to Virgo discussed in those posts include:

- the fact that she is often depicted as riding on a lion (Virgo follows Leo across the sky, and hence the Goddess in widely dispersed mythologies is very often associated with a lion or with lions, sometimes riding a lion, or riding in a chariot drawn by a lion or by lions, or seated on a throne flanked by one or more lions),
- the fact that she is often depicted with an outstretched arm which is one of the most characteristic features of the constellation Virgo, and
- the distinctive bend in the hip depicted in some statues and reliefs featuring Durga, which corresponds to the outline of the constellation Virgo, and which can be seen in other artwork featuring Virgo-goddesses from ancient Greece and Rome as well.

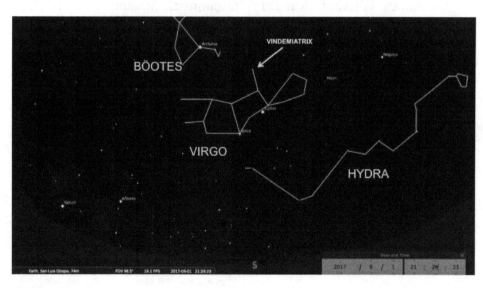

Below is an image of Durga from a bas-relief in which her outstretched arm appears to be holding a bow. This outstretched arm is marked in the sky by the bright star Vindemiatrix, labeled in the diagram above. In other depictions, she may be holding a sword or other weapon in this outstretched arm. Female figures corresponding to Virgo often have bows, sometimes bows with miraculous powers -- see for example the story of "The old man and his daughter" from the Native American people of North America (in this case, from the northwest coastal island region near the modern-day state of Washington in the US and British Columbia in Canada, discussed in *Hamlet's Mill* and also in my book *Star Myths of the World, Volume One*).

Note also in the bas-relief below that Durga is shown in the act of defeating the buffalo-bull-headed Mahishasura.

We can see additional indications that Durga is associated with Virgo in the image at the top of this post. There, a lion is again prominently featured, and if we wanted to take the time to do so we could draw direct correspondences between the posture of that particular lion and the outline of the constellation Leo in the sky, who is so close to Virgo that he can be seen in the star-chart above, although his outline has not been drawn in (his stars are seen directly to the right of the word "Vindemiatrix" in that star chart).

Another clue that Durga is an aspect of Virgo can be seen in the fact that in the modern Durga Festival depiction of Durga at the top of this post, she is holding a serpent in the lowest of her hands on the right side of the image as we look at it (it is actually a cobra). This hand holding the cobra is on the opposite side of Durga's body from the hand that is holding the long Vajra-shaped scepter-weapon (the one that ends on top of the bloody buffalo-head).

Once again, this depiction is astronomically correct for the constellation Virgo. In the star chart above which shows the outline of the constellation, we can see that Virgo's outstretched arm (represented in the top image by the Vajra-scepter) is on the opposite side of her body from the long constellation of the serpent Hydra (whose starry outline does in fact resemble a cobra, if the circle of stars at the far right of his body as we look at it corresponds to the cobra's hood).

But perhaps the most important clues that we are correct in connecting Durga and Virgo are seen in the timing of her festival, and the fact that the festival celebrates her triumph over the buffalo-bull demon. The festival's timing is tied to the complex lunar-sidereal Hindu calendar, but it basically falls near the autumnal equinox and the part of the year which has anciently been associated with the sign of Virgo (particularly in the Age of Aries).

The reason Virgo is associated with this part of the year is that, during the Age of Aries, she was the constellation and zodiac sign seen above the eastern horizon just prior to the rising sun. Even today, although the background of stars has been delayed due to the ages-long motion of precession, the sun's rising at this time of year is in the sign of Virgo, a fact you can readily see for yourself by going to the excellent online, browser-based, free planetarium app at Neave.com [or by downloading the even more robust open-source planetarium app at Stellarium.org].

There, if you simply leave the "location" at its default point, and dial up the app for today (October 2), you can swing the sky around to the east and then "dial back the hours" using the "upper arrow" on the date-time window (be sure to dial back the hours, not the days) until you see the sun rising on the eastern horizon. You will quite clearly see that the sun is in the midst of the constellation Virgo as it rises.

Now, if you "freeze" at the point where the sun is just below the horizon (and Virgo is already charging up over the eastern horizon), swing the view back along the horizon from the east (where the sun is rising) to the west (where the stars are setting). If you then shift your view upwards a bit, you can find the distinctive outline of Taurus, the Bull, who contains the V-shaped Hyades and is just "above" (actually "to the right of") the constellation Orion. Taurus is getting ready to set in the west as Virgo (Durga) rises in the east. Dial the hours forward and Taurus will be banished below the western horizon, as Virgo rises more and more fully into the sky in the east.

This is the meaning of Durga's slaying of the buffalo-bull demon Mahishasura.

At least, it is according to my interpretation of the celestial clues. And the number of clues in this case is pretty overwhelming. Thus, the Festival of Durga is another example of the fact that nearly all the ancient sacred myths and scriptures and traditions are built upon a common system of celestial metaphor.

The implications of this commonality are profound and far-reaching, but one of them is that humanity is actually united and not divided by these scriptures and traditions -- and thus we can all in some measure celebrate the Durga Puja, if we choose to do so!

R. Smirke R.A. pinxit.

A. Smith A.R.A. sculpsit.

THE GENIUS OF THE LAMP APPEARING
TO ALADDIN AND HIS MOTHER.

LONDON. PUBLISHED BY WILLIAM MILLER, OLD BOND STREET, APRIL 1802.

The Arabian Nights: can you unlock
their celestial metaphors?

2014 October 24

Richard Francis Burton (1821 - 1890) "was one of those Victorians whose energy and achievements make any modern man quail," in the words of the novelist A. S. Byatt in the introduction to Burton's translation of the *Thousand Nights and a Night*, also commonly known as the *Thousand and One Nights*, or the *Arabian Nights* (xv). A partial list of examples ensues, of course:

> He lived like one of his own heroes, travelling in Goa, Equatorial Africa, Brazil, India, and the Middle East. He took part in the Crimean war. He went with J. H. Speke to find the source of the Nile and discovered Lake Tanganyika. He disguised in himself as an Afghan dervish and doctor and went on pilgrimage to the sacred cities of Mecca and Medina -- a journey where unmasking would have cost him his life. He wrote books on swordsmanship and geology. According to Borges he dreamed in seventeen languages and spoke thirty-five -- other sources say forty. xv.

When he died on October 20, 1890, we are told that, "alarmed by the sexually explicit content of her husband's papers, Isabel Burton burned almost all of his notes, diaries, and manuscripts -- an immeasurable loss to history" (vii -- this quotation from the publisher and not from A. S. Byatt's introduction, which begins on page xiii). That could be what happened, or it could be a convenient cover-story -- we will probably never know.

In any case, Burton's translation of the *Nights* was begun in the 1850s and finally published in the 1880s in sixteen volumes. The introduction by A. S. Byatt cited above declares that of all the translations of the *Nights*, "the most accessible complete translation remains Burton's extraordinary translation" along with its "immense apparatus of extraordinary footnotes" (xv). Of the massive work Burton himself said:

This work, laborious as it may appear, has been to me a labour of love, an unfailing source of solace and satisfaction. During my long years of official banishment to the luxuriant and deadly deserts of Western Africa, and to the dull and dreary half-clearings of South America, it proved itself a charm, a talisman against ennui and despondency. Impossible even to open the pages without a vision staring into view [. . .] Arabia, a region so familiar to my mind that even at first sight, it seemed a reminiscence of some by-gone metempsychic life in the distant Past [. . .] air glorious as ether, whose every breath raises men's spirits like sparkling wine [. . .] while the reremouse flitted overhead with his tiny shriek, and the rave of the jackal resounded through deepening glooms, and -- most musical of music -- the palm-trees answered the whispers of the night-breeze with the softest tones of falling water. xxiii - xxiv.

Burton's translation -- and his voluminous endnotes -- are famous for their extremely sexually explicit nature, especially during the period that they first appeared, as a private printing of one thousand copies to subscribers only. Modern readers will find that their content (and perhaps their translation) also appears on the surface to be highly objectionable in terms of being both sexist and racist -- so much so, in fact, that they may prove difficult or even impossible for some to actually read.

And yet, as with other ancient tales, I would argue that the tales which made their way into the *Thousand Nights and a Night* are almost certainly deeply esoteric in nature, and that to read them only on a literal level is as mistaken as reading Herman Melville's *Moby Dick* as a story about whaling (this concept is discussed in my most recent interview on *Truth Warrior* with David Whitehead, beginning at about 0:17:00 and continuing through to 0:24:00, as well as in the essay I wrote for Jacob Karlin's meditation and *Selfless Self-Help* site entitled "Clothing spirit with matter and raising it up again: How metaphor transcends and transforms the material realm").

312

The themes of the *Thousand Nights and a Night* ostensibly center around the differences between men and women, and their different "powers," and this is the approach to these fabulous tales that is most commonly employed today (simply search for them on the internet for a host of examples). In the world of the *Nights*, women appear on the surface to be less powerful in the extremely patriarchal (and violent) society that is depicted, and yet they ultimately prove to be far more powerful.

In fact, the entire tension of the story is established by the deflation experienced by first one royal brother, Shah Zaman, and then his brother, King Shahryar, when their wives "get the better of them," each of their frustrations being relived in turn only when each successively encounters an example even more egregious than his own humiliation (their humiliation is only relieved by the even greater humiliation of another man by his wife). Their humiliation leads to a predictably (if excessively) "male" response, the rule that sets the stage for the "thousand and one nights," an extreme and violent "solution" which is finally subverted and corrected by the wisdom, patience, grace, charm, wit, circularity, and feminine power of Shahrazad (or Sheherezad in some translations), assisted by her sister Dunyazad.

Throughout the tales, the power of women can be destructive and devouring, or it can be constructive and restorative, but it is almost always ultimately far more formidable than that of men, despite the latter's excessive bluster, arbitrary ultimatums, and readiness to try to solve most problems by immediately swinging at them wildly with a scimitar.

While the above theme of the "power" of women versus the "power" of men is undeniably present throughout the *Nights*, I would still argue that to read them on this fairly literal level, or to approach them as a sort of "women's studies" about how women "were treated" in some historical society and how they dealt with and overcame that treatment, is actually a mistake, in that it fails to see the *Nights* as deeply esoteric and as almost certainly metaphorical, not literal. The same can be said for the extremely racist episodes and descriptions in some of the tales: while the racist elements are highly objectionable and regrettable, and one would prefer that some other metaphor had been employed (the same could be said for some of the sexual content as well), it is likely that the real meaning of the tales is on a level other than the literal, and that the fantastical and often bizarre events and episodes which are related were originally intended to highlight aspects of our universal human condition, or were descended from ancient myths whose original intent was to do so (it is possible that the more racist elements came in later, perhaps during medieval times).

And this is the key: if the *Nights* in all their incredible tales and transformations and encounters with fire-beings such as jinns and janns and ifrits are actually describing a vision of the soul in its incarnations, and a vision of the universe as shamanic and holographic in nature, then they are *not* primarily about the division of humanity into men versus women, or this "race" against that one. When a wife is depicted as leaving an almost-ideal husband to chase after rag-bound and filthy and abusive adulterous lovers in illicit affairs, this can be seen as an esoteric depiction of our incarnate condition, in which we can so easily forget our innate (but hidden) spiritual or even divine component and embrace too thoroughly our "animal" or physical nature: a metaphor which applies equally to incarnate men as to incarnate women (see the many similar examples in the scriptures of both the Old Testament and the New Testament, including that of the Prodigal Son, who ends up eating husks among the swine before he remembers his true origin).

In other words, if we read the *Nights* on a literal level, they will almost certainly appear to *divide* humanity, along "racial" or "ethnic" or "gender" lines. They will also be quite disturbing and even revolting to many readers, or at least deeply offensive to their sensibilities -- even degrading to the human condition and destructive of human dignity. However, if we read them on a metaphorical and esoteric level, they can actually be seen as

314

teaching a unifying and an uplifting and even a dignifying message -- because they show how our descent into the material realm (the very words *matter* and *material* being feminine in connotation, related to the Latin word *mater* or "mother") exposes us to death, to "beatings," to a type of enslavement, to oppressions, to exigencies beyond our control, to transformations, and subjugations, and yet opens the door for exaltation and transformation and even to a transformation that benefits others and enables them to be transformed as well (all of which Shahrazad experiences and demonstrates throughout the *Nights*).

See previous posts such as the post from 10/14/2014 for more on this concept of *unifying* rather than *dividing*.

When profound truths put on the garments of metaphor, they descend from the spiritual realm to the material, in order to enable our matter-bound minds to see, through them, that spiritual realm which we have forgotten -- and then these metaphors leap back upwards to the spiritual realms from whence they came, and drag our consciousness along with them. This is what Melville's *Moby Dick* demonstrates,[66] when deep spiritual subjects come down to put on the rough garments of a whaling vessel, and it is what the *Thousand Nights and One Night* demonstrate when profound matters of human incarnation and the nature of our spirit-infused universe are clothed in the often gratuitously violent and sexually explicit situations depicted in those tales.[310]

This motion of "metaphor itself" in descending from the "realms of the ideal" into the physical trappings of the vehicle chosen to house or to clothe the metaphor in familiar material form, for the purpose of elevating our consciousness and pointing us back towards the spiritual and helping us to transcend the physical and material *can be seen to mirror our own experience in this human incarnation*. We descend from the realm of spirit into material and physical vehicles, with the purpose of somehow transforming and transcending and returning with new understanding, and

315

elevating and "dragging along" and reawakening the spiritual which is hidden inside the material world in the process.

This esoteric understanding of the *Nights* is supported by an aspect of the tales that has rarely, if ever, been explored, and that is the fact that -- like the ancient sacred scriptures and mythologies of the human race, they frequently employ clear celestial metaphor, using the exact same system which underlies other myths the world over.

To demonstrate, I will here offer just two of the many hundreds of possible examples. However, at the request of an extremely insightful and astute correspondent who wrote to me about these interpretations, I will give my interpretation of the constellations underlying these two episodes from the *Nights* in a future installment of this blog in a couple of days -- enabling you, gentle reader, to work them out for yourself in the interim!

Feel free to post or message your "celestial interpretations" of these two passages, naming the constellations that you believe correspond to each important character (or object, in the case of the second of the two episodes).

Currently, the best places to post (or message, if you wish to be more private and less public) your interpretations are either Twitter (yes, you can fit your explanation in a single tweet or two -- you can just say "X = this constellation; Y = that one") or Facebook. If neither Twitter nor Facebook work for you, send me a message on one of those two channels and suggest a better place to communicate. I will look forward to reading your submissions, if you wish to post them, and then I will put up my own interpretations (which I have already formulated for myself -- obviously I'm not going to offer examples which I am not already fairly confident contain clear celestial correspondences which people can work out: that wouldn't be very helpful).

To get yourself warmed up, feel free to check out the many examples of star myths and their explanations listed elsewhere in this collection.

316

Here are the two episodes from *The Arabian Nights*, as translated by Richard Francis Burton:

First episode:
the adulterous affair that started the whole story.

Shah Zaman, the younger brother of King Shahryar, is invited to go visit his brother after many years of separation (in which each ruled their own domain with great "equity and fair-dealing," but as Zaman begins to go, he returns for something he forgot. Here is how he begins to describe what took place:

> "Know then, O my brother," rejoined Shah Zaman, "that when thou sentest thy Wazir with the invitation to place myself between thy hands, I made ready and marched out of my city; but presently I minded me having left behind me in the palace a string of jewels intended as a gift to thee. I returned for it alone, and found my wife [. . .]. 9.

Finding his wife with another, he says, Shah Zaman "drew his scimitar and, cutting the two into four pieces with a single blow, left them on the carpet and returned presently to his camp without letting anyone know of what had happened" (5).

Can you determine which celestial inhabitants might correspond to Shah Zaman, his adulterous wife, her adulterous lover, his scimitar, and the string of jewels that he forgot to take with him?

Second episode: the Fisherman and the Jinni.

This is the first story in which a Jinni comes forth out of a lamp. There is a story prior to this one which features a Jinni (and a beautiful and formidable woman, who proceeds to exercise absolute power over both Shah Zaman and his brother King Shahryar), but that one strides up out of the ocean onto the shore, and does not emanate from an ancient lamp. The Tale of the Fisherman and the Jinni is presented as the very first tale Shahrazad tells to King Shahryar on her first night with him, and it is long and involved and contains many "stories within stories within stories," but the first part of the action involves an old fisherman and his wondrous catch. Listen as Shahrazad begins her tale:

> It hath reached me, O auspicious King, that there was a Fisherman well stricken in years who had a wife and three children, and withal was of poor condition. Now it was his custom to cast his net every day four times, and no more. On a day he went forth about noontide to the sea shore, where he laid down his basket; and, tucking up his shirt and plunging into the water, made a cast with his net and waited till it settled to the bottom. Then he gathered the cords together and haled away at it, but found it weighty; and however much he drew it landwards, he could not pull it up; so he carried the ends ashore and drove a stake into the ground and made the net fast to it. Then he stripped and dived into the water all about the net, and left not off working hard until he had brought it up. He rejoiced thereat and, donning his clothes, went to the net, when he found in it a dead jackass which had torn the meshes. 25.

The Fisherman is grieved at this development, but he gets it clear of his net and casts again, but with similar results. After a great deal of effort, he gets the net in a second time: this time we are told "found he in it a large earthern pitcher which was full of sand and mud; and seeing this he was greatly troubled" (26). So he has another go, but only brings up "potsherds and broken glass" (26).

Finally, he goes through the motions one last time, after first "raising his eyes heavenwards" and imploring "O my God! verily Thou wottest that I cast not my net each day save four times; the third is done and as yet Thou hast vouchsafed me nothing. So this time, O my God, deign give me my daily bread" (26). This time, we are told, he pulls up an old jar or lamp of yellow copper, with a seal stopping its mouth with a leaden cap. Removing the seal with great effort, we watch along with the Fisherman in amazement as:

318

presently there came forth from the jar a smoke which spired heavenwards into ether (wherat he again marveled with a mighty marvel), and which trailed along earth's surface till presently, having reached its full height, the thick vapour condensed, and became an Ifrit, huge of bulk, whose crest touched the clouds while his feet were on the ground. His head was as a dome, his hands like pitchforks, his legs long as masts and his mouth big as a cave; his teeth were like large stones, his nostrils ewers, his eyes two lamps and his look was fierce and lowering. Now when the fisherman saw the Ifrit his side muscles quivered, his teeth chattered, his spittle dried up and he became blind about what to do. 27.

Can you identify the <u>net</u>, *the* <u>dead jackass</u>, *the "*<u>earthern pot</u>,*" and the* <u>magic lamp</u>? *If so, you will probably be able to guess at who is likely to play the Fisherman in this tale. How about the smoke which pours from the lamp and spirals upwards? The* <u>Ifrit</u> *is a bit tricky, and could be one of a couple different figures, but you may want to give him a try as well.*

Enjoy!

Star Myths in the Arabian Nights!

2014 October 25

SPOILER ALERT: This blog post will reveal my interpretations of the celestial foundations underlying two episodes from the incredible *Thousand Nights and a Night* (otherwise known as the *1,001 Nights*, or the *Arabian Nights*).

320

These two episodes were introduced in the previous essay entitled "The Arabian Nights: can you unlock their celestial metaphors?" If you want to go back and try to unlock them for yourself before you read the following explanation, just turn back to page 310 before reading any further, and come back after you're done!

Here we go . . .

In the first episode, which really launches the entire dynamic of the Nights and sets up the horrific situation in which a king (King Shahryar) decides to enjoy a new virgin bride each night and then slay her in the morning, and the king's brother Shah Zaman is invited to visit -- but as Zaman leaves his palace, he remembers that he has left behind him a string of jewels he intended to give to his brother Shahryar.

He returns home for the string of jewels, only to find his wife on the bed in the arms of an adulterous lover. Drawing his scimitar, he immediately cuts them both in half, leaving them in four pieces. He then proceeds to fall into depression, refusing to eat and languishing in self-pity . . . and the story proceeds from there.

The *Arabian Nights* can be graphic, violent, and even horrifying -- but I believe that, just like other remnants of the ancient wisdom bequeathed to humanity, the literal stories are only the vessels used to contain ineffable spiritual truth, and that to focus only on the literal action is to "miss all that heavenly glory" towards which they are pointing us.

While they are certainly fascinating and entertaining and moving and memorable as literal stories, the *Nights* also function as profound spiritual metaphors regarding the nature of our human condition as incarnate spiritual beings, and regarding the nature of this apparently physical universe, which itself is actually infused with and interpenetrated by an unseen world.

This metaphorical spiritual message can also be found in the sacred texts and mythologies of nearly every other culture on earth, and which actually unites the world's sacred traditions, as discussed in numerous previous blog posts and in my most-recent book, *The Undying Stars.*

One of the biggest indicators that the *Thousand Nights and a Night* should be interpreted esoterically is the fact that, like the sacred mythologies found around the world, they are built upon the same common system of celestial metaphor which can be seen operating in "star myths" of ancient Egypt, of ancient Greece, of Japan, or North America, or northern Europe, or Africa, or Australia, or China, or the surviving texts of the Maya, and even in the scriptures of what are commonly called the Old and New Testaments of the Bible. For hundreds of examples establishing this undeniable fact, see the books and videos I have published since this post was first written in 2014.

The story of Shah Zaman returning for a string of jewels and catching his wife *in flagrante* can clearly be seen to correspond to a specific set of familiar constellations in the night sky. The "string of jewels" is an important clue, and one with which readers will be familiar if they remember the explanation I offered of the irresistible necklace of Freya from the Norse myths.[282] There, we saw that this necklace corresponds to the *Corona Borealis*, or "Northern Crown," a beautiful feature of the northern sky and one which appears over and over in the world's mythology, playing many different roles.

From the Northern Crown, we can fairly easily identify the rest of the constellations in the story of Shah Zaman and the two illicit lovers. Below is a screen-shot of the region of the sky containing the Northern Crown, taken from the excellent application Stellarium.org (free and open-source and available on the web).

Note that in the default setting on most planetarium apps, including Stellarium, the constellation outlines are provided, but they are *not* the constellation outlines proposed by H. A. Rey in his outstanding books on the stars, and therefore they are not very helpful. The following image, with outlines showing my

322

interpretation of the incident with Shah Zaman and his wife, provides my own added full-color outlines which follow the H. A. Rey system.

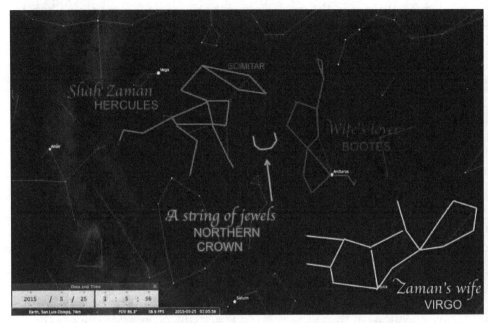

In the above Stellarium image, the circlet of stars which make up the Northern Crown are seen as a "U"-shaped constellation near the center of the screen-shot. This is the "string of jewels" which Shah Zaman forgets. To the right of the Northern Crown is the important constellation of Boötes, the Herdsman, and below him is the even more important constellation of Virgo the Virgin, with the bright star Spica on her hip (labeled, near the bottom of the image and on the right-hand-side of the screen as we look at it on the page).

To the left of the Northern Crown is the constellation Hercules, a mighty warrior and a hulking gigantic figure in the night sky. Hercules, of course, is brandishing his favorite weapon, his Club -- but in the story of Shah Zaman and his unfaithful wife, the Club of Hercules becomes a scimitar!

In the lower-right of the image, we see Shah Zaman's wife, played by the zodiac constellation of Virgo, and outlined in yellow in my diagram. Just above her is the adulterous lover, outlined in red and played by Boötes the Herdsman, who often plays the role of the consort of Virgo in various myths around the globe. To the left of Boötes is the pesky string of jewels, forgotten by Shah Zaman when he headed out to visit his brother, and they are outlined in a kind of lavender color. Finally, to the left of these we see the constellation Hercules, representative of Shah Zaman charging in upon the surprised couple, raising his dreadful scimitar and preparing to cut them down. He is outlined in green.

The number of celestial clues that have been worked into the story as related in the *Arabian Nights* really leaves little doubt that the story corresponds directly to the heavenly drama, as do so many other myths and sacred stories from humanity's ancient past.

[Thank you, by the way, to those of you who sent me your interpretations, even if you did not want to share those publicly! I hope you enjoyed the exercise of exploring the possible celestial connections in the *Nights*! No one actually proposed either of the interpretations shown here for the two tales, but that doesn't mean nobody out there came up with some version of these interpretations -- I'm sure most people who gave it a try just decided to participate "in private." Also, my interpretations are not "the answer," of course -- these are open to debate and discussion. One reader sent in a thought which had not occurred to me, which was that this "cutting in half" of the two lovers might have to do with an equinox, which is a very good observation! I would argue that if it *does* have to do with an equinox, then it would most likely be the fall equinox, and not the spring equinox, since Virgo is associated with the fall equinox in the northern hemisphere – but if it is an equinox, then we can further speculate that perhaps Shah Zaman represents one half of the year, probably the "lower half," and his brother King Sharyar represents the other, and probably the "upper half." Great observation and thanks for sharing it!]

Turning now to the story of the Fisherman and the Jinni, we encounter an absolutely fabulous tale and the one with which the beautiful, courageous, and intelligent Shahrazad opens her thousand-and-one nights of storytelling, with which she will save her life -- and, by extension, the lives of all the other young unmarried women of the kingdom including her own sister, and with which she will ultimately save King Shahryar from his own madness and self-destructive jealousy and pride.

The Fisherman and the Jinni is a tale-within-a-tale-within-a-tale and it contains several more "nested" and interwoven tales within it, but it opens with the account of a poor old Fisherman who casts his net into the waters each day, and one day pulls up a series of strange catches beginning with a dead jackass, followed by an earthen jar (Richard Francis Burton calls it an "earthern pitcher"), followed by some potsherds and broken glass, and finally by a lamp containing a genie (or Jinni -- and one who in this story is identified as an Ifrit, and who pours from the lamp in a towering column of smoke spiraling up to the heavens).

What could be the celestial counterparts to this fantastic opening to the series of stories contained in the tale of the Fisherman and the Jinni?

Well, there are a number of clues in the story to help us, not least of them a lamp next to a column of smoke -- which almost certainly corresponds to the "Teapot" portion of the constellation Sagittarius, which is located right next to the rising "smoke" of the Milky Way galaxy, as discussed in this previous blog post from 07/06/2012) regarding Revelation chapter 9 (which also refers to the Milky Way that rises between Sagittarius and Scorpio as a rising smoke).

Another powerful clue is the Fisherman's net itself, which may suggest to the minds of readers familiar with the recent discussion of the celestial foundations of the story of Shem, Ham and Japheth (the sons of Noah) the Great Square of Pegasus, which appears in that story as a sheet carried backwards over the shoulders of Shem and Japheth. That distinctive Square in the sky could also be the net of the Fisherman, which keeps bringing up everything except fish from the briny deep.

The connection to the Great Square in the story found in the *Arabian Nights* is strengthened by the story's repetition of the fact that the Fisherman only casts his net into the waters *four* times per day, and never more than that: if we are looking for a celestial counterpart to the net, the repetition of the number four is certain to suggest to us the mighty celestial Square, which after all is a figure containing four corners and the constellation that might come to mind most readily in connection with that particular number.

From there, we can readily identify the other details of the Fisherman's tale, and there are quite enough of them to make the correspondence more than certain. Below is a screen-shot showing the region of the sky which corresponds to the start of the tale of the Fisherman and the Jinni:

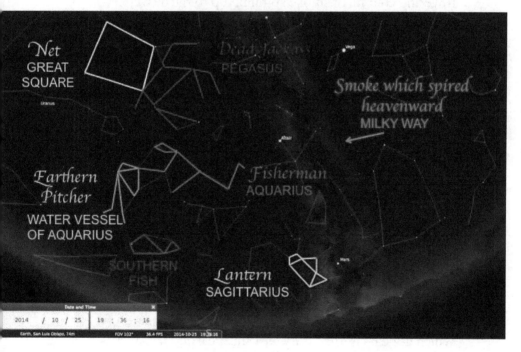

The beautiful towering "column of smoke" of the Milky Way galaxy can clearly be seen rising up out of the southern horizon, and just to the left of it as we look at the image can be seen the "teapot" portion of the zodiac constellation of Sagittarius (just to the left of the planet Mars, which is labeled and which just happens to be located in the center-line of the Milky Way in this particular screenshot for this particular date and time and year -- Mars is not always located there, by any means).

Beginning from the top-left of the sky, we see the Fisherman's Net, played by the Great Square of Pegasus and outlined in white. Just to the right of the square we see Pegasus himself, that celestial winged horse (the Square represents his wings), but in this particular story he is playing a decidedly more ignominious role as the Dead Jackass which the Fisherman first hauls up with his Net. Pegasus in the above image looks about "right-side up," but at other points during his journey across the sky (particularly when the Great Square is just rising up in the east, for viewers in the northern hemisphere), he is kind of positioned "upside-down," and this fact no doubt accounts for the depiction in this story of the outline of Pegasus as a dead donkey, with his four feet pointing up in the air.

Just below Pegasus we see the constellation Aquarius, outlined in green. I believe that Aquarius plays the role of the Fisherman in this particular story, primarily because Aquarius is located in close proximity to the Net, and also because directly below Aquarius there is a constellation known as *Piscus Austrinus*, or the "Southern Fish." This constellation is rather faint, but contains the brilliant star Fomalhaut which is very easy to spot in the night sky below Aquarius (you can see it in the tip of the nose of the Fish even in the above diagram).

The second thing that the Fisherman dredges up with his Net in the tale of the Fisherman and the Jinni is an earthen pitcher (Burton calls it an "earthern pitcher," and it is thus labeled in the diagram above). This object clearly corresponds to the jar or water-vessel of Aquarius, which is really part of the constellation

Aquarius but which I have outlined in light blue in the image above, so that you can see it more easily.

The third haul of the Net in the tale brings up "potsherds and glass," which really could be anything and which I am not exactly certain about identifying definitively with any particular stars or groups of stars. My most-likely candidate for these potsherds and glass would probably be the glittering trails of stars located at the bottom of each of the two "streams" of water you see depicted coming out of the water-pitcher of Aquarius. These are very distinctive and easy to spot in the actual night sky, although they don't show up very well in the screen-shots above.

Below, I have "zoomed-in" on Aquarius and his water-vessel in order to try to show these little glittering trails at the bottom of each (imagined) stream of water pouring out of the vessel. These little curves of stars are quite beautiful, and they actually "create" the stream of water that we imagine coming out of the pitcher of Aquarius, since the two "streams" themselves have no stars in them: the streams are entirely imaginary, and are created when we "connect" using our mind's eye the pitcher with these two little "curved lines" of stars.

Here is a closeup of Aquarius and his jar, with the two lines of water coming out of the jar ending in the two glittering curves of stars (colors inverted so that you can see them more easily):

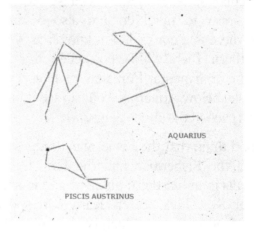

AQUARIUS

PISCIS AUSTRINUS

Finally, we now come to the "fourth catch" of the day -- the one which will ultimately change the Fisherman's fortunes forever. He utters a prayer before sending his Net one more time into the deep, noting that so far he has brought back nothing which he or his wife can eat, and asking that he might please be granted his daily bread.

When he brings back the Net this time, there is a copper-colored lamp, its mouth sealed with a leaden seal upon which is fixed "the stamp of the seal-ring of our Lord Sulayman son of David," whom we would commonly refer to as King Solomon (27). As we might expect, these being the *Arabian Nights*, when the Fisherman removes the seal, what should pour forth from the lamp but a spiring column of smoke reaching to the heavens, which ultimately resolves into a powerful Jinni, who promptly informs the poor Fisherman that he must now kill him within the hour, although he will allow the Fisherman to choose the manner of his death.

And the story proceeds from there -- it is a remarkable tale, and one with which many modern readers may not be familiar. Be sure to take the time to check it out (there are various places on the web to read translations of the *Nights*, including Burton's translation in its entirety, but of course it is my fixed opinion that *The Arabian Nights* belongs on everyone's bookshelf in its physical paper form, if it is at all possible for you to obtain it).

In any event, the constellation that plays the part of the genie's magic lantern in this tale is fairly easy to spot, and it is the distinctive outline of the brightest stars in the zodiac constellation of Sagittarius, shown in the full-story star-chart diagram above as an outline of yellow lines and labeled "Lantern." The fact that its "spout" points right into the glowing column of the rising "smoke" of the Milky Way galaxy makes this identification of the celestial counterpart to the story almost certain.

In fact, the wealth of detail in the story which corresponds directly to the constellations surrounding the "Fisherman" of

Aquarius makes the above interpretation a very strong hypothesis, in my opinion. The fact that literally hundreds of other myths and sacred stories from around the world are built upon this very type of celestial metaphor makes the celestial correspondence that I am here proposing for the *Thousand and One Nights* even more likely.

In fact, it should be pointed out that I did not even know these correspondences existed when I revisited the *Arabian Nights* recently (although I strongly suspected the *Nights* would be full of them).

The fact that familiarity with the system of celestial metaphor enables us to discover the same metaphorical system in operation in other myths or stories not previously examined (such as just demonstrated with the *Arabian Nights* -- and many more examples from the *Nights* could be offered) argues very strongly that the existence of this ancient and worldwide system of celestial metaphor is no mere figment of the imagination. The number of correspondences to the details of the story offered in the two explanations above shows that these celestial metaphors were actually part of the tales: they are not "subjective interpretation," because the details are actually present in the constellations of the night sky.

The ramifications of this fact are profound, and have the potential to change our understanding of sacred literature, of the connections between all the various branches of the human family, and of the very history of mankind. Where did this nearly universal system come from, and how does it turn up over and over again in the treasured stories and myths of humanity around the globe?

Perhaps if, like the Fisherman in the story, we persevere in putting our Net out into the deep waters -- and if we accompany our efforts with a heartfelt prayer -- we will one day receive an answer.

Winter solstice, 2014: the Stable and the Manger

2014 December 20

The earth will pass through the point of December solstice this year on December 21st at 2303 Greenwich time (now referred to as UTC), which is 1503 Pacific time and 1803 Eastern time for those in North America (numerous sites on the web can help you determine the time at your location if the references above aren't enough to zero-in on it).

As has been remarked upon in many other discussions, the word "solstice" descends from a combination of the Latin noun *sol* ("the sun") with a form of the Latin verb *sistere* ("to stand"), and thus means "sun-standing," as in "standing *still*." We find another example containing a derivation of *sistere* in the word "interstitial," which describes the "boundary space" in between two larger spaces -- the border-zone, the threshold region, the "standing-in-the-middle" place.

331

When the earth is hurtling towards the December solstice, it causes the sun's apparent path to observers on earth to move further and further south each day. As a consequence, ever since we passed the June solstice, the sun has been rising on the eastern horizon at a point further and further south, and arcing across the sky on a path that is further and further towards the southern horizon, and then setting at a point along the western horizon at a point that is further and further south each day.

At the solstice, the sun seems to "stand still" before it turns back around and reverses the process. The reason for this standstill is discussed in this previous post involving the metaphor of a mighty sailing ship with the bowsprit acting as the north pole.[91]

For those observers in the northern hemisphere, where the sun's steady progress towards the south has caused its rays to be less and less direct, and the warming effects less and less effective, plunging the world deeper and deeper into winter, as the days grow shorter and shorter and the nights longer and longer, the anticipation of that turnaround is tremendous. It seems as if life itself hangs in the balance, and the time in which the sun finally grinds to a halt in its southward progression and stands still before finally turning back towards the north feels like a breathless pause in which the entire world freezes in place to see if the life-giving orb will actually "make the turn."

It is this moment, when all the world collectively "holds its breath" (figuratively speaking), that Alvin Boyd Kuhn says is commemorated in the concept of the "Silent Night," the stillness that is celebrated in the Christmas tradition, with carols which proclaim: "O little town of Bethlehem, how *still* we see thee lie," and (in "*It came upon a midnight clear*"), "The world in *solemn stillness* lay to hear the angels sing" (13).

In a lecture entitled *The Stable and the Manger*, delivered in 1936, Alvin Boyd Kuhn elucidates the connections between the elements of the Christmas story and the significance of the winter solstice as a spiritual allegory, in which spirit which has been plunged deep into matter begins its "upward turn," but prior to doing so there is a pregnant pause in which all is in perfect stillness, and the tension between the two creates a moment of equipoise in which "all is calm."

Outlining the framework of the metaphor, he explains:

> The sun in its apparent passage from the high glory of summer to its enfeebled power in the solstice of winter exactly symbolizes, because it repeats, the experience of the soul in its alternating swing from the heights of spiritual purity in disembodiment -- in summer -- to the depths of diminished shining in the lowest arc of its immersion in a body, its night, its winter. 11.

And, tying this concept to the Christmas story, he explains that it is this commingling of the spark of spirit plunged into the body of matter which gives birth to the "third principle," the higher self, the Christ within. Kuhn says:

> Suffice it to say for the moment that obviously if a higher and a lower force are to meet and unite at the point midway between their status of being, they must so meet as the result of the ascent of the one and the descent of the other. Nature could not well arrange such a meeting in any other way. That nature has so arranged the matter is one of the bits of knowledge furnished us by the ancient wisdom. When God or Life at the beginning of each period of its activity bifurcates into the polarization of spirit and matter, the two forms of being move toward each other, meet in the middle ground, so to say, effect their conjunction and interplay, and at the end of the cycle retire into latency again. For the earth evolution that point of middle distance between the two is the body and life of man. here is where the "marriage" takes place and the Son, the Christ, is born. And when the two forces meet at this point, they counteract each other's energies and bring each other to a standstill. Spirit descending came to a stop in the arms of matter, for the inertia of matter stilled the vibrations of spirit. 9.

333

Thus, he notes, it is highly appropriate that the ancient scriptures describe the birth of the Christ as taking place in a *stable* -- the word itself means "steady" and "standing upon a base," appropriate for this story that takes place at the very base of the year, the bottom of the zodiac wheel shown below, and appropriate to the point where as Kuhn says "spirit and matter, soul and body, are 'stabilized' in relation to each other" (12).

DAYS LONGER THAN NIGHTS:
Heaven, Promised Land, Greece, etc.

Horizontal Column:
The Djed cast down

Vertical Column:
The Djed raised up

NIGHTS LONGER THAN DAYS:
Hell, Egypt, Troy, etc.

He further points out that the stable is the place "where animals come to stand for the night," and a place where the animal nature connects with the benevolent care of the higher human intellect (which presumably designed and constructed the stable, to shelter and protect the animal), and which thus may symbolize this point where "the brute kingdom is elevated by the grace of mankind, as mankind in turn is exalted by the grace of the gods" (12).

But that is not all -- for, as Kuhn goes on to explain, the Christ-child who is born at this point of tension between spirit and matter, where spirit has descended to its deepest place in the cycle, is then *laid in a manger* -- the place where the animals are fed! The animal nature must be fed and nourished and ultimately elevated by their participation with the Christ nature (17 - 19).

Astronomically, we have seen that the sign of Virgo, standing as she does at the autumn equinox where days begin to be shorter than nights, presides over the plunge of the spirit from the higher realm into the material realm (see the image of Virgo, wearing the crown of the "Queen of Heaven," located just above the horizontal line before the "crossing point" indicated by the red "X" on the right-hand side of the zodiac circle as we face it in the diagram above). Virgo appears in the ancient Egyptian myth-cycle as Isis, holding the divine Horus on her lap in exactly the same way that she appears in the New Testament accounts as the Virgin Mary, holding the divine Jesus:

The identification of Isis, and Mary, with Virgo is evident from an examination of the outline of the constellation itself, but also from the fact that Virgo is associated with wheat and with grain, and that in fact the constellation is often depicted as holding a sheaf of wheat, and in fact the name of her brightest star, Spica, comes from a Latin reference to an "ear of grain" (and in Arabic this star is called *Sumbalet* which also means "an ear of wheat"). The fact that she lays her divine son in the *manger*, where the *grains* are fed to the animals, should cement this identification between the heavenly queen and the Virgin in the story found in the gospel account. See also the discussion of Mary and Virgo, and the visit of the Magi, in my video entitled "Star Myths: a thousand times more precious . . . ".

There are many more astonishing connections to be found in the lecture of Alvin Boyd Kuhn, and in consideration of the spiritual symbology present in the point of the winter solstice with all its implications. The reader is encouraged to consult the full text of that lecture, and what better time to do so than this portentous point on the year, when all the world stands still at the December solstice?

But, perhaps the most important part of Kuhn's entire lecture is found before he actually begins to elucidate the details of the solstice-scene at all, when he explains that these exquisite metaphors are meant to convey a drama of which the central player is *each and every human being*. He asserts:

> Bible stories are in no sense a record of what happened to a man or a people as historical occurrence. As such they would have little significance for mankind. They would be the experience of people not ourselves, and would not bear a relation to our life. But they are a record, under pictorial forms, of that which is ever occurring as a reality of the present in all lives. They mean nothing as outward events; but they mean everything as picturizations of that which is our living experience at all times. The actors are not old kings, priests and warriors; the one actor in every portrayal,

in every scene, is the human soul. The Bible is the drama of our history here and now; and it is not apprehended in its full force and applicability until every reader discerns himself [or herself] to be the central figure in it! The Bible is about the mystery of human life. Instead of relating to the incidents of a remote epoch in temporal history, it deals with the reality of the living present in the life of every soul on earth. 4.

The Bodhi Tree

2015 May 09

The Buddha is traditionally said to have attained enlightenment while sitting and meditating underneath the bo tree, or bodhi tree.

The term *bodhi* is one word for enlightenment, and does not mean a specific type of tree: however, the bodhi tree itself is traditionally understood to have been a *ficus religiosa* or "sacred fig," also known as a *pipal* (in Hindi) and an *ashwanth* (in Sanskrit). Buddhist monasteries in parts of the world in which this tree can prosper will almost invariably have one as one of their most sacred treasures.

Additionally, in order to be designated a bodhi tree today, a tree is supposed to be descended from that original tree by direct

338

propagation from it or one of its descendants. There are several such bodhi trees said to be descended in a direct line from the original bodhi tree under which the Buddha achieved enlightenment; one of those is pictured above.

The sacred fig or ashwanth has a distinctive heart-shaped leaf, clearly visible in the statue of the Buddha under the tree shown below (from the first century AD):

The shape of this leaf is so deeply associated with the achievement of this blessed state, and so imbued with meaning in Buddhist culture that this shape appears in stylized form even with no additional "explanation" necessary:

Now, what I find extraordinarily interesting and significant is the fact that the ashwanth or sacred fig, the very tree associated with the bodhi tree under which the Buddha achieves enlightenment, is associated in the ancient Vedic tradition of India with a specific celestial pair of stars, designated together by the name Pushya.

You can see this ancient association between certain important *Nakshatras* (stars) and specific tree species attested to in various texts, for example in the scholarly publication of the *Proceedings of the Seventh International Congress of Ethnobiology* for 2002, and particularly on page 90 of that collection.

Now, you might be asking yourself which specific star or stars are associated with the Nakshatra known as Pushya! (*Self...*)

Astonishingly enough, Pushya is associated with two stars: the Northern and Southern Colts, *Asellus Borealis* and *Asellus Australis*, which flank the beautiful Beehive Cluster in the zodiac constellation of Cancer, and which we have already seen to have been associated with the Manger in which the Christ is born and the Triumphal Entry into Jerusalem in the New Testament scriptures (see videos I have posted on these subjects on the web).

We have also seen that the zodiac sign of Cancer the Crab is located at the very "top of the year" on the zodiac wheel, beginning immediately following the point of summer solstice, and that it is thus associated with the upraised Djed column and all that that powerful symbol was intended to convey, including the "raising up" of the invisible and divine spirit within the individual and within all of the material-spiritual cosmos through which we sojourn in this incarnate life.

Due to this positioning at the "top of the cycle" which the great zodiac wheel symbolizes in its entirety, the upraised arms of the Crab (visible in the constellation itself) were associated in ancient symbolic art and in ancient myth with the upraised arms of the sacred Scarab, with the upraised arms of the ancient Egyptian god of the air (Shu), with the upraised arms of Moses when signaling victory, and with the upraised arms depicted on the sacred Ankh above the vertical Djed column, such as in one famous image from the Book of the Going Forth by Day (also more commonly known as the Egyptian Book of the Dead, or in previous centuries sometimes referred to simply as *the Ritual*) found in the Papyrus of Ani.

Now, the association of the bodhi tree of the Buddha with the stars of the zodiac sign of Cancer the Crab thus becomes incredibly important, and powerfully resonant with all the other manifestations of this same concept in the ancient wisdom of the world -- the concept which I usually refer to as the "raising of the Djed" with all of its myriad layers of significance.

This association means that, in addition to all else that this "vertical element" in the great cross of the year represents (all that is "vertical" or spirit-elevating in our individual journey and all that brings forth the invisible spirit world that infuses and animates everything in the universe around us), it is also directly related to the concept of enlightenment, of transcendence of the "cast down" condition we experience when we enter into incarnate form and of profound connection with the infinite.

The bodhi tree can thus also be seen to have connections to the World Tree which Odin ascends and upon which he must hang until he is suddenly granted a vision into the invisible realm of the infinite, and to the tree which the shaman ascends literally in cultures around the world as part of the ecstatic journey.

Ultimately, this is a journey undertaken not just by Odin or the Buddha but in fact by every single human soul. I believe (and have quoted Alvin Boyd Kuhn[33] on this specific point several times in the past) no ancient myth or cycle "is apprehended in its full force and applicability until every reader discerns himself or herself to be the central figure in it!"

One need not journey to a specific location where an external Buddha is said to have achieved his enlightenment, nor visit a specific tree reputed to be descended from the very tree under which he sat when he achieved this union with the infinite (although there is nothing wrong with doing so, and it would indeed be a beautiful experience to be in the presence of one of the sacred ficus trees revered and lovingly tended by so many generations of fellow-journeyers through this vale of tears). The bodhi tree, and enlightenment, are in fact inside us at all times (see the tremendously helpful perspective shed upon this concept by Peter Kingsley, discussed *in his superlative book, In the Dark Places of Wisdom*).

We can each sit under that very tree at any time, no matter where in the universe we happen to be.

Namaste.

DAYS LONGER THAN NIGHTS:
Heaven, Promised Land, Greece, etc.

NIGHTS LONGER THAN DAYS:
Hell, Egypt, Troy, etc.

(*Note the two "small celestials" to either side of the Buddha, each of which I have indicated by a red arrow. I believe the Buddha and the bodhi tree in this image clearly relate to the "vertical line" running up from the winter solstice through the summer solstice, while the two flanking figures represent the two equinoxes and the horizontal line between them: the line of being "cast down" into incarnation, which the Buddha and the enlightenment under the fig tree overcome with the "raising back up" of the Djed. In this interpretation, the two flanking figures thus play the same role that Isis and Nephthys play in the Papyrus of Ani image shown on page 127, while the Buddha and the Tree play the same role as the Djed column and the Ankh with upraised arms in that Papyrus of Ani image. This role is also played by Cautes and Cautopates in the Mithraic symbology discussed here*).*

343

The sacred fig tree, continued: Jonah and the gourd

Jonna am erften capitel und ſie namen Jonna und erwurffen in ins meer da ſtund das meer ſtil von ſiunem wüten und die leute forchten den herren ſer und thoten dem herren opffer und gelübde aber der herre verſch afft ein groſſen viſch Jona zu verſchlingen und Jonka war im leibe des fiſches drey tag und drey necht

2015 May 12

The previous post on The Bodhi Tree[338] examined the very strong evidence that the imagery of the sacred fig tree under which the Buddha is described as attaining enlightenment has powerful points of resonance with the "vertical Djed" symbology found throughout the mythology of the world, and associated with the invisible, divine, spirit-component in human beings and indeed in all the universe.

This "vertical component" symbology can be shown to be directly related to the "vertical component" of the great cross of the year which runs from the winter solstice (at the "bottom of the year") straight up to the summer solstice (at the very "summit" of the year), in contrast to the "horizontal component" that connects the two points of equinox and which represent the "crossing points" between the worlds of spirit and matter. In contrast to the vertical spirit-component of this great cross, the horizontal component almost always pictures the physical, animal, material nature into which we are "cast down" when we incarnate in this mortal life, during which time we are "crossed" in the human

344

condition of being simultaneously spirit and matter, divine and animal, vertical and horizontal.

Hence, the vertical-component symbology of the bodhi tree under which the Buddha achieves enlightenment can be shown to be related to the reconnection with the divine and the transcendence of the dual and conflicted condition in which we find ourselves: a spiritual transcendence which can only be achieved by actually entering into the lower or material realm (in much the same way that plants and trees which grow up towards the heavens must first begin as seeds planted in the "lower realm" of the earthy soil, as Alvin Boyd Kuhn frequently explains in his writings on the subject).

Readers who are familiar by now with the thesis that a common system of celestial allegory can be shown to run through virtually all of the world's ancient myth and sacred tradition may have already begun to question whether this sacred fig tree under which the Buddha achieves the height of divine consciousness has any echoes in other sacred traditions around the world -- and indeed we would probably be very surprised if a symbol of such central importance did not have echoes in other world mythology.

Students of classical literature, and especially those who love the Odyssey of Homer, might immediately think of the fig tree which saves Odysseus from certain destruction between the whirlpool of Charybdis and the ravenous snaking heads of the monster Scylla, in the Odyssey's Book 12 (particularly lines 464 - 478). This fig tree is almost certainly connected to the fig tree of the Buddha – because I believe that in addition to being associated with the vertical "Djed column" which runs through the great circle of the year from the lowest point at winter solstice up to the highest point at summer solstice, the "fig tree" of sacred tradition can be shown to be associated with a very prominent feature of the starry heavens, the same feature that runs between Scylla and Charybdis, to which Odysseus is described as clinging to "like a bat" in order to escape being sucked down into the vortex.

Students of the Hebrew Scriptures may have read the previous post about the Buddha sitting beneath the sacred fig of the bodhi tree[338] and been reminded of the numerous passages in which the promise that "every man should dwell safely . . . under his vine and under his fig tree" is given as a formula that describes the golden age under King Solomon in 1 Kings 4:25 and which is referenced in many other passages in the books of the prophets, including the scrolls of Isaiah and Micah and Zechariah.

Students of the New Testament scriptures may have considered the discussion of the Buddha underneath the bo tree and been suddenly reminded of the passage found only in the gospel according to John, in which Jesus calls Nathanael and tells Nathanael that he saw him "when thou wast under the fig tree," before Philip had told Nathanael to come and see Jesus (John 1:46 - 51).

In other words, fig trees feature prominently in myths and sacred stories around the world! There are many more like these, including from sacred stories in the Americas, some of which are examined in *Hamlet's Mill* (1969). Many readers will also have thought immediately of Adam and Eve, whose story certainly

346

involves a central tree, and who are specifically described as making coverings for themselves out of fig leaves in Genesis 3:7.

What celestial feature might be playing the role of the fig tree in all of these celestial allegories?

Perhaps the most revealing passage which helps to decode this vitally important symbol, and one which was the first one that I myself thought of when reflecting on the image of the bodhi tree, is the story in the book of Jonah, which describes Jonah as taking shelter beneath a friendly *kikajon* or vine, translated as a "gourd" in the 1611 English translation.

There, in the fourth chapter of Jonah, after Jonah has been persuaded (by a stint in the belly of the fish) to preach to the Ninevites (whom he begrudged God's grace and did not want to see spared), we read:

> 5 So Jonah went out of the city, and sat on the east side of the city, and there made him a booth, and sat under it in the shadow, till he might see what would become of the city.
> 6 And the LORD God prepared a gourd, and made it to come up over Jonah, that it might be a shadow over his head, to deliver him from his grief. So Jonah was exceedingly glad of the gourd.
> 7 But God prepared a worm when the morning rose the next day, and it smote the gourd that it withered.
> 8 And it came to pass, when the sun did arise, that God prepared a vehement east wind; and the sun beat upon the head of Jonah, that he fainted, and wished in himself to die, and said, It is better for me to die than to live.
> 9 And God said to Jonah, Doest thou well to be angry for the gourd? And he said, I do well to be angry, even unto death.
> 10 Then said the LORD, Thou has had pity on the gourd, for the which thou has not laboured, neither maddest it grow; which came up in a night, and perished in a night:
> 11 And should I not spare Nineveh, that great city, wherein

are more than sixscore thousand persons that cannot discern between their right hand and their left hand; and also much cattle?

And on that note the book of Jonah ends.

There are certainly deep subjects being treated here in these passages, but it also seems that Jonah sitting under his gourd has some points of resonance with the Buddha sitting under the sacred bo tree, even though the vine that shelters Jonah is not specifically described as a fig (although other passages in the Old and New Testaments specifically indicate a fig and characters who sit underneath one, as we have already seen).

As with so many other sacred myths around the world, and so many other passages based on celestial allegory in the passages of the scriptures of the Old and New Testaments, enough "clues" have been included in the passage above for us to determine with some confidence just which celestial figures this ancient sacred story brings down to earth and clothes in "terrestrial form," so to speak.

Perhaps the feature of this story that does the most to unlock its celestial correlatives is the figure of "the worm" in verse seven, which is depicted as gnawing at or "smiting" the sheltering vine and causing its demise. If you are familiar with the night sky, you might immediately recognize this "worm" at the base of a glorious vertical tree or vine in the heavens as the sinuous constellation Scorpio, one of the most beautiful constellations in the heavens and one that is situated right at the very "base" of the thickest and brightest part of the shining band of the Milky Way galaxy, as it rises out of the southern horizon during the summer months (for observers in the northern hemisphere).

Below is my interpretation of the celestial figures depicted in the events of Jonah chapter 4, beginning with the "worm" of Scorpio, and working around to the rest of the events depicted in the chapter:

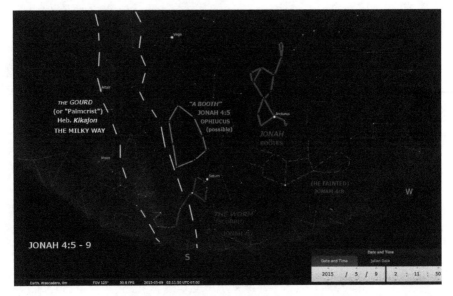

THE GOURD
(or "Palmcrist")
Heb. *Kikajon*
THE MILKY WAY

"A BOOTH"
JONAH 4:5
OPHIUCUS
(possible)

JONAH
BOÖTES

(HE FAINTED)
JONAH 4:8

THE WORM
SCORPIO

JONAH 4:7

JONAH 4:5 - 9

This is a modified Stellarium screen-shot of the night sky as it looks to an observer at a latitude of about 35 north, looking towards the southern horizon (almost due south), such that east will be to the left and west to the right. There, stretching upwards like a mighty tree, is the shining "trunk" of the Milky Way galaxy, and directly at its base or its "root" we can see the dreaded worm, in the zodiac constellation of Scorpio.

Just above Scorpio is a constellation we have not previously discussed on this blog, and we won't really discuss it at length in this post either, except to remark that its outline may well be the explanation for the line in Jonah 4:5 cited above in which we see that "Jonah made him a booth," in which to get a little shade as he sat looking towards Ninevah. The outline of Ophiucus is indeed somewhat suggestive of a "booth" or a narrow peaked tent, and although the interpretation of Jonah 4 does not stand or fall on the identification of Jonah's "booth" with the outline of Ophiucus, this correspondence appears to be a strong possibility.

Just outside the "booth" (if that's indeed what it is), we see Jonah himself, sitting with his back to the vine. It is almost certain that the constellation of Boötes the Herdsman is playing the role of the seated (and sulking) prophet Jonah in this chapter, and you can see that the constellation Boötes itself does indeed have a seated posture. In fact, the same seated posture can also be envisioned as being a posture of kneeling, or of sitting "cross-legged" or even in a "lotus position," if we envision a horizontal line connecting the two lowest points on the constellation as shown above.

We have already seen strong evidence that the constellation Boötes plays the role of the kneeling sage Bodhidharma or Da Mo, who knelt against a wall for nine years without moving (in some versions of the story, without even blinking), as discussed in a previous post entitled "Bodhidharma, Shen Guang, and the Shaolin Temple" (not included in this collection but available online and also discussed at length in *Star Myths of the World, Volume One*).

I believe it is very likely that the seated prophet Jonah, the kneeling sage Da Mo, and the meditating figure of the Buddha underneath the bodhi tree, are all manifestations of one and the same celestial figure in the sky, the constellation Boötes beside the glorious vertical column of the Milky Way.

This identification, at least in the case of Jonah, is strengthened by the events described in verse 8, in which the worm has destroyed the gourd, and the sun comes up and beats upon the unprotected head of Jonah, who then faints. While the constellation Virgo located below Boötes figures in numerous Star Myths around the world as the wife or lover of the figure played by Boötes, such as in the story of Adam and Eve in which Boötes is almost certainly Adam and Virgo is almost certainly Eve, in this particular passage it seems quite likely that the figure of Virgo stretched out below Boötes represents Jonah having fainted from the sun beating down upon his unprotected head (and indeed Boötes does have a prominent and rather bulbous head, based upon the outline of the stars themselves in the constellation). The many places in Jonah chapter 4 in which Jonah says he might as well die or he is angry "unto death" would seem to add support to this identification in this particular part of

350

the Jonah story.

Further confirmation that the fig tree of the world's sacred myths is indeed identified with this portion of the Milky Way can be obtained by considering again the story of Odysseus escaping from Scylla and Charybdis: in this story, Scylla is undoubtedly Scorpio, which appears to have multiple long heads emerging from its body on snaky necks, while the "top" of the Milky Way stretches towards the point of the north celestial pole, around which the entire "starry ocean" of the northern celestial sky appears to turn, just like a whirlpool.

Between these two mortal threats, Odysseus is rescued by the friendly fig tree, to which he clings "like a bat" -- and you can easily confirm for yourself that just above the Scorpion in the shining path of the Milky Way there are two great bird-constellations, Aquila the Eagle and Cygnus the Swan, either of which might be playing the role of the hapless hero Odysseus, clinging for dear life to the fig tree in order to avoid being sucked down into the vortex of Charybdis (a vortex which is actually located in the "up" direction, for observers on earth, but not for players upon the great stage of the heavens, where "up" and "down" can take on different meanings in order to make the poetry work).

Still further confirmation is provided by the fact that the head of the constellation Boötes actually appears to resemble a "gourd," and is so described or depicted in many another Star Myth around the world. See for example the illustration of Da Mo shown on this page, (scroll down to the image in which Da Mo has a crooked staff over his shoulder, from which a gourd can be seen to dangle), or the image of Daikoku and Otafuku from Japanese myth shown and discussed in a previous post not included in this collection but available online and also explored in detail in *Star Myths of the World, Volume One*, involving the story of Amaterasu in which the following image is displayed, which shows Daikoku and Otafuku, and in which Daikoku represents Boötes and holds an enormous gourd, while Otafuku represents Virgo and holds a wand in one hand, just as Virgo's outline has a distinctive outstretched arm, marked by Vindemiatrix, which is often portrayed in the world's myth as a stick, a sword, or even an arm holding a bow):

Thus, we have fairly strong evidence from literally around the globe to support the identification of Boötes with Jonah when Jonah is sitting "under the gourd," and fairly strong evidence from many of the world's myths to support the identification of the ubiquitous fig tree with the "vertical trunk" of the Milky Way as it rises up from the horizon.

[Later note: during the writing of Star Myths of the World, Volume Two, which deals almost entirely with the mythology of ancient Greece, I became aware of the very strong probability that many of these trees in ancient myth under which individuals sit are in fact derived from the "spinning-form" outline of the constellation Hercules, located immediately behind the head of Boötes and which can be envisioned as a branching tree growing up from the "head" of the serpent held by Ophiucus; later work on Star Myths of the World, Volume Three, which deals almost entirely with episodes and characters from the Old and New Testaments of the Bible, confirmed that this interpretation is almost certainly accurate for many of the important trees in the Biblical scriptures, including for the fig tree which Jesus curses during the events of the Passion week. This newer interpretation is presented in the modified diagram below]:

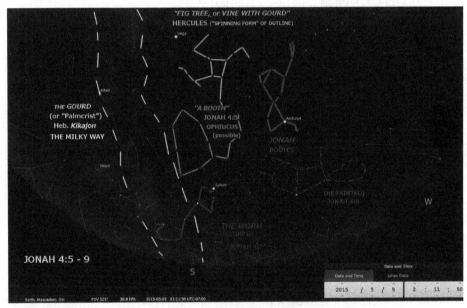

Of course, the figure of the Buddha sitting under the bo tree achieving the state of highest divine consciousness, and the figure of Jonah petulantly nursing his anger that the LORD God has shown mercy to Ninevah could not present a greater contrast.

But note: the scroll of Jonah ends abruptly with the verses quoted above. We are not told anything more about Jonah. We only see

353

that he is being admonished for his failure to have pity upon the people of Ninevah, whom he apparently hates because they are of a different family of humanity than he is -- and the divine voice tells Jonah in no uncertain terms that Jonah is wrong to think of them in this way.

We do not know at all whether or not Jonah ever achieved enlightenment, like the Buddha who likewise sat beneath the same celestial tree.

And here once again we must return to the incredibly helpful quotation from Alvin Boyd Kuhn[331], who reminds us that these stories are not about an external figure but that they are in fact about each and every man and woman on earth, and the experience of each and every human soul.

In other words, we are both Jonah and the Buddha.

The depiction in one story describes one aspect of our journey, while the depiction in the other depicts another part of our ultimate experience. We should not spend too much time wondering about whether Jonah ever changes, and spend perhaps more time considering our own state of mind and consciousness.

As well as our concern for our fellow human beings, whether they live in Ninevah or elsewhere.

Blessing and not cursing.[550]

Ultimately, these stories point us towards the concept of "raising the Djed" (or "the fig tree") and all that this concept appears to have entailed, in the ancient system of sacred wisdom imparted to the human race.

Mukasa, the Guardian of the Lake

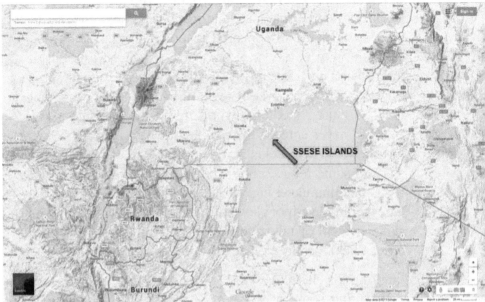

The Ssese Islands, in Lake Victoria, indicated by the red arrow. Google Maps.

2015 May 18

Among the Baganda people of eastern-central Africa, whose land in their own language is called Buganda but in the Swahili language is called Uganda, one of the central figures of the spirit world is Mukasa, the Guardian of the Lake.

Of this powerful entity we read in *African Mythology* by Geoffrey Parrinder (1967) that:

> The greatest of the demi-gods of Buganda, Mukasa, was a great giver of oracles, a kindly deity who never asked for human sacrifice. Myths say that when Mukasa was a child he refused to eat ordinary food and disappeared from home, later being found on an island sitting under a large tree. A man who saw him there took him to a garden and lifted him onto a rock. People were afraid to take him into their houses, thinking he was a spirit, so they built a hut for him on the rock. They did not know what to give him to eat, for he refused all their food, but when they killed an ox he asked

for its blood, liver and heart. Then people knew he was a god and consulted him in any trouble. Mukasa lived on the island for many years, married three wives, was cared for by priests, and at last disappeared as suddenly as he had come. His temple was a conical reed hut, which was rebuilt at intervals on the express orders of the king. Originally it is said that Mukasa spoke his will directly to the priests, but later they used mediums who uttered his messages. The medium never entered the temple but had a special hut in front of it. When seeking to know the will of Mukasa she smoked some tobacco until the spirit came upon her, and then she announced in a shrill voice what was to be done. The medium was not allowed to marry, or walk about in the sight of men, or talk to any man but the priest, and once chosen held the office till death. 89-90.

This information is remarkable on several levels, and may immediately ring some bells for readers who have studied the previous two posts in which I presented arguments to support my theory that the details of the story of the Buddha underneath the bodhi tree, as well as the story of Jonah underneath the vine or "the gourd" or the palmcrist or the *kikajon* found in Jonah chapter 4, are based upon the celestial figure of Boötes the Herdsman sitting with his back to the glorious column of the Milky Way galaxy -- see "The Bodhi Tree"[338] and "The sacred fig tree, continued: Jonah and the gourd."[344]

The general details regarding Mukasa presented above are corroborated in other accounts of the Baganda. A page from the webiste uganda.com, for example, discusses the understanding of a spirit world beyond this one, and Mukasa as one of the most important of the *Lubaale* or "Guardians" who dwell in the invisible realm. There, we see that the location of the oracle where the medium (or *mandwa*) obtained messages from Mukasa was located on Bubembe island, one of a chain of over eighty islands known as the Ssese Islands (after the tsetse flies which swarm there) in Lake Victoria.

See the map above for the location of Lake Victoria -- which lake is known in the Luganda language of the Baganda as *Nalubaale*, or "Lake of the *Lubaale*" -- and the Ssese Island archipelago in that great lake. Nalubaale is the second-largest freshwater lake on earth, with a surface area of 26,600 miles, second only to Lake Superior in size measured by surface area (the subterranean freshwater lake of Lake Vostok in Antarctica has a surface area of "only" 4,800 miles although it is so massive that it contains roughly 1,300 cubic miles of water, compared to Nalubaale's 660 cubic miles and Lake Superior's 2,900 cubic miles and Lake Baikal's 5,700 cubic miles).

It is actually somewhat difficult to find a good detailed map labeling all the Ssese Islands and especially Bubembe island, the location of the oracle and primary temple of Mukasa, but I believe Bubembe is the island that I have indicated with an arrow in the map below, which "zooms in" on the Ssese archipelago from the map shown above:

BUBEMBE ISLAND

The details regarding Mukasa given in the quotation above are further supported by accounts found in *The Baganda: An account of their native customs and beliefs*, by John Roscoe (originally published in 1911). There, we learn more information

357

regarding the *mandwa* and her entering into a state of trance or ecstasy in order to receive information from the spirit world:

> When she was about to seek an interview with the god, or to become possessed, she dressed like one of the priests with two bark-cloths knotted over each shoulder, and eighteen small white goat-skins round her waist. She first smoked a pipe of tobacco until the god came upon her; she then commenced speaking in a shrill voice, and announced what was to be done. She sat over a sacred fire when giving the oracle, perspired very freely, and foamed at the mouth. After the oracle had been delivered, and the god had left her, she was very fatigued and lay prostrate for some time. While giving the oracle, she held a stick in her hand with which she struck the ground to emphasize her words. 297-298.

Again, these details are extremely significant and noteworthy. First, they provide yet another example of a concept that can be seen to be absolutely ubiquitous around the world -- the understanding of the the existence of a spirit world with which it is possible to communicate and to which it is possible to journey even during this life, and the importance of doing so in order to obtain information or effect change which impacts aspects of this material world, which is intimately connected to and in fact can be said to be "interpenetrated by" and even "projected from" the spirit world in a very real sense. We have examined the importance of this concept in numerous previous discussions.

Second, they again demonstrate that the actual techniques with which human beings may enter into a state of ecstatic trance or contact with the invisible realm are incredibly diverse, a fact borne out by the encyclopedic research presented by Mircea Eliade in the landmark text *Shamanism: Archaic techniques of ecstasy* (first published in 1951), and discussed in the previous post entitled "How many ways are there of contacting the hidden realm?"[208]

358

But perhaps most importantly and most strikingly, the details provided above illustrate powerful and undeniable points of resonance with other sacred traditions from different cultures around the globe, and what is more these points of resonance can -- I argue -- be seen to be distinctly celestial in nature, relating very clearly to specific important constellations which are used in other cultures and other traditions to point the way to the importance of the realm of spirit within and around us, just as they do in the sacred traditions of the Baganda.

Let us examine some of those details more closely.

First, we see that Mukasa shares very clear points of correspondence with the story of the life of the Buddha: he seated himself under a tree, he refused ordinary food, he was against sacrifice (in the case of Mukasa, he was specifically against human sacrifice).

Further, the temple of Mukasa is described as a "conical reed hut," and the *mandwa* herself also dwelt in a special hut near the conical temple or shrine of Mukasa, although she did not enter it herself, even when she communed with the *Lubaale* himself, but instead smoked a pipe of tobacco in her hut and sat over a fire, perspiring and even foaming at the mouth. John Roscoe shows an image of one of the conical shrines of the Baganda in his 1911 book, and it looks very much like the image shown below of one of the sacred tombs of the Baganda:

We also see in the accounts that the mandwa is always a woman, that she begins her contact with the god by sitting above a fire and smoking a pipe, but that at the end she falls down exhausted, and lies prostrate for some time.

All of these details have very powerful correspondences to the specific details of the constellation Boötes the Herdsman and the other surrounding constellations and celestial bodies near Boötes, which the previous posts on the Buddha and the bo tree[338] and on Jonah and the gourd[344] have argued to be the foundation of those sacred stories as well.

The clear celestial connection of the story of the Buddha, the story of Jonah, and the details of the powerful Mukasa of the Baganda is extremely significant, and extremely powerful evidence supporting the actual celestial *connection* of all of the world's ancient sacred wisdom.

Let's spell out those celestial correspondences (which will be illustrated in the planetarium image below):

- The sitting figure of Mukasa on the rock, the Buddha under the bodhi tree, and Jonah under his gourd are all related to the constellation Boötes, who can clearly seen to be seated in the sky (and can also be envisioned to be kneeling). In fact, the figure of Bodhidharma who is known as Da Mo in China and who traditional legends describe as bringing Buddhism to China and kneeling in front of a stone wall for nine years without moving, and in some cases to have originated the martial arts as a way of strengthening the monks and giving them a physical-spiritual practice that would function as a kind of "moving meditation," can also be shown to be connected to Boötes, as I have demonstrated in previous posts such as the post discussing Shen Guang, from February 27, 2015.
- The beautiful tree arching over their heads is the shining column of the Milky Way, which rises up behind the sitting or kneeling figure of Boötes in the heavens.

- The "conical hut" (or the "booth" that Jonah makes under the gourd) is most likely the outline of the constellation Ophiucus.

The diagram below shows the major players in these Star Myths. The constellation Scorpio is also outlined, latching on to the base of the Milky Way, because Scorpio almost certainly plays the role of the worm who smites the vine that shelters Jonah, and causes it to wither away, much to Jonah's frustration and anger.

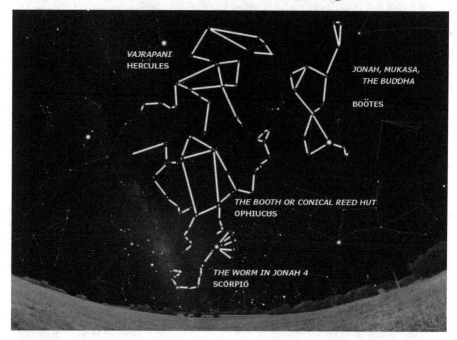

Note that in the diagram, the gigantic constellation of Hercules with his raised club is also outlined. This constellation plays a role in the legend of Da Mo (where, I argue, Hercules represents Shen Guang, the faithful follower and first disciple of Da Mo). Interestingly enough, the proximity of Hercules to the seated figure of Boötes provides an important confirmatory piece of evidence that this celestial interpretation is correct for the story of the Buddha as well.

The image below, from the 2d century AD, shows the unmistakeable figure of Hercules (or Vajrapani) standing behind the seated figure of the Buddha underneath the bo tree, exactly as the constellation of Hercules can be seen to stand behind the seated figure of Boötes in the night sky. This confirms beyond a shadow of a doubt that the ancients knew the connection between the Buddha and the celestial figures of Hercules and Boötes:

What is perhaps most striking in the sacred Baganda tradition surrounding Mukasa is the way in which the *mandwa* herself enacts the postures of the celestial constellations when she makes contact with the spirit world: first she sits above the fire smoking a pipe, just as Boötes can be seen to be "smoking a pipe" in the outline shown above, and then she falls down prostrate just as the constellation Virgo (who is located directly below Boötes and whose outline is shown in the image below from the Jonah story) can be said to be "lying prostrate and exhausted" in the way the constellation is arranged in the sky:

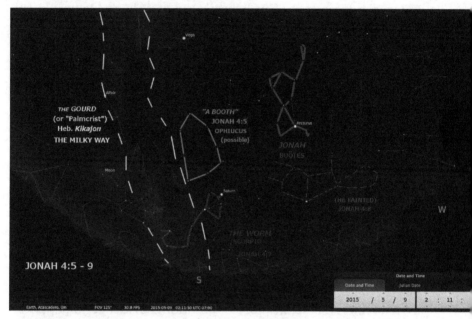

Note also that the *mandwa* carries a stick with which she strikes the ground for emphasis while reciting the message from the spirit world during her trance. The constellation Virgo can be seen to have a distinctive "outstretched arm" (marked by the star Vindemiatrix), which in some legends from around the world becomes a stick (and in other world myths it is a sword, a bow, or another implement connected to the story in question).

She is thus enacting, in the most direct way imaginable, the concept of "as above, so below," which conveys a number of deep teachings, one of them the fact that every single man and woman *embodies* within themselves, *contains*, and *connects to* the infinite universe itself: that we are each a *microcosm* which reflects and which in fact is not separate from the infinite *macrocosm* around and above us.

363

It is also extremely noteworthy that the famous Pythia who sat in the tripod at the oracle at Delphi can also be shown to reflect the constellation Virgo, who herself is in a seated position and who is directly above a celestial serpent, the constellation Hydra (corresponding to the dead carcass of the Python who was supposedly entombed deep beneath the temple at Delphi [later note: Scorpio is the more likely candidate for the Python, and the fumes or smoke from the carcass would then correspond to the Milky Way -- if so, then Sagittarius would correspond to the god who slays the Python with his arrows?]). In other words, the priestess at Delphi also entered into a state of ecstasy and communion with the gods by actually imitating the constellation Virgo, and embodying the concept of "as above, so below" and the microcosm/macrocosm.

Thus, we see that the sacred traditions surrounding the benevolent deity Mukasa of central Africa share extremely close and significant correspondences with the sacred traditions at the heart of Buddhism, ancient Greece, the scrolls of the Hebrew Scriptures and specifically of the prophet Jonah, and the legend of Da Mo in China, and that they thus provide an extremely powerful and significant piece of additional evidence to support the thesis that the world's sacred myths, scriptures, and traditions all share a *common celestial foundation*.

This fact, if true (and I believe the evidence is overwhelming and nearly beyond dispute; dozens more examples are discussed in other posts and in my previous books) is of incredible significance for world history, and for our lives today.

Some of the implications might be:

- That the sacred myths, scriptures and traditions of the world are not literal but that they are sophisticated celestial metaphors and that they use the celestial realm to convey the reality of the invisible realm of spirit.
- That we are not in fact separate from the realm of spirit, but that we are intimately connected to it at all times, and

364

that it is also within us at all times (as above, so below: microcosm and macrocosm).

- That if the various myths and sacred traditions teach that we are "descended" from figures in Star Myths, they are talking about our spiritual nature, and that such stories are not intended to be used to divide people on the basis of ancestry (or supposed ancestry) -- in fact, since they teach the existence and importance of the infinite spiritual nature inside each man and woman, this can be seen to supersede the far less important external distinctions which people have used to set men and women against each other based on external differences.

- That we are all deeply connected to one another and in fact to all beings and even to the universe itself.

- That on this basis, it is wrong to kill other beings, and especially that human sacrifice is profoundly wrong -- in fact, Mukasa's ordinance against human sacrifice can be seen as teaching that it is wrong to take the life of another man or woman, and that one cannot even use "religious devotion" as an excuse to harm another man or woman.

- That the ancients clearly understood these sacred myths to be connected to the constellations over our heads, and that they consciously depicted this understanding in their art and in their ecstatic practices and techniques.

- That this ancient understanding has been subverted, and that it has in fact been overturned or "stood on its head," such that for at least seventeen hundred years it has been taught that sacred traditions are only meaningful if taken literally.

- That literalism tends to invert the original meaning of the myths themselves, including all of the points outlined above.

- Literalism tends towards creating divisions between different people and different groups based on supposed descent from figures in stories that were originally intended to be understood as celestial metaphor.

- Literalism has often been used to "excuse" (or, it should be said, only "supposedly excuse," since it does not in fact excuse) violence against other men and women.
- Ultimately, all of these sacred traditions point us towards the importance of the spiritual realm, and especially the importance of the spiritual realm within ourselves and within everyone around us: the importance of recognizing and elevating and evoking the spiritual and the divine side of ourselves and of the cosmos, rather than demeaning and debasing and brutalizing and denying the spiritual and the divine in ourselves and in others and in the world around us.

And there are many other implications, in addition to those listed here.

Namaste.

The Hymn to Durga in the Mahabharata

2015 June 28

Immediately prior to the great battle of Kurukshetra, Lord Krishna urges the great warrior Arjuna to utter a hymn to Durga.

Tthis direction from Krishna to seek out the great goddess Durga helps confirm that the great battle in that ancient epic is indeed a metaphor for the endless interplay and "struggle" between the visible material world and the invisible world of spirit -- a struggle which is taking place in the universe around us and indeed within us at all times in this incarnate human existence.

There are abundant clues throughout the Mahabharata that the entire epic uses the endless cycles of the heavenly bodies -- the sun, the moon, the visible planets, and especially the stars -- to convey profound truths about the nature of our incarnation in this material plane, and about the existence and importance of the unseen realm.

Just as the Bhagavad Gita itself is presented as the song and counsel of the divine Lord Krishna to the semi-divine bowman Arjuna prior to descending into the great struggle, in the two sections of the Mahabharata immediately prior to the Bhagavad Gita we see Krishna telling Arjuna to utter his hymn to Durga -- and it can be conclusively shown that the goddess Durga is replete with imagery associated with the sign and constellation of Virgo the Virgin, the very sign which is located immediately prior to the autumnal or fall equinox on the great wheel of the zodiac: the point at which the sun's arc "crosses down" into the lower half of the year, towards the winter months and the December solstice, the half of the year in which darkness reigns and nights are longer than days, the half of the year associated with incarnation in this "lower world" of matter, when the soul clothes itself in bodies made of the lower elements of earth and water.

Thus the sign of Virgo (outlined in blue on the zodiac wheel shown below) truly does stand at the very "eve of battle" -- the final position before the plunge into the struggle of incarnate existence:

DAYS LONGER THAN NIGHTS:
Heaven, Promised Land, Greece, etc.

NIGHTS LONGER THAN DAYS:
Hell, Egypt, Troy, etc.

The goddess Durga, whom we can see to be associated with the sign and constellation of Virgo using the superabundant clues and references provided in the Hymn to Durga uttered by Arjuna at Krishna's request in Mahabharata Book 6 and Section 23, thus can be seen as preparing the soul for incarnation, sending the soul into battle, and (as we see in the events described in this section, in which Durga herself appears to Arjuna and gives him blessing and encouragement for the struggle) as the one who guides the soul along the difficult path and promises that the struggle will not be in vain.

More than that, however, the contents of the hymn identify the goddess Durga as "identical with Brahman," and the one who supports the Sun and the Moon and makes them shine: in other words, as the infinite and undifferentiated and eternal Cosmic Principle, the undefinable and the un-namable -- just as we see in the Bhagavad Gita the Lord Krishna declares himself (and reveals himself) to be.

And yet she is immediately available to Arjuna, and appears when he utters his hymn to the Goddess. This indicates that we, the human soul embarked upon this journey of incarnation, in actual fact are in the presence of the ultimate and the infinite at all times -- and that we have access to the supreme and undifferentiated and un-definable at all times as well.

And perhaps this is why at the end of the section describing the directive from Krishna to Arjuna to utter his hymn to Durga, and giving the contents of the hymn itself and the results (the appearance of Durga to Arjuna, and her promise to him that he shall conquer, that he is in fact invincible, and that he is incapable of being defeated by his foes), the text of the Mahabharata tells us to recite this same hymn every day, and to do so when we rise, "at dawn."

In doing so, we are focusing upon the infinite and connecting with the infinite: transcending the "chatter" of the mind and the senses (which are endlessly defining, and partitioning, and assessing, and evaluating -- all important and necessary functions, but functions that can keep us from being in contact with that undifferentiated and un-definable infinite which we in fact can and do have access to at all times and in all places, even in our incarnate situation).

By beginning each new day connecting with this ultimate principle, who is in fact always with us, the Mahabharata promises that we "can have no enemies," and "no fear," freedom from animals that attack with their teeth -- and "also from kings" -- victory in all disputes, freedom from all bonds, from thieves, and the enjoyment of victory in every struggle.

Balaam and the Ass

2015 October 10

The position of the earth on its annual journey around the sun is currently bringing the part of the heavens into view which I believe forms the basis for the fascinating ancient scriptural incident of Balaam and his ass (or donkey).

The account of Balaam is found in chapters 22 through 24 of the Old Testament book of Numbers, and it involves a number of important themes, chief among them the theme of blessing versus cursing.

The story of Balaam probably does not get much focus from those devoted to a literalistic reading of the scriptures these days (and my own personal experience during the nearly twenty years I was devoted to such an understanding was attending churches teaching a literalist understanding supports that assertion), due to the fact that it poses some fairly significant difficulties for those trying to read it literally.

Chief among these problems is undoubtedly the climax of the story, in which Balaam's donkey turns around and speaks to him to complain about Balaam's inhumane treatment. Balaam doesn't help things, because he answers right back to the donkey as if it is the most natural thing in the world do be accosted by one's mount while out for a ride. The two get into a conversation.

This is actually not the biggest difficulty in the text, as we shall see. The biggest problem is probably the fact that God appears to become angry with Balaam even after he explicitly tells Balaam to go ahead and travel to Moab, as we'll see in the text below.

Another factor which has probably led to the decline in focus on this story is the fact that the older translations consistently refer to Balaam's mount as an ass, which is what it is, because it was apparently not until some time in the 1700s that the word donkey was even used in English to refer to one particular sub-variety of ass. The 1611 King James translation, which had an enormous impact on literature and culture, thus refers to the animal as an ass, and the story has generally been referred to through the centuries in English-speaking cultures as the incident of "Balaam's ass."

However, if we can just get past those superficial problems, we can see in this story yet another example of the incredible worldwide system by which the same celestial foundations were dressed up in *myth* after *myth* after *myth*, in order to convey profound truths to us for our benefit during this earthly sojourn.

Unfortunately, trying to force the ancient scriptures into a literalistic-historical framework can cause us to miss their beautiful message altogether, or to distort it into something that means the exact opposite of what they were actually intended to convey.

The story of Balaam begins in Numbers chapter 22:

> 1 And the children of Israel set forward, and pitched in the plain of Moab on this side Jordan by Jericho.
>
> 2 And Balak the son of Zippor saw all that Israel had done to the Amorites.
>
> 3 And Moab was sore afraid of the people, because they were many: and Moab was distressed because of the children of Israel.
>
> 4 And Moab said unto the elders of Midian, Now shall this company lick up all that are round about us, as the ox licketh up the grass of the field. And Balak the son of Zippor was king of the Moabites at that time.
>
> 5 He sent messengers therefore unto Balaam the son of Beor to Pethor, which is by the river of the land of the children of his people, to call him, saying, Behold, there is a people come out from Egypt: behold, they cover the face [literally "the eye"] of the earth, and they abide over against me:
>
> 6 Come now therefore, I pray thee, curse me this people; for they are too mighty for me: peradventure I shall prevail, that we may smite them, and that I may drive them out of the land: for I wot that he whom thou belssest is blessed, and he whom thou cursest is cursed.

The messengers from Balak come to Balaam and convey the message, but Balaam consults with God and is told in verse 12 not to go with them nor to curse the people, "for they are blessed."

Disappointed, Balak sends yet more princes to Balaam, even more honorable than the first messengers, and this time offers great honor and says that Balaam can name his reward if he agrees to come.

Balaam is again visited by God at night who tells Balaam that if the messengers ask Balaam to go with them, he should rise up and go, but only say the word which God gives to him (verse 20).

This brings us to the most famous part of the story (still in Numbers chapter 22):

22 And God's anger was kindled because he went: and the angel of the LORD stood in the way for an adversary against him. Now he was riding upon his ass, and his two servants were with him.

23 And the ass saw the angel of the LORD standing in the way, and his sword drawn in his hand: and the ass turned aside out of the way, and went into the field: and Balaam smote the ass, to turn her into the way.

24 But the angel of the LORD stood in a path of the vineyards, a wall being on this side, and a wall on that side.

25 And when the ass saw the angel of the LORD, she thrust herself unto the wall, and crushed Balaam's foot against the wall: and he smote her again.

26 And the angel of the LORD went further, and stood in a narrow place, where was no way to turn either to the right hand or to the left.

27 And when the ass saw the angel of the LORD, she fell down under Balaam: and Balaam's anger was kindled, and he smote the ass with a staff.

28 And the LORD opened the mouth of the ass, and she said unto Balaam, What have I done unto thee, that thou hast smitten me these three times?

29 And Balaam said unto the ass, Because thou hast mocked me: I would there were a sword in mine hand, for now would I kill thee.

30 And the ass said unto Balaam, Am not I thine ass, upon which thou hast ridden ever since I was thine unto this day? was I ever won't to do so unto thee? And he said, Nay.

31 Then the LORD opened the eyes of Balaam, and he saw the angel of the LORD standing in the way, and his sword drawn in his hand: and he bowed down his head, and fell flat on his face.

The angel then informs Balaam that, had it not been for the fact that the ass perceived the presence of the angel, the angel would have slain Balaam. Balaam offers to go back home, but the angel tells him to continue, repeating the previous admonition from verse 20 that Balaam is only to say what is given to him to speak.

So Balaam continues, and joins Balak, who takes him "up into the high places of Baal" (verse 41). Balaam instructs Balak to have seven altars prepared, for seven bulls and seven rams, which are made into a burnt offering (Numbers 23: 1 - 6). But when the time comes that Balak expects Balaam to pronounce a great curse, Balaam announces that he cannot curse what God hath not cursed, and concludes with words of blessing (23: 7 - 12).

Balak is upset, but Balaam notes that he had said from the very start when first approached by Balak's messengers that he could only say what was given to him by God for Balaam to say.

Balak doesn't give up, however, and suggests they try another location, where seven altars are again constructed for seven bulls and seven rams. But the LORD meets Balaam and tells him exactly what to say, resulting in an even more eloquent blessing than before (this time replete with celestial imagery, particularly of a great lion). Balak isn't very happy about this and asks Balaam if he can just say nothing if he's not going to pronounce a curse, but Balaam explains that he must say what the LORD tells him to say (23: 25 - 26).

Balak decides to try one more time, and seven more altars are built as before, with similar results. This time the blessing is even more elaborate and takes up the first part of Numbers 24 (verses 5 - 9). The text also tells us that to deliver this message, Balaam falls into a trance, in which his eyes are open but in which he was given a vision of the Almighty (Numbers 24: 4).

After this, Balak tells Balaam to flee back to his home, but Balaam asks Balak if he wouldn't like to know more, and goes into another trance to give more predictions -- all of which I believe have to do with the celestial realms and to have spiritual meaning for our lives here on earth, but which could be (and often are) misinterpreted as literal predictions of things that would happen in earthly history. After delivering this message, Balaam returns to his place (Numbers 24: 25).

Now, how can we be reasonably certain that this event, preserved in ancient scripture, is allegorical and not literal and historical?

Setting aside the fact that donkeys cannot actually carry on conversations with humans as Balaam's ass is literally described as doing, there are abundant clues in the story which indicate the exact set of constellations involved.

The best place to start is with Balaam himself. The specific detail that he has his foot crushed by his donkey's efforts to avoid the awe-inducing presence of the angel (Numbers 22: 25), gives us our first clue as to his identity – and it is a very important one. There is one specific constellation who appears to have a severely twisted foot, and that constellation is currently rising brightly in the east during the "prime-time" viewing hours after the sun goes down: the constellation Perseus.

Below is a star diagram looking generally south and east, in which I have drawn in the outline of the constellation Perseus and several of the accompanying constellations surrounding Perseus which may also play a role in this story.

I've labeled Perseus as playing the role of Balaam in this story, and noted the location of the foot that was injured (ouch -- that looks pretty bad):

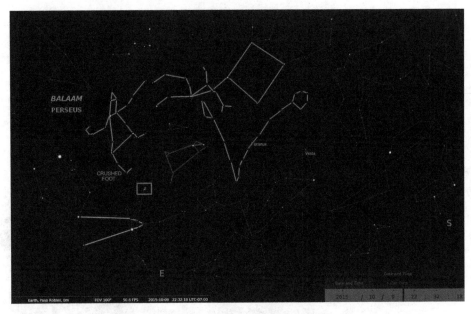

Now, if we're correct in identifying Balaam with Perseus (primarily on the basis of the crushed foot in the story, although there is plenty of other corroborating evidence that we will find shortly), then we need to find out which constellation is playing the role of Balaam's mistreated beast of burden in the story: the ass.

It just so happens that, directly beneath the figure of Perseus is the zodiac constellation of Taurus the Bull. Now, we know that this story has not come down through history as the famous tale of Balaam's Bull but rather of Balaam's Ass, so how can we possibly assert that Taurus could be playing the role of an ass in this story?

Well, as you can see from the diagram above (and the labeled diagram below, both of which indicate the outline formed by the brightest stars of the constellation Taurus using orange lines), the zodiac constellation of the Bull primarily consists of the brilliant V-shaped Hyades, and then there are two stars much further out above each of the "prongs" of the "V" which enable us to trace a long line in our imagination from the top of the Hyades to the ends of two mighty bull-horns.

377

These "horns" could also be envisioned as the ears of an ass.

The ass as a species can have some pretty long and impressive ears, as shown in the image collection below

Looking again at the stars of the constellation Taurus, it is not hard to understand why the formulators of the world's ancient Star Myths sometimes chose to envision this outline as a long-eared ass:

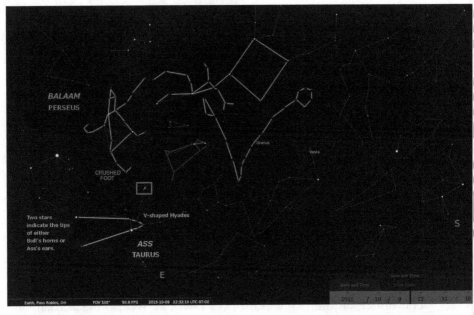

In the diagram, I've indicated the location of the V-shaped Hyades, and then if you look directly to the "left" of the "V" you can see the two stars which form the tips of the horns (if playing the role of the Bull) or the tips of the ears (if playing the role of an ass, as in the story of Balaam).

But in addition to the fact that the outline of the brightest stars in Taurus can very easily be envisioned as fitting a story with an ass or donkey, there is also plenty of evidence from other myths which help to confirm that our interpretation of the story of Balaam is on the right track so far.

Perhaps the most powerful piece of confirmatory evidence comes from elsewhere in the Hebrew scriptures themselves, for the V-shaped Hyades feature prominently in another Star Myth which I have outlined and discussed in some detail: the Samson cycle of myths.

In the story of Samson, of course, Samson's chosen weapon for slaying thousands of Philistines is the famous "jawbone of an ass," which does not seem to make much sense if the story is taken as literal history. Perhaps Samson might use such an implement in a

379

hurry for one or two opponents, but it hardly seems likely that he would continue to employ it over and over against literally a thousand: wouldn't he decide to pick up one of their weapons after slaying a few enemies who had swords or spears? (Unless, that is, all his opponents were also using jawbones as weapons that day, which seems unlikely).

The account is recorded in the scroll of Judges, chapter 15 and verse 15. I have explained in previous posts and in a video that the story of Samson is clearly not intended to be understood literally, but that it was almost certainly intended to convey powerful esoteric truths regarding our experience in this physical incarnate life (Samson was not a literal-historic character but in fact represents aspects of the incarnation of each and every human soul: in a very real sense, the story of Samson is all about *you*).

The understanding that Samson's jawbone-weapon is actually a group of stars -- that this jawbone is, in fact, the very specific V-shaped formation of the Hyades -- was one of the first breakthroughs in my own understanding that the stories in the Bible are built upon the very same celestial foundation that underlies all the other myths found in virtually every culture and every corner of our planet. This conclusion is explained by Hertha von Dechend and Giorgio de Santillana in their groundbreaking 1969 text, *Hamlet's Mill*, in which they present evidence that jawbone-weapons are described in myths from the Americas and from the Pacific Islands as well, and all of them relate to the Hyades (the Hyades are located above the constellation Orion, who can be seen "reaching out" towards them, just as Samson is described as "putting forth his hand" to grasp the jawbone in the book of Judges -- you can actually see a few stars of Orion peeking above the horizon in the star-diagrams presented here).

If the Hyades can function as a jawbone-weapon, and if that jawbone is described as "the jawbone of an ass" rather than "the jawbone of a bull" (as we might expect, since the Hyades are in Taurus), then this is very strong confirmatory evidence to support

380

the proposition that Taurus is functioning as the ass in the story of Balaam as well.

Interestingly enough, as can be seen from the included diagrams here, Perseus is reaching out with one arm in the direction of another important constellation: the beautiful maiden Andromeda, whom Perseus rescues in the Greek myth based upon these same stars. In a moment, we will see that Andromeda is playing the role of the powerful angel in this Old Testament story, but first let us briefly note another important confirmatory piece of evidence from Greek myths which also involves the theme of "ass's ears," and that is the story of King Midas.

In that story, of course, Midas reaches out towards his daughter (played, I am convinced, by the same constellation of Andromeda who plays the heroine in the story of Perseus). It is very noteworthy that Midas was later given ass's ears as a sign of his foolishness, given the above discussion regarding the likelihood that Taurus functions as the ass in the story of Balaam in the Old Testament. The existence of another myth involving Perseus and Andromeda, and featuring ass's ears, indicates that myths involving Perseus and Andromeda can also feature nearby Taurus, but as an ass rather than as a bull in some cases.

Note also that there seems to be an element of greed or of overstepping proper bounds due to *temptation of money* in both the story of Balaam and (much more clearly) the story of Midas.

All of this evidence appears to indicate that we are on the right track in our analysis of the Balaam story. Let's proceed to the identity of the angel.

In the scriptural text, we are told that an angel blocks the path of Balaam, and that specifically (in Numbers 22: 24) the angel "stood in a path of the vineyards, a wall being on this side, and a wall on that side."

Andromeda is positioned between Perseus and the Great Square of Pegasus, and she is actually touching one corner of the Square itself. If the Square represents the vineyards that are mentioned in verse 24, then it is quite evident that indeed she has a wall on "this side" of her, and a wall on "that side" of her. In fact, I believe this is exactly what the scriptures intend us to understand (it is very common for Star Myths all around the world to contain this very kind of super-abundant evidence, pointing us towards a fairly clear understanding of which constellations they represent).

Just in case we are still in confusion, we can also take a look at verse 22, where the angel is first mentioned, and see that in that verse we are told that Perseus is traveling with "his two servants with him." Just beneath the Great Square of Pegasus is one of the notable "dual constellations" in the zodiac wheel: Pisces. I would argue that the twin fishes of Pisces are probably the "two servants" of Balaam, traveling along the road with him (the road, in this case, following the zodiac through the heavens, up from Taurus to Aries to Pisces to Aquarius).

In fact, I have previously outlined another very important Biblical Star Myth in which Andromeda plays the role of an intercepting angel: the story of Abraham and Isaac. In that story, Perseus plays Abraham about to sacrifice his son, and Andromeda is the angel who stays his hand and points the way to the substitute: the Ram of Aries (located below Andromeda). In fact, the artist who drew this image included in that previous post does a very good job of depicting the characters as they are arranged in the sky -- Abraham standing with his arms out like Perseus, the angel flying in with outstretched arm in the location that Andromeda is found in the heavens, and the Ram trapped in the thicket just about where Aries is actually seen in the sky as well.

The fact that Andromeda plays an intercepting angel in another Biblical scripture is very strong confirmatory evidence that our interpretation of the Balaam story is on track.

Let's have a look at the analysis of the Balaam story thus far:

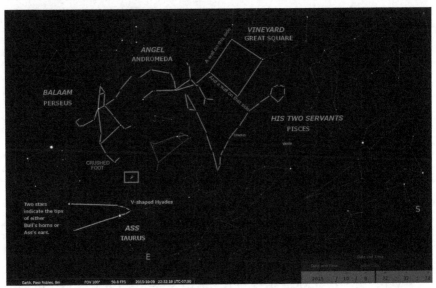

383

All in all, the amount of details included in the scriptural account provide overwhelming evidence that the story of Balaam is a celestial allegory, and that it is specifically a celestial allegory involving the region of the heavens containing the constellations Perseus, Taurus, Andromeda, Pisces and the Great Square. To hold that all these celestial correspondences are "merely coincidental" and that the story is really supposed to be understood as a literal-historical account of someone named Balaam (who also happens to have a literal conversation with his donkey using spoken human language, when his foot is crushed because the animal sees an angel blocking the path) seems to be a very unlikely hypothesis at this point, because *the texts themselves* provide us with abundant evidence that they want to be read as celestial metaphor.

One more set of clues from the text is worth a brief mention, which is the construction of seven altars for seven burnt offerings, which Balaam requests to have built each time Balak takes him up to a high place. Of course, the number seven is fraught with many layers of significance and may be present in the story because of some other aspect of its numerical and symbolical import. However, a very strong argument can be made that the presence of *seven altars* in this story (a detail repeated over and over) is one more textual clue regarding the celestial origin of this episode.

Just beneath the twisted foot of the constellation Perseus can be found one of the most beautiful celestial formations in the heavens: the brilliant Pleiades. The importance of the Pleiades to cultures around the world is very well known, and has been explored in numerous previous posts on this blog over the years.

The Pleiades is a dazzling cluster of bright and beautiful stars, unmistakeable once you know how to locate it in the sky. While the number of stars in the Pleiades cluster which can be visible to the naked eye under good conditions number far more than seven, the Pleiades in many myth-systems of the world are depicted as "Seven Sisters" or as related to the number seven (the brightest of the Pleiades are six in number, and sometimes there are stories about the "missing sister" as well, although as you can see from the NASA images and my own hand-drawn diagrams in the blog posts above, there are more than seven stars that you can probably identify for yourself in the Pleiades cluster).

Because of the strong connection between the Pleiades and the number seven, and because the Pleiades are located very near to Perseus (Balaam) and are in fact technically part of Taurus (the ass in the story), I believe it is very possible that the seven altars which are built in the Balaam story are a reference to the Pleiades.

This possibility gains further traction from the fact that we are told that the altars are the site of *burnt* offerings -- very appropriate for a cluster of glowing stars.

Additionally, we are told that the burnt offerings are bulls and rams. Of course, the two zodiac constellations in this part of the sky are Taurus and Aries.

Below is our now-familiar diagram of the Perseus - Andromeda region of the sky, with a few final labels added to round out the details we've discovered in our analysis of this Star Myth:

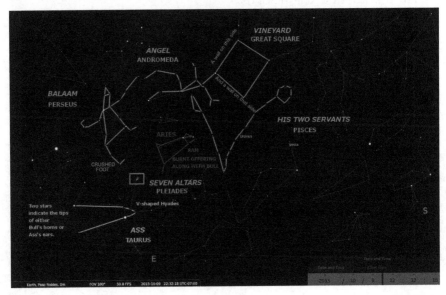

With this many details, I believe we can make a very strong case to argue that the incident of Balaam and the Angel is entirely celestial in nature, and that its message is thus *allegorical* and not literal-historical.

But what does it all mean? That, of course, is open to interpretation, but previous posts have cited the assertion of Alvin Boyd Kuhn to the effect that the ancient myths are *not* about fabulous kings, powerful warriors, or even enlightened sages and mystics, but are actually about the experience of each and every human soul in this incarnate life (see for instance here, here and here). In an important 1936 lecture entitled "The Stable and the Manger," Kuhn said:

> The one actor in every portrayal, in every scene, is the human soul. The Bible is the drama of our history here and now; and it is not apprehended in its full force and applicability until every reader discerns himself [or herself] to be the central figure in it!

That means that we don't have to try to imagine an external literal-historical figure named Balaam having a conversation with his donkey -- the story is not really about anyone named Balaam at all! It is about each one of us.

But we will not be able to figure out what it is trying to convey to us if we try to force the text to be about a literal-historical figure named Balaam. In fact, as we will see shortly, doing so risks inverting the esoteric message entirely.

To understand what I think the story of Balaam is intended to convey (or at least part of what it is intended to convey -- there is no doubt much more to this very deep metaphor, the depths of which each reader is invited to plumb on his or her own), we must understand that the specific part of the heavens which we have been examining in our analysis is very significant due to the sun's rising in the sign of Aries at the point of the spring equinox during the Age of Aries during which many ancient myths (and especially Biblical myths in the Hebrew scriptures) are set.

386

This is the point of "crossing upwards" into the upper half of the year, when hours of daylight begin to dominate again over hours of darkness, after the long winter months in which darkness dominated over day.

DAYS LONGER THAN NIGHTS:
Heaven, Promised Land, Greece, etc.

NIGHTS LONGER THAN DAYS:
Hell, Egypt, Troy, etc.

The constant interplay between the "lower half" and the "upper half" -- between the forces of "darkness" and the forces of "light" -- were anciently allegorized in myths around the world as a great struggle or battle. Previous examinations of the Mahabharata and the Bhagavad Gita, for instance, have discussed the evidence for this assertion, and the stories of the Trojan War in the Iliad as well as the crossing of the Red Sea in the Old Testament can be shown to relate to this same interplay.

But the myths are not "just" about the natural cycles of the year: they can be definitively shown to have used the great cycles to convey knowledge about *spiritual* truths. In other words, the myths use the most majestic physical models conceivable -- the mighty cycles of the heavens, the turning of the stars through the night, the progression of the zodiac signs and the planets through the year, the interplay of the seasons and the sun's path from equinox to solstice and back, the phases of the moon, and even the longer cycles of planetary conjunctions and the titanic precessional mechanism that grinds out the ages over the course of thousands and thousands of years -- in order to convey to our understanding truths about *invisible* matters.

In terms of the great zodiac wheel, at least on one level of metaphor, the upper half of the year is associated in myth with the invisible realm of spirit but also the spiritual and divine aspect in each and every incarnate human being, just as the lower half of the wheel is associated with our physical, material, animal, corporeal nature, into which we are plunged upon incarnation.

Much of the purpose of the myths of the world appears to have been to remind us that we are not merely physical, to awaken the spiritual within and point us towards the truth of our divine inner nature. Previous posts have connected this awakening of the "spiritual component" in ourselves, others, and in all the universe around us, with the concept of *blessing*.

The opposite impulse, of course, denies the spiritual, seeks to degrade, debase, brutalize and otherwise reduce to the physical and the animal (which is why violence is so wrong, on any level). All forms of cursing can be seen to be connected to this opposite "physicalizing" and "brutalizing" impulse.

In the story of Balaam, the concept of blessing and cursing is clearly central to the narrative. In the allegory presented, Moab and her king Balak are representative of the lower half of the wheel, and the forces of darkness. The king, Balak, specifically wants cursing, and seeks to hire Balaam to do it.

The children of Israel in the metaphor are representative of the upper half of the wheel. In one part of the Biblical passage quoted, the text tells us that they "cover the face [literally the 'eye'] of the earth" (Numbers 22: 5). In other words, they are associated with the sun (the "eye of the earth") and with the half of the year in which the hours of daylight cover more and more of the

388

hemisphere in question (the summer months, the upper side of the wheel).

The upper half of the year metaphorically represents the realm of spirit, and the re-establishment and re-affirming and uplifting of the divine present at all times in men and women (and in all of creation). It is the same concept expressed by the raising-up of the Djed Column in ancient Egyptian myth-systems discussed in many previous posts and videos, such as here.[42]

It should not have to be repeated at this point, but because literalism has so firmly entrenched itself in the cultural consciousness of the west for the past seventeen hundred years, it must be stressed that the *children of Israel* in this story do not represent historical or literal personages, any more than does Balaam (or, for that matter, King Midas). The text is a *spiritual allegory*. The children of Israel in this story represent a spiritual aspect of reality that is present in *all of us* -- not a group of literal or historical people (the allegorical understanding is inclusionary, not exclusionary as the literalistic understanding tends to be).

They (like the Danaans in the Trojan War) represent *the upper half* of the zodiac wheel, and allegorically the realm of spirit and the uplifting of the divine spark present in all human beings (and all nature as well). This is made clear in some of the "blessings" pronounced by Balaam in Numbers 22 - 24 (see for example the mention of the Lion in Numbers 23: 24, which is undoubtedly a reference to the sign of Leo, strongly associated with summer and the "upper half" of the zodiac wheel). None of us are *literal* descendants of any constellation -- but the idea of being descended from the stars conveys an allegorical truth about our spiritual condition.

Moab and Balak represent the lower half of the wheel. The story is about *spiritual* matters, and not about historical and literal battles between different physical branches of the human family.

Thus, when Balaam is asked to *curse* the allegorical representatives of the divine spark, the invisible realm of spirit --

the very aspect of our dual human nature that we are supposed to be lifting up and calling forth – he is being asked to deny the spiritual, the divine, and everything associated with the invisible realm.

Doing so would be to send the message that we are nothing but physical, animal, brutal beings, with no invisible, spiritual, divine component.

Of course, whenever Balaam gets in touch with the realm of spirit, with the realm of the divine, by going into a state of trance, he is strongly warned *not* to convey such a brutalizing, cursing message. He is instead given a message that raises up the spiritual -- and indeed a message that predicts the eventual and inevitable triumph of spirit over the brutal, the physical, the debasing and the degrading aspects of our physical incarnate condition.

Whenever Balaam is on the way to cooperate with the king of Moab, he is opposed by the angel, representative of the invisible realm (and indeed, invisible to Balaam until his eyes are opened). We watch as he grows more and more angry at his beast, more and more violent, more and more brutal, until his ass with her just questions appears to be at least as human as he is.

She is more in touch with the spiritual realm than he is, and she saves him from destruction even though he beats her for it.

Clearly, Balaam in this story is representative of our own human condition. And this helps us to understand one of the aspects of the scriptural passage which could give literalist readers major difficulties -- the fact that God told Balaam to go along with the messengers of the king of Moab, and then sent the angel to oppose Balaam (literalist interpreters often try to construe some kind of culpable motive to Balaam in his going along, even though he has just been told in a dream to do so).

If Balaam is representative of some aspect of our own soul's condition, here in this incarnate life, then our entry into incarnation is akin to "going into the kingdom of Moab" and it is

390

ultimately for our own good and in thus in accordance with the divine will. In other words, we descend into this life from the realm of spirit for our own benefit. But our mission here is not to become brutal, not to become violent, not to become bestial, but rather to bless and to uplift and to reconnect with that upper half of our nature – our spiritual and divine True Self.

When we understand this allegorical system, then the story begins to make sense in a way that it does not when we try to force a literal reading on the text. It is a story of hope and of the dignity and divinity inherent in each and every human being. We all are a combination of physical and spiritual, but we are told that the spirit will eventually and inevitably triumph, no matter how ugly the physical circumstances and situations may become, and no matter how our own spiritual blindness will often lead us to do stupid and even self-destructive things as we go up the path.

When we understand the story as esoteric and allegorical, then we see that it applies to each and every person, and that it teaches us to work to lift up the spiritual in ourselves and in others, and not to put them down.

But when it is taken as literal and historical, this message can become distorted, because when it is externalized then it can be mistakenly seen as a message which lifts up some groups and puts down others.

In fact, by externalizing the text, a literal reading can lead to some conclusions that are "180 degrees out" from the interpretation just offered. A "physical" message, so to speak, instead of a spiritual one.

But, when we see the clear and overwhelming evidence that the text describes the motions of the stars, it becomes clear that the literal and historical reading – already very difficult to maintain in light of the incidents in this particular episode -- is almost certainly *not* the intended message of the ancient text.

The same exercise can be performed with virtually *every single* other story in the scriptures included in what we today refer to as "the Bible" (both the "Old" and "New" Testaments), and indeed with virtually every other myth and sacred story from around the world.

Leaving us with what I believe are several inescapable conclusions, among them:

> *that we are all connected,*

> *that we are all primarily spiritual and that thus the external and physical should not be used to divide us from one another,*

> *that we should pay attention to the invisible realm* (as Balaam learned "the hard way" in the story, but as we ourselves also generally "learn the hard way" in this life),

> *that we should bless and not curse,*

> *that we should lift up and otherwise draw forward the divine spark in others and, as much as possible, in the part of the cosmos that we can impact around us* (including by planting gardens, opposing degrading treatment of animals, and opposing the pollution of the air and land and waters around us),

> and *that the side of uplifting will ultimately and inevitably win out, and that those who are on the side of cursing and debasing and brutalizing may seem to be powerful now but that in fact* they are not.

2015 December 31

Now that the day of Christmas has passed, and now that the full moon that was in the sky during that time has begun to rise later and later each night as it wanes towards new moon (enabling better star-gazing), those wishing to contemplate the celestial motions on which the familiar story is almost certainly patterned have a wonderful opportunity to go outside and watch the heavenly figures in action for themselves.

Seeing it take place in person can -- I believe -- open up an entirely new and personal level of apprehension (a word that has as its root a verb meaning "to seize" or "to grasp") of the powerful knowledge that the ancient story was intended to convey.

I have previously published a short video which details some of the abundant evidence suggesting that the stories found in the scriptures that became what we commonly call the Old and New Testaments of the Bible are based upon the motions of the stars.

I'm going to repeat a bit of the argument covered in that video here, with some new diagrams, in order to explain how you can go out and observe the celestial actors for yourself, and also in order to offer a few brief suggestions as to what these ancient texts might be trying to tell us.

There are specific details in the texts from which we derive the Christmas story which argue very strongly that the texts themselves were never intended to be understood as describing events which happened in literal, terrestrial history. One such indication from the texts themselves is the familiar story of the visit of the "wise men" or Magi.

The event is described in the gospel according to Matthew, where we read in the first verses of chapter 2 that "when Jesus was born in Bethlehem of Judaea in the days of Herod the king, behold, there came wise me from the east to Jerusalem" (Matthew 2: 1). The text tells us that they "saw his star in the east" and had therefore come to worship him (verse 2), and later that "the star, which they saw in the east, went before them, till it came and stood over where the young child was" (verse 9).

These verses cause something of a geographic problem, for those who wish to interpret the text as describing an event which took place in terrestrial history (but no problem at all if those texts are describing an event which takes place in the celestial realms above, as we will see momentarily). As typically understood, the Magi came *from* the east, and followed a star which they *saw in the east*, and which "stood over where the young child was," thus leading them to the place where they would find the divine child.

No matter where on the planet you choose to try to make these verses work, they require some significant contortions if they are interpreted as describing a journey on the planet's surface, as the diagram at the top of this post attempts to illustrate.

The geographic problem, as may perhaps be best perceived by looking at the map above on which north is "up" as we look at the page, east is to our right and west is to our left as we look at it, should be immediately clear if we try to imagine the Magi traveling *from the east* while simultaneously following a star which they have seen *in the east.*

The arrow shows a possible terrestrial route or general direction of travel for the wise men in the story (but note that it does not really matter where on the planet we draw this arrow -- the "from the east" problem will still remain). If they come *from the east* while following a star seen *in the east,* they will not get to Jerusalem or Bethlehem or anywhere else that is to the *west,* unless they go east and keep on going east for a very long time and circle the globe.

However, this "geographic problem" is only a problem if we try to understand the text as if it were speaking in the language of literal, terrestrial history. If the text is instead understood to be speaking in the language of the stars and constellations and heavenly cycles, then the problem resolves itself most satisfactorily.

As Robert Taylor (1784 - 1844) proposes in a series of talks entitled "The Star of Bethlehem (parts I, II and III)" and delivered over three weeks in November of 1830, all of which were later published in a collection of his lectures entitled *The Devil's Pulpit*), the Magi in the story, who have traditionally been referred to as the "three kings" since ancient times, may be identified with the three glorious belt-stars of the constellation Orion, a constellation who dominates the night sky during the winter months (see especially pages 43 - 44).

These three stars that make up Orion's belt are among the brightest and most-recognizable groupings of stars in the entire celestial panoply -- and in their dignified motion across the sky they do indeed begin *in the east,* as do all the other celestial objects including our sun, due to the motion of earth's daily rotation on its axis.

As they pass the zenith point in their progression across the sky, and begin to arc back downwards towards the west (where they will eventually set), the constellation Virgo the Virgin will begin to rise in the east -- and the star which marks her outstretched arm was interpreted in many Star Myths from around the world as a young child either nursing at her breast or sitting upon her lap.

Because the Virgin (and the star that marks her child) is rising *in the east* even as Orion is beginning to go down *in the west*, it is entirely appropriate to say that the "three kings" of Orion's belt (who came across the sky *from the east*) now look and see the child's star *in the east.*

This neatly resolves any dilemma with the text -- and shows that the scriptures were not at all mixed up in their description of the directions, and that they did not intend for those directions to be applied to terrestrial events, but rather to celestial ones.

The diagram below shows the scene as it appears in the sky at this very time of year. The three belt-stars of the constellation Orion are framed with a bright yellow line above and below, and the direction that they have traveled from the east is indicated by red arrows. The extended arm of Virgo is indicated by a purple arrow, pointing to the bright star Vindemiatrix which was sometimes envisioned as a child in her arms or on her lap:

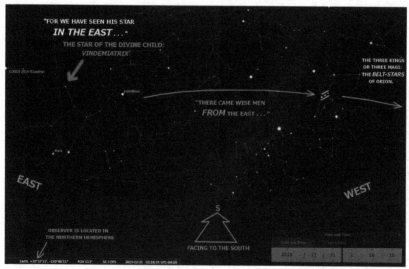

Note that in the above diagram, we are facing towards the south, because an observer in the northern hemisphere, which this image replicates using the outstanding open-source planetarium app *Stellarium,* must look towards the south in order to see the zodiac constellations such as Virgo, and in order to see the belt-stars of Orion, which are located almost exactly upon the celestial equator, which is an imaginary line found "ninety degrees down" from the north celestial pole (located "behind our back" in this illustration) or "ninety degrees up" from the south celestial pole (if you are in the southern hemisphere).

Of course, because we are facing south in the illustration above, the eastern horizon is to our left and the western horizon is to our right (and that is indicated in the diagram).

You may be familiar enough with the stars and constellations to envision the outlines of Orion and Virgo in the diagram above, but in order to help out, I have added their outlines in the diagram below, which is identical to the one shown above but with a few additional lines and labels:

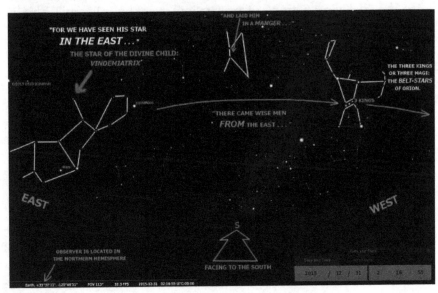

Virgo is shown on the left, and you can see that both the moon and the bright planet Jupiter are currently near the crown of her head (the moon will continue rising later and later each evening, and will move out of the area of Virgo in a couple more days).

The diagram above also adds the outline of Cancer the Crab, which is straight up over the due-south direction (at its highest point on its arcing path across the sky, a point known as "transit" and also as its "culmination" and its "zenith"). As Robert Taylor also explains in the lectures referenced above, the constellation Cancer contains the beautiful and very significant cluster of stars known as the Beehive Cluster, which was anciently also referred to as *Praesepe* -- "the manger" (probably because it is in between two stars known as the Northern Ass or Donkey-colt, and the Southern Ass or Donkey-colt).

As is well known, a different scripture text (found in the gospel according to Luke) says that the Virgin laid the child *in a manger.* As you can see from the time-marker in the lower-right part of the screen in the image above, the Beehive is crossing the zenith point just after two in the morning in the modern epoch – but due to the "delaying action" of precession over the millennia, this is "behind schedule" compared to the time that it would have been crossing the zenith point thousands of years ago.

There was a time in a previous epoch at which the "manger" in the Crab was crossing the zenith right at midnight on the nights surrounding winter solstice (instead of around two in the morning as it does today, due to the delay) -- and Robert Taylor believed that this explains the aspect of the story of the baby being born in a manger, because it is at the winter solstice that the year or the sun is reborn.

The point of winter solstice, as explained in some detail in previous posts such as this one,[150] was anciently used as a metaphor for the awakening of the consciousness to the existence of the connection to the Infinite, which actually permeates our entire universe and is (according to many ancient teachings) the

398

"real world behind this one" from which the manifest world originates.

This awakening is powerfully expressed to us in the story of the birth of the divine child in the manger – and expressed in many other forms and guises in countless other Star Myths from around the world. As the discussion above should convincingly indicate, the story is not intended to be taken as literal, terrestrial history -- which would make it a story about events (however wonderful, moving and mysterious) that happened to *someone else* in another time and place – but rather to be understood as teaching you and me about something we need to *know for ourselves.*

It describes the birth of awareness of the reality of and possibility for connection and communication with our own Higher Self (see discussions of the identity of Doubting Thomas[499] for more on this subject). It is a birth that was described in many myths from around the world using the metaphor of a twin, but a twin so close that they are part of us – "closer than a brother." And the profound teachings contained in this ancient celestial metaphor no doubt go on for layers far deeper than anything I can express in a written discussion, but must be experienced and grasped and felt by each person for himself or herself.

Perhaps one of the most singular lines in the Biblical text, and one on which I have not hitherto remarked, is found in the Matthew account, in verse 12 of the same chapter: "And being warned of God in a dream that they should not return to Herod, they departed into their own country another way."

Now that we understand the celestial foundations of this famous episode, we might be able to bring in an entirely new perspective to that phrase "another way"! Obviously, if they crossed the sky from east to west in an arcing path above the horizon at approximately zero degrees right ascension (that is to say, along the path of the celestial equator), the stars in question will return to "their own country" – the east – by an entirely different path!

That is to say, they will do so (from our perspective) "under the earth," where neither Herod nor anyone else will be able to see them.

Celestially, this makes perfect sense, and adds another extremely satisfying piece of evidence to support the conclusion that the scriptural text itself is telling us that it is speaking in the language of the heavenly cycles, describing the motions of stars.

Spiritually, it opens up even more new paths of consideration for the apprehension of deep knowledge of profound benefit for our daily life in the here and now.

First, this "underworld path" was used in the ancient Star Myths of the world to symbolize our condition here in this incarnate form, where we find ourselves at this moment.

The motions of the stars (including the star we know as the sun) present us with a perfect metaphor for the interplay between the realm of the infinite (the realm of the gods, the realm of pure potentiality in the language of quantum physics, in which finite boundedness has not yet manifested, which physicist David Bohm referred to as the "implicate order") and the realm of the incarnate, the material realm (in which "infinite possibility" has manifested into one of its potential forms, "temporarily unfolding" in the terminology with which David Bohm described it, into the "explicate order").

The stars arc across the sky -- traveling through a realm which is, in a very real sense, infinite. And yet, with the exception of those close enough to the two central hubs or celestial poles around which the sky appears to revolve (only one of which is visible to us at any given point on the globe, unless we happen to be located somewhere on the line of the equator), the stars cross that infinite realm only to dip down into contact with our horizon (note that if you are located on the line of the equator, none of the stars will fail to dip down below the horizon, although if you are far enough to the north, the stars making a circle fairly close to the point of

the north celestial pole will not dip down below the horizon --
they are referred to as the "undying stars" in the texts of ancient
Egypt).

When the stars sink below the horizon, they appear to leave the
realm of infinity and plunge into the lower elements of matter --
disappearing into either earth or water, depending upon what
you see when you look towards the western horizon from your
location on our planet. Alvin Boyd Kuhn devotes several chapters
in his masterful *Lost Light* (1940) to the spiritual symbolism that
the ancient wisdom attaches to each of the so-called "four
elements," and also argues that the ancients knew very well that
there are not "only four elements" but that they used this system
primarily for its outstanding capacity for conveying esoteric
spiritual knowledge.

Thus, when the stars sink below the horizon and seem, from our
perspective on the planet, to be "plowing through the
underworld" on their way back to the eastern horizon, they
symbolize rather perfectly a teaching about our own incarnate
condition, plunged as we are into a body composed of earth and
water (or *clay*, as Genesis describes the material from which
Adam is fashioned). When the "three kings" go back to their own
country by "another way," they are symbolizing our own sojourn
through this apparently material world, this "explicate order."

And yet, in the very same verse, we see a very clear hint that the
Magi have a very direct connection to the Infinite: they are
"warned of God *in a dream*." In an altered state of consciousness,
and one into which we enter basically every single day (every
single rotation of this planet of ours, that is), the Magi receive
messages from the divine.

Note that this teaching, contained in that very important verse
Matthew 2: 12, resonates very powerfully with the message
described as being given to another king, in another text of
ancient scripture collected into what today we call the Bible – the
vision given to Solomon in a dream, in which he receives the gift

of God-given wisdom, in the first fifteen verses of I Kings chapter 3. This powerful dream is recounted in the ancient scriptural text *immediately* prior to the famous episode of the living baby and the two mothers (hmmmm . . . many of the metaphors regarding the inception of the divinely-given awakening seem to have to do with a birth of a baby, the incarnation into this explicate order).

Thus, that amazing verse not only provides a powerful additional clue that the familiar story is built upon a celestial pattern, but it also provides a very strong indication that the story has to do with our own incarnation in this "explicate" order in which we find ourselves, but also with the necessity of our realization that while here we do have access -- not only through dreams but through a wide variety of other techniques and avenues, a variety that itself seems to be nearly infinite -- to the infinite realm, the realm of the divine, the realm of the unbounded, the "implicate order."

The way in which the story of the visit of the Magi is framed also makes it clear that access to that realm can be *absolutely essential* to the very practical questions of the path or way that we choose to follow "here below" as well.

As stated at the very beginning of this discussion, now is an ideal time to go outside and gaze into the infinity of the night sky to meditate upon these celestial cycles for yourself, if it is at all possible for you to do so. The glorious outline of the constellation Orion will be immediately visible to you as you go outside into the night at any time after sundown right now -- and may in fact take your breath away, as you first turn towards that part of the sky in which Orion is moving alongside the brilliant star Sirius (Sothis) and among the circle of other bright stars often referred to as the "winter circle."

In order to see Virgo rising with Vindemiatrix (her outstretched arm) and Spica (her brightest star, currently also accompanied by the planet Mars), along with the moon and Jupiter, you will have to wait until the early morning hours, after midnight. Virgo should begin to rise into view above the eastern horizon near one in the morning, depending on the skyline of your eastern horizon where you are.

You will have to wait until shortly after two in the morning to see Cancer the Crab with the beautiful Beehive Cluster climb all the way to the transit point above the due-south line (if you are in the northern hemisphere), but you can actually observe the Beehive long before it reaches its transit or zenith point. The Beehive and the faint constellation of Cancer the Crab are located between the Twins of Gemini (who stretch out horizontally from the direction of Orion's trailing shoulder, and are part of the "winter circle") and the mouth of the majestic constellation Leo the Lion.

Note that finding the Beehive does require a dark sky, so you will want to try to find it before the moon rises into view and if possible will want to get to a place where there is little or no ambient light from city streets or buildings.

But, no matter how many of the stars and heavenly figures you can actually identify, the very act of going out and gazing into the night sky can be conducive to a closer and deeper apprehension of the knowledge preserved for us in the ancient wisdom of the human race. When you stare out into space, you are in fact staring out into infinity. And, even if the only stars you can confidently identify are the three great belt-stars of Orion, you can gaze at them and think about the message of the turning-point of the year and the awakening to the reality of the infinite realm that this great pivot-point represents, and the connection we have to it, a connection which in fact is always present.

And, as we consider the evidence that the stories in the Bible follow the same system of celestial metaphor upon which virtually all the other sacred stories, scriptures, and myths of humanity are also founded, it should also become very clear that it is very likely that they were trying to tell us the same thing, and that there is no basis for disrespecting one expression of the ancient wisdom or trying to supplant it with another expression of the same ancient system.

I hope that you have an opportunity to go out and spend some time with the stars at this turning-point of the year, if it is at all possible for you to do so – and I send my very sincere wishes for positive renewal and growth to each and every reader out there (wherever you are on this terrestrial ball) at the beginning of a new cycle.

Salmacis and Hermaphroditus

2016 March 20

It is probably safe to say that there is no actual fountain on earth which literally possesses the power to cause any man who sets foot in it to emerge from the waters half-man and half-woman.

And yet Ovid, in the fourth book of his *Metamorphoses*, relates the story of the son of Hermes and Aphrodite whose fateful encounter with the nymph Salmacis imparted this power to the waters of the spring, as though the location and effects of that place were actually well-known in his day.

Ovid actually tells the story as a "story-within-a-story" in his poem, during an extended episode in which the daughters of Minyas refuse to set aside their work and join in the rituals of the god Dionysus, but instead continue weaving -- and as they do so, they relate stories of various interactions with the divine realm, debating amongst themselves as they do so whether or not the gods could really perform all the wonders described (a question which is answered at the end of the tale, when the impious daughters who failed to recognize the divinity of Dionysus are transformed into chattering bats).

The final story they tell before this fate befalls them is the story of Salmacis and Hermaphroditus. Alcithoe, one of the daughters of Minyas, begins:

> I will explain the way in which the fountain
> of Salmacis, whose enervating waters
> effeminate the limbs of any man
> who bathes in it, came by its reputation,
> for though the fountain's ill effects are famous,
> their cause has never been revealed before.
> *Metamorphoses* 4. 396 - 401.

Thus in the excellent translation of Charles Martin published in 2005.

I find that very literal translations can be the most helpful for examining Star Myths for the celestial clues that may have been included in the original but which may have been "lost in translation" if the translator does not pick them up and bring them across into the new language. With literal translations, the clues are usually carried over, because the translator is trying to render the words of the original as closely as possible into the new language, even if the result sounds a little unusual.

Here is one such a version, from the nineteenth century scholar Roscoe Mongan. The account of the encounter between Salmacis and Hermaphroditus at the pool which ever after bore its unique powers (and ever after was named after the nymph herself, becoming "the fountain of Salmacis") is translated there as follows (Alcithoe is speaking as she and her sisters weave at their loom):

Learn, then, from what cause Salmacis became notorious, and why, with its enfeebling waters, it unnerves the limbs bathed in it. The cause lies hid, but the power of the spring is very well known. The Naiads nursed, in the caves of Ida, a boy, born to Mercury from the Cythereian goddess, whose face was of that kind in which both father and mother might be recognised; he also obtained his name from them. As soon as he had completed thrice five years, he forsook his native mountains and, leaving Ida, that had been his nurse, he loved to wander about in unknown places, and to see unknown rivers, his curiosity lessening the fatigue. He proceeds to the Lycian cities also, and to the Carians that border upon Lycia. He sees here a pool of water, clear even to the very ground below. There are not here any fenny reeds, nor barren sedges, nor rushes with sharp points. The water is transparent, yet the borders of the pool are fringed with fresh turf, and with plants perpetually blooming. A nymph dwells there, but one who is not suited either for the chase, nor one who is won't to bend the bow, nor one who is to compete in the foot-race, and she alone, of all the naiads, was not known to the swift Diana. The report is, that her sisters often said to her: "Salmacis, do take either a javelin, or a painted quiver, and combine they leisure time with the toilsome chase." She does not take either a javelin or a painted quiver, and she does not combine her hours, spent in leisure, with the toilsome chase; but at one time she bathes her beautiful limbs in her own fountain; often she smooths down her tresses with a comb of Cytorian box-wood; and consults the waters which she looks into [to see] what is most becoming to herself [i.e., she looks into the pool to see which way of arranging her hair is the most beautiful on her (DWM note -- all others are from Mongan)]. And, at another time, having her person enveloped in a transparent garment, she reclines either upon the soft leaves, or upon the soft grass. She often gathered flowers, and now, also, by chance, she was gathering them when she saw the youth, and wished to possess him as soon as she beheld him. However, although she was hastening to approach him, she did not actually approach to him until she had arranged herself, and until she had looked at her raiment, and had assumed her [most captivating] aspect, and deserved to appear beautiful.

Then thus she began to speak: "O boy most worthy to be believed to be a god! if thou art a god, thou mayst be Cupid; or, if thou art a mortal, happy are they who gave thee birth. [. . .] If thou hast any spouse, let my pleasure be secretly enjoyed; or, if thou hast none, let me be [thy consort], and let us enter the same bridal chamber." After these words the naiad became silent. A blush suffused the features of the young. He knows not what love is, but even the very act of blushing was becoming to him. Such a colour is in apples hanging upon a tree exposed to the sun, or in painted ivory, or in the moon blushing beneath her brightness, when the auxiliary brazen cymbals resound in vain.

To the nymph soliciting, without cessation, at least such kisses as he might give to a sister, and to her now advancing her arms to his neck, as white as ivory, he says: "Wilt thou cease? or must I fly and leave these places, along with thyself also?" Salmacis was alarmed, and said: "I surrender these places free to thee, O stranger!" and, with a retreating step, she pretends to depart. But then, also looking back and being concealed in a thicket of shrubs, she lay hid, and placed on the ground her bended knees. But he, as being only a boy, and as if being unobserved, goes hither and thither on the lonely sward, and dips in the playful ripples [first] the soles of his feet, and [afterwards] his feet as far as the ankles. Nor is there any delay; being delighted with the temperature of the gentle waters, he throws off from his tender person his soft garments. But then, indeed, Salmacis was amazed, and became excited with desire for his unrobed beauty; the eyes, too, of the nymph burn, no otherwise than the sun, when shining most brilliantly with a clear disk, it is reflected from the opposite image of a mirror. With difficulty can she endure delay; and now with difficulty can she defer her joy. Now she desires to embrace him; and now, distracted with love, she can scarcely restrain herself. He, striking his body with his hands bent inwards, swiftly plunges into the stream, and throwing out his arms alternately, shines in the clear waters, just as if any one were to enclose ivory figures, or white lilies, within clear glass.

"We have conquered!" exclaims the naiad, "lo, he is mine!" and, throwing all her garments far away, she plunges into the midst of the waters, and seizes him, resisting her, and snatches kisses in the struggle, and puts down her hand and touches his breast much against his will; and clings around the youth, sometimes in one direction, sometimes in another. Finally she entangles him struggling hard against her, and anxious to escape from her, like a serpent, which the royal bird takes up and carries away aloft it, as it hangs suspended, holds fast his head and feet and entangles his expanded wings with its tail.

And [she clung to him as closely] as the tendrils of the ivy are wont to entwine themselves around the tall trunks of trees, and as the polypus, but letting down his sucker on all sides, grasps his enemy captured beneath the water. The descendant of Atlas persists, and denies to the nymph her hoped-for joy. She presses him closely, and as she was clinging to him with her entire person she said: "Although thou mayest struggle, O thou obstinate being! notwithstanding this thou shalt not escape. May ye so ordain it, O ye gods! and let no length of time separate him from me or me from him!" These supplications obtained the favour of the [*lit.* their own] deities, for the persons of these two, becoming incorporated, are united together, and one form includes both of them, just as if anyone should see from beneath a bark formed over both of them, branches to become united in their growth and to spring up equally. Thus, after their bodies were united in a firm embrace, they are no longer two bodies; but yet the form of them is two-fold; so that it could be called neither woman nor boy; it seems to be neither, and yet both.

Wherefore, when Hermaphroditus sees that the clear waters, into which he had descended as a man, had rendered him only half a male, and that his limbs were becoming softened in them; holding up his hands, he says, but now not with the voice of a man: "O both father and mother! grant this favour to your son who has the name of you both. Whosoever comes as a man to these streams, let him go out thence as half a man, and let him suddenly become effeminate in the waters that he touches." Both parents being moved, confirmed the words of their double-shaped son, and tinged the fountain with a drug that renders sex ambiguous. II - 14.

Where is this famous fountain, whose properties were apparently well-known? Is it possible that it actually existed in ancient times, or that its waters still possess such properties to this day?

I believe in fact that this fountain actually *does* exist -- but that it is located in the celestial realms, and not on earth. The pool is found at the widest, brightest section of the Milky Way band, where the two zodiac constellations of Scorpio and Sagittarius are stationed on either side (visible this time of year in the hours prior to sunrise).

The clearest indication that this is the section of the night sky to which this myth is giving reference is the extended metaphor in which the poet compares the clinging of Salmacis to the person of Hermaphroditus to a serpent being carried upwards by an eagle, and twisting and wrapping about the bird. This metaphor clearly points to the two Milky Way constellations of Aquila and Scorpio, the Eagle of Aquila being located above the Scorpion in this brightest portion of the Milky Way band.

In fact, I believe that from the clues in the ancient poem itself (the best extant version of this particular myth, although it is also referenced by the earlier historian Diodorus Siculus, and obviously has an origin much earlier in the mythology of ancient Greece rather than Rome, since the boy's name is a combination of the Greek names of the god Hermes and goddess Aphrodite, rather than the Roman versions of the same, although Ovid of course uses their Roman names Mercury and Venus), the youth who dips his feet into the waters is played by the constellation Ophiucus, which is flanked by serpents on either side -- just as Salmacis is described as clinging to him with her entire body, like a serpent, first on one side and then on the other.

In the diagram below, you can see that Ophiucus is "dipping his feet" into the pool (the widest and brightest part of the Milky Way band):

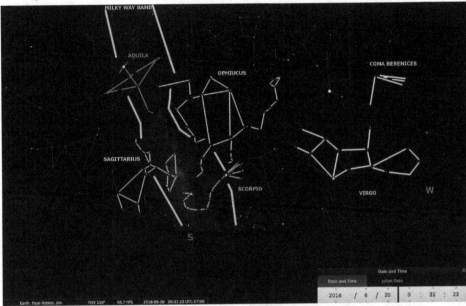

The nymph Salmacis is probably Scorpio, crouching in a thicket before she rushes out to wrap herself around Hermaphroditus, although earlier in the poem some of the description suggests Virgo (particularly the part about gathering flowers when she first spies Hermaphroditus – can you see what celestial features might play a role in this part of the account?)

The struggle that ensues contains an extended metaphor involving an eagle ("the royal bird") and a snake – again, this probably refers to Aquila and Scorpio, and is a pattern found in many myths and traditions involving this part of the sky.

After Hermaphroditus emerges from the pool, now merged with Salmacis and sharing the gender of both man and woman, the waters from then on have the power of effecting the same change upon those with whom they come in contact.

The constellation Sagittarius, which in ancient Greek myth frequently plays characters of either male or female sex (as you will see if you examine the evidence discussed in my latest book, *Star Myths of the World and how to interpret them, Volume Two*) may play the role of Hermaphroditus emerging from the stream, now changed.

Thus, the "action" of the myth can be said to commence on the right of our screen as we look at the star chart above (the west) and proceed towards the east -- a very significant direction of movement, from a spiritual point of view (this is also discussed in the latest book). The two figures of Hermaphroditus and Salmacis begin on the right side of the sky (as we look towards the south in the northern-hemisphere view above), then they go into the fountain and there is a struggle (described in terms of a serpent entwining about an eagle) and emerge on the other side as one hermaphroditic figure (Sagittarius).

From the above analysis, we can be confident that the above encounter never actually took place on this earth between literal, historical figures who merged into one being.

But that does not mean that the myth itself is "not true."

In fact, I believe the myths *are* actually true, and on many profound levels (perhaps not just "many levels" but rather *infinite* levels, descending deeper and deeper without end).

One of the ways that they are true is that they describe the experience of our human soul, "plunged down" into this material realm, a realm characterized by the "lower elements" of earth and *water* (as opposed to the "upper elements" of air and fire).

When our invisible spirit takes on a material body, we become for a time a "blended being" composed of both divine soul and physical form.

Myths having to do with the "plunge" into the waters of incarnation often do involve the constellation Virgo, who stands at the edge of the "lower half" of the year, at the point of autumnal equinox (see discussion in this previous post[50]).

But the plunge down into matter often involves (at first) the loss of awareness of our spiritual or divine inner spark, as we sink more and more into sensual enjoyment of the body (and Salmacis is described as basically spending all of her time in such enjoyment, looking at her reflection in the water, combing her hair and bathing her limbs, and lying around in the grass wearing diaphanous garments). At a certain point, there is a "spiritual turn" at which we begin to have an awakening of awareness of our spiritual nature -- and I believe that this myth actually depicts that very point of awakening, when Salmacis sees the child of Hermes and Aphrodite and exclaims that he must be of divine origin, and that she must have him.

The integration of the two natures is actually the point of this famous incident, I believe -- portrayed here in the frank sexual imagery sometimes employed in ancient myth, but actually using the sexes as a way of expressing spiritual concepts in allegorical or metaphorical form: to "clothe" the truths of the invisible reality in the physical forms of nature, to better convey them to our deeper understanding.

As we begin to understand how to interpret the myths in the language which they are actually speaking, the language in which they actually ask us to listen to them, we can begin to hear a message that we might otherwise have totally missed.

Each and every ancient myth is worthy of deep and careful contemplation, and the above explication of the myth of Salmacis and Hermaphroditus may serve as an example of the sort of examination and meditation we can profit by applying to the myths of the portions of the corpus of ancient wisdom which draws each of us most strongly (some will perhaps find themselves drawn to the myths of ancient Greece, others to the myths of ancient India or ancient Japan, or of the cultures spread across the vast Pacific, or the continents of Africa or Australia or the Americas, and so forth).

413

In fact, the above discussion only barely ripples the surface (so to speak) of the deep pool of the fountain of Salmacis: one could meditate upon this tiny portion of the stories in Ovid's work, and in the wider context of the daughters of Minyas, for years on end and probably never exhaust the amazing lessons that it might hold for him or her.

The bad judgment of King Midas, and what it teaches us

·MIDAS·DAUGHTER·TURNED·TO·GOLD·

2016 October 22

King Midas is a well-known figure from ancient Greek mythology famed for his bad judgment.

He is most remembered for his request, when granted one wish by the god Dionysus, that everything he touched would turn to gold – a request which, when granted, made him so giddy with happiness that he could hardly believe what he thought to be his good fortune. As everyone knows, however, he soon came to regret that awful request.

There is another episode from ancient myth in which Midas again displays his bad judgment, this time when he was asked to judge a competition of musical skill between Apollo -- the very god of music who is referred to in some ancient texts as Apollo *Musagetes*, a title which signifies "leader of the Muses" (see for instance Diodorus Siculus Book I and chapter 18, fifth sentence, which you can read online here) -- and a satyr (in some accounts a satyr named Marsyas, and in others the god Pan himself).

Apollo of course played upon a lyre, and the satyr upon the pan-pipes, and in some accounts Midas, the King of Phrygia, was appointed to be the judge of the contest, while in other accounts it was the mountain of Timolus itself (or the god of that mountain) which was to be the judge. In those accounts, Timolus wisely judged that Apollo was the winner, but Midas loudly disagreed with him and indicated that the satyr was the more skilled, while in the accounts in which Midas alone was the judge, he also unwisely selected the satyr as the winner of the contest -- and as a punishment, Apollo gave Midas the ears of an ass, saying that the dull judgment of Midas and his lack of discernment in hearing should from then onwards be visible for all to see.

Both of these episodes have clear celestial foundations, and add to the overwhelming body of evidence which supports the conclusion that virtually all the world's ancient myths, scriptures, and sacred stories are built upon a system of celestial metaphor, in order to impart deep knowledge about the simultaneously "material-spiritual" universe in which we find ourselves, as well as *our own* inherently dual material-spiritual nature as men and women.

In fact, not only do I believe that overwhelming evidence points to the fact that virtually all the world's ancient myths are built upon celestial metaphor involving the constellations and heavenly cycles, but I also believe the evidence indicates that they are all built upon *the same system* of celestial metaphor -- a common, worldwide system which appears to indicate that they all somehow share the same common source.

This common system unites the ancient myths and sacred stories of all the varied cultures from around the world and across the

416

millennia -- and should in fact be seen as uniting us all as men and women sharing an incredible common inheritance of tremendous value.

In fact, in both of the above episodes involving King Midas and his terrible judgment, we can see very clear echoes to two other well known "judgment myths" or sacred stories involving very much the same theme: the famous "Judgment of Paris," which ultimately leads to the Trojan War, and the equally-famous "Judgment of Solomon," in which -- unlike both Midas and Paris -- King Solomon displays right judgment when presented with a very similar choice.

In the Judgment of Solomon episode, the most famous aspect of the story involves two mothers and two babies, one of them alive and one of them dead, and Solomon's wisdom in solving the dilemma with which he is faced, in which each of the mothers claim that the living child belongs to her. However, as other posts discuss in more detail (06/23/2016), the famous scene with the two mothers as told in the text of I Kings chapter 3 actually follows immediately from a previous episode in the same chapter, found immediately preceding the two mothers scene, in which Solomon in a dream is visited by the Most High, who asks Solomon what he would like to be given.

This offer very much parallels the offer made to Midas in the myths of ancient Greece, in which Dionysus also offers to grant one request to Midas. Midas unwisely asks for unlimited riches -- specifically, the power to turn everything he touches into pure gold. In contrast, Solomon asks for a wise and understanding heart, so that he can be a better ruler on behalf of the people -- and the text tells us that this request pleases the Lord, who says:

> Because thou hast asked this thing, and hast not asked for thyself long life [*literally: "many days"*]; neither hast asked riches for thyself, nor hast asked the life of thine enemies; but hast asked for thyself understanding to discern judgment;

Behold, I have done according to thy words [...]
And I have also given thee that which thou hast not asked, both riches, and honor: so that there shall not be any among the kings like unto thee all thy days. (I Kings 3: 11 - 13).

Note that Solomon's request for wisdom and discerning judgment is contrasted with other possible choices, including riches, honor, long life, or power over his enemies. Clearly, this story has points of resonance with with the story of Midas, who unwisely asked the divine Dionysus for the equivalent of riches -- with disastrous results.

In similar manner, in the episode from Greek myth known as the Judgment of Paris, the youth of the same name (Paris, a prince of Troy) is presented with a contest of beauty among three goddesses, each of whom offers him a reward if he will select her. The rewards offered to Paris by the three goddesses include rulership and power (offered by Hera), heroism and fame (offered by Athena), and the most beautiful woman in the world to be his bride (offered by Aphrodite).

As we know, Paris selected Aphrodite and in doing so launched the Trojan War, because the most beautiful woman in the world, Helen, was already married to a king of the Achaeans, and all the other Achaean kings and heroes had previously promised to defend whichever among them would be so fortunate as to have won the right to marry Helen.

This disastrous decision by Paris again has clear echoes with the judgment offered to Solomon -- who decided *not* to request riches or honor, but rather asked for wisdom, and who was told that because of this choice, he would also be given those things for which he did not ask, such as riches and honor.

The episode in the Midas story which perhaps resembles the Judgment of Paris even more closely is the episode in which Midas must judge the musical contest between Apollo and either the satyr Marsyas or the god Pan -- because in both of those myths, there is an actual contest involved. Midas, clearly an

418

exemplar of bad judgment, fails to recognize the god Apollo as the winner -- the very deity from whom all skill and talent in music proceeds in the first place. Thus, Midas inverts the proper order of things, disrespecting the divine source, and is punished by being given the ears of an animal (in this case, the long hairy ears of an ass or donkey).

Interestingly enough, this punishment brings to mind a masterpiece of esoteric fiction written by the later Roman author Apuleius (who appears to have been an initiate into the Mysteries of Isis, as well as perhaps other mystery schools). That story was originally called the *Metamorphoses* (not to be confused with the more famous work of the same title by Ovid), but the Metamorphoses of Apuleius is more commonly known as *The Golden Tale of the Ass*, or simply *The Golden Ass*. In that story, the narrator (Lucius) is himself transformed into an ass, and undergoes a series of outrageous adventures before he is restored to his original form by the goddess Isis herself.

Not only is the condition of Lucius when transformed into an ass reminiscent of the fate of Midas who is given long ass-ears for his lack of judgment, but the restoration of Lucius comes not long after a climactic episode in which Lucius witnesses a re-enactment in a Roman arena of the mythical episode of the Judgment of Paris itself! Thus, it would appear that the theme of "judging or discerning rightly or wrongly" is very much central to the tale of Lucius in *The Golden Ass* -- and that Apuleius himself understood the important thematic connection between the Judgment of Midas (in which King Midas ends up receiving donkey-ears, just as Lucius in the story is turned into a donkey) and the Judgment of Paris (the very episode Lucius sees enacted just before his own restoration, and an episode in which Paris brought "damnation upon mankind" by his desire to possess another man's wife, in the words of Apuleius -- an interesting way of viewing the story of the Judgment of Paris).

The fact that Paris in that beauty contest selects the winner by giving her an apple is extremely interesting -- especially because the mystery initiate Apuleius says that the disastrous choice of Paris brought "damnation upon mankind." We can all probably think of another ancient episode involving an apple (or other unspecified fruit) which was said to have brought damnation upon mankind as well.

I believe that all these ancient mythological episodes can be shown to be built upon celestial metaphor -- and therefore to be esoteric in nature, designed to impart knowledge to us about our own inner connection to the infinite realm: our own spiritual nature, even encased as we are in a material (animal) body of flesh during this life.

The story of King Midas can be shown to relate to very specific constellations in our night sky (and constellations which are in fact visible at this very time of year, during the end of the month of October).

In the story of the disastrous request for the gift of the golden touch, several ancient sources tell us that Midas was at first overjoyed at the granting of his request, but soon realized to his horror that he could neither eat nor drink anything without it also turning to gold (a situation which would soon end in his own death as well).

In some versions of the story, the king's own daughter runs up to embrace her father and is herself transformed into solid gold. This particular aspect of the story does not seem to be present in many of the most ancient accounts, but it is perhaps the most well-known part of the King Midas story today.

In almost every ancient version of the myth, Midas prays to heaven (in some versions to Dionysus, who had originally granted Midas one request, and in other versions to Apollo) to have the curse of the golden touch taken away from him, and is told to go immerse himself at the source of the river Pactolus (which is found at the aforementioned Mount Timolus). In some versions of the story, Midas is to immerse his head three times in the river at its source. Thereupon, all the things which had been turned to gold by Midas after his terrible choice were restored to their original condition -- and the river Pactolus from then on had golden sands which often yielded up gold flakes or gold nuggets.

For a variety of reasons, I believe it is almost certain that this story of King Midas is founded upon the constellation Perseus, who is presently rising above the eastern horizon in the hours after midnight. Perseus is a constellation who is located near the very "top" of the Milky Way band as it arches across the sky, on the far side of the galactic trail from the brightest and widest part found between Sagittarius and Scorpio (the galactic core). Thus, it can be envisioned as the "upper reaches" of the galactic river -- allegorized in the myth as the upper source of the river Pactolus.

There, Perseus can be clearly seen to be immersing himself in the river -- or even dunking his head in it! Here is a star chart with the constellation Perseus outlined in yellow (on the left as we face the page, which is the east), as well as the constellation Sagittarius, who will play a role in the story discussed later on:

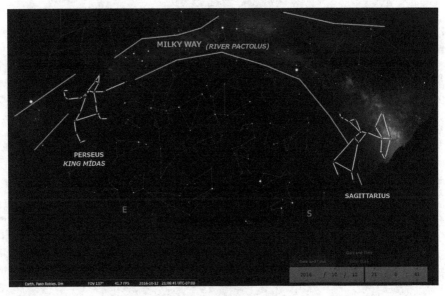

Can you see the brilliant arc of the Milky Way band? In real life, if you go outside into the night sky, you will see this arc going over your head, beginning at the western horizon (on the right side of the image above) and crossing the center of the sky towards the east where you will see Perseus (on the left side of the image above). This image is from the perspective of a viewer in the northern hemisphere, looking towards the south.

From the star charts, we can see why the myth tells us that Midas (played by Perseus) went to the source of the river Pactolus, and there he immersed his head (or his entire body, depending on the ancient account).

The most dramatic part of the episode of Midas and his golden touch, of course, comes when his daughter runs to him and is herself turned to gold, to the king's horror. This aspect of the myth is almost certainly inspired by the outstretched arm of Perseus on the western side of the constellation (the right side as we face the image above). That part of the constellation reaches out towards and almost touches the constellation Andromeda, representing a beautiful maiden in many myths:

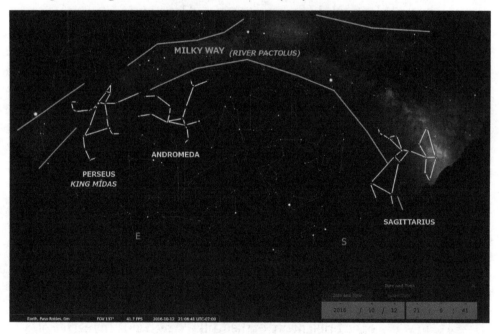

Can you see how the story of Midas touching his daughter can be clearly seen in the above constellations? We can be very thankful

422

that the god allowed Midas to change his mind and restore all that he had previously turned to gold, by immersing himself in the river!

I believe that the very same constellations that form the basis for the disastrous golden touch episode also play the main parts in the episode of the Judgment of Midas between the music of the god Apollo and either the satyr Marsyas or the god Pan.

Now, instead of playing the beautiful daughter of the king, the constellation Andromeda actually plays the role of the satyr, with arching tail and pan-pipes. Can you look at the image above and see which parts of the constellation play the role of the tail of the satyr, and the pan-pipes?

In the star-chart below, I have labeled the parts of the constellation Andromeda which (I believe) give rise to the connection between this constellation and the satyrs:

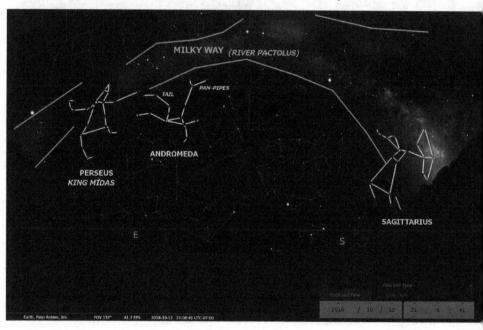

It is notable that the contest takes place in the vicinity of Mount Timolus (or Tmolus), which is also the source of the river

423

Pactolus – indicating that we are still in the same place in the sky (because Midas went to Pactolus' source at Timolus to dunk his head in the stream).

Look at some of the classical paintings from later centuries shown below, where we can see that the posture and attitude of Midas is very reminiscent of the constellation Perseus with its outstretched arm (sometimes envisioned as a sword or a wand):

In the above image, Midas can be easily identified by his pointed ass-ears. He is reaching towards Pan with his wand to indicate that he believes Pan to be the winner of the contest rather than the god Apollo. Apollo can be seen playing a viola instead of his usual ancient lyre in this painting (from the early 1600s). Note that on the left side of the painting, the artist has depicted one of the nymphs or young women holding an urn or jug – a celestial detail indicative of the constellation Aquarius, and signaling the awareness on some level of the constellational connections in these myths among at least some of the traditions which were preserved within the schools of fine art in Europe. Apollo is almost certainly associated with a constellation not far from

Aquarius, as we will presently discuss.

Below is another example, from the late 1800s:

Note in the above image that the pan-pipes are held aloft, in a manner very reminiscent of the way the constellation Andromeda holds up the part of the constellation that I believe can be identified as the pan-pipes in the constellation (when playing the role of the satyr Marsyas or the god Pan). Note also that Apollo in this painting exhibits the characteristic of "walking away while looking back" -- a very distinctive feature of the constellation Sagittarius found in countless ancient myths from around the world. I believe for a great number of reasons that Sagittarius frequently plays the role of the god Apollo -- and I discuss these reasons in detail in *Star Myths of the World, Volume Two* (which focuses almost entirely on the myths of ancient Greece).

The constellation Sagittarius holds what is usually envisioned as a bow, on the west side of the constellation and pointing towards the west, but this could also be envisioned as a lyre (or even a viola -- and notice that in the earlier painting from the 1600s in which Apollo is depicted with a viola, the musical instrument is held in the same general position that Sagittarius holds the bow in the heavens). Additionally, in the painting from the 1800s, the god Apollo is depicted standing on the edge of a pool -- just as Sagittarius is located at the edge of the widening of the Milky Way (at its brightest and thickest point -- the Galactic Core), which was anciently envisioned as a pool in many myths.

For those still not fully convinced that the constellation Andromeda, usually envisioned in myth as a beautiful maiden, can also play the role of a satyr (as the constellation does in this episode of the Judgment of Midas), please observe the characteristic "arching tail" of the satyr in the ancient Greek artwork below, which corresponds very well to the "upper leg" of the constellation Andromeda (which is labeled as "tail" in the star-chart above):

Recall also that in the story of King Midas, after his poor judgment in the music contest involving the god of music himself, the hapless king receives ass's ears as a punishment and a sign of his brutish lack of discernment. In almost every ancient version of the story, the detail is included that the king usually tried to hide his deformity beneath a tall cap -- known specifically as a "Phrygian cap" (King Midas, after all, was the King of Phrygia).

Note that the constellation of Perseus, as outlined in the star-charts above and as labeled in the chart below, does indeed feature a tall, peaked cap! This is only evident if you follow the inspired outlines suggested by H. A. Rey and discussed in previous posts and also my published books.

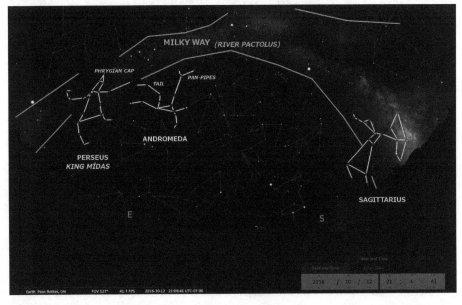

This celestial detail should pretty much cement the identification of Midas King of Phrygia with the constellation Perseus in these mythical episodes.

Below is an image of a classic Phrygian cap, for those not familiar with them. Note that in many cases, the cap has "ear flaps" which can perhaps be envisioned in the constellation outline as seen above.

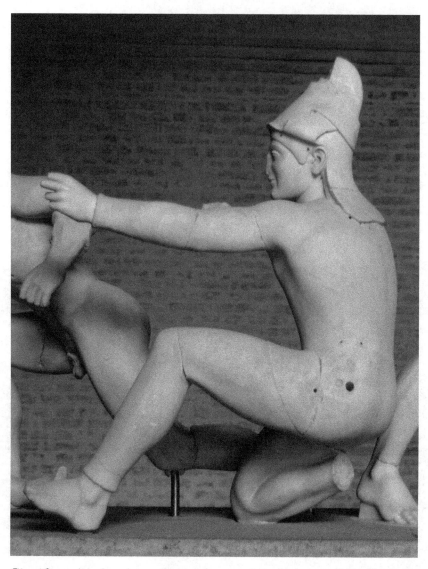

Significantly, the above figure from an ancient temple is thought to represent Paris -- the very character responsible for the disastrous judgment that led to the Trojan War! The ear-flaps on his Phrygian cap appear to be "tied up" so that they will not hang down, in this particular case.

And, if all of the above discussion does not thoroughly establish that Midas in these myths is associated with Perseus the constellation, there is also the fact that Perseus is almost directly above the constellation Taurus in the sky -- and that the long horns of the constellation Taurus could sometimes in ancient myth be envisioned as the long ears of an ass instead of the long ears of a bull. For this reason Perseus-characters in many myths are described as riding upon an ass or donkey -- as Balaam is described as doing in another Old Testament story, this time from the book of Judges.

This connection adds still more weight to the mountain of evidence we have already discovered linking Midas to Perseus, and linking the Midas episodes described above to the region of the sky surrounding Perseus.

This discussion should help to firmly establish the argument that the world's myths are all in fact closely related, and built upon a common system of celestial metaphor. The purpose of this ancient system is, I am convinced, to impart to our deep understanding the true nature of the simultaneously material and spiritual universe in which we find ourselves, and the simultaneously material and spiritual nature of the human condition.

In the Midas story, we see powerful illustrations of the teaching that we are not advised to contact the invisible realm for the purpose of acquiring wealth or riches. Midas is given a wish by the god Dionysus, and can ask for anything he chooses -- and Midas chooses unwisely.

In contrast, Solomon is approached by the Lord in a dream and similarly offered anything Solomon wishes to request -- and Solomon does not ask for riches, or for harm to his enemies, or long life. I believe that one way to interpret this story is that we are strongly advised *not* to use our contact with the invisible or the divine realm in order to try to obtain riches for ourselves, or power over others.

Instead, Solomon asks for wisdom in order to help others: this request *is* in fact a proper request to make of the divine and the infinite.

Midas with his animal ears is a picture of our own condition in this world, entangled with an "animal nature" and prone to becoming seduced by that which is material and that which is lower -- instead of seeking that which is spiritual, which is in fact the true source and fount of everything which we see manifested in the material realm. The mistaken judgment of Midas in the contest with Apollo is a dramatic example of this failure to acknowledge the true divine source from which everything flows and has its fount (in this case, Apollo is the divine source of music, but Midas fails to acknowledge this truth).

Failing to acknowledge and properly value the infinite realm, the spirit world -- or trying to use it for personal gain or for destroying one's opponents -- leads to objectification of oneself and others, turning us and them into *objects*, as Midas ends up doing to his own daughter after his disastrous wish for gold.

The ancient myths provide us with powerful teaching to help us to overcome the "Midas condition" and elevate our spiritual awareness, and to put us in touch with the true divine source which we should acknowledge and recognize and revere and uplift.

I am convinced that our ability to hear their powerful message is greatly enhanced when we begin to understand the celestial language that they are speaking.

Shango and Oya of the Yoruba

2016 December 04

Abundant evidence from myths found around the world, on every single inhabited continent on our planet -- as well as the inhabited islands of the vast Pacific Ocean -- points to the incredible conclusion that these ancient myths all appear to be built upon a common system of celestial metaphor.

This same worldwide system underlies the stories of what we call the Old and New Testaments of the Bible, as well as the myths, scriptures and sacred traditions of ancient India, ancient Greece, ancient China, ancient Japan, ancient Egypt, ancient Mesopotamia, the peoples of various part of Europe, and of the Americas, and other parts of Asia, and the islands of Polynesia and Micronesia, and the same system can even be seen to form the basis for myths and sacred traditions found in Australia and in Africa.

Some of the myths and sacred traditions of Africa are explored in *Star Myths of the World, Volume One*, which seeks to provide an overview of representative Star Myths from numerous cultures on different continents (whereas *Star Myths of the World, Volume Two* and *Star Myths of the World, Volume Three* focus more deeply on myths from Ancient Greece and from the Bible, respectively).

Because Volume One tries to give a broad introduction to the vast scope of this ancient worldwide system, only a few myths from each different continent could be highlighted. Entire multi-volume sets could of course be written on the Star Myths of each of the different traditions, showing ways that the myths of all these different cultures appear to be based on the motions of the constellations and other heavenly bodies and heavenly cycles.

Many more myths and sacred stories from the continent of Africa and its many different cultures and myth-systems could be explored in addition to those featured in Volume One of the *Star Myths of the World* series.

One of the myth-cycles that could be explored would be the myths and traditions surrounding two important Yoruba deities or Orisha: Shango and Oya.

Shango is a powerful Orisha of fire and of thunder and lightning.

In his 1980 study of Yoruba oral tradition and divination entitled *Sixteen Cowries*, William Bascom writes of Shango (sometimes also spelled Xango):

> Shango is a God of Thunder. Living in the sky he hurls thunderstones to earth, killing those who offend him or setting their houses afire. His thunderbolts are prehistoric stone celts which farmers sometimes find while hoeing their fields; they are taken to Shango's priests, who keep them at

his shrine in a plate supported by an inverted mortar, which also serves as a stool when the heads of initiates are shaved (cf. Bascom 1972: 6). The stones in Shango's sacrifices may be an allusion to his thunderbolts, and in one verse Shango kills a leopard by putting an inverted mortar over it. [. . .]

He was noted for his magical powers and was feared because when he spoke, fire came out of his mouth. One verse has Shango lighting a fire in his mouth with itufu, oil-soaked fibers from the pericarp of the oil palm, which is used in making torches and starting fires. In a state of possession it is said that a Shango worshiper may eat fire, possibly using itufu, carry a pot of live coals on his head, or put his hand into live coals without apparent harm. 44.

Shango is a formidable deity or Orisha -- but so is his favorite consort, the goddess Oya. William Bascom describes her thusly:

Oya is the favorite wife of Shango, the only wife who remained true to him until the end, leaving Oyo with him and becoming a deity when he did. She is Goddess of the Niger River, which is called the River Oya (odo Oya), but she anifests herself as the strong wind that precedes a thunderstorm. When Shango wishes to fight with lightning, he sends his wife ahead of him to fight with wind. She blows roofs off houses, knocks down large trees, and fans the fires set by Shango's thunderbolts into a high blaze. When Oya comes, people know that Shango is not far behind, and it is said that without her, Shango cannot fight. The verses tell that Oya is the wife of Shango, "The wife who is fiercer than the husband." Her town is Ira, which is said to be near Ofa. 45.

Bascom also notes that Oya is associated with buffalo's horns, and that a set of buffalo horns will be rubbed with cam wood to make them red and placed on Oya's shrine. In another book discussing the mythology of the Yoruba, *Yoruba Myths* by Ulli and Georgina Beier (1980), we learn that one time, when Shango and Oya were having a fight,

> she charged him with mighty horns. But Shango appeased her by placing a big dish of akara (bean cakes) in front of her. Pleased by the offering of her favourite food, Oya made peace with Shango and gave him her two horns. When he was in need, he only had to beat these horns one against the other and she would come to his aid. 32 - 33.

Based on these details from the different sacred traditions involving Shango and Oya, I believe we can very confidently identify Shango and Oya with the constellations Hercules and Virgo. Below is a star-chart showing some of the features of these constellations which correspond to aspects of the mythology of Shango and Oya:

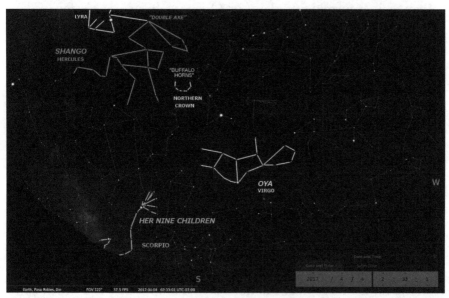

The details of the stories may have already tipped you off to this conclusion, if you have worked your way through previous Star Myth examinations presented on this blog or in the "Myths" section of the *Star Myth World* website, as well as some of the "Videos" on the same website, and especially if you have worked your way through any of the volumes of the *Star Myths of the World* series of books.

In nearly every ancient myth-system, the powerful figure who wields a thunderbolt weapon will be associated with the figure of Hercules in the sky, whether that thunderbolt weapons is wielded by a god in the Maya account contained in the Popol Vuh, or by a god in the myths of ancient Greece and Rome, or in the myths of the Norse.

Images of Shango and symbolic scepters sacred to Shango usually feature a double-axe motif, a potent symbol which is also found around the world. The carved wooden image of Shango shown at top features a wide double-axe above the figure's head, as well as two more smaller double-axes placed in front of the image in the carving.

It is possible that the great weapon held menacingly aloft by the constellation Hercules in the sky, which in some myths becomes a club or a sword, can also be seen as an enormous double-axe shape, especially if the "blade-shaped" outline of Lyra the Lyre nearby is also envisioned as being part of the same weapon (see star-chart above).

There are other details in the myths which give added certainty to the identification of Shango with the constellation Hercules, which we will examine in a moment. First, however, let's look at the identity of the goddess Oya, who is so powerful that Shango cannot fight without her, and who is described as *going ahead of Shango* in everything.

I am convinced that Oya is associated with the constellation Virgo: can you see how this arrangement gives rise to the tradition that Oya always precedes Shango? The motion of the stars each night is from east to west (just like the motion of the sun each day -- both are caused by the rotation of the earth towards the east on its daily rotation). In the star chart above, which looks towards the south, east is on the left and west is on the right, and the constellations move from left-to-right in the diagrams.

The definitive clue that Oya is associated with Virgo is the fact that she is sometimes called the "Mother of Nine" (*Iyansan*, or *'Yansan*) in Yoruba tradition (Bascom, 45). The constellation Virgo, as we have seen in many myths from around the world, is often envisioned as a mother about to give birth, due to her posture in the sky, lying on her back with feet elevated.

Virgo is sometimes envisioned as giving birth to the multi-headed figure of Scorpio, which follows Virgo in the sky. Scorpio, as seen in many of the discussions in the *Star Myths of the World* books, is sometimes envisioned as having *nine heads*. The fact that Oya is called "Mother of Nine" pretty much seals her association with the constellation Virgo in the heavens.

You can also see the "buffalo horns" which Oya gave to Shango, almost certainly identified with the beautiful arc of stars known as the Northern Crown (or Corona Borealis), very close to Shango-Hercules in the sky and included in the diagram above.

What about the details of the story in which Shango breathes fire out of his mouth? The star chart below shows that the "lower arm" of the constellation Hercules (the arm not holding a club or weapon) can be envisioned as proceeding out of the mouth of the constellation. I believe this is very likely the source of the association of "breathing fire" with this particular Orisha.

There is also a "torch" in the sky not far from Hercules and Virgo, in the form of the constellation Coma Berenices, which actually plays the role of a torch in many other Star Myths (some of them discussed in the *Star Myths of the World* books). This may be the itufa torch that appears in the myths of Shango:

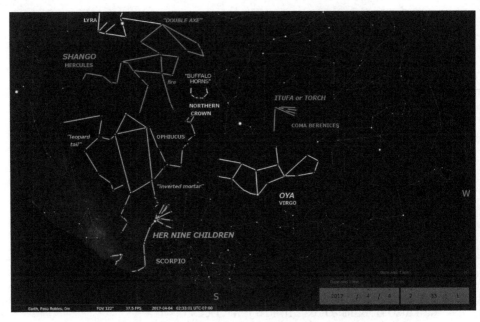

The aspect of the myth in which Shango is described as killing a leopard by crushing it beneath "an inverted mortar" no doubt have to do with the constellation Ophiucus, directly beneath the constellation Hercules. The body of Ophiucus has a distinctive oblong shape with triangle at top (as outlined by the ingenious outlining system proposed by H. A. Rey). This almost certainly represents the inverted mortar (a mortar and pestle are tools for crushing up grains and spices: the mortar usually a stone bowl with a depression or hole in the center, and this shape also gives its name to the later weapon known as a mortar, which shoots shells out of a tube – Ophiucus could be envisioned as a tall mortar, turned upside down so that its conical base is at the top).

Note that the head and tail of the unfortunate leopard can be seen protruding from either side of the upturned mortar of Ophiucus!

Note also that Shango is sometimes described as defeating his enemies with a cudgel, which is another weapon very closely associated with the outline of the constellation Hercules (and Hercules-figures throughout the world will often carry a club or cudgel as their favorite weapon). William Bascom cites Yoruba

437

verses in which Shango uses a cudgel in the verse labeled "L1" in *Sixteen Cowries*, and he mentions this fact on page 44 as well. This cudgel is yet another clue that Shango corresponds to the constellation Hercules – in addition to all the other clues, I believe we can be quite confident in associating Shango with Hercules, and Oya with Virgo.

In another set of verses cited by William Bascom, we learn of Shango that: "He drove away the hartebeeste that had been eating the children of the people of Ijagba, and became the deity that all the people of Ijagba worshipped" (45).

A hartebeeste is a large African ungulate, with majestic curving horns. As you can see in the image below, it is very possible that this hartebeest which Shango drives away might be associated with the outline of the horned figure of Taurus the Bull, which can also be said to resemble a hartebeeste:

438

As the constellations Virgo and Hercules rise over the horizon in the east, the constellation Taurus can be seen to be sinking down into the west. This perfectly describes the situation in which Shango (Hercules) "drives away" the hartebeest (Taurus). There are other Star Myths from around the world in which the arrival of a god or goddess associated with Hercules or Virgo signifies the demise of a figure associated with Taurus (for instance, the traditions associated with the goddess Durga[367]).

Below is a star-chart showing Hercules rising in the east, as Taurus sinks down in the west:

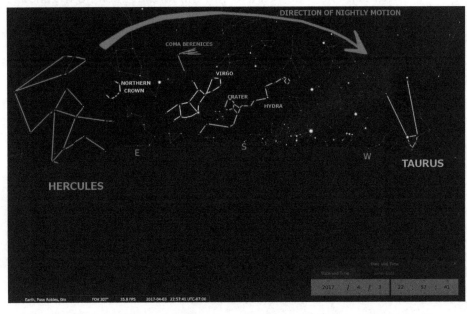

Note that in the diagram, the planetarium app distorts the size of constellations along the east and west, in order to simulate the "wraparound" effect you would see outside (the left and right of the image represent turning towards the eastern and western horizons, respectively; looking to the center of the image represents looking towards the southern horizon: the planetarium app from Stellarium.org makes constellations look smaller when they are in the middle of the diagram, and larger when they are near the east or the west to your left and right).

In Yoruba Myths, we read that the goddess Oya was originally an antelope who periodically took off her antelope skin to reveal a beautiful woman:

> Oya was an antelope who transformed herself into a woman. Every five days, when she came to the market in town, she took off her skin in the forest and hid it under a shrub.

> One day Shango met her in the market, was struck by her beauty, and followed her into the forest. Then he watched, as she donned the skin and turned back into an antelope.

> The following day Shango hid himself in the forest, and when Oya had changed into a woman and gone to market he picked up the skin, took it home and hid it in the rafters. 33.

We learn that Shango's other two wives become jealous of Oya, who bears Shango twins, and they tell Oya where to find her skin, hanging in the rafters. She dons the antelope form again and disappears into the forest.

I believe that this story is also based upon the same celestial mechanics shown in the star-chart above. When Virgo takes off her antelope skin and hides it beneath a bush, Taurus is sinking down into the horizon (into the bushes of the horizon, you might say).

The part about hanging the skin up in the rafters resonates with a very common theme in Star Myths around the world – for instance, in the Maui myths of the Pacific, Maui's grandfather hangs Maui up in the rafters when he is a baby! In that case, the grandfather is undoubtedly Hercules, and Maui the infant is almost certainly Corona Borealis (the Northern Crown). This identification is discussed in greater detail in *Star Myths Volume One*. The Northern Crown, oddly enough, plays a baby in many other Star Myths.

440

In the story of Shango and Oya and the hanging of Oya's skin in the rafters, I believe the constellation Coma Berenices fits the identification of the skin better than the nearby Northern Crown. There are other Star Myths we could look at which make this identification the likely answer to the celestial source of this sacred Yoruba story.

When the other wives tell Oya where her skin is hanging (they do this using a chant, in which they sing about its location), she resumes her antelope form and bounds away into the forest so that Shango cannot find her. Again, this detail probably stems from the fact that, when Hercules rises in the east, Taurus sinks down out of sight in the west.

For those who wonder whether Taurus could play the role of the female antelope which is one of the shapes of the goddess Oya, note that in Africa the female of the many species of antelope often has horns, in addition to the horns of the male antelope.

Below is an illustration of the female Hirola, an antelope found in the areas where the Yoruba cultures traditionally have lived for thousands of years:

There are still more stories of Oya and Shango which point to a celestial foundation associated with the constellations Virgo and

Hercules. One of the stories involves the mother of Shango, the goddess of the River Yemoja (*odo Yemoja*), which flows through Yoruba lands. In one version of the story, Yemoja was pursued by her husband Okere (who is not the father of Shango) and he knocks her down, causing her to turn into a river which flows out of pots of water she was also carrying (Bascom, 46).

This story may also involve the constellation Virgo, which is located in the sky adjacent to the constellations Crater the Cubp and Hydra the Snake (see star-chart above). When Shango's mother Yemoja falls down (and note that Virgo is recumbent), she may become the river which is associated with the flowing form of Hydra, directly beneath Virgo and beneath the water-cup-like outline of the constellation Crater.

Shango and Oya are very important deities in the Yoruba mythology, with many devotees around the world to this day. Their clear celestial parallels provide still more evidence which argues that the system of celestial metaphor which we can see operating in the stories of the Bible and in the other myths of the world, is in fact a common system which somehow provides the underlying bedrock upon which all the world's ancient traditions have their foundation.

Shango is a god of fire. I believe that the world's Star Myths convey powerful truths regarding the Invisible World – the realm of spirit, the realm of the gods, the Infinite Realm.

One of the lessons that they teach is that, just as the stars themselves can be seen to rotate down to sink into the western horizon, so also we ourselves came down to this incarnate realm from a spirit realm -- and that we all contain a divine spark, an internal divine fire, through which we have immediate access to that Invisible Realm at all times, if we learn how to become re-acquainted with that aspect of our nature.

I am convinced that the ancient myths, scriptures and sacred stories which were entrusted to humanity the world over are here to help show us how to do that.

442

Revelation chapter 22 and
the Egyptian Book of the Dead

2017 March 18

My most recent book, *Star Myths of the World and how to interpret them, Volume Three* (*Star Myths of the Bible*), examines hundreds of stories, characters, events and themes in the Old and New Testaments of the Bible to show that they are based on the same system of celestial metaphor which forms the foundation for virtually all of the other ancient myths, scriptures, and sacred stories from around the world.

The book totals 766 pages, with over two hundred seventy full color images, including well over one hundred star charts showing connections between specific constellations and aspects of various episodes in the scriptures. Even so, due to the limitations of space, many important events and episodes had to be left out.

The apocalyptic literature in the book of Revelation is touched upon in one section, which deals specifically with events described in Revelation chapters 9, 10 and 12. An entire book could of course be written about the celestial metaphor present in the book of Revelation alone -- in fact, it would probably take up more than one volume. Therefore, I did not deal with the imagery found in the final chapter of Revelation, chapter 22 -- although it is of course full of important material worthy of study.

Recently, Moe Bedard, host of the *Gnostic Warrior* podcast and website, invited me back for another interview (you can find the previous interview, which was published in October of 2014, online). Moe asked me about a specific passage in Revelation 22, into which I had not previously delved deeply (as I had selected other parts of Revelation, which have more obvious and unambiguous connections to specific constellations, to study more closely instead, when preparing *Star Myths of the Bible*).

However, because it is a very important chapter, and because listeners to the podcast might be wondering what my analysis of Revelation 22 would contain, I promised to look into it further and follow up with some thoughts on the contents of the passage in question.

The final chapter of Revelation 22 can be found online in many places; one good resource where you can view this text in various translations is the *Blue Letter Bible* project online.

The verse in chapter 22 which Moe focused on in his question is verse 16, which reads: "I Jesus have sent mine angel to testify unto you these things in the churches. I am the root and offspring of David, *and* the bright and morning star." The translators of the King James version, which is cited here, italicize words which they added based on their assessment of the best syntax to convey in English what they felt the text was conveying in the original language -- thus, in the final sentence, the actual text does not contain the word for "and" in between the two titles, but simply declares, "I am the root and offspring of David, the bright and morning star." That's why the final "and" in the sentence is written in italics.

The imagery in Revelation 22, like the imagery in the rest of the book of Revelation, can be shown to incorporate celestial imagery which points to specific regions of the night sky. For examples which support the argument that the text of Revelation consists of celestial metaphor, see a previous post from 07/06/2012 discussing some of the imagery in Revelation 9, or this previous post and accompanying video from January of this year (10th of January, 2017), discussing Revelation 12.

For example, at the beginning of the chapter, the narrator is describing what is being shown to him by the angel described in the previous chapter's ninth verse. The text begins:

"And he showed me a river of water of life, clear as crystal, proceeding out of the throne of God and of the Lamb. In the midst of the street of it, and on either side of the river, *was there* the tree of life, which bare twelve *manner of* fruits, and yielded her fruit every month: and the leaves of the tree *were* for the healing of the nations" (Revelation 22: 1 - 2).

This river of water of life is almost certainly the shining band of the Milky Way galaxy, which features prominently in many of the metaphors contained in the book of Revelation, and which is described as a river in many other stories in the Bible (including the river in which Jacob wrestles at the time that his name is changed to Israel, in Genesis 32, which is discussed at length in *Star Myths of the Bible*), as well as in many other Star Myths from other cultures around the world.

For evidence that the Milky Way plays the role of a river in certain ancient myths and stories, see for example the discussion of the story of King Midas in the myths of ancient Greece found in this previous post from October 2016, in which I present evidence that the River Pactolus described in the Midas myths can be identified with the Milky Way band.[415]

The river is described in verse 1 as proceeding out of the throne of God and of the Lamb – both of which can be identified with specific constellations discussed in *Star Myths of the Bible*, beside which the Milky Way flows: the constellations Hercules and Aries.

The tree in the midst of the river of water of life, described as being both "in the midst of the street of it" and also "on either side of the river," may well describe the constellation Ophiucus, which can be seen to be "standing" with one foot in the river in the star-chart below. The serpent which is held by Ophiucus, and which stretches out to either side of the constellation, is very frequently associated with a tree in numerous Star Myths from around the world – especially the upper part of the half of the serpent on the west side of the body of Ophiucus (the "head-end" of the serpent, on the right side as we face the image below, in which the view is taken as though facing towards the south from a point in the northern hemisphere):

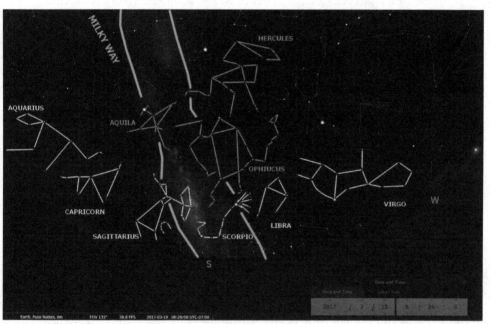

Many such examples of the upper part of the western half ("head-end") of the serpent of Ophiucus being associated with descriptions of a tree in various myths can be found in the *Star Myths of the World* series -- especially in Volumes Two (Greek mythology) and Three (the Bible).

446

However, as we saw a moment ago, the text of Revelation 22: 2 states very specifically that the tree in the "midst" of the river is also "on either side of the river," and so part of it must be envisioned as being on the east side of the Milky Way (that is to say, the left side in the image above).

If Ophiucus is indeed playing the role of a branching tree which stretches to either side of the river, then what stars in the chart above could represent the branches of the tree on the east of the river? I would propose that some of the stars of the constellation Aquila the Eagle may be envisioned as being connected to the eastern half of the serpent that Ophiucus is holding.

Note that the upper part of the Ophiucus serpent comes very close to the outline of Aquila -- and note that of course there are no actual "connecting lines" between the stars in the sky, which means that we can imagine a line connecting Ophiucus and Aquila if we want to do so, or imagine no line between them. There are many Star Myths which I have analyzed in which an "optional line" appears to have been envisioned by the originators of the ancient myths, such as those in which a connection is envisioned between the "head-end" of the serpent of Ophiucus (on the west or right side of the image above) and the "downward-reaching arm" of the constellation Hercules.

Note also that the rest of the description in Revelation 22: 2 goes on to describe the fruits of the tree as being twelve in number or in type, and to be produced "every month." Any time we find a reference to the number *twelve* such as this reference in Revelation 22, a possible explanation to consider would of course be the idea that the passage may be referring to the stations of the zodiac -- and indeed, I believe that this is a good explanation for this reference to the twelve fruits of the tree.

The confirmatory detail which the text provides in order to increase our confidence in this conclusion is the statement that the tried "yielded her fruit every month." The earth's annual orbit around the sun causes the sun to move through the twelve signs

447

of the zodiac as we go through the year, spending about one month in each. Thus, the tree's twelve types of fruit, which are yielded each month, are almost certainly a reference to the zodiac band, which stretches out to either side from the base of the tree – almost like fruit that has fallen down periodically, once per month.

Note in the image above that the zodiac band does indeed stretch outwards on either side of the upright form of Ophiucus. The constellations of the zodiac visible in the chart above are outlined in yellow, beginning with Aquarius on the left side of the image (the east side) and proceeding to Capricorn, Sagittarius, Scorpio, Libra and Virgo. These zodiac constellations, I'm convinced, represent the "fruit" that the tree of life has "yielded" (or "given up") each month, described in Revelation 22: 2.

There are many other examples of celestial metaphor in this chapter of Revelation, many of them using the very same constellations but clothing them in different metaphor. For example, in verse 8, the text tells us: "And I John saw these things and heard *them*. And when I had heard and seen, I fell down to worship before the feet of the angel which showed me these things" (Revelation 22: 8).

In this particular part of the chapter, I believe that the same constellation Ophiucus, which was just a moment before playing the role of the tree of life in the midst of the pure river of water of life, is now playing the role of the angel, before whose feet John says he fell down to worship. Note that just below the feet of the outline of Ophiucus we find the constellation Scorpio, a constellation with a "bowing down" posture, and one which will sometimes play just such a role (of a person bowing down) in the world's various myths.

Immediately after this, the angel tells John not to bow down in this way, and then goes on to make a series of declarations, including the declaration in verse 13: "I am the Alpha and Omega, the beginning and the end, the first and the last," as well as the

448

later declaration in verse 16 cited above: "I Jesus have sent mine angel to testify unto you these things in the churches. I am the root and offspring of David, *and* the bright and morning star."

This series of declarations seems a little confusing, since at its outset it is clearly the angel who is addressing the narrator, but the unbroken series of proclamations then continues to the statement "I am the Alpha and Omega" and the declaration "I Jesus have sent mine angel to testify."

However, I believe any difficulty is resolved by the understanding that we are still dealing with the constellation Ophiucus, a constellation which actually does have a few of its stars crossing the ecliptic path of the sun, thus qualifying it as being part of the zodiac band, even though Ophiucus is not normally associated with the twelve signs of the zodiac. For this reason, Ophiucus is sometimes described as being the "thirteenth zodiac sign."

This placement of Ophiucus, above the majority of the constellations arrayed along the ecliptic path, but still crossing with some of its lower stars into the zodiac band, undoubtedly accounts for statements such as "I am the Alpha and Omega, the beginning and the end, the first and the last." All of these statements are appropriate for a point on an endless ring or circle, which describes the zodiac itself. Any point on a circle can be described as being both a "beginning and end," or as being both "first and last" in a procession of figures arranged in a great circle.

When the one speaking to John then identifies himself by the specific name of Jesus and gives his descriptions as "the root and offspring of David" as well as "the bright and morning star," I believe we should also look to the constellation Ophiucus.

The constellation Ophiucus can indeed be seen as both "the root and offspring of David," if we understand that the figure of King David in the Old Testament is almost always associated with the powerful outline of the constellation Hercules, located immediately above Ophiucus. The evidence to support this assertion is provided in the lengthy examination of the figure of

David found in *Star Myths of the Bible*, particularly in pages 145 through 157 and pages 515 through 542.

The image below, showing a painting by Flemish master Peter Paul Rubens (1577 - 1640), shows David in a posture that is almost identical to the outline of the constellation Hercules in the sky (and reveals that the connection between the figure of David and the constellation Hercules must have been known in previous centuries):

The constellation Ophiucus, being directly beneath Hercules, can be said to be the "root" of that constellation in the same way that a root of a tree is directly beneath a tree, and supports it from under the ground. Alternately, the constellation Ophiucus can also be seen as the "offspring" of the constellation Hercules, in that it is directly below Hercules in the sky and thus "descended" from it, metaphorically speaking.

Thus, the heavenly speaker showing these visions to the narrator in the 22nd chapter of Revelation, who in verse 16 explicitly identifies himself as "I Jesus," is also one and the same with the tree of life, which stands in the midst of the pure river of the Milky Way, and whose "fruit" is yielded up each month and identified with the circular band of the zodiac, and who is both "the first and the last" of that same zodiac ring, and whose leaves are "for the healing of the nations."

But what about the declaration that accompanies the description in verse 16 of the "root and offspring of David," the proclamation that the speaker is also "the bright and morning star"? There certainly do not seem to be any bright morning stars in the constellation Ophiucus, which is a constellation without any extremely bright stars, and one that is not extremely easy to locate (although once you know how, it is not too difficult, but instead extremely satisfying, to locate Ophiucus – a post from 12/26/2016 gives some tips, although you will need to look for Ophiucus during a time when the other constellations in the star chart shown above are also visible in the sky).

For help with this question, I turned to the indispensable Alvin Boyd Kuhn (1880 - 1963), who in *Lost Light* (1940) argues that this phrase "the morning star" was anciently identified by the Egyptians with the star Sothis – the Greek version of their name for the star we call Sirius, brightest of all the stars in the heavens (other than the sun itself, of course).

There, on page 545, Kuhn declares:

> The morning star (at one time) was Sothis: the watch-dog
> that barked to announce the coming of the Day-Star from
> on high, as the ape clicked at the rising sun.

We know from ancient texts that the ancient Egyptians eagerly
awaited the first morning re-appearance of the star Sothis or
Sirius (the heliacal rise of Sirius) and began their year with it.

The fact that the rising of Sirius *in the morning* marked the
beginning of a new year in ancient Egypt (as well as in some other
ancient civilizations), is a strong argument for the possibility that
Sirius is indeed the "bright and morning star" mentioned in the
text. Not only was Sirius eagerly awaited in the morning, but
Sirius is also the brightest of stars in our heavens, and what's
more, the heliacal rise of Sirius marked *a new year.*

It should be quite obvious that the start of a new year can be
accurately described as "a beginning and end."

It is also perhaps significant that the constellation Ophiucus and
the star Sirius are almost 180 degrees apart from one another in
the void of space, meaning that they will rise about twelve hours
apart from one another. By saying that he is both "the root and
offspring of David" (associated with Ophiucus) and also "the
bright and morning star" (probably associated with Sirius), the
speaker in this passage is identifying not only with one particular
point on the great wheel of the year, but rather with the whole
thing (both halves of the ring).

But there is still more significance to the assertion by Alvin Boyd
Kuhn that the identity of the morning star is the bright star
Sirius, the morning star of ancient Egypt whose appearance
marked the end of one year and the beginning of the next.
Because the presence of that reference, in a passage in the New
Testament which also contains a reference to "the beginning and
the end, the first and the last," betrays a clear connection to the
sacred traditions of ancient Egypt – and the possibility that the

452

imagery in this "New" Testament text has roots that are much, much older than is commonly understood.

And, on the very same page that he makes the identification of the morning star with Sothis, Alvin Boyd Kuhn cites a passage from a translation of the Book of the Dead as found on the papyrus of Nebseny (also commonly rendered "Nebseni") which contains clear resonance with verses in Revelation 22.

Nebseny was a scribe and probably a priest of ancient Egypt, believed to have lived during Eighteenth Dynasty of ancient Egypt, which is thought to have stretched from about 1550 BC to 1298 BC and to include famous rulers such as Hatshepsut, Amenhotep I, Amenhotep II, Amenhotep III, Akhenaten, and Tutankhamun, among others. Some scholars believe that Nebseny was probably writing during the period around the time of the reign of Thutmose IV, whose reign is believed to have lasted from about 1398 BC to 1388 BC.

The texts which we know today as the Egyptian Book of the Dead were not actually fixed in number or in order or sequence in ancient times, but appear to have been selected from a larger corpus of possible sayings or spells. The Pyramid Texts found in the pyramid of Unas, for example (who is thought to have reigned from 2375 BC to 2345 BC), which form some of the earliest texts which have been preserved to modern times, are actually versions of this same corpus.

The version of the Book of the Dead which was buried with Nebseny contains illustrations of very high artistic quality. There is some evidence for scholars to conclude that Nebseny might have prepared these extensive texts himself, which makes them especially interesting and poignant. They contain illustrations of Nebseny as well as his wife Senseneb, and three of his children (see discussion on page 279 in *Journey Through the Afterlife*, edited by John H. Taylor).

The papyrus of Nebseny is now located in the British Museum in London (I wonder what he would have thought about that), and has been since the 1830s. It was one of the first versions of the Book of the Dead to be translated. The first translation by Samuel Birch (1813 - 1885), from the 1860s, can be found online at http://www.masseiana.org/ritual.htm. A translation into the French language from 1885 can be found online at : http://gallica.bnf.fr/ark:/12148/bpt6k5803525d/f2.image

The passage from the papyrus of Nebseny cited in *Lost Light* on page 545 comes from chapter (or "spell") 149. It reads (in the translation by Birch):

> Made the day of the month of festival of the Sixth and the festival of the Fifth, of the festival of the Lintel, that of Thoth, that of the birthday of Osiris, of Skhem, and the night of the festival of Haker, the mysteries of the Gate, and of traversing the secret place in Hell, prevailing against the Evil, pawing the secret valleys, the mouth and path of which are unknown, corroborating the Spirit who stretches his legs, to go his journey correctly or making a hole in it to pass through it with the God. No man sees it except a king and a priest, no slave's face looks at it. Every Spirit for whom this book has been made having come and gone round, his Soul comes away on the day with the living, he has prevailed as the Gods do, he is not stopped in true linen for a million or times. The Gods, they approach him, they touch him, for he is like one of them; he lets [them] know what he has done in this secret book of truth. There is not known any such anywhere or ever; no men have spoken it, no eye has perceived it, no ear has heard it, not any other face has looked in it to learn it. Do not though multiply its chapters, or do not thou let any face except thy own [see it] and eat thy heart, doing it in the midst of the Hall of Clothes, it is put forth by the God with all his power. It is a true secret; when it is known, all the providers in all places supply the Spirits in Hades, food is given to his Soul on earth, he is made to live for ever, nothing prevails against him.

The wording as quoted by Kuhn is slightly different, and comes from a version contained in publications by Gerald Massey (1828 - 1907), who lived before Kuhn and whose work Kuhn held in high regard (though Kuhn differed with Massey on a few important points). I suspect the version cited by Massey and Kuhn might have been Massey's own translation of Nebseny. It reads:

> By this book the soul of the deceased shall make its exodus with the living and prevail amongst, or as, the gods. By this book he shall know the secrets of that which happened in the beginning. No one else has ever known this mystical book or any part of it. It has not been spoken by men. No eye hath deciphered it. No ear hath heard of it. It must only be seen by thee and the man who unfolded its secrets to thee. Do not add to its chapters or make commentaries on it from the imagination or from memory. Carry it out in the judgment hall. This is a true Mystery unknown anywhere to those who are uninitiated.

Massey had already noted, on page 726 of his *Ancient Egypt, the Light of the World*, the startling similarity between the language of this ancient text from the Book of the Dead and the passage in the Apocalypse of John (the book of Revelation included in the New Testament canon), chapter 22 and verses 18 and 19. Alvin Boyd Kuhn adds the further connection to the language of the writer who calls himself Paul in Ephesians chapter 3, and who (Kuhn says): "speaks in quite similar terms of a mystery made known to him 'which in other ages was not made known unto the sons of men; as it is now revealed unto his holy apostles and prophets by the spirit'" (545).

Kuhn also should have added that in 1 Corinthians 2: 9, the same writer Paul (citing Isaiah 64: 4) says, "But as it is written, Eye hath not seen, nor ear heard, neither have entered into the heart of man, the things which God hath prepared for them that love them," both passages being clear derivatives from the Egyptian Book of the Dead as cited in the passage above from the papyrus of Nebseny.

It would seem from these connections that the wisdom encoded in the so-called "New" Testament is in fact extremely ancient indeed!

And that is by no means the end of the parallels between the celestial imagery in the book of Revelation and that found in the Book of the Dead. There are many others -- and parallels to other parts of the New Testament as well. For example, in the version found in the papyrus of Ani, an Egyptian priest and scribe thought to have lived around 1250 BC (over a hundred years after Nebseny), the 149th chapter describes the ship which rows in the Field of Reeds, and declares, "I know the gate in the middle of the Field of Reeds from which Re goes out into the east of the sky, of which the south is the Lake of Waterfowl and the north is the Waters of Geese," and also says, "I know those two trees of turquoise between which Re goes forth, and which have grown up at the Supports of Shu at that door of the Lord of the East from which Re goes forth" (translation by Dr. Raymond O. Faulkner, found on page 121 of the complete reproduction and translation of Papyrus of Ani cited in the Bibliography of this text).

This imagery is celestial, and has clear parallels to the celestial explication of Revelation 22 we have just been exploring. The "gate" in the "middle of the Field of Reeds" is the same constellation Ophiucus we see described as standing in the "midst" of the crystal river in Revelation. If you look at the star-chart above, you can see why Ophiucus often plays the role of a gate in ancient myth (many examples can be found in *Star Myths of the World*, especially Volumes Two and Three). The description of the "Waters of Geese" which are located to the north undoubtedly describe the portion of the Milky Way in which the constellation Cygnus the Swan can be seen to be flying, just a little ways above (that is, just north) of the constellation Aquila in the star-chart. And, "those two trees of turquoise between which Re goes forth" are undoubtedly related to the tree of life described in the vision of John the Revelator, as well as to trees described in the visions in the Old Testament scriptures, such as those of Daniel and of Ezekiel.

What, then, could be the message of Revelation 22? What are all these celestial metaphors trying to convey?

I personally believe that the best approach to the ancient wisdom contained in the scriptures, myths, and sacred stories of the world would be to seek those answers in the myths themselves, rather than to rely upon the interpretation of any individual. I believe that the myths (or our Higher Self) will tell us the answer, if we approach them in the right way (and I also believe it helps if we understand the celestial and metaphorical language that they are speaking).

I also believe that the message has many layers, and may be literally bottomless in its profundity, for any very ancient scripture or myth.

Nevertheless, in order to try to helpfully point readers what I believe to be the right direction regarding the verses found in Revelation 22, I would begin with some of the assertions Alvin Boyd Kuhn makes in the pages surrounding the observations cited in the above discussion.

He says, for example, that:

> The Messianic Son came ever as the manifested and witness for the father, who had sunk his life in matter to reproduce himself in his next generation. According to Herodotus (2: 43) the Egyptian Jesus with the title of Iu-em-hetep was one of the eight great gods who were in the papyri twenty thousand years ago! He bore a different name according to the cult [in other words, each different "cult" or culture gave him a different name]. To the sages of old time the coming was a constantly recurring and only typical event. The ancient Messiah was a representative figure coming from age to age, cycle to cycle. He came "each day"

in the Ritual; he came periodically; he came "regularly and continuously." He came once through the cycle; but his solar and lunar and natural *types* came cyclically and in eternal renewal. The Egyptian Messiah was one whose historical coming was not expected at any date, at any epoch. The type of his coming was manifest in some phenomenon repeated as often as the day, the year, or the lunation came around. The constant repetition of type was the assurance of its unfailing fulfillment. [. . .]

The coming was taking place in the life of every man at all times. Each man had his evolutionary solstice, his Christmas; and he would have his Easter. The symbols were annuals; the actual events they typed in mankind's history were perennials. In nature every process is but typical and repetitive. But it is typical of all other process of life in its entirety.

Horus, a form of Iu-em-hetep, was not an individual historical person. For he says: "I am Horus, the Prince of Eternity." Jesus was with the Father before the foundation of the worlds. Horus calls himself "the persistent traveler on the highways of heaven," and "the everlasting one." "I am Horus who steppeth onward through eternity." Here is wisdom to nourish the mind and lead it out of its infantile stage into maturity of view. Horus declares himself forever above the character of a time-bound personage, an indestructible spirit that advances onward through one embodiment after another to endless days. 546 - 547.

All of these themes can clearly be seen to be present in the passages of Revelation 22 we've been exploring, with their declarations of "the beginning and the end" and of the "bright and morning star" which marks the renewal of the year.

It should be quite clear that the book of the Apocalypse is not about a literal "end of the world" but rather about renewal, which is figured or "typed" in all the heavenly cycles, and that it is not just about the heavenly cycles themselves but about the way in which these majestic and awe-inspiring heavenly motions convey to us truths about our own cycle of descent of the spirit or divine nature into matter (in our incarnation) and its elevation and restoration, which involves transfiguration and transformation (related to the central theme of alchemy).

As Alvin Boyd Kuhn reminds us in a related passage in *Who is this King of Glory?* (1944):

> It is ever to be remembered that the "deceased" in the Egyptian Ritual is the living mortal, not the earthly defunct; and therefore its making its exodus among the living is a reference to its coming to full development in the life on earth. The great Mystery is of course the whole import and reality of life in the cycles, the secret wisdom that the soul picks up throughout its whole peregrination through the kingdoms of organic existence. It unfolds in course as the cycling spiral of experience extends. 408.

Our examination of Revelation 22 has thus furnished yet further powerful evidence that the ancient myths of the world are all closely related, and based upon a now largely-forgotten system of celestial metaphor.

They are designed to impart profound spiritual wisdom for our benefit in this life -- for our soul's uplifting, its renewal, its blessing, and its transformation.

Self and Higher Self

Plotinus and the upward way

A recent post discussed the well-known myth of the Judgment of Paris, and the fact that the second century writer Apuleius (born c. AD 125) somehow saw the "damnation of mankind" as connected to Paris' selling of his vote in that contest for the "lucre of lust."

The philosopher Plotinus (c. AD 204 - AD 270) may be able to shed some light on this interesting comment from Apuleius. We last heard from Plotinus in the post entitled "giving forth, without any change in itself, images or likenesses of itself, like one face caught by many mirrors," after a line from the First Ennead of Plotinus (*Enn.*I, 1:8).

This idea of a mirror was clearly central to Plotinus' teaching on the nature of human existence. Later, in Ennead IV, 3:12, he writes:

> The souls of men, seeing their images in the mirror of Dionysus as it were, have entered into that realm in a leap downward from the Supreme: yet even they are not cut off from their origin, from the divine Intellect; it is not that they

have come bringing the Intellectual Principle down in their fall; it is that though they have descended even to earth, yet their higher part holds for ever above the heavens.
[Ttranslation by Stephen MacKenna and B.S. Page, 148].

Thus, Plotinus is teaching that the mirror is a good metaphor for the relationship between the Intellectual Principle and the souls of men. In an essay entitled "Judaism, Judaic Christianity, and Gnosis," Professor Gilles Quispel explains:

> The mirror is a powerful symbol in Greek and Gnostic religion. Narcissus is said to have jumped into the water and to have embraced his own shadow and to have drowned, when he looked into the water and saw his own shadow and fell in love with it. This is not true. For he was not suffocated in the water but he contemplated in the transient and passing nature of his material body, his own shadow, namely the body, which is the basest *eidolon* of the real soul. Desiring to embrace this, he became enamoured with life according to that shadow. Therefore he drowned and suffocated his real soul and a real and true life. Therefore the proverb says, 'Fear your own shadow.' This story teaches you to fear the inclination to prize inferior things as the highest, because that leads man to the loss of his soul and the annihilation of the true Gnosis of ultimate reality. Thus the *Anonymus de incredibilibus* IX.

> Nonnus tells us that the young Dionysus was looking in a mirror when the Titans tore him into pieces [. . .]. 5

Professor Quispel notes that French philosopher Jean Pepin (1924 - 2005) points to the Plotinus passage quoted above as the first conflation of the mirror myth of Dionysus and the reflection myth of Narcissus, which Plotinus combines in *Enn.* IV, 3:12 to illustrate the condition of the souls of men and women in this world. Certain ancient traditions appear to have taught that the fall of mankind could be understood through the metaphor of

Narcissus (or Dionysus), becoming enamored with a reflection in a mirror.

It is important to note that Plotinus does not teach that love of beauty is bad – quite the contrary. In his discussion of "the Upward Way," he notes that there are three paths which lead to the upward way: that of the musician, that of the "born lover," and that of the metaphysician (the philosopher). In his description of the born lover, Plotinus writes:

> The born lover, to whose degree the musician also may attain -- and then either come to a stand or pass beyond -- has a certain memory of beauty but, severed from it now, he no longer comprehends it: spellbound by visible loveliness he clings amazed about that. His lesson must be to fall down no longer in bewildered delight before some, one embodied form; he must be led, under a system of mental discipline, to beauty everywhere and made to discern the One Principle underlying all, a Principle apart from the material forms, springing from another source, and elsewhere more truly present. *Enn.* I, 3:2.

Thus, Plotinus seems to teach that love of beauty is an entry-gate to the upward way, but that the "lesson" for the lover of beauty is to learn to disentangle from being enamored with one specific embodied form (whatever form that lover of beauty is enamored with) and to see that specific form of beauty as a pointer to "beauty everywhere" (this being the very opposite of Narcissus, who could only see beauty in himself), and ultimately to the "One Principle underlying all." Plotinus says that from there, "thence onward, he treads the upward way."

In other words, although enrapture with the reflection of beauty led to the fall ("a leap downward from the Supreme," Plotinus calls it), love of beauty can lead back upwards, if the process can be somehow reversed (directing the gaze from love of the specific image back to the underlying One Principle).

464

These passages from Plotinus appear to shed light on the work of Apuleius, and help us to understand what he meant when he said that the Judgment of Paris was somehow the fall of mankind.

Self, the senses, and the mind

2015 June 09

In the introduction to his famous *Light on Yoga*, Sri B. K. S. Iyengar quotes a passage from the sacred Vedic Upanishad *Katha Upanishad*, or *Kathopanishad*, regarding what Sri Iyengar calls the well co-ordinated functioning of "body, senses, mind, reason and Self" (30).

466

The passage he quotes from that Upanishad comes from the third chapter of Part I, beginning in the third verse -- you can read the entire *Kathopanishad* online at

http://www.gayathrimanthra.com/contents/documents/Vedic-related/katha_upanishad.pdf

in an English translation, and find the passage in question beginning on page nine of fifteen in that file (the page itself bears the page number "7" at the bottom of the image of the page). There, we read:

> 3 Know the atman to be the master of the chariot; the body, chariot; the intellect, the charioteer; and the mind, the reins.
> 4 The senses, they say, are the horses; the objects, the roads. The wise call the atman -- united with the body, the sense and the mind -- the enjoyer.
> 5 If the buddhi, being related to a mind that is always distracted, loses its discriminations, then the senses become uncontrolled, like the vicious horses of a charioteer.
> 6 But if the buddhi, being related to a mind that is always restrained, possesses discrimination, then the senses come under control, like the good horses of a charioteer.

In the translation found in *Light on Yoga* (also given below for comparison), the first mention of Atman is capitalized, and next to Atman in parenthesis the text gives as a "gloss" or synonym the capitalized word: "Self."

Clearly, in this passage, there is a clear distinction being made between "body, sense, and mind" and "the atman -- the enjoyer" (or the Self) which is somehow separate not just from body and senses (which is fairly easy to understand) but also from "mind" -- which is a lot less intuitive.

We don't have much difficulty making a distinction between our "Self" and our body or our senses. However, we are usually accustomed to thinking about "ourselves" as being identical or co-equal with our mind. For example, someone might say they appreciate a man or a woman not for his or her physical beauty or for his or her body, but rather for his or her *mind* – meaning, we usually think, who they *really* are, who they are *inside*.

Why is this passage apparently making a distinction between our mind and the Self? Is this passage from the Vedas teaching us that our true Self is somehow distinct from our mind as well as from our body and our senses?

The distinction is even more evident in the English version of the same passage given in the introduction to *Light on Yoga*:

> Know the Atman (Self) as the Lord in a chariot, reason as the charioteer and mind as the reins. The senses, they say, are the horses, and their objects of desire are the pastures. The Self, when united with the senses and the mind, the wise call the Enjoyer (Bhoktr). The undiscriminating can never rein in his mind; his senses are like the vicious horses of a charioteer. The discriminating ever controls his mind; his senses are like disciplined horses. 30.

How can we understand that we are not the same as our mind? Or, to put it another way, if we are accustomed to thinking of ourselves as co-existent with our "mind," then what definition of "mind" are the above ancient scriptures using, since they obviously are *not* using the word "mind" to mean our True Self?

Obviously, what they are calling "mind" is something from which the Self stands apart -- like a charioteer. The mind is compared to the reins, which the Self uses to control the horses, which themselves are connected to the senses. The mind somehow makes all the difference between being carried away by the power of the senses, and guiding the senses "like disciplined horses."

Recently, I saw a video of a talk given by clinical psychologist and author Dr. Darrah Westrup at the mindbodygreen "Revitalize 2015" conference, in which she discusses another metaphor that (to me) sheds a lot of light on this distinction between the Self and the mind which we see operating in the ancient Vedic scriptures.

Dr. Westrup's talk can be found as the last talk in the segment from "Friday Morning Session One" and it begins at approximately 1:03:00 on that video segment.

In the beginning of her interview, which has been titled "Why Stress is a Healthy Part of a Meaningful Life," Dr. Westrup makes a very interesting comment in response to a question about "dealing with stress."

At about the 1:05:30 mark in the video, when asked about the source of stress and suffering, she identifies a culprit we might not have expected, when she says: "It turns out that our ability to develop and use language is a key player in this."

Interviewer and mindbodygreen CEO Jason Wachob then asks, "Like, just poor communication?"

But that's not what she is pointing towards at all: Dr. Westrup clarifies, "No – just language itself."

It turns out, she says, that while language has tremendous benefits, it is also directly related to what the *Kathopanishad* calls "mind," and it can and does threaten to "run away with us" like the "vicious horses" described in the metaphor above.

With language, we are able to analyze, criticize, evaluate, and project. We can speculate about the future, and we can brood about the past. In fact, long before the invention of "computers" and "virtual space," we could create our own "virtual worlds" with language, in which we can test out ideas or think about future and past events and analyze them from every angle in a "virtual space," in much the same way that a modern aircraft design team might "construct" a jet airplane inside of a virtual "computer-modeled" space, in order to test out its strengths and weaknesses before the actual airplane is ever built in the physical world.

This ability to analyze actions and events from every different angle inside of the "virtual reality" of language, Dr. Westrup says, is an incredibly powerful and potentially beneficial ability and this

469

aspect of language itself must be appreciated in order to understand how it can also lead us astray.

With our internalization of language itself, Dr. Westrup says, which she calls a sort of "verbal virtual reality," we create the virtual-world concepts of "future" and "past," neither of which actually exists in the present. And, while this ability is something we cannot actually get rid of (nor would we want to, she says), it is also the cause of stress and suffering. She says:

> All those concepts – all those concepts -- *I'm inadequate, I'm too fat, I should've, if only* – those are all language-based. As far as we know, only humans have the ability to create *constructs* like that with words -- and then we carry them around -- and it causes a huge amount of suffering. We get ideas about what we should and shouldn't be experiencing; what is and isn't OK.

Again, Dr. Westrup never says that this ability to create such verbal mental constructs is not a good ability: it is vital and necessary to our lives in a myriad of different ways.

Some examples I might offer would be that through language, we can ask ourselves (about the future), "Should I go to that event this weekend, or should I work on the other project that I've been meaning to finish?"

We can ask ourselves (about the past), "Did I turn off the stove burner in the house when I left an hour ago?"

These can be helpful and useful and appropriate things to run through our minds -- as long as they don't get out of control.

But, as Dr. Westrup explains beginning at about the 1:08:00 mark, that is exactly what tends to happen, and why this incredible ability to create "verbal virtual reality" can become a problem, if not understood. And it is here that she offers a metaphor which may shed light on the passage from the *Kathopanishad* quoted earlier, and the distinction between Self and mind.

470

In response to a question from interviewer Jason Wachob about how we can "deal with stressful situations," Dr. Westrup says:

> In my book, *Advanced ACT*, I talk about this metaphor called "the over-eager assistant." So this is the idea of that assistant that's *really, really* trying to help -- so we all probably may have encountered an assistant like this -- always in there, full of ideas, full of suggestions, commentary -- and, you know, a lot of times just not that helpful. And I think of my mind like that: my mind is on overdrive right now, for instance. She's -- my assistant -- is handing me commentary, writing little post-it notes, weighing in, telling me how I'm doing, grading, all of that, OK? And *struggling* with that assistant is not going to do -- it's not going to make her go away. I can't get her to stop -- she's not going anywhere. But what I can do is understand that she's doing her job. My mind in a stressful moment is doing what our minds are supposed to be doing, means well. And so understanding that, kind of allowing that to be there, it's kind of like, "Yeah, I know you mean well," but that allows me to be in this moment, vital and engaged here -- I'm not trying to get that to go away."

In this metaphor -- of the mind as the over-eager office assistant -- we can suddenly begin (I believe) to understand what the ancient Vedic scriptures mean when they describe Self as being separate not only from the body and the senses, but also *from the mind as well* (in her book on page 213, she also notes that she attributes this metaphor to a fellow ACT practitioner, Jeremy Goldberg).

In this metaphor, she explains, what she is referring to as "mind" is not the same as "who we are" but rather mind is there to serve us, and it does the best it can, but in many situations the observations or suggestions that come from this eager assistant are "just not that helpful."

We might think of a character from the long-running television comedy *The Office*, in which the character who may in fact resemble the above description of "mind" might be Michael Scott himself: usually well-meaning, but often *not that helpful*, and in many cases making the situation worse with his constant desire to give suggestions, commentary, and analysis -- even when his "help" is not needed.

You may prefer to think of another character from film or literature (or from *The Office*) who better exemplifies to you this concept of the "over-eager assistant," but Michael Scott may in fact be the perfect example, because Dr. Westrup has identified *mind* with the facility of *language*, and with the ability to create "verbal virtual reality" through language itself -- something that actually characterizes Michael Scott in *The Office* to an extraordinary degree.

And yet, as well-meaning as Michael Scott is, and as funny he can be, we really don't want to let him "run the office" completely unchecked -- and that is why Dr. Westrup explains that we have to cultivate the ability to sort of "stand apart from" the constant chatter of the mind, and analyze what it is doing and saying and suggesting, without letting it take things in whatever direction it wants to take them.

Towards the end of her discussion, in response to a question about what we should not do when we find ourselves in a stressful situation, Dr. Westrup suggests that one should *not*:

> Really buy everything your mind tells you about it, which is, "This is not OK," "I can't tolerate this," "This means my life isn't working," "I'm never going to figure it out." Notice I'm not telling you not to have those thoughts -- good luck! But rather, when they show up, that doesn't mean that they're True with a capital T.

And this brings us back to the distinction that the passage from the Upanishad cited above is trying to articulate, between the True Self or Atman and the body, senses, and mind. It stands to reason that if mind is somehow related to the construct of language, and that its greatest strength (and greatest weakness) is its ability to create "virtual worlds" or "virtual realities" in which it can analyze, critique, judge, describe, compare, and contrast, then the higher Self, the True Self, the Atman must somehow exist *beyond* all of that.

It stands to reason, in other words, that the True Self is not constructed of language, or of modifiers and descriptors and adjectives and judgments and labels.

And this is exactly how Sri B. K. S. Iyengar -- and the sacred Vedas and other texts in the same tradition – describe the transcendence of the mind and the achievement of samadhi. At the end of the introductory section, Sri Iyengar writes that in this state:

> The mind cannot find words to describe the state and the tongue fails to utter them. Comparing the experience of samadhi with other experiences, the sages say: 'Neti! Neti!' -- 'It is not this! It is not this!' The state can only be expressed by profound silence. The yogi has departed from the material world and is merged in the Eternal. There is then no duality between the knower and the known for they are merged like camphor and the flame. 52.

Note the important observation that the state of samadhi is characterized by the complete absence of description, of modification by language: it can only be said that it is "Neti! Neti!" -- it is *not* whatever one wants to compare it to or describe it as.

It should be apparent that this is identical to the famous opening lines of the Tao Te Ching (at least as traditionally arranged for the past 1000 years) in which it is said of the Tao or the Way that

> The ways that can be walked are not the eternal Way;
> The names that can be named are not the eternal name.
> > [Translation by Victor H. Mair, p59].

The actual Tao is beyond description – as soon as it is "named," we know that we are not actually dealing with "the eternal name." The very concept of "the Eternal" (which is mentioned in the Sri Iyengar quotation immediately above as well) means in a state which is still pure possibility, all possibility, containing all options, and thus not manifested in one form or another, and thus not able to be "pinned down" or labeled.

We might note that the divine name in the Biblical scriptures implies the same rejection of modification or description, the same Eternal potentiality and Eternal present (past and future being constructs of language, as we have just seen).

To evoke the same ineffable concept, Sri Iyengar quotes the Song of Sankaracharya, the *Atma Shatkam* or Song of the Soul (Sankaracharya, pictured above, is also known as Adi Shankara and his song is also known as *Nirvanashatkam*):

> I cannot be heard nor cast into words, nor by smell nor sight ever caught [. . .]
> I have no speech [. . .]
> Neither knowable, knowledge, nor knower am I, formless is my form,
> I dwell within the senses but they are not my home;
> Ever serenely balanced, I am neither free nor bound --
> Consciousness and joy am I, and Bliss is where I am found.
> 53.

Note that in the image above showing a shrine to Adi Shankara, the statue of
Shankaracharya in its alcove is flanked on either side by twin female deities,
each of whom is carrying a torch in her inside hand, pointed downwards.* This
quite clearly links to the concepts discussed in the previous post entitled Isis
and Nephthys: March Equinox 2015,[127] as well as to equinox-and-torch
symbology discussed here. Thus Adi Shankara and his message can be clearly
linked to the central "Djed-column raised up" which is also depicted in
between the equinoxes (and which is associated with the vertical column
connecting winter and summer solstices), and all that it represents.

* <u>Later note</u>: Special additional thanks to correspondent Ramakrishnan T.,
who points out that these figures are known as Dwarapalakas, are found
throughout India, are almost exclusively male, and are carrying a mace and not
a torch! However, what is very interesting to me is that, while he is certainly
correct, these figures do sometimes appear to have characteristics that are
slightly androgynous, but even more interesting is that they often have their
legs crossed in a very distinctive manner reminiscent of the equinoctial figures
(who are also male) discussed in the link included in the above paragraph at
the word here. And, there is also no doubt that the mace seen in the
Dwarapalaka symbology is similar in form to the torch found in the symbology
further west, raising interesting questions about possible common origin or
cultural diffusion on this particular symbol.

Thank you to Dr. Darrah Westrup and Jason Wachob for
sharing their helpful discussion of this extremely important
subject!

475

The Gospel of Thomas and the Divine Twin

2015 July 06

Why was an entire "library" of ancient texts carefully sealed in a large storage jar at the base of the steep cliffs of the massif known today as the Jabal al-Tarif, along the banks of the Nile River in Egypt not far from the ancient city of Thebes, sometime during the second half of what we label today as the fourth century AD (the fourth century being the years in the 300s, since the *first* century AD consists of the years with numbers *below* 100, such as for example AD 60 or AD 70, causing all the subsequent centuries to have numbers "one higher" than the "hundred multiple" on the year-numbers, which is why the years in the 1900s were the "twentieth century")?

476

What would be the purpose of carefully sealing an upside-down bowl over the top of the large jar containing these texts, and burying them some distance from the city, underneath the talus at the base of the cliffs?

What was so important about the texts that someone would want to bury them? Were they worried about the texts being stolen? Or was there some other reason?

After this ancient jar was rediscovered in the 1940s (more details about that, along with some maps showing the location of the discovery, are in this previous post[628]), and scholars began to decipher the ancient manuscripts, one possible reason these texts were buried began to suggest itself: these were ancient texts that were *not* included on the lists of approved writings that church authorities began to publish in the second half of that same fourth century -- and texts that did not make it onto the list of approved writings were no longer safe to have in one's possession (often texts excluded from the approved list were specifically denounced as heretical and spurious by the authorities).

Thus, it is quite possible that someone or some group who personally treasured these texts and their teachings, but did not feel it was safe to keep them in their immediate possession as the pressure against "heretical" texts ratcheted up during the second half of the fourth century, took them up the Nile to the cliffs away from the city and buried them there, fully intending to come back to them at some point in the future.

Apparently they never got the opportunity to go back.

These ancient texts, along with some others that have come to light in more recent discoveries, as well as a very few other fragments and manuscripts that had been found or preserved prior to those found in the jar at the Nag Hammadi, suggest to some researchers a very different history of the early centuries of the Christian church than has traditionally been taught. Some of the evidence can be interpreted as indicating that early teachings very different from what we today think of as "Christian teaching" were forcibly suppressed and driven underground (literally driven "under ground" in the case of the texts buried at Nag Hammadi) during the second, third, and especially fourth centuries, and replaced by an "approved list" of texts and teachings, which were to be interpreted from a primarily literalist perspective.

In the next few posts, let's briefly examine a few of the ancient texts that were pretty much lost to history for nearly 1,600 years, surviving (as far as we know) only inside that sealed jar buried under the earth beneath the cliffs of Nag Hammadi and safely out of the way for the spread of literalist teachings until that jar was unearthed again in the twentieth century.

When we do so, we will find some teachings which seem to strongly resonate with some of the themes we have been examining recently in our examination of some of the "Star Myths" in the Mahabharata of ancient India, and in the Bhagavad Gita that is part of the Mahabharata. In fact, we will find teachings in some of those long-buried Nag Hammadi texts that I believe have clear affinity with much that is found in the ancient wisdom preserved in myth and sacred stories literally around the world -- and indeed, that is even found in the texts of what we think of today as the Bible (the texts that *did* make it onto those approved lists), but which are more evident in those Biblical texts when they are understood as esoteric allegory rather than as literal accounts.

Previous posts have presented evidence that the stories of the Bible were not intended to be understood as literal history but as esoteric allegory, and that forcing a literal reading onto them has resulted in an interpretation that is pretty much the *polar opposite* of their intended teaching -- see, for example, the many discussions of Bible characters and episodes available in my online blog and videos, as well as in my book (published the year after this post was published) entitled *Star Myths of the World, and how to interpret them, Volume Three* (*Star Myths of the Bible*).

The entire "library" of texts that have survived from the discovery of that jar at Nag Hammadi (apparently, not all of the texts found in the jar survived, because when they were first found a few of the texts were actually burned as fuel for a cooking fire, according to stories surrounding the discovery) can be found online at http://gnosis.org/naghamm/nhl.html, as well as in print form in various translations and collections (such as the collection edited by Nag Hammadi scholar and translator Marvin Meyer).

Out of that collection, we'll just look at a few passages from a couple of texts over the next few days or weeks. However, those interested in learning more can go straight to the Nag Hammadi texts themselves -- although the passages often appear cryptic at first, sometimes quite strange and alien, and even downright off-putting, remember that they are intended to be understood (I believe) as esoteric allegory and that as such they are intended to convey spiritual truths which our literal or rational mind would "choke on" or reject, but which can often be best absorbed through powerful stories or metaphors.

Remember also that these texts were considered precious enough by someone living in ancient times to bury them, possibly at some risk to themselves, because they couldn't bear to see them destroyed -- and remember as well that the teachings in these texts was apparently considered so dangerous by those trying to spread a different system that these specific texts were literally unavailable after a certain point; they were completely or nearly completely eradicated.

And, it should be noted, these texts were not marginal or unimportant texts: some of them (such as the one we will discuss in a moment) were mentioned quite often by ancient authors (including literalist Christian authorities, who were denouncing the texts), and so their titles were know to modern scholars even though -- until the discovery of the Nag Hammadi library -- their contents could not be consulted (except, in a few very limited cases, in a few fragments that survived, including in one case fragments which survived in a rubbish heap).

One of the most well-known and important of the texts found in that long-buried jar from Nag Hammadi is the text known as *The Gospel of Thomas*, which introduces itself as a record of the "secret sayings that the living Jesus spoke and Didymos Judas Thomas recorded" (this is the translation version found at http://gnosis.org/naghamm/gosthom.html, by Stephen Patterson and Marvin Meyer; there are several other versions of English translations available and linked from that location, and it is interesting to read the different translations to try to get additional perspectives on the ancient text).

This opening line itself offers us some extremely important insights, based on the name "Didymos Judas Thomas" – the title "Didymos" or "Didymus" for Thomas is also found in the canonical gospel of John (in chapters 11, 20, and 21) and it means "Twin" (as does the name Thomas itself, apparently, but *Didymos* comes from the Greek word for "Twin" and *Thomas* comes from the Aramaic word for "Twin").

Of course, a character specifically identified as a Twin might suggest a connection to the Twins of Gemini, to those who have become familiar with the patterns found in Star Myths around the world, and it is certainly possible that the Thomas character has some connection to the zodiac constellation of Gemini.

However, it is also quite possible that something even more interesting is at work here, something related to the previous discussion entitled "Why divinities can appear in an instant: The inner connection to the Infinite."[580] That post argued that the ancient Star Myths are intended to convey the knowledge to us that even in this incarnate existence, we have inside of us a connection to the infinite: a connection to the divine, what is also described as the "hidden divine spark" or the "god within" (and see other related discussions on this very important subject, such as "Namaste and Amen,"[544] or any of the previous posts about Osiris and the casting down and raising-up-again of the Djed).

How does the character of "Didymos Judas Thomas" convey a related message? The answer comes when we ask, "if Thomas is a twin, who is the other twin in the pair?" After all, that is a natural question to ask if we are reading a story and we are told that a character is a twin, but we are not immediately introduced to the other twin.

Interestingly enough, in another of the Nag Hammadi texts -- and in fact in a text which was bound up together with the Gospel of Thomas in the book-form or "codex" known to Nag Hammadi scholars as "Codex II" -- a text called The Book of Thomas the Contender, we get a startling answer as to who the other twin of Thomas might be (in the esoteric allegory).

In Section II of The Book of Thomas the Contender, which is called "Dialogue between Thomas and the Savior," we read these words in a sub-section regarding the subject of ignorance and self-knowledge:

> The savior said, "Brother Thomas, while you have time in the world, listen to me and I will reveal to you the things you have pondered in your mind. Now, since it has been said that you are my twin and true companion, examine yourself, and learn who you are, in what way you exist, and how you will come to be. Since you will be called my brother, it is not fitting that you be ignorant of yourself . . ."

Stop! What?
Did this text just say that the Savior addressed Thomas as "my twin"?

Yes, that is what was asserted in this text. Now, if you are one who wants to *interpret literally* ancient texts about what the Savior said, then you are probably going to reject this text as being heretical. If you try to take this text literally, it will cause big problems with other texts, such as the scriptures describing the birth of the Savior (in which it is never said that he was born as one in a set of twins, for instance).

481

But, if you are not troubled with a need to force every ancient scripture into a literal mold, and if you believe that they were not intended to be understood that way, then you can ask yourself what this assertion that Thomas was the twin of the Savior might mean -- what it might have been intended to convey.

As you did so, you might remember that in other ancient mythologies, most notably perhaps in Greek myth, there are sets of twins in which one twin is divine or immortal, and the other twin is human and mortal. These Thomas narratives in the Nag Hammadi texts seem to be resorting to this same metaphor: we have a divine twin ("the living Jesus" as he is called in the opening line of the Gospel of Thomas, and "the Savior" as he is called in the Book of Thomas the Contender), and we have the mortal counterpart, the human twin: Thomas, the one who writes down the sayings for us, which he received from the divine twin.

Now, as we saw at the end of the preceding discussion regarding the "inner connection to the Infinite,"[603] there is a passage in the wisdom-book of Proverbs which declares "there is a friend that sticketh closer than a brother." As that post argued, and presented evidence from myth (particularly myths in which a god or divine being appears instantly, which also happens to be one of the characteristics of the risen Christ) this teaching may well be trying to convey to us the knowledge that our connection to the infinite, to the realm of the gods, is not external to us: it is *within* us already.

The metaphor of a divine twin and a human twin, such as the Gemini Twins in Greek mythology of Pollux (divine) and Castor (human), may well be referring to just such a concept or teaching. Expressing it in this way can convey this truth to us in a powerful, metaphorical, *esoteric* manner.

If that is the case, then what we see here in the Gospel of Thomas (and in the Book of Thomas the Contender) may well be conveying the very same truth, just in a slightly different form than it is found in (for example) the Greek myth of Pollux and Castor. In the Nag Hammadi texts mentioned here, Jesus is the divine twin and Thomas is the human twin, but *they are not in fact two different entities*. This is a teaching about the "Christ within" (which is a teaching also found in the writings of the apostle who called himself *Paul*, a name which the Reverend Robert Taylor points out is very much linguistically related to *Pollux* and to *Apollo*).

We are already, perhaps, getting a sense as to why these texts ended up buried in a large jar in a secret location, where the authorities who had declared such teachings to be "heretical" could not find them and destroy them.

There is much within the *Gospel of Thomas* itself to back up the interpretation that has been suggested above. In future posts we may have occasion to examine a few more of them, but for now let's just look at another metaphor, offered as a saying of Jesus, found in section 109 of the *Gospel of Thomas*.

There, in the translation of Stephen Patterson and Marvin Meyer, Jesus says:

> The (Father's) kingdom is like a person who had a treasure hidden in his field but did not know it. And [when] he died he left it to his [son]. The son [did] not know about it either. He took over the field and sold it. The buyer went plowing, [discovered] the treasure, and began to lend money at interest to whomever he wished.

This is a very interesting metaphor, and one that suggests that the "treasure" of the infinite is buried away deep inside us like the treasure in the story that lies buried under a field, which can remain there our entire lives without our knowing it. But it is something which we actually already have, if we just knew.

The scriptures appear to be trying to break through our ignorance on this subject, to tell us that we already are connected to something that is actually inexpressible in its infinity (that cannot be quantified or defined or even named, as the opening lines of the Tao Te Ching declare, and that thus lies beyond all the quantifying and labeling and chattering of the part of us that we call our mind).

Thomas is telling us the words of Jesus, but perhaps "Thomas" received these sayings from a divine source that was not external to him (though none the less divine and none the less real for that). In fact, we should not think of the Gospel of Thomas as being about some "twin" who lived thousands of years ago: as Alvin Boyd Kuhn advised us in a passage quoted in several previous posts,[331] we won't understand ancient texts unless we realize that they are about us. Each and every individual soul that incarnates in this world is, according to such a reading, like Thomas: a twin to a living infinite inner divinity, possessed of a friend that sticketh closer than any "external twin" (as close as literal twins are to one another, this twin is even closer).

This teaching is also portrayed in the Mahabharata and the Bhagavad Gita, with Arjuna and his companion and divine charioteer, the Lord Krishna (as well as in the episode in which Durga appears[367] before the battle: see videos here and here and additional discussion here).

These are *not* the messages that are traditionally drawn from the scriptures of the Bible when they are approached with a literalist hermeneutic (because literalist readings necessarily start off by seeing the characters in the text as primarily *external* to us, since those characters are understood to be literal-historical figures). But they are messages which resonate strongly with all the other myths and sacred traditions of the world -- and they are in fact the messages which I believe these texts were intended to convey to us, before something happened and that message was all but wiped out, around the period of time that the Nag Hammadi library was being sealed away.

484

The Gospel of Thomas and the Everlasting Spring

2015 July 09

We're currently engaged in an examination of some of the ancient texts found buried at the base of the Jabal al-Tarif near Nag Hammadi in Egypt, for evidence of teachings which resonate with the teachings conveyed by other Star Myths around the world.

The previous post examined the Gospel of Thomas,[476] found in Nag Hammadi codex 2, and argued that it is using a powerful esoteric metaphor to teach us that we are beings composed of two natures, that we are like a "set of twins," but contained within one being. We have our human, incarnate, doubting side -- but one privileged with the gift of direct access to and intimate communication with the divine, the Christ within, who declares in another manuscript contained in Nag Hammadi codex 2 that Thomas is indeed his *twin*, his *true companion*, and the one who *will be called his brother.*

In section 13 of the Gospel of Thomas, we find the following exchange:

> Jesus said to his disciples, "Compare me to something and tell me what I am like."
> Simon Peter said to him, "You are like a just messenger."
> Matthew said to him, "You are like a wise philosopher."
> Thomas said to him, "Teacher, my mouth is utterly unable to say what you are like."
> Jesus said, "I am not your teacher. Because you have drunk, you have become intoxicated from the spring that I have tended."
> And he took him, and withdrew, and spoke three sayings to him. When Thomas came back to his friends they asked him, "What did Jesus say to you?"
> Thomas said to them, "If I tell you one of the sayings he spoke to me, you will pick up rocks and stone me, and fire will come from the rocks and devour you." [link to this translation: http://gnosis.org/naghamm/gosthom.html].

This passage is noteworthy for many reasons.

First, it is very clearly a parallel to an episode found in the canonical gospels (those which, unlike those texts buried at Nag Hammadi, *were* included in the "approved list" of texts that eventually came to be called the "New Testament"): specifically, the mountaintop experience recounted in Matthew 16, Mark 8 and Luke 9, in which Jesus asks "Whom do men say that I the Son of Man am?" and then "But whom say ye that I am?"

In the versions included in the canonical gospels, it is Simon Peter who gives an answer that Jesus approves. In this version, it is Thomas -- and the answer that Thomas gives is different from that given by Peter in the canonical gospels. Thomas here says, in answer to the question, that his mouth is "utterly unable to say" what Jesus is like.

This answer is actually very profound, in that it is expressing the idea that the one with whom Thomas is conversing cannot be

486

defined, cannot be labeled, cannot be delineated: he is utterly unable to be framed or contained by the faculty of language. This answer immediately points to the previous discussion in the posts: "Self, the senses and the mind,"[466] in which a distinction is made between the mind (with its endless attempts to define and describe and discriminate and delineate) and the infinite and ineffable Supreme Source which is behind and above mind, and of which Sri B. K. S. Iyengar, in commenting upon the teaching of the Vedas upon this subject, declares:

> The mind cannot find words to describe the state and the tongue fails to utter them. Comparing the experience of samadhi with other experiences, the sages say: 'Neti! Neti!' -- 'It is not this! It is not this!' The state can only be expressed by profound silence. The yogi has departed from the material world and is merged with the Eternal. There is then no duality between the knower and the known for they are merged like camphor and flame. *Light on Yoga*, 52.

Note how well the above statement reflects the sentiment dramatized in the Thomas Gospel. Thomas declares that he can only say, "I am unable to say!" In other words, he must declare "Neti! Neti!" like the sages described by B. K. S. Iyengar and the teachings of ancient India.

Further, in the description of Sri Iyengar, we see the assertion that there is in fact a merging of the yogi with the Eternal: there is no duality between the two; they merge like camphor and flame. The previous post[476] makes the argument that the Nag Hammadi texts express this same idea by declaring that Thomas and the one who is ineffable, who cannot be described, are in fact *twins*. They are, in some mysterious sense, merged. There is no duality between them. In the words of the Hebrew scriptures in the book of Proverbs, the heavenly friend is the one who "sticketh closer than a brother" (Proverbs 18: 24).

In other words, the Nag Hammadi text of the Thomas Gospel is trying to convey to us that in our incarnate condition we are like

Thomas: we are intimately connected to the infinite, the ineffable, the Eternal -- so closely that we are "twinned;" we are "merged like camphor and flame."

And, this one with whom we are so close is in fact the un-namable, the undefinable: the Ultimate. In the Bhagavad Gita, this is expressed by Arjuna's divine companion and confidant, the Lord Krishna, who declares that: "The entire universe is pervaded by me" (section 9), "I am the origin of all. Everything emanates from me. [. . .] There is no end of my divine manifestations" (section 10). Krishna then displays his ultimate form, and shows Arjuna that his divine companion is indeed unbounded, unlimited, unable to be described with words, endless and infinite.

The same is declared in the Hymn to Durga which is found in the Mahabharata immediately prior to the Bhagavad Gita,[367] in which the goddess Durga is also declared to be "identical with Brahman [. . .] the unconsciousness [. . .] the beauty of all creatures [. . .]" (Book I, section 23).

The fact that she appears to Arjuna immediately upon his meditation upon her and his hymn of praise to her indicates the same teaching that we have been exploring above: there is so little distance between the human being and the deity that they are as close as the camphor and the flame, they are closer than even the closest of brothers, they are twinned: the mortal with the immortal (like Castor and Pollux).

After Thomas declares that his mouth is utterly unable to say what the divine one is like, Jesus then declares to him that: 1) Jesus is not his teacher, and 2) that Thomas has drunk from the spring which Jesus has tended, and it is this spring which has made Thomas drunk.

This aspect of the passage is also extremely noteworthy. Thomas began his "confession" by saying "Teacher," but Jesus in a sense rebukes him and says "I am not your teacher." This might be interpreted as telling us that he is not separate from Thomas:

there is not an external one to whom Thomas must look for guidance. The divine is within Thomas himself.

This interpretation might be seen as comporting very well with the declaration of Paul in the epistle to the Galatians, in chapter 1 and verse 16, in which Paul can be interpreted as saying that when God revealed the Christ in him, he did not confer with any teacher.

This interpretation is strengthened by the next metaphor, in which Jesus declares that Thomas has obtained this insight because Thomas has drunk from the spring which Jesus has tended. In other words, according to this passage in the Nag Hammadi text, Jesus is here declaring that his role is as the one who tends to the spring (almost like a barista who tends to the coffee that is given to those who come looking for it).

This declaration is very interesting in light of the passage in the canonical gospel of John describing the episode of Jesus' encounter with the Samaritan woman at the well (at Jacob's well, in fact): Jesus tells her that he has living water that one can have and never thirst again, and then explains that this living water is within her. In verse 14 of John 4, Jesus says that this water can be in anyone a well of living water (unending water, unlimited water), "springing up into everlasting life." In other words, he is tending a spring which is infinite in nature, but which is available to each person *internally.*

Based on this declaration found in John 4:14, and the declaration found here in the Thomas Gospel section 13 that Jesus is *tending* the spring, it does not seem too far of a stretch to conclude that the spring from which Thomas has drunk is the everlasting or infinite and Eternal spring *within himself* (within Thomas). Thomas is connected with the infinite, not externally but in himself.

Again, it bears repeating that this passage of ancient scripture is not intended to be understood as describing some ancient enlightened being named Thomas, who was different from ourselves. It is intended to convey to us a truth about *each and every human soul* who comes into this material life: we, like the "Thomas" in the text, are actually a composite being, a dual being -- a "set of twins," in which we usually identify with only the human aspect but which has a hidden or forgotten connection to the divine or the infinite or the eternal: a "divine twin," but our divine twin is not external to ourselves.

Switching to a different metaphor, the text shows us that the divine or infinite or Eternal is already within us, like an everlasting or unending spring, from which we can drink.

In the beginning of the Gospel of Thomas, Jesus declares: "When you know yourself, then you will be known, and you will understand that you are children of the living Father. But if you do not know yourselves, then you live in poverty, and you are the poverty" (section 3). This parallels the metaphor discussed in the previous post regarding the Gospel of Thomas,[476] which says we are like one who has a field but is unaware of the treasure buried within that field.

Ultimately, then, the purpose of this ancient text seems to be identical to the famous dictum of the temple at Delphi:

"Know thyself."[6]

Note that the temple at Delphi was closed under the reign of the Emperor Theodosius,[693] after the literalists took control of the Roman Empire, in the year AD 390 -- during the same second half of the fourth century AD in which scholars believe the Nag Hammadi texts were hidden and buried, possibly due to persecution by the ascendant literalist hierarchy.

The Gospel of Thomas is telling us that if we know the truth, we are actually connected intimately with the ultimate -- with the divine. We are, like Thomas, a twin: a twin to divinity (like Castor was a twin to Pollux). We contain within us a bubbling spring which is connected to Eternity. But, if we remain in ignorance of this fact, we are like the one who had a treasure buried in his or her own field, and never knew about that treasure.

Peter Kingsley, who writes about the ancient knowledge of this internal connection to the infinite, says that when we are disconnected from that infinite source, we become impoverished indeed -- filled with a longing we can never satisfy, and with a hollowness that drives us to chase after substitute after substitute for what we perceive to be missing. This hollowness and chasing after substitutes, not surprisingly, characterizes western civilization (because western civilization almost by definition is directly descended from those cultures that are heir to the Roman Empire which had shut down the temple at Delphi and declared heretical the texts buried at Nag Hammadi).

As the Gospel of Thomas tells us, if we do not know this truth, this treasure, then we will live in poverty, and will in fact *be* that poverty (clearly describing spiritual poverty, rather than material poverty, since the rushing after substitutes which Peter Kingsley describes can in many cases produce material wealth, although without corresponding release from the spiritual hollowness).

The worst part about this situation is that the actual solution is already within our grasp: the bubbling spring is already available to each of us. It is that Tao which cannot be named, that Krishna who declares that there is no end to his manifestations, that one of whom Thomas says the mouth is utterly incapable of describing or defining.

But that is also the best part, as well.

The Gospel of Philip and the Zodiac Wheel

DAYS LONGER THAN NIGHTS:
Heaven, Promised Land, Greece, etc.

NIGHTS LONGER THAN DAYS:
Hell, Egypt, Troy, etc.

2015 July 14

The previous two posts have attempted to demonstrate that ancient texts buried beneath a cliff near modern-day Nag Hammadi, likely placed there during the second half of the fourth century AD after authorities promoting what can generally be called a literalist approach as opposed to a gnostic approach had declared these texts to be heretical and suppressed their teachings, can be shown to be using esoteric metaphors to convey the very same ancient wisdom found in other myth-systems the world over.

In particular, the preceding posts argued that specific metaphors in the Gospel of Thomas, an extremely important text found in Codex 2 when the Nag Hammadi codexes were unearthed in the twentieth century, after spending perhaps sixteen centuries beneath the ground, are conveying the same message found in the ancient Sanskrit scriptures of the Bhagavad Gita and the Mahabharata concerning the nature of human incarnation, the constant interplay between the material realm and the realm of spirit, and the reality of each individual to have inner access to the infinite -- the higher self, the supreme self, the Atman -- at all times.

For those discussions, please see the previous posts entitled

- "The Gospel of Thomas and the Divine Twin,"[476]
- "The Gospel of Thomas and the Everlasting Spring,"[485]
- "Star Myths of the World: the Bhagavad Gita,"
- "Star Myths of the World: the Hymn to Durga in the Mahabharata,"[367] and
- "Why divinities can appear in an instant: The inner connection to the Infinite."[580]

These discussions can be seen to be related to the larger pattern of the world's ancient myths, all of which can be shown to be very deliberately and intentionally using the celestial cycles to convey profound spiritual truths, most often within the framework of the great wheel of the zodiac and the great solar cross formed by the "horizontal" line running between the equinoxes (which generally relates to the "casting down" of the spirit into material incarnation in this life) and the "vertical" line running between the solstices (which generally relates to the "raising up" or "calling forth" of the spiritual aspect present -- though often hidden or forgotten -- in ourselves and indeed within every aspect of the apparently physical universe). Numerous previous posts have discussed this overall pattern -- often relating it to the ancient Egyptian metaphor of the "casting down" and the "raising back up" of the Djed column.

Very significantly, there are passages in the Nag Hammadi texts which I would argue can be shown to explicitly declare the major outline of this very same mythological zodiac metaphor: the metaphor which forms the foundation for Star Myths from virtually every continent and culture around the globe.

In another important text from the same collection, the Gospel of Philip, which was also contained in codex 2 of the texts buried in the large jar beneath the cliffs near Nag Hammadi along the Nile River in Egypt, there is a specific passage in the subsection labeled (for ease of reference) as "Sowing and Reaping" by translator Marvin Meyer, which plainly tells us:

> Whoever sows in winter reaps in summer. Winter is the world, summer is the other, eternal realm. Let us sow in the world to reap in summer.

This passage is completely consistent with the metaphor-system which previous posts have alleged can be seen to be operating in myths literally around the world, stretching across time from the civilizations of ancient Egypt and Sumer and Babylon, all the way up through the present day in cultures where the connection to the ancient wisdom remains to some degree intact.

The system uses the "lower half" of the cycle of the heavenly bodies (from the daily cycle created by the rotation of the earth on its axis, to the monthly cycle of the moon and the yearly cycle created by earth's annual path around the great cross of the year, as well as some other cycles which are even longer than these) to describe our incarnation in "this world" -- that is to say, in the familiar, visible, material realm.

The system uses the "upper half" of the same cycles to describe "the other, eternal realm" -- the invisible realm, the realm of spirit.

Each day the turning of the earth causes the stars (including our own sun, the Day Star) to appear to rise up out of the eastern horizon and arc their way into the celestial realm: the realm of the air, the realm of celestial fire -- a perfect metaphor for the realm of spirit, the invisible realm. But the same turning of the earth also causes the stars (including the sun) to plunge down again into the western horizon, disappearing into the "lower elements" of earth and water -- a perfect metaphor for this "lower realm" of matter, in which we find ourselves in this incarnate life.

494

And, using the annual cycle of the year (which has certain advantages over the daily cycle, because it is conveniently broken up into much smaller sub-sections which can be conveniently discussed using the twelve subdivisions of the zodiac signs which precisely indicate very specific parts of the annual cycle) we can use the same general metaphor. This time, the "lower half" of the year -- the half which runs from the autumnal equinox down through the winter solstice and up to the crossing point of the spring equinox -- represents the same thing that night-time represents for the daily cycle: the incarnate realm, the material realm, the imprisonment in a body of earth and water, plowing through the "underworld" of the physical universe.

The "upper half" of the year -- the half which runs from the spring equinox up through the summer solstice and down again to the autumnal equinox -- represents the realm of spirit, the invisible realm, all that is eternal, unbounded and infinite.

The ancient Egyptian myth cycles depicted this same principle using the gods Osiris and Horus. Osiris, god of the dead, ruler of the underworld, represents the sun in the "lower half" of the cycle: when it is plowing through the lower realm of incarnate matter, "cast down" into incarnation. Horus represents the "upper half" of the cycle, when the sun soars upwards "between the two horizons" into the celestial realms of air and fire -- the realm of spirit.

Here in the Gospel of Philip, buried for those long centuries among the other texts in the Nag Hammadi collection, we find an explicit confirmation of this pattern: "Winter is the world, summer is the other, eternal realm."

It could hardly be more clear if the text were to tell us: "The lower half of the wheel represents this world: this material realm -- the upper half, or the summer months on the annual circuit, are used as a metaphor for the other realm, the invisible realm, the eternal realm, the realm of spirit."

This in itself is remarkable, and it has tremendous implications for our understanding of the scriptures included in what today is called the Bible, but all of it might still be (mistakenly) dismissed by some as being of limited practical value.

"So what?" they might ask. "How does this matter to my daily life?"

The answer, according to the Nag Hammadi texts themselves, is: "plenty."

Because, just as we have seen in the previous examinations of the Bhagavad Gita or the Mahabharata, and just as Peter Kingsley has argued in his powerful book *In the Dark Places of Wisdom*, the ancient texts which were literally "driven underground" and buried in the urn at Nag Hammadi tell us something remarkable about the location of this eternal realm, and where we need to go in order to have access to it.

In section 3 of the Gospel of Thomas, for example, we find another explicit statement which can perhaps be profitably juxtaposed with this "zodiac wheel explanation" from the Gospel of Philip. There, giving the words which "Thomas" the twin has heard from his divine counterpart Jesus, the scripture tells us:

> Jesus said, "If your leaders say to you, 'Look, the kingdom is in the sky,' then the birds of the sky will precede you. If they say to you, 'It is in the sea,' then the fish will precede you. Rather, the kingdom is within you and it is outside you.

Notice how this passage can be interpreted, in light of all we have discussed above, as telling us that both halves of the cycle -- the upper half of the "sky" and the lower half of the "sea" -- are talking about something that really has nothing to do with physical location (neither sky nor sea). What is being discussed is the invisible realm with which we already have intimate contact, right inside of us.

496

And, this same invisible realm with which we already have contact (within) is also present within and behind every single molecule of the seemingly physical realm all around us as well -- it is both "within you" and it is "outside you," the Thomas Gospel tells us.

And this is knowledge with absolutely world-changing implications for each of us. Because, as Peter Kingsley explains so powerfully in the beginning sections of *In the Dark Places of Wisdom*, western civilization has somehow been cut off from that truth (very likely, I would argue, by ancient events that were part of the very same chain of events which led to the burying of the Nag Hammadi texts that we have just now been considering), and because of being cut off from that truth has spent the better part of the past sixteen or seventeen centuries trying to find external substitutes for something that is already internally accessible, right now, in "the peace of utter stillness" (and see further discussion of this concept in the previous post entitled "Two Visions"[770]).

The previous posts and accompanying videos exploring the significance of the invocation of the goddess Durga[367] in the Mahabharata (immediately prior to the Bhagavad Gita) and the significance of the relationship between Arjuna and his divine charioteer, who is none other than Lord Krishna whose form is shown to be without limits, impossible to define or delineate or describe or bound with words, also indicates the practical impact that this ancient wisdom can have on our daily lives.

Because it would argue that we can have access to this divine higher self literally every day, at any time (and the passage in the Mahabharata containing the Hymn to Durga[367] specifically advises making the calling upon her divine presence a daily habit -- first thing each day, in fact). For more discussion of this subject, see previous posts such as "Self, the senses, and the mind"[466] and "The Bodhi Tree."[338]

Below is a famous statue from ancient Egypt of the king Khefren or Khafra (who probably reigned for over two decades around the year 2560 BC), showing the king with the falcon-god Horus spreading his wings over and behind his head.

It is a powerful image, and one which can be interpreted as depicting the very teaching conveyed by the Gospels of Thomas and Philip above, as well as by the section of the Mahabharata dramatizing invocation of Durga or the Bhagavad Gita's dramatization of Arjuna and Krishna in the chariot, prior to the battle of Kurukshetra.

It appears to indicate the state in which we are in contact with, in communion with, in harmony with, and under the guidance and protection of the higher self, the supreme soul, the infinite and unbounded principle which both Durga and Krishna declare themselves and reveal themselves to be, and which the Gospel of Philip plainly says is symbolized by the "upper half" of the great annual wheel: the summer half, the Horus half.

The infinite to which we each have access, within ourselves, in the peace of utter stillness, without going anywhere.

This is the truth of which the world's ancient scriptures and myths all testify.

Who is Doubting Thomas?

2015 September 07

One of the more famous episodes in the New Testament resurrection story is the account of "Doubting Thomas," also referred to as "The Incredulity of Thomas" ("incredulity" meaning literally "the not-believing" of Thomas, or the "not-giving-credit [i.e., trust]" by Thomas).

The account of this episode is found in the Gospel According to John (and only there, out of the texts that were included in what came to be the accepted texts of the "canon"), and is there described as follows (in the 20th chapter of the Gospel According to John):

> 24 But Thomas, one of the twelve, called Didymus, was not with them when Jesus came.
> 25 The other disciples therefore said unto him, We have seen the Lord. But he said unto them, Except I shall see in his hands the print of the nails, and put my finger into the

print of the nails, and thrust my hand into his side, I will not believe.

26 And after eight days again his disciples were within, and Thomas with them; then came Jesus, the doors being shut, and stood in the midst, and said, Peace be unto you.

27 Then saith he to Thomas, Reach hither thy finger, and behold my hands; and reach hither thy hand, and thrust it into my side: and be not faithless, but believing.

28 And Thomas answered and said unto him, My Lord and my God.

29 Jesus saith unto him, Thomas, because thou hast seen me, thou hast believed: blessed are they that have not seen, and yet believed.

This passage is often interpreted as being about *belief* or *faith*, particularly by those who assert that the texts were intended to be understood literally and historically (that is to say, those who believe the texts were intended to be understood as describing an encounter that took place in literal history between two literal and historical figures).

But what if that is not what this episode is actually about at all?

If the text *is* describing a literal event that took place after a literal resurrection, then it *would* make sense to understand this encounter with Thomas as being about believing that the literal resurrection happened in the manner described.

But we have already examined some evidence that the character of Thomas has a meaning that goes far beyond the common understanding of Thomas as a particular individual who lived a long time ago and had a particularly "incredulous" disposition and a particularly blunt way of expressing himself.

A major clue to the critical importance and true identity of this character "Doubting Thomas" is in fact found in the very passage cited above from the "canonical" text of the Gospel According to John: the information given in verse 24 of John 20 that Thomas (one of the twelve) was also *called Didymus.*

The word "didymus" is Greek in origin and means "twin" (the prefix *di-* is still found in many English words, many scientific in nature, which carry the meaning of "twin" or "two" or "twinned," such as a *diode* or a *dipole* or a *diplodocus* or a *dichotomy* or even a *diploma* -- diplomas apparently being so named because they were originally "folded in two" or "doubled" instead of being rolled up in a cardboard tube the way they are today).

The Gospel According to John is the only text among those admitted to the New Testament canon which uses the word Didymus (or *didymos* in the New Testament Greek) or reveals that Thomas was either an actual twin or was for some reason called "the twin" (even though Thomas is listed in the naming of the twelve apostles found in the books of Matthew, Mark and Luke, as well as in Acts of the Apostles). Neither John nor the others ever explain why Thomas is called that, or who his other twin might be.

But, as has already been discussed at some length in the previous post entitled "The Gospel of Thomas and the Divine Twin,"[476] there were other "New Testament era" texts which were specifically excluded from the New Testament canon but which were apparently preserved in a large sealed jar which in ancient times (in fact, during the same century that the current canon was being established and other texts not included in the canon were being marginalized or even outlawed) was buried beneath the sands at the base of a cliff near the modern-day village of Nag Hammadi in Egypt -- and one of these texts has Jesus addressing Thomas as "my twin and true companion."

This remarkable statement opens up an entirely different interpretation of the so-called "Incredulity of Thomas" episode -- and indeed of the identity and meaning of the character of Thomas altogether.

The statement simply cannot be understood literally, as in referring to a literal-historical twin of Jesus, since such an interpretation would then undermine a literal-historical interpretation of the descriptions of the birth of a single child (not a twin) found elsewhere in the New Testament scriptures.

But just because something is not *literal* does not mean that it is not *true.*

If we are not meant to interpret these scriptural passages as literal-historical, then how else could we be intended to interpret them? If the passage is not intended to describe a literal individual named Thomas with an incredulous disposition and a gruff manner of speaking, then what are we supposed to learn from it?

The answer is: Something of tremendous importance and applicability to our daily lives -- something which makes Thomas a character of immediate and ongoing relevance to each of our individual journeys through this world, every single day (in a way, I would argue, that a Thomas who lived a couple of thousand years ago might not be).

In fact, I would argue that even those who take the scriptures as literal and historical probably do not find themselves thinking about Thomas and his importance to their lives multiple times every day.

But I would submit that after reading the esoteric interpretation of Thomas offered below, you might (at least I do).

Because if Jesus and Thomas are twins, and if out of the two of them Jesus represents the "divine twin" in the pairing (and, as we explored in that previous post on Thomas,[476] there are many such "twins" in ancient mythology, including Castor and Pollux in Greek mythology and Gilgamesh and Enkidu in the mythology of ancient Sumer and Babylon), then what does that imply about the identity of Thomas?

Why, it could imply that Thomas is "the human twin."

And which one would that make us?

502

(This is a trick question).

The answer, of course, is: "both of them."

We are running around in this incarnate life with both of these "twinned" natures within us at all times (in fact, as has been explored at some length, the very symbol of the cross itself can be seen to represent the "crossing" of two natures in each and every human being – a horizontal nature and a vertical nature, so to speak).

To put it very plainly, I believe that the episode of "Doubting Thomas" is intended to teach us to *get in touch with* the divine Infinite.

And our "Thomas nature" – while serving a very necessary function – can be an obstacle to that connection with the Infinite, at least when overcome by doubt.

Having offered that interpretation, let's now take another look at the text itself to see if it is possible to find any support for such an assertion.

In verse 25, the other disciples say to Thomas: "We have seen the Lord."

Thomas replies in the same verse: "Except I shall see in his hands the print of the nails, and put my finger into the print of the nails, and thrust my hand into his side, I will not believe."

Let's just think about that for a second.

As noted above, the episode of the "Incredulity of Thomas" is usually interpreted as instructing *belief* or *faith*, and not doubt – and hence a "negative spin" is imputed to this "incredulity" of Thomas (this failing to "extend credit" or trust to the account of the other disciples, on the part of "Doubting Thomas").

But is this statement from Thomas really something that we are meant to see in a negative light?

He did not say, "Even if I see the print of the nails, I will not believe."

He did not say, "Even if I thrust my hand into his side, I will not believe."

Thomas is actually displaying critical thinking, a desire to check things out and examine the evidence which supports one or another theory, or which might disprove one or another theory . . . even what we might call "the scientific method."

And this kind of thinking is actually indispensable in our daily life, from one moment to the next in this physical world (in fact, it is essential to our very *survival* from one moment to the next).

If a traffic light turns green, telling you that it is safe to proceed into an intersection, or a railroad crossing signal tells you that it is safe to proceed across the railroad tracks, and you don't exercise at least a very little bit of what Thomas here displays when he receives the report of the other disciples, there may come a day when those signals are telling you an untruth that could be extremely dangerous to you. It is advisable to just swivel your head to glance quickly up and down a cross-street or a train-track as you approach it, to "see for yourself" in the same way that Thomas might advise you to do.

In other words, critical thinking is critically important to anyone living in the material world.

It is the same critical thinking that enables us to categorize things into one category or another ("this" and "not that"), to communicate using language (which is built upon definitions of "this" and "not that," the very word *definition* meaning "to put a boundary or a limit around something"), and to analyze our situation and come up with possible hypotheses to explain what we see, and then examine the evidence that could help us accept or reject the different possible explanations or hypotheses.

All that being said, the scriptural passage itself does indeed appear to be telling us that all of this critical thought, while

504

essential, can have a negative side (like any other good thing, especially when there is "too much of a good thing").

The very same *essential* and *indispensable* faculty that enables us to categorize, to hypothesize, and even to criticize ("this is good" versus "that was not so good" or even "that was a disaster") is exactly the same faculty that makes possible self-doubt, self-criticism, and even what we might term "self-imposed isolation from the divine twin."

If you haven't watched it already or don't remember the details of this previous post[466] discussing the excellent conversation with Dr. Darrah Westrup at the mindbodygreen "Revitalize 2015" conference (her talk can be seen in this video clip beginning at about the 1:03:00 mark), please check it out or give it a re-look.

Because in her talk, after pointing out that animals do not typically walk around wracked with self-doubt, and that even if a cat makes a terrible failure of trying to leap somewhere, it doesn't seem to reduce its self-image or cause it to wonder if it is going to be a failure at it the next time, Dr. Westrup states that it is through language (and thus, I would argue, through the entire facility of defining into "this" and "not that") that we can let our minds "run away with us" with negative results.

In the same presentation, she explains that ancient practices such as meditation and ancient scriptures such as the Vedas seem to teach that what we call our mind is not the whole of who we are, but rather a very useful and indeed indispensable tool, one which we should view as occasionally detrimental: a sort of "over-eager office assistant" that will sometimes make absolutely terrible recommendations, from which we can learn to "stand aside" or "stand above" through disciplines and methods which were known to the ancients and which can put us in touch with something altogether different.

In the scripture passage from John chapter 20, the remedy or solution given to Thomas does not involve thinking or talking or reasoning at all: it involves feeling and seeing and experiencing and knowing. And it involves getting in touch with the divine twin.

Note that this does not mean "getting rid of the Thomas" -- as Dr. Westrup says in her talk about the "over-eager office assistant," we actually cannot get rid of that assistant, nor would we really want to.

Once we have the faculty of defining and critically thinking (and hence of criticizing and also of doubting) then we cannot ever get rid of that, nor would it be good to do so: but we can get in touch with something which is beyond *defining*, which *cannot* be "*defined*": something which is in fact In-finite (non-boundaried, non-bounded, non-finite).

Something which other traditions (such as the Vedic texts and epics and commentaries) call variously the Higher Self, the Supreme Self, the Brahman -- which is just as much a part of who we are as is the part we might call our Thomas-self. That's why they are described as *twins*. You can't separate them: they are both part of our identity.

In other words, the relationship between Thomas and Jesus implied by the word Didymus may be intended to convey the very same thing that the relationship between Arjuna and the Lord Krishna in the Bhagavad Gita is intended to convey.

And note that at the beginning of the Bhagavad Gita, Arjuna (who corresponds to Thomas) is racked by *doubt.*

Not doubt about the existence or divinity of his divine charioteer, Krishna, but doubt about himself, his worthiness, and whether it is right or not for him to engage in the upcoming battle of Kurukshetra.

And as we see in the Bhagavad Gita, Krishna reveals himself to be unbounded and infinite (just as the goddess Durga revealed herself to be unbounded and infinite immediately prior to the Bhagavad Gita, and in fact was addressed as identical to *the Brahman*, in the hymn to Durga uttered by Arjuna).

506

In those Vedic texts, which I believe were designed to convey the very same message being conveyed by the episode of "Doubting Thomas," the metaphor of a chariot is used, in which the horses are the senses and the desires, and the mind is compared to the reins, but the driver is the "divine charioteer," who in the Bhagavad Gita is Lord Krishna himself. Here, mind is shown to be an essential tool, but it must be guided by the divine charioteer, held in the hands of the divine charioteer.

In other words, I believe we need our critical-thinking "Thomas-faculty" nearly all the time during our waking hours, but there is a very real sense in which this aspect of our humanity gets in the way of our accessing something much deeper, something that is in fact infinite, and that can actually be properly described as divine (and that is described as divine in ancient sacred texts and traditions, including those of the New Testament, as discussed in previous posts such as "Namaste and Amen"[544] and previous examinations[248] of the teachings of the person called Paul).

And we are actually designed to be in touch with the divine Infinite in this life.

Many of us have in fact experienced moments when we seem to suddenly touch something that is beyond or beneath all of the mental chatter, perhaps in a sports situation when (looking back later) we realize we were playing "out of our head."

(Conversely, we can also probably recall situations in sports or other areas of endeavor in which we seemed to "self-sabotage" – through a sudden onset of "doubting Thomas" self-talk – a play or a catch that we normally would have been able to easily make).

Examples from daily life which we might put into the "uncontroversial" category could include parallel parking perfectly on the first try (even into a very difficult spot), or fetching the exact right amount of water to pour into a coffee-maker to come exactly up to the "max-fill" line without measuring (in an unmarked jug or pitcher that you use to fetch it), or even looking at the clock exactly at 3:33 on several different days, without even thinking about it (we might wonder what exactly was "ticking" in the back of your mind that seemed to be keeping track of the time, since it is clear in this example that it was not the conscious part of the mind that "reasoned out" the exact right moment to glance over at the clock on those different days).

But there are other examples that are far from "mundane" and which seem to evidence a sudden manifestation of the "hidden divine within" or the "unconscious connection with the Supreme Self," such as the incredible displays of timing caught on camera in popular videos such as the "Greatest 'Dad saves' ever" shown here https://www.youtube.com/watch?v=nj85L2r17iw (and there are many other collections along the same lines -- many showing situations that are clearly not staged, unless people are deliberately hazarding their infants to make these movies).

It should be pointed out that in nearly every one of these "Dad saves," the injury-saving action is completely unpremeditated and even apparently "unconscious" (without conscious thought). In some of them, the "save" even appears to be *literally* "unconscious," as in "he was half-asleep (or more than just half) and his arm reached out to save the baby."

It should also be pointed out that these kinds of difficult-to-explain displays of unconscious genius are not limited to "Dads," although saving a baby or a child does seem to be a common denominator. For instance, there was an incident in my own experience (known to me personally) in which a mother was in line at the grocery store, facing the clerk, and reached completely behind her back to grab the shopping cart and stop it from tipping over as her older son climbed onto the side of it while her younger son (an infant at the time) was inside of it. She was not looking in that direction at all when this took place: it was behind her and she was about to say "hi" to the clerk in anticipation of moving up to the check-out point.

As difficult to explain as such examples appear to be, there are some who would argue that even these displays of human response -- admittedly beyond our "day-to-day" way of behaving or reacting -- are still explainable within the realm of the "natural, material world" and do not require descriptions involving the words "divine" or "infinite" or connections to anything non-material or super-natural.

Perhaps they are just manifestations of highly-developed instinctual abilities on the same level as those which animals routinely display (untroubled as they are by anything resembling the "Thomas-mind" and the self-doubt that comes along with being able to think critically and maintain inner dialogues), and which we usually forget in our civilized setting, but which "pop up" from time-to-time when they are most necessary (a kind of "animal-like survival instinct" that is usually forgotten but occasionally awakens).

That is certainly a possible explanation, and one that our critical-thinking, scientific-method-following minds should consider.

But even if that is a valid explanation for *some* manifestations of behavior (like the "Dad saves" shown above) that fall completely outside of what we usually experience in what might be called "ordinary reality," there are *other* examples of human beings apparently accessing the fabric of non-ordinary reality for which even *that* explanation (already a stretch) seems to be completely inadequate.

For example, in a post from all the way back in January 2012 (not included in this collection but available online here: http://mathisencorollary.blogspot.com/2012/01/heartfelt-portrait-of-john-blofeld-from.html), we examined an account of a daughter who was visited in dreams and who received information about the existence of a Buddhist monastery the existence of which she had previously been unaware, but which upon visiting she learned from the presiding abbott that her father had helped found that particular monastery, years before she had even been born.

It is difficult to explain that account as an example of "highly-developed human ability or instinct," because it involved

509

information that came to a person (while unconscious, it should be noted) who could not be expected to know that information at all – even subconsciously.

Or, see for another example the situation described in the account of Norman Ollestad in his book *Crazy for the Storm*, in which as a young eleven-year old boy, he had to make his way down a steep and icy mountain in what can only be described as a life-or-death situation.

In that book, we see an excellent real-life example of the "Doubting Thomas" phenomenon: young Ollestad must overcome his own fears, anxieties, self-criticisms and self-doubt – both on the mountain and in the challenging situations he faced while growing up in the canyons and suburbs around Los Angeles and the California coast during the 1970s.

In order to overcome those doubts, he relies on the uplifting influence of his father, and on reserves of courage and resourcefulness inside himself that at first the boy might not even have known or realized were there.

However, that is not all that helps him survive, as those who have read the book (or who will read the book after this) see by the end. Indeed, in order to eventually make his way off the mountain, several events (including something that he is able to "see" which he later realizes he would not have been able to see based on actual terrain and line-of-sight) and "coincidences" took place which directly contributed to the author's survival on that awful day in 1979.

Although they might not be as dramatic, many of us can also think of "coincidences" or "synchronicities" in our own experience in which people who could not possibly have known that we were thinking about something or considering some course of action suddenly contacted us with information or suggestions that make it seem as though something from outside of ordinary reality is at work.

It is my belief that the episode involving the encounter of "Doubting Thomas" and "the risen Lord" is intended to describe this exact dynamic in our human experience: the fact that we ourselves are endowed with an important facility of critical thinking, which is well-suited for many aspects of day-to-day life (and which is in fact indispensable for our survival), but which can also be a hindrance to us, to the extent that it can lead to self-doubt, self-sabotage, self-destruction in extreme cases, and self-imposed separation from someone we are actually supposed to rely upon as absolutely vital to our experience in this life: our Higher Self, the "divine charioteer," the Christ within.

Indeed, while some readers may remain unconvinced by the analysis and examples offered so far (and especially those who are especially committed to a literal-historical interpretation of the sacred texts of the New Testament) -- even though I believe that the discussion so far should already be fairly convincing -- I believe there is actually a whole additional line of evidence which makes the above interpretation not only "likely" but nearly "indisputable."

The more I have studied the ancient mythology of humanity, the more evidence I have found that virtually all of it, from every single inhabited continent on our globe, and from millennia in the past right up to living traditions which have remained in practice into the present day, is built upon a common system of celestial metaphor, the purpose of which is to convey exactly the type of knowledge that we have been examining above regarding the human condition and the makeup of the natural world and the cosmos in which we find ourselves.

Knowledge regarding its dual material-spiritual composition: the existence of a Spirit World or an Infinite Realm which interpenetrates this material realm at all times and at all points, and with which we are actually in contact all the time ourselves, through our own inner divine spark, our own inner connection to the Infinite.

This inner connection may be often neglected, or even completely forgotten, but (as the embedded video and some of the other examples discussed above make clear) it is very real, and it is very powerful.

It absolutely transcends and blows away our limited understanding of what we ordinary think of as "reality."

But our normal facilities of thinking and understanding and analyzing (the "Thomas side" of our "twinned" existence) tend to doubt the very existence or reality of the divine nature, and when we listen to them enough we can miss out on something that is actually a huge part of who we really are.

A nice "contemporary film allegory" for this self-doubt and self-sabotage which keeps us from reaching our "non-ordinary potential" is the famous exchange between "doubting Skywalker" and Yoda in the famous "X-wing in the swamp" scene from *The Empire Strikes Back* (1980): in that scene, Yoda exclaims (when confronted by Luke's constant *doubting*), "Do, or do not -- there is no *try*."

One way that we can help to confirm that the "Doubting Thomas" episode in the Gospel According to John was intended to be understood as an esoteric metaphor and not as a literal account of an event which took place in terrestrial earthly history is the fact that, like so many other events related in the scriptures of the Old and New Testaments of the Bible, it incorporates clearly-identifiable celestial components.

In fact, I can find enough celestial components to this story and those which precede and follow it in John's gospel to amply confirm to my own satisfaction that it is describing of the heavenly cycles of the sun, moon, stars and planets (which serve throughout the world's mythology as an allegorical system which relies upon some of the most majestic and awe-inspiring aspects of our *physical, material* universe to discuss and explain aspects of the *invisible, spiritual* world) and not a description of anything that took place in terrestrial human history.

512

Very briefly, the Reverend Robert Taylor (1784 - 1844), who lived well before the discovery of the Nag Hammadi texts, and who spent considerable time discussing the identity of Thomas in a collection of his public sermons or lectures published in 1854 (ten years after his death) and entitled *The Devil's Pulpit*, makes much of the fact that the traditional observance of St. Thomas' Day was held on December 21st, the point of winter solstice and the point on the zodiac wheel which during the Age of Aries marked the beginning of the sign of Capricorn.

Since at the points of solstice ("sun-station" or "sun-stand-still") the sun appears to "pause" and rise at roughly the same point on the horizon for about three days before turning back around and moving in the other direction again, which is why the "birth" of the solar child is celebrated three days later (at midnight on the 24th of December, rather than on the solstice-day of the 21st of December), Robert Taylor argued that the 21st is a sort of "day of maximum doubt," when the sun has been rising successively further and further south since summer solstice in June, and tracing an arc that is lower and lower across the sky, and the hours of daylight each day have been getting shorter and shorter relative to the hours of darkness – and that "unbelieving Thomas" is thus the figure who doubts that the sun will ever turn around again (*Devil's Pulpit*, 42).

Interestingly enough, there are also traditions which associate the feast-day of Thomas with the 3rd of July, which is in the sign of Cancer the Crab, the sign which follows the sun's point of maximum arc and its northmost rising- and setting-points (as well as with other days of the year).

Robert Taylor incorporates all these details into his explanation, in which he argues that Thomas is associated both with the Goat of Capricorn (beginning at the sun's lowest point) and with the Crab of Cancer (beginning just after its highest) -- and also the related fact that he is called "the twin."

DAYS LONGER THAN NIGHTS:
Heaven, Promised Land, Greece, etc.

Horizontal Column:
The Djed cast down

Vertical Column:
The Djed raised up

NIGHTS LONGER THAN DAYS:
Hell, Egypt, Troy, etc.

Zodiac wheel with positions of the signs of Capricorn (green) and Cancer (red) indicated, one beginning at the point of winter solstice (Capricorn) and the other beginning at the point of summer solstice (Cancer).

One look at the zodiac image above should be enough to perceive just how ingeniously the ancient myths (including those in the Bible) were crafted to impart their esoteric message, and how the majestic cycles of the celestial realms were employed in order to convey knowledge of spiritual truths -- in this case, the truth allegorized by the metaphor of Thomas and the Divine Twin, one enmeshed in the doubts and definitions of the "practical" struggles of the finite world, and the other completely free of the bounds of earth (passing easily through locked doors) and of the endless defining and analyzing of the "Thomas" side of our nature: the divine nature, at home in and representative of the realm of the Infinite.

Images of Thomas in this famous encounter with the Lord painted in previous centuries have in fact emphasized his Capricornian nature.

514

Now is a good time of year to observe the Goat of Capricorn in the sky (look to the west of the Great Square of Pegasus, or to the east of the distinctive "teapot" outline in the constellation of Sagittarius, which is currently still easily visible looking towards the south during the prime stargazing hours after sunset and before midnight, at the base of the rising column of the Milky Way).

Below is an image of the night sky as it looks to an observer in the northern hemisphere in the temperate latitudes, and looking towards the southern horizon (where the zodiac constellations make their nightly procession):

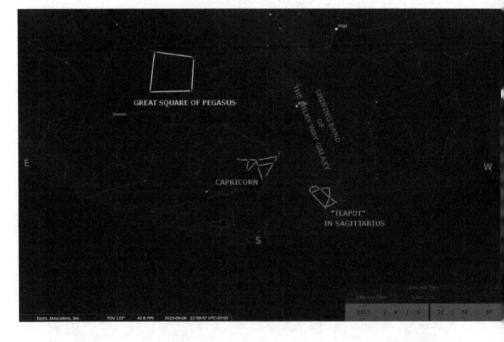

In the above image, you can see that the zodiac constellation of Capricorn the Goat (or the Sea-Goat) is actually reaching its highest point (its *transit point*) as it turns through the due-south celestial meridian-line (the highest point on its arc through the sky between rising in the east on the left and setting to the west on the right) right around 11pm.

Note that the constellation suggests the shape of two "point-downward" triangles: one for the head of the Goat, and the other formed by Capricorn's two feet, which come together in a near-point, as if he is a rock-hopping mountain goat instead of a Sea-Goat as he is often portrayed.

It is also notable that he has some fairly formidable "goat-horns" pointing almost straight forward from his head, which are distinctly two in number: there are two stars to mark the tips of the Goat's horns (one is labeled in the image above, Deneb Algedi, and the other is just a bit further to the right and on a line slightly below Deneb Algedi in the sky).

Below is the same "zoomed-in" image of Capricorn, this time with green outlines to help make perfectly clear the line of the horns and the "two triangles" shape of the constellation:

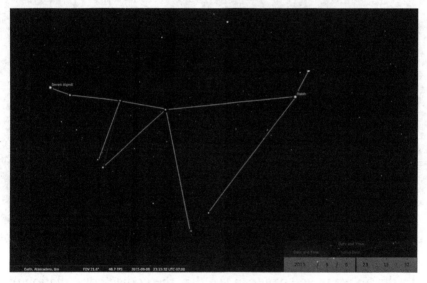

Having familiarized ourselves with the outline of Capricorn, let us now take a look at some of the images created by master artists over the centuries depicting the famous encounter between Thomas and the risen Lord in the episode of "The Incredulity of Thomas."

The first (and perhaps most revealing) is from Giovanni del Giglio, who lived from some time in the late 1400s through approximately 1557. It is entitled *L'incredulita di San Tommaso*:

Take a close look at the hands of Thomas and the divine twin (the risen Lord).

If you have studied the images of the constellation Capricorn presented above, you will find that the unmistakeable features of the heavenly symbol are reproduced in this drawing and are associated with the probing fingers of Thomas (the horns of the Goat), the downward-facing triangle of the hand of Jesus (the head of the Goat), the bend of the arm of Thomas below the elbow of the risen Lord (the feet of the Goat), and the distinctive hand-symbol being displayed by the woman in the image (the tail of the Goat).

517

If you are having trouble seeing the correspondence between the image and the constellation, it is outlined in the zoomed-in detail from the identical image, with Capricorn added:

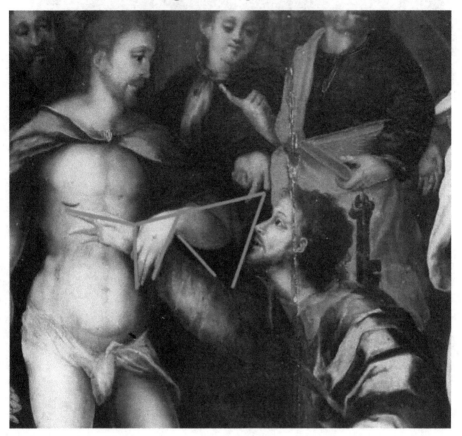

Below is another example, much more recent, from Tissot (1836 - 1902), which envisions the same scene but instead appears to use the bearded, downward-bowed head of Thomas himself to evoke the idea of the head of Capricorn, and the down-stretched arm and one leg of the apostle to suggest the front and back legs of the constellation which are nearly together in the outline of Capricorn in the actual night sky [later note -- I believe it is more likely that Tissot is using the "triangle shapes" created by the portion of Christ's garment above Thomas's arm, and the trianglular space above Thomas's head, as well as the arms of the

518

woman just above his neck and shoulder, to create the "two triangles" of the constellation Capricorn in this depiction of the "Doubting Thomas" scene]:

The examples could be multiplied on and on: the reader is invited to examine them for himself or herself to decide whether or not the identification of Thomas with the constellation Capricorn is valid in these examples, based on what we know of the outline of the stars themselves in the sky.

There is actually much more which could be added to the celestial metaphors at work in this particular scriptural event, and in the events which surround it in the John gospel, which act to confirm even more powerfully the fact that this story was originally intended to be understood esoterically rather than literally as an event taking place in earthly history.

One other important piece of evidence which Robert Taylor offers in his extensive analysis of the identity of Thomas is the fact that his name itself points to the connection between Capricorn and Cancer in this story (the signs marking the celestial low-point and the celestial high-point).

The name Thomas, he alleges (and others have made the same assertion) is related to the name Tammuz, which is both the name of an ancient deity and also of the fourth month of some ancient calendars (including the Hebrew calendar still in use today).

If you look again at the zodiac wheel reproduced above, and count to the fourth sign after the point of spring equinox (the beginning of the year in many ancient cultures), you will find that this count brings you to the sign of Cancer the Crab (1 - Aries; 2 - Taurus; 3 - Gemini; 4 - Cancer).

In other words, Thomas is associated with *both* Capricorn and Cancer: both the "doubting twin" and with the exalted Supreme Self (and some have even noted that his confession or exclamation "My Lord and my God" appears to refer to both human kingship and divinity, an expression of the dual nature of the Christ).

All of this appears to rather strongly confirm the powerful insight of Alvin Boyd Kuhn, quoted many times in previous posts,[33] that the ancient myths of the world (including those in the Bible) are not about ancient history but about our experience "here and now;" that they are not about "old kings, priests and warriors" but rather that in every scene they treat the experience of "the human soul."

"The Bible is about the mystery of human life," he says, "[. . .] and it is not apprehended in its full force and applicability until every reader discerns himself [or herself] to be the central figure in it!"

520

(Note that the two halves of the foregoing quotation are from different sentences in the same lecture by Alvin Boyd Kuhn, but by quoting them in this way I have not altered the sense of what he is asserting).

Indeed, when it comes to the story of Thomas the Twin (Didymus), we might alter Alvin Boyd Kuhn's quotation a bit further and say that this particular story "is not apprehended in its full force and applicability until every reader discerns himself or herself to be a twin in exactly the same way!"

The metaphor of Thomas and the divine twin is a metaphor to teach us a profound truth. It could be taught a different way, using a different metaphor – such as the metaphor of Arjuna and the divine charioteer, in the Bhagavad Gita. In fact, there are endless different ways of expressing the same concept, found throughout the myths of the world, which collectively are the precious inheritance of humanity, intended for our benefit and use in this life.

We are each a twin in exactly the same way that Thomas is a twin: permanently 'twinned' with the 'divine twin,' who can appear in an instant no matter where we are or in what circumstance we find ourselves. No locked door can prevent the appearance of the divine twin, for we ourselves have within us -- always and in every circumstance -- an inner connection to the Infinite; we ourselves contain both Capricorn and Cancer: both twins, simultaneously. We are prone to doubting and to forgetting -- to saying with Luke in the swamp, 'I'll give it a try' -- cutting ourselves off from unlimited potential, when in reality we have access to all of it, all the time.

The fact that the story is a metaphor in no way means that it is "not true" (it just is not, at least in my understanding of it based on the evidence that I have seen for myself, literal or historical).

In fact, I believe it is *profoundly* true, and that it has daily practical applications for us in virtually every field of our human experience.

There are ways to learn these truths other than through the exquisite metaphors found in the world's ancient myths -- but when we have this incredible treasure which has been imparted to us for our good, it would seem to be a terrible waste to ignore these ancient teachings, or to turn them into something which they quite plainly are not (especially if we know what they are).

Who is "Doubting Thomas"?

Well, obviously, we have him with us every day.

But if we recognize his good aspects (incredulity, after all, can be a good quality), while avoiding the negative side of incredulity (self-doubt, over-criticism, over-haste in labeling defeat or failure, self-sabotage, and disconnection with the divine nature which is as much a part of who we are as is the Thomas-nature), we can touch the Infinite. Every day.

Namaste.

You may have a Higher Self
(at least, according to the ancient myths and scriptures), and He or She wants you to know it

2016 August 24

If the world's ancient scriptures and myths are not literal but rather allegorical, then it is quite likely that attempts at literal interpretation risk serious mis-interpretation.

For instance, I have written previously[499] that the dramatic encounter described in the Fourth Gospel, in which Thomas (often referred to as "Doubting Thomas") encounters the risen Lord, may well represent a teaching about the existence of a Higher Self, a Divine Twin -- and that the two figures portrayed as two different persons in that story may actually have been intended to convey to our understanding the true situation of each and every man or woman in this incarnate life.

They are not two separate persons: that is a mistaken interpretation which comes from reading the story literally.

In another example, previous posts have also explored the possibility that the Battle of Kurukshetra described in the ancient Sanskrit epic of the Mahabharata may be celestial allegory and not literal, terrestrial history (or even a "mythologized" or "supernaturally enhanced" version of literal, terrestrial history) -- meant to describe the experience of each and every human soul which descends into the "battlefield" of this incarnate life.

Some of those discussions have examined the possibility that the depiction of the semi-divine hero Arjuna with his divine charioteer being none other than the Lord Krishna himself may also be a dramatization of the proper relationship with what some translations of the teachings of that tradition have labeled the Higher Self or the Supreme Self: see for example this previous post entitled "Self, the senses, and the mind"[466] in which quotations from the *Katha Upanishad* or *Kathopanishad* were cited which use the image of the chariot to illustrate the goal of bringing the senses and even the mind under the direction of the Higher Self.

The *Kathopanishad* says that the senses are like powerful horses, which if not properly guided by the mind (which acts as the reins) under the control of the Higher Self can run off after their desires, out of control.

In the illustration from the Bhagavad Gita (a portion of the Mahabharata detailing the discussion between the Lord Krishna and Arjuna just prior to the start of the great battle) shown below, we see that the Divine Charioteer (Lord Krishna) is between the horses and Arjuna, and we see Arjuna placing his palms together in recognition and acknowledgement of the divinity of Krishna:

It is very easy for the "horses" to run away with us, so to speak, and we have all experienced this first-hand (probably later regretting what happened). For example, if in a discussion or a debate, if someone makes a pointed insult (as a deliberate tactic to incite emotion such as anger in the other person), all clear thinking can go out the window as the horses stampede (blood rises to our head, we might even begin to literally "see red," and what we say or do at that point may be more driven by anger or emotion than anything else -- that is, if we are holding the reins ourself, without the Higher Self in between).

Another common example might be our performance in a sporting event, in which we are about to take a crucial shot at the goal, and the mind is suddenly seized with doubt. Through training, we can actually learn to observe the impulse of the horses in a more dispassionate way, saying in effect, "I see these emotions arising -- that's interesting, but I am not going to let them take me wherever they want to go."

These are mundane examples (although quite important ones -- in which letting our doubts or our anger have full control can lead to various levels of disaster). The ancient myths of the world, however, demonstrate over and over that the existence of a Higher Self goes *far beyond* these examples (as helpful as those examples are for understanding the concept). The world's ancient myths, scriptures and sacred stories appear to show us that integration with the Higher Self -- the divine or Supreme Self -- is one of *the* critical missions in this incarnate life, if not *the* critical mission (period).

However, if we insist on trying to read the stories literally, we may well miss that message altogether. Because if we read the above episodes literally, we will mistakenly conclude that the duo of Thomas and the risen Lord, or the duo of Arjuna and Lord Krishna, are *separate* personages -- and that the human figures at least (Thomas in the gospel account and Arjuna in the Mahabharata) represent people who lived thousands of years ago, and who probably have little or nothing to do with us and our own personal situation, whatever that may happen to be.

The message is so important, in fact, that the myths of the world present it to us in hundreds or thousands of different ways. And, although many previous posts have cited illustrations which use "male" characters to illustrate the principle, that is by no means always the case either. Several important and dramatic illustrations employ female figures to illustrate very much the same message -- for example, the Sophia cycle, or the memorable myth of Psyche and her divine lover, Eros (or Cupid), relayed for us in written form in the *Metamorphoses* of Apuleius (most commonly called *The Golden Tale of the Ass,* or more simply, "*The Golden Ass*"), although earlier statues and mosaics and shorter references from earlier texts show us that the myth predates Apuleius himself by at least some centuries.

The story of Cupid and Psyche occupies a very prominent position in the tale of Apuleius, who himself appears to have been an initiate of the Mysteries of Isis, and whose Metamorphoses should be read very carefully, because it appears to be the work of someone who understood the deep truths which his seemingly "idle tales" were designed to convey. A 1924 edition is available online in its entirety here, which contains the original Latin text alongside the English translation, although I personally am partial to the 1960 translation by Jack Lindsay, which I believe is far superior, although it does not provide the Latin text for comparison, which is a definite advantage of the online version linked above.

I would personally recommend obtaining a physical copy of the Lindsay translation for your own library, if possible, and then the Latin version can be seen online if desired for comparison (Latin scholars or Apuleius aficionados may want to obtain the original Latin in hardcopy as well).

The Psyche story deserves to be read as Apuleius tells it – I will not spoil it by giving a summary here. I will say, however, that I believe the story is *spiritual allegory*, and that we should be extremely careful about forcing it into service to make modern points about supposed relationships between genders or sexes (the entire text of the *Metamorphoses* of Apuleius consists of stories within stories within stories, all related to us by a narrator who spends most of the book transformed into the outward appearance of an ass – which itself can be seen as a metaphor relating to our incarnate experience in this material plane of existence).

However (and if you wish to stop here and read the story as told by Apuleius first, now would be a good time to do that), I will point out a few aspects of the story which clearly relate it to the discussion above.

First, the story clearly has parallels to the story of "Doubting Thomas." In the story, Psyche lets her "doubts," so to speak, run away with her -- to her great sorrow, and temporary estrangement from her divine lover.

The reconciliation of that relationship between Psyche and Cupid (or Eros, as he is called in Greek myth, a name which Gerald Massey hints may well be related to the Egyptian deity

Horus, and who may play the same role in the story that Horus plays in the myth-cycle of Isis, Osiris and Horus – this observation is found in Massey's *Ancient Egypt: the Light of the World, Volume One* on page 223) takes up the majority of the tension in the memorable tale.

In all of the mythical illustrations – Thomas and the risen Christ, Arjuna and the divine Krishna, Psyche and Eros -- the proper relationship is seen when the divine and the human are in harmony, and when "control" is turned over the divine twin in the relationship.

The story of Cupid and Psyche also contains two "awakenings" -- first, Psyche awakens Cupid (shown in the illustration above, as well as in numerous ancient and Renaissance depictions of the myth), when she lets the doubts and insinuations of her two sisters get the better of her. In the illustration, Psyche is about to accidentally spill burning oil from a small lamp or cruse onto the god's sleeping form.

Later, however, Cupid awakens Psyche, from a sleep that is described as a deathlike sleep. This precedes the ultimate union of the two in a divine marriage at which all of the assembled gods and goddesses are present.

Alvin Boyd Kuhn, following and citing the arguments made earlier by Gerald Massey, says plainly in *Lost Light* that the story of Cupid and Psyche represents the "welding at last in blissful harmony of the mortal and immortal elements" (587).

Discerning readers might be wondering at this point, however, if we are supposed to understand *all* of the gods and goddesses found in the ancient myths as representative of our Higher Self, our divine invisible nature, our Christ within.

The answer, I believe, is *no*.

Psyche got into trouble, at the beginning of her story, when everyone for miles around began to pay more attention to her than to Venus herself, the *actual* goddess of love and beauty. So it would probably be a mistake to conclude that the myths are teaching us that the gods and goddesses (including Krishna) are always representative of our own Higher Self.

However, Kuhn does make the very important argument on page 550 of *Lost Light*, based on passages found in the Pyramid Texts (specifically some of the Pyramid Texts from Teta or Teti, which can be read in the Budge translation on page 139 of *Osiris and the Egyptian Resurrection*) that:

> man is to summarize in himself the qualities of the whole scale of being, denominated gods. All their powers and virtue have to be embodied in man's organic wholeness to make him, like the resuscitated Osiris, "Neb-er-ter, the god entire."

Obviously, although Kuhn (writing in 1940) uses the word "man" (as was common in previous generations to indicate "humanity in general"), he means *men and women* (and he says "men and women" and "male and female" explicitly in other parts of the same discussion, such as on pages 551 and 587).

The point being made is absolutely critical, and worthy of deep consideration, as a guide to what we are supposed to be doing here in this life in the (seemingly) material world in which we find ourselves. But we are apt to miss this message altogether if we attempt to read the ancient myths and scriptures as if they were intended to be taken literally as opposed to esoterically.

Note that the presence of all the gods and goddesses at the conclusion of the story of Psyche and Cupid can be seen as a visual dramatization of the very teaching which Kuhn articulates above.

Finally, it must be noted that the entire story of Psyche in *The Golden Ass* is presented with what I would argue to be very clear indicators of celestial allegory. Readers of the series *Star Myths of the World, and how to interpret them* (particularly Volumes Two and Three) will find extensive discussion (and illustration) of

529

other myths and sacred stories which involve high cliffs and personified divine winds, such as are found in the tale related by Apuleius.

Note also that the critical moment in the story, in which Psyche spills hot oil upon the sleeping god (at a true low-point in the narrative, when she has succumbed to her all-too-human doubts), can be seen to have very specific celestial correspondences. The cruze of oil is found in many Biblical stories (discussed in *Star Myths of the World, Volume Three*) and it has powerful spiritual meaning, as well as clear connections to the outline of the important zodiac constellation of Sagittarius, as discussed in *Volume Three*.

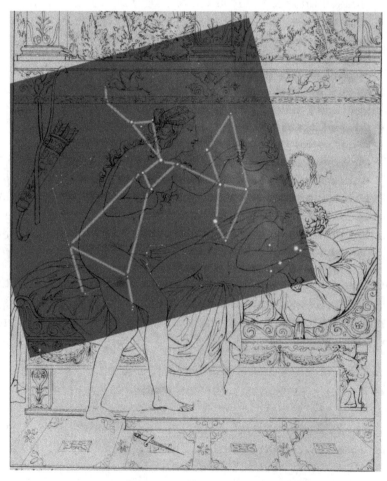

I believe that the understanding of *where* certain scenes can be found on the Great Wheel of the zodiac gives us additional insight into the deeper meaning of the myths and sacred stories (see previous discussion here[156]). The fact that this "low-point" (which is also a "turning point" in the story, and in the life of Psyche) takes place at Sagittarius – at the bottom of the Wheel – has deep spiritual significance (also outlined in depth by Alvin Boyd Kuhn, most notably in a 1936 lecture entitled *Easter: the Birth-day of the Gods*).

The absolutely profound importance of the fact that the ancient myths and scriptures of the world are in fact speaking to us in the language of celestial metaphor may sometimes be difficult to relate to our individual lives -- but I believe that the above discussion should help demonstrate the awe-inspiring and potentially life-changing (as well as exciting) message these ancient myths are telling us: they seem to be saying quite clearly that each and every man and woman does indeed have a Higher Self, and that he or she wants you to *wake up* to that fact.

The Inner Connection to the Infinite

A brief examination of the importance of chakras and singing praises

2011 July 31

In the 1950s, Frank Waters (1902 - 1995) and Oswald White Bear Fredericks (1905 - 1990s) spent three years tape recording the sacred traditions of the Hopi people as told to them by twenty-seven Hopi elders, who had consented to preserving their wisdom in a form other than oral tradition for the very first time.

From the recordings, Mr. Waters and Mr. Fredericks created a manuscript, which the elders reviewed and approved (an important point, as there are some who allege that Mr. Waters treated the Hopi tradition in a freewheeling fashion and changed what he was told into what he wanted it to be, which is a baseless allegation). This text became the *Book of the Hopi*, published in 1963.

One of the most striking features of the Hopi creation story, as related to Mr. Waters, is the belief that we are currently living in the Fourth World, the previous three having been destroyed by fire, ice, and water, respectively. As is now widely known, the Maya also believed that we are living in the fourth age of the world, which they called the Fourth Sun, the previous Suns having been destroyed, as this one will inevitably be as well. It is quite fascinating to examine the reason given by the Hopi elders for the destruction of the previous three worlds.

In the *Book of the Hopi*, we learn that the Creator gave the people two clear commands at the beginning of each new age: "First, respect me and one another. And second, sing in harmony from the tops of the hills. When I do not hear you singing praises to your Creator I will know you have gone back to evil again" (16).

Clearly, singing to the Hopi was a matter of very great importance. Further, the elders revealed their understanding that when men and women were first created, they began to multiply and spread throughout the earth, but that "This did not matter, for they were so close together in spirit they could see and talk to each other from the center on top of the head. Because this door was still open, they felt close to Sóktunang and they sang joyful praises to the Creator, Taiowa" (15).

This "center" or "door" on top of the head is very significant. As related by the Hopi elders, it was the uppermost of the vibratory centers inside every human, which corresponded to the vibratory centers in the earth itself:

> The living body of man and the living body of the earth were constructed in the same way. Through each ran an axis, man's axis being the backbone, the vertebral column, which controlled the equilibrium of his movements and his functions. Along this axis were several vibratory centers which echoed the primordial sound of life throughout the universe or sounded a warning if anything went wrong.

The first of these in man lay at the top of the head. Here, when he was born, was the soft spot, *kópavi*, the "open door" through which he received his life and communicated with his Creator. For with every breath the soft spot moved up and down with a gentle vibration that was communicated to the Creator. 9 - 10. Other vibratory centers included those at the brain, the throat, the heart, and the navel. The Hopi tradition makes clear that the successive destruction of the first three worlds took place after the majority of men began to cease their singing of praise to the Creator and stopped communicating with him through their first center.

What is most remarkable is that these vibratory centers described by the Hopi elders correspond quite closely to the *chakras* described in ancient Hindu texts and persisting in Tibetan, Hindu and Buddhist beliefs to this day, a similarity which Frank Waters points out in a footnote to his manuscript (pages 10 and 11). While most eastern traditions hold that the *chakras* are seven in number (and ten in number in some Tibetan traditions), the corresponding location of the crown of the head, the forehead (or brain), the throat, the heart, and the navel in both the eastern and the Hopi traditions is quite fascinating.

It is also fascinating that the Hopi described these important "doors" as vibratory, in light of the subjects touched upon in a discussion published in a post from July 19, 2011 of the importance of wavelengths and energy as they relate to human beings through music and (as John Anthony West has described in his books on ancient Egypt) the proportions of architecture and art. We discussed the importance of this concept further in another more recent blog post here.

Finally, it is quite interesting that the Christian faith also places great importance on singing (as well as chanting, in some traditions, and singing without accompanying instruments in some traditions as well).

Even more intriguing is the command found in 1 Corinthians 11:4 which says:

> "Every man praying or prophesying, having *his* head covered, dishonoureth his head."

The reasons for this strict admonition are worthwhile to contemplate.

Does the direction you lay your head down to sleep matter?

2011 September 19

Does it matter which direction your body is oriented when you lay your head down to sleep?

Some ancient civilizations apparently thought so. The Mahabharata, one of the ancient texts of India, was probably composed in the 8th or 9th centuries BC but possibly much older than that (certainly the authors of *Hamlet's Mill* demonstrate throughout their book that the Mahabharata contains encoded ancient knowledge which is likely from the same source as the ancient knowledge that turns up in ancient Egypt and ancient Sumer and other civilizations around the world). The Mahabharata instructs its readers:

> They that are wise should never see themselves in an unpolished or dirty mirror. One should never have sexual congress with a woman that is unknown or with one that is quick with child. One should never sleep with head turned

538

towards the north or the west. [Book 13 (Anusasana Parva), section 104].

This very ancient injunction against sleeping with the head to the north or to the west has been preserved among Indian culture, according to some sources. For example, Swami Buaji (who lived to quite an advanced age) apparently believed the same thing about the importance of the direction of the head while sleeping. In this passage, which is cited around the internet in various places, he says:

"Never lie down to sleep with your head northward or westward" is a common injunction given from time immemorial by the Indian mother to her children. Almost every Hindu- orthodox or heterodox- observes this dictum of his ancestors, but he doesn't know the rationale or significance behind the dictum, although it has been handed down to him through generations. For example, Vishnu Purana says: "O King! It is beneficial to lie down with the head placed eastward or southward. The man who lies down with his head placed in contrary directions becomes diseased." The Varshaadi Nool says: "Sleeping eastward is good; sleeping southward prolongs life; sleeping westward and northward brings ruin." The Mahabharata says: "Men become wise by sleeping eastward and southward." There are two Tamil proverbs which run thus: "Vaaraatha Vashvu Vanthaalum Vadakkae Thalai Vaikkakkuudathu", meaning; " Even in the heyday of sudden fortune, one should not lie down with head to the north", and " Vidakkeiyayinum Vadakkaakaathu", meaning: "Even the head of the dried fish should not be placed northward." The Ayurvedic physician seats his patients facing eastward before diagnosing the disease or administering his medicine. Brides and bridegrooms are always seated facing eastward on the wedding day. Even corpses are placed down with the head southward.

In addition to cited numerous other texts and proverbs beyond the Mahabharata, Swami Buaji also gives some explanation as to why he believed the direction of the body at rest was important. According to that Yogi, human beings reflect the planet earth in having a north pole and a south pole, and the alignment of our body while sleeping matters because the flow of energy through the earth effects our own magnetic field.

This unrelated website
http://www.spiritualresearchfoundation.org/articles/?id = spiritualresearch/spiritualscience/how_to_sleep_well/sleepingwellfeetfacingwest
appears to counsel very much the same thing, arguing the importance of the direction of the body during sleep and proposing that head facing east and feet facing west is beneficial.

We have seen in previous posts that ancient cultures appear to have been aware of the low-frequency underground electric currents called telluric energy which flow through the earth. It is quite likely that the ancients were aware of some of the influence that the earth's energy has on the human body (some researchers point to evidence that this energy also impacts seeds and the positive growth of plants).

Even in our supposedly advanced modern civilization, we know very little about sleep. In fact, the Division of Sleep Medicine at Harvard Medical School admits that we still do not even really know why we and other animals require sleep at all! Recent studies have found a strong link between quality of sleep and successful aging.

Based upon the fact that Swami Buaji was healthy and active at a very advanced age (according to his followers, he was still teaching at ages over 110, determining his age by photographic evidence from various times in his life), it may be prudent to consider carefully his advice on the direction of the head and body during sleep. The fact that his teaching is backed up by very ancient texts, from civilizations that appear to have known more about the earth's energy field and its impact on life on earth than we do today, would appear to make his case even stronger.

540

How much time do you spend chanting praises?

2011 October 11

In this previous post,[535] we considered the teachings of the Hopi elders who consented to pass on their sacred traditions in the 1950s for the benefit of others, which Frank Waters and Oswald White Bear Fredericks recorded in *The Book of the Hopi* (1963). We saw that, according to the oral tradition, the world had been destroyed three times in the past, and that at the commencement of each new age after each successive cataclysm, the Creator gave to his people the same two commands:

> First, respect me and one another. And second, sing in harmony from the tops of the hills. When I do not hear you singing praises to your Creator I will know you have gone back to evil again. 16.

What is it that is so important about singing praises to the Creator in harmony from the tops of the hills that this would be one of two strict commands, the neglect of which would bring as a consequence another world-ending catastrophe? If this command is so important, should we perhaps be paying more attention to it?

It turns out that singing praises in harmony to the Creator is held to be extremely important in other places where very ancient teachings are remembered as well. Here is the link to a video of the late Swami Buaji (whom we met in this previous post[538]):
https://www.youtube.com/watch?v=D3ol9SxbtJs

In it, he begins a very interesting chant, which contains words which can certainly be described as "singing praises to your Creator," to wit:

> Relax your body!
> Relax your mind!
> Mentally massage your body.
> Give a mental massage!
> Pray to God, Almighty.
> The Creator of the universe.

Pray to him to give you your long life
Your healthy life
Your disease-free body
Health and strength
Strength and stamina
Vigor and white energy (?)
Peace and prosperity
He is the doctor of doctors
He is the father of fathers
He is the architect of this human being (?)
He is the engineer of this human machine
He alone can give you whatever good you require
Therefore, pray to him
Surrender to him
Take shelter under him
Seek perfection under him
Seek asylum under him
He is the Creator
He is the protector
He is the (?)
He is the survivor
He is the sustainer
He is the preserver
He is the designer
He is the (?)
He is the requirer of all wisdom (?)
He alone can give you whatever good (?) you require
Therefore, remember him always
Forget him not [. . .]

Not only are these words themselves noteworthy, but the tone in which they are sung or chanted is extremely noteworthy as well. They are not simply spoken, but rather intoned. The sing-song pattern of this "singing" or "chanting" appears to be important, because it is found in other ancient sacred chanting as well. Compare the pattern to the singing or chanting in the following videos.

542

Byzantine liturgy:
https://www.youtube.com/watch?v=jbOJNHrhPbE

Native American chanting or singing:
https://www.youtube.com/watch?v=8JI20-nxHd8

Tibetan monks chanting in a new home:
https://www.youtube.com/watch?v=P2xqE1fVL1M

It appears that some people are remembering the importance of chanting (and specifically, "singing in harmony" and "singing praises to your Creator," as expressed by the Hopi elders), while many of us have forgotten it.

Without rushing to any conclusions presumptuously, it is at least possible to say that in light of all of the above, it might be wise to consider the possibility of the ongoing importance of chanting. It might also be possible to conclude that the actual language one uses is not as important as other aspects of this kind of singing – the video of Swami Buaji indicates that it can even be done in English!

It is certainly possible that these similar patterns of expression sprang up independently around the world. On the other hand, it is also possible that they preserve some common heritage of mankind that is very ancient (we have seen previously that John Anthony West provides extensive evidence that "harmonies" – including audible harmonies – were considered to be of great importance by the ancient Egyptians as well).

This seems to be a subject worthy of further exploration.

Namaste and Amen

2014 July 10

In the previous post entitled "Amen and Amenta: the hidden god and the hidden realm" (July 5, 2014), we examined enough evidence from ancient mythology (including stories found in both the Old and New Testaments) to conclude "that buried within the scriptural metaphors is the concept of the 'hidden god,' the divine spark which resides within each man or woman, practically unknown and even forgotten."

Alvin Boyd Kuhn devotes an entire chapter in *Lost Light* to this concept and to the mythological metaphors which were meant to convey this teaching of the hidden spark of divinity within humanity. In "Colonists from Heaven" (pages 105 - 127), Kuhn demonstrates that myths involving a god or goddess who is cast down into the "underworld" or imprisoned there or cut up and scattered across the earth or lost and searched for by another god or goddess (usually by a goddess) may all be seen as illustrating this teaching.

Among these myths, Kuhn lists the dismemberment of Osiris and scattering of the pieces of his body across the earth, the search by Isis for Osiris, the search by Demeter for her daughter Persephone, the descent of divine Eros to wed the mortal Psyche, the bringing down by Prometheus of fire inside a smoldering Thyrsus reed to humanity, the casting down of Lucifer and the fallen angels into the lake of fire, the descent of Hercules into the underworld to drag up the three-headed dog Cerberus, the New Testament parable of the prodigal son, the command given to Abram in the Old Testament to depart from Ur of the Chaldees and go to a new land, the various descents into Egypt found in the Old and New Testaments, and many others.

In addition to all the evidence Kuhn himself musters in his argument, we could add the evidence found in ancient sources, including the hints given to us by the important philosopher and historian Plutarch, (c. AD 46 - AD 120) who describes the searching for dismembered body of the god Osiris by the goddess Isis in his famous essay "On Isis and Osiris" in *Moralia.*

Plutarch begins the account of the murder of Osiris by Set and the subsequent searchings and wanderings of Isis in section 12 of that essay, but the things he says, seemingly unrelated banter, in the lead-up to the Isis and Osiris myth itself should not be dismissed and were no doubt intended by Plutarch to help to enlighten his readers upon the true meaning of what follows. In section 9, Plutarch takes up the idea of the deepest philosophy being "veiled in myths and in words containing dim reflexions and adumbrations of the truth," and then he takes up the concept of the hidden god "Ammon" or "Amoun," saying:

> Moreover, most people believe that Amoun is the name given to Zeus in the land of the Egyptians, a name which we, with a slight alteration, pronounce Ammon. But Manetho of Sebennytus thinks that the meaning "concealed" or "concealment" lies in this word. Hecataeus of Abdera, however, says that the Egyptians use this expression one to another whenever they call to anyone, for

the word is a form of address. When they, therefore, address the supreme god, whom they believe to be the same as the Universe, as if he were invisible and concealed, and implore him to make himself visible and manifest to them, they use the word "Amoun"; so great, then, was the circumspection of the Egyptians in their wisdom touching all that had to do with the gods.

This little discourse by Plutarch (on the subject of concealing) is very revealing! No sooner has he given us a reference to a famous authority that the name of the Egyptian god Amun or Amen (whose name is found in the names of Egyptian kings such as Tutankhamun or Amenhotep) than he veers off to the seemingly unrelated observation that "the Egyptians use this expression one to another whenever they call to anyone, for the word is a form of address."

What on earth is Plutarch doing? Is he just participating in some early "stream-of-consciousness" literary technique, nineteen centuries before it would become popular? Is he showing off his vast knowledge of trivial matters? What does the fact that the Egyptians use the word "Amun" to greet one another have to do with the question of whether the hidden god Amen of Egypt is or is not their version of the great god Zeus of the Greeks?

I believe that Plutarch is here giving away the entire key to the philosophy which, in the preceding sentences, he has just finished telling us has always been "veiled in myths" -- a veil which (according to the inscription upon the statue of Isis which Plutarch also cites) "no mortal has yet uncovered."

In other words, he is telling us that the very name of the "hidden god" is also used as a form of address by the Egyptians (and I quote) "whenever they call to *anyone*." In other words, Plutarch is telling us (without coming out and saying it directly): "The hidden god is found in every single person -- the Egyptians, about whom I have just declared 'so great was their circumspection in their wisdom touching all that had to do with the gods,' find it appropriate to use the name of the hidden god Amun whenever they are calling out to anyone . . . not just some extremely special people, but *anyone*."

Plutarch immediately goes on to say that when the Egyptians address this hidden god, "whom they believe to be the same as the Universe," they "implore him to make himself visible and manifest to them." In other words, they are performing the same function depicted in the myths of Isis searching for the hidden and scattered Osiris: they are declaring the truth that divinity is buried in everyone they meet (and within themselves as well), but that we have to remind ourselves of this truth, because it is not intuitive and it is not always obvious! It is, in fact, quite easy to forget and to start acting like the prodigal son in the parable, eating among the swine and descending into the condition of a brute.

The Egyptians, Plutarch notes, thus greet everyone with the name that signifies "the hidden divinity" (whom they believe to be the same as the Universe), treating him "as if he were invisible and concealed, and implore him to make himself visible and manifest to them."

It is very interesting to note that this practice of the Egyptians, preserved for us in the writings of Plutarch, seems to parallel quite closely the practice which survives to this day in India and Tibet of the salutation *Namaste* or *Namaskaram*, a very ancient practice and one described in the Vedas.

The expression means literally "I bow to you" and is usually explained to mean "I bow to in recognition of the divine in you," or even "the divine in me recognizes the divine in you." For example, in this explanation, we read that in performing this gesture:

> you recognize the source of creation within them. This is the intention behind doing namaskaram. [. . .] And it is a constant reminder that the source of creation is within you too. If you recognize this, you are paving the way towards

547

your ultimate nature every time you do namaskaram.
http://www.ishafoundation.org/what-is-the-meaning-of-namaskar.isa

Note the emphasis here upon a "constant reminder" of a truth which is a hidden truth -- one we need to reminded ourselves regarding.

Here is another explanation online about namaste or namaskaram. There again, we read that:

> The reason why we do namaste has a deeper spiritual significance. It recognizes the belief that the life force, the divinity, the Self or the God in me is the same in all. Acknowledging this oneness with the meeting of the palms, we honor the god in the person we meet.
> http://hinduism.about.com/od/artculture/p/namaste.htm

Clearly, the parallels to what was still a teaching and a practice among the Egyptians during the time of Plutarch and to the practice of doing namaste or namaskaram are quite pronounced. We could almost say that the descriptions of the meaning of namaste cited above could apply equally to the ancient Egyptian practice of greeting with the word "Amun" or "Amen."

And here we encounter an unexpected parallel which strikes us with all the force of one hand clapping (so to speak). And that is the fact that all the descriptions of namaste include the fact that this greeting is always accompanied by the mudra or hand posture of placing the palms together (anjali mudra), generally in front of the heart chakra, and that in fact the gesture itself is often performed without the verbal greeting, although the verbal greeting is not performed without the hand gesture.

In light of the evidence that we have been finding which suggests that the ancient Egyptian greeting "Amen" carries the same meaning as the "namaste" or "namaskaram" (both "Amen" and "namaste" recognize the hidden god in ourselves and in other men and women and children), is it any wonder that the word "amen" in Christianity is often accompanied by the exact same hand gesture? That is to say, are prayers which are concluded

548

with the word "Amen" not also accompanied in many instances by the mudra of the hands placed palms together, at the level of the heart chakra?

This fact is powerful evidence that the greeting Plutarch is describing, which has to do with the concept of the hidden god (Amen or Amun in ancient Egypt) is in actual fact closely related to the salutation namaste or namaskaram in India – they are not just similar concepts that express the same idea but they are in fact related in some way (one is either descended from the other, or both are descended from some even more ancient common ancestor). The fact that the Christian "Amen" also uses the hands-together gesture shows that it, too, is directly related to the salutation of namaste or namaskaram, as well as to the hidden god Amun of ancient Egypt.

This is yet further evidence that all the world's ancient sacred traditions are in fact deeply connected.

_/_ *Namaste*

Resolved: blessing and not cursing

2014 December 31

My resolution as we contemplate the end of one year and the arrival of another is to engage in the act of blessing and not cursing.

As discussed in a post from 09/19/2014, the concept of *blessing* can be conceived of as the act of:

- *recognizing* the spirit world which is behind all we can perceive here in the material world, and which in some sense can be said to generate everything we perceive here in the material world

and

- *awakening* and bringing out that hidden, veiled, invisible spirit dwelling within everything and everyone we encounter here in this material world.

I am grateful to Sandra Ingerman and Hank Wesselman for articulating this wonderful definition of the act of blessing in the book *Awakening to the Spirit World* (pages 25 - 26). As they explain in that definition, it appears that our ongoing mission in this material world may well be the continuous act of recognizing and acknowledging and waking up and calling forth this hidden spark of spirit within ourselves and the rest of the material world around us:

> the physical plane appears to most as a camouflage universe where Spirit does not appear to exist [. . .]

> many of us respond to the physical world by assuming a deep hypnosis, a deep sleep where we no longer recognize that Spirit is present [. . .]

> So it is our job to wake up and to awaken all that is around us. This act of waking up could be called "blessing the world." 26.

Previous posts have spent a great deal of time examining the symbology found in the ancient wisdom around the world in various forms using various metaphors describing the "casting down" of spirit into matter and the subsequent "raising up" of spirit again. Symbols describing this dynamic include:

- the "casting down of the Djed-column" and the "raising it back up again,"
- the entombment of Osiris in a sarcophagus and the subsequent standing back upright of the god,
- the ascent upon a central tree which is a foundational image in shamanic cultures around the world and can also be found in the Norse myth of Odin and in both the Old and New Testaments of the Bible,
- the symbol of the kundalini serpent rising along the central column of the body, which is also related to the symbol of the caduceus
- the Vajra or Thunderbolt,
- the widely-known symbol of the cross with its horizontal component (the spirit cast down into matter) and its vertical component (the spirit awakened and ascendant)
- the similar and related symbol of the Ankh, as well as the symbol of the Scarab, and
- the annual "cross of the year" formed upon the zodiac wheel by the horizontal line of the equinoxes (between which the spirit is cast down into matter) and the vertical line of the solstices (topped by the sign of Cancer the Crab, whose upraised arms resemble the upraised arms of the Scarab beetle and serve the same symbolic function).
- The concept of walking the north-south red road, which is crossed with the east-west black road in the Sacred Hoop of the Lakota.

The fundamental importance of these symbols in the sacred traditions found around the globe testifies to the profound centrality of the continuous process of acknowledging and recognizing and then calling forth and elevating the spiritual which has been veiled and hidden beneath or within this material covering that we perceive with our physical senses.

In other words, the act of blessing appears to be our central ongoing task, according to the world's ancient wisdom!

And yet how often and how easily this physical world can get us to lose sight of the world of spirit pulsing just beneath the surface of everything we see -- and how easily the sharp and sometimes painful exigency of the material realm can cause us to reverse the process just described, and fall into cursing when we are supposed to be blessing!

If cursing is the opposite of blessing, then the definition of blessing just discussed would seem to lead to a definition of cursing which involves the *denial* of the spiritual within ourselves and other beings around us, and flowing just beneath the surface of everything here in the physical realm. Instead of elevating spirit, cursing denigrates it, or degrades it, or diminishes it, or denies its existence altogether. Instead of seeing ourselves and others as spiritual beings immersed for a time within physical bodies, cursing objectifies, physicalizes, and profanes.

The downward direction of cursing, driving down the spiritual instead of elevating it, seems to be very much related to the definition of violence offered by Simone Weil in her powerful 1940 essay entitled "The Iliad, or the Poem of Force," in which she famously defined physical violence as "that x that turns anybody who is subjected to it into a thing."

The curse words and phrases I am familiar with tend to emphasize violence, physicality, and the animal aspect of human existence -- they tend to focus on the carnal in a way that is stripped of any accompanying deeper meaning of spirit, to emphasize the bodily functions of the human body in a way that "turns anybody who is subjected to [incarnation] into a thing," and in doing so they obscure, or deny, or attempt to take our mind away from the dual spiritual-physical reality of human existence. It is a reality that we are always prone to forgetting or

552

ignoring, as the quotation from Sandra Ingerman and Hank Wesselman cited above makes clear -- and as the ancient wisdom expressed in the teaching of the "hidden god" buried inside the material realm but hidden from sight and easily overlooked (see discussions here and here).

If you are anything like me, you know how easily the daily frustrations offered up by life inside the material realm and its unforgiving laws of physics can cause "cursing" in some form to erupt almost spontaneously, whether in thought or in actual expression. And yet how damaging this tendency is to our true mission of blessing rather than cursing.

This tension is expressed in many ancient scriptures -- the scriptures in the Old and New Testaments, as well as ancient sacred texts which were left out by the literalists when they assembled the New Testament, enjoin blessing rather than cursing, and warn us against the constant temptation to fall into patterns of cursing.

The text known as the general epistle of James memorably declares: "Out of the same mouth proceedeth blessing and cursing. My brethren, these things ought not so to be. Doth a fountain send forth at the same place sweet water and bitter?" (James 3:10-11).

Proverbs 11:11 declares: "By the blessing of the upright the city is exalted: but it is overthrown by the mouth of the wicked" (it is interesting to note that this passage associates blessing with "*the upright*" and that "uprightness" is associated in the ancient symbology catalogued above with the Djed-column raised-up, the Osiris raised to the vertical position, and the vertical portion of the universal cross -- all of them symbols of the elevation of spirit and hence with the concept of *blessing*).

In both texts known as the Gospel according to Matthew and and the Gospel according to Mark, Jesus is recorded as saying that it is what comes out of our mouths that can defile us, rather than what we put into our mouth, saying: "But those things which proceed out of the mouth come forth from the heart; and they defile the man [. . .] These are the things which defile a man: but to eat with unwashen hands defileth not a man" (Matthew 15:18, 20).

The same teaching is expressed in a text known as the Gospel of Thomas (an important Gnostic text found in the Nag Hammadi library), in which Jesus tells his listeners: "After all, what goes into your mouth will not defile you; rather, it's what comes out of your mouth that will defile you" (14).

These passages tend to support the definition of "cursing" which we have derived above from our definition of "blessing" -- the concept of "defiling" means making profane, denying the sacred aspect, driving out the sense of the spiritual and emphasizing all that is most associated with the solely physical aspects of our incarnate existence. These passages seem to be enjoining us to be constantly blessing and not cursing: to be seeing the sacred and the spiritual in ourselves and in everyone and everything around us, and to try to bring it out – as opposed to doing the opposite.

Of course, as we all know, maintaining this focus is not easy (if it were, there probably would not be so many passages in ancient scriptures enjoining us to do it).

That's why it is a resolution of mine, and something I hope to do more of in the coming year!

Blessings to you in this new solar year. Namaste.

Ecstasy every day: Qigong energy work

2015 April 18

A post from 04/15/2014 discussing the central concept of "raising the Djed" contained the assertion that this vital symbol -- which shows up in virtually *all* of the world's sacred traditions, based as they are upon Star Myths and incorporating as they do the symbology pointing to the Great Cross of the Year, which is allegorized in ancient Egypt as the "casting down" and then "raising up" of the Djed, but shows up in other mythologies in different guises -- most certainly relates to the overall cycle of our lives (and what happens before and after life in this body), but it also and with equal importance relates to an important part of this life, and a practice that we should find ways to incorporate on a regular basis, maybe even every single day.

That post said of this cycle of "casting down" and "raising up" the spiritual, invisible Djed-component that it

> describes a process that is meant to be part of our life here and now: the connection with the realm of spirit, the raising of the spiritual component inside ourselves and the spiritual-material world around us, and the entry into the state of ecstasy *on a regular basis.*

It also contained a link to a video which concluded with the assertion that the concept of raising the Djed "can be incorporated into our lives every single day . . . and maybe, even, every single minute."
https://www.youtube.com/watch?v=SNm674gLcPw&feature=youtu.be

But *how* can we incorporate the concept of "raising the Djed" in our lives on a regular basis -- every single day or even more regularly than that?

Perhaps it would be helpful to spend a few posts discussing just a few specific possible answers to that question (out of many, many more possibilities). We'll title this little "mini-series" *Ecstasy every day.*

The good news is that, as has been explored in many previous posts, human beings appear to have almost unlimited options for practices which invoke the invisible world and enable it to shine through into this material world.

The simple (but incredibly challenging) cultivation of an attitude of blessing rather than cursing can become a perspective that changes nearly every aspect of our daily lives, and which involves us in the constant practice of "raising up" rather than "casting down."

But, the concept of ecstasy actually involves experiencing first-hand the power of the non-material, and of going beyond or breaking free from the normal bounds of our physical "static"

556

nature. The word itself, *ex stasis*, means "out of" the "static or unchanging" -- and thus involves making contact with the realm which is not just "beyond the physical" but which is also "beyond the static" -- the realm of pure dynamism, of pure potential or pure "potentiality."

We can obtain glimpses of the state where everything is "non-static" in some stages of dreaming, especially when we are just falling asleep and reach a stage where our thoughts feel like they are painting images on their own but we can almost just barely change them or influence the shapes that they take on -- they are purely dynamic and have the potential really to change into anything or go in virtually any direction. Such "dynamism" seems to be incorporated in the word "ex-stasis."

Even though this material realm is not really "static" in the sense that static means "unchanging" (our physical bodies change as they age, of course, as does almost everything else around us over time), it is "static" in that in this realm things have already "manifested" -- they are no longer in the realm of "pure potentiality" (the famous foundational experiments of quantum physics come to mind here).

Pure potentiality and pure dynamism are clearly related to the concept of pure creativity: creative ideas which have never before been manifested in this world seem to come from some other place, some "other side" (and many artists and inventors and scientists and discoverers have attested to the fact that new ideas or new solutions or new songs have "come to them" while they were in a dream-state or other state of contact with the realm of pure potentiality, the realm of the non-static or ecstatic).

And, in some forms of ecstatic experience, our consciousness can actually travel beyond the body -- in "ecstatic transport" where our we enter into an actual state of trance or journeying into the spirit realm.

Some of the many ways that human beings in many different environments and cultures have used in order to achieve states of

557

ecstasy are listed in the previous post entitled "How many ways are there to contact the hidden realm?"[208] That post shows that while many cultures have used various substances including special plants and mushrooms to induce ecstatic states, these are by no means the only possibilities, and that in fact human beings appear to have a natural innate ability to access the spirit world -- which is exactly what we would expect if, as the ancient wisdom attests, we are beings who are composed of a "cross" between spirit and matter, and who actually enter into the material realm temporarily, with our spirit nature being our true nature.

Previous posts have also discussed the fact that shamanic drumming is one of the most widely-used and effective forms of regularly accessing the Invisible World, and that longtime shamanic drumming practitioners and teachers have attested to the fact that nearly everyone can enter a state of ecstatic transport *on their first try* after fifteen minutes using the right techniques.

But the methods which men and women have used down through the centuries for raising the invisible and spiritual force inside themselves and calling it forth in the world around them do not stop there -- and it may be beneficial to visit a few others, which may have an incredibly positive impact for those who choose to incorporate them into their daily lives.

Some people may find that they are more drawn towards or more comfortable with one type of ecstatic discipline rather than another, or that one seems to "raise their spiritual force" more easily or more reliably than other methods they have tried.

And, it may well also be that some or all individuals would benefit from cultivating more than one such practice on a regular basis.

One method which can almost certainly be seen as calling forth and raising-up the invisible spirit present in each man and woman is the ancient Chinese discipline which today is known most widely by the name *qigong* or *chi gung*:

氣功

The first of these symbols means "breath" or "spirit" or "energy" and is pronounced "chi" or "chee" in Mandarin and "hei" in Cantonese, and the second and third symbols together are the symbol for "work" or "skill" or "acquired power" or "practice" and which are pronounced "gung" or "gong" (and which is the same word found in the first half of "gung fu" or "kung fu").

The name *qigong* or *chi gung* itself appears to be a fairly recent label, appearing in the 1940s or 1950s as part of a rather incredible phenomenon of sudden promotion and widespread adoption of qigong practice in China, which had previously been transmitted and practiced more secretively. That story is discussed in a book called *Qigong Fever*, by David A. Palmer.

However, it is also indisputable that what is now most widely known by this newer name has in fact been practiced for centuries and perhaps for millennia, and it is not only very real but also can have very real health benefits for those who take the time to learn and practice this ancient technique as part of their daily lives.

A brief definition for those who may be unfamiliar with this form of internal practice is that qigong involves the practice of physical motions designed to increase consciousness and to foster the greater and greater recognition of the spiritual energy within each individual and indeed within the universe itself, and the ability to feel, and direct, and move that energy around -- in a way that is actually "tangible."

One teacher and long-time practitioner of qigong, who has spent decades studying with some of the amazing masters of different forms of this art, is Taoist Master Bruce Frantzis, whose description of qigong makes some very intriguing points which resonate with concepts we have encountered before in different contexts. In a page entitled "What is Qigong?" which is worth reading in its entirety, he writes:

> According to Taoism, every human being contains "the three treasures" -- jing (sperm/ovary energy, or the essence of the physical body), chi (energy, including the thoughts

559

and emotions), and shen (spirit or spiritual power). Wu (emptiness) gives birth to and integrates the three treasures.

[...]

Popular opinion has it that once you have reached a state of emptiness, you stay there, but this idea is false. You merely become increasingly familiar with this state and learn how to spend more and more time there. As longs you live in a physical body, physical needs continue to exert demands, and dwelling completely in emptiness is not possible. Taoism has developed advanced techniques to work with the energy of wu.
http://www.energyarts.com/what-qigong

This description is extremely interesting on many levels and for many reasons -- one of which is the fact that this description appears to resonate very strongly with assertions made by philosopher and scholar Peter Kingsley, in his exploration of the lost wisdom of the pre-Socratic philosophers of the Greek islands and Mediterranean settlements, particularly those on the Italian peninsula and particularly that lineage of which Parmenides (or Parmeneides) was an important figure.

Dr. Kingsley's discoveries on this subject are published in his 1999 text *In the Dark Places of Wisdom*, and discussed in a March 9, 2015 blog post entitled "The peace of utter stillness . . . "

The similarities between the concept of "incubation" or the deliberate cultivation of "the peace of utter stillness" that the ancient philosophers of Parmeneides' day appear to have been practicing, and the concept of "wu" in the description of qigong given above, should be quite evident.

The same qigong definition page also asserts that there is a strong connection between qigong and meditation (another important technique that can be used to access the invisible realm and which can be incorporated into daily life), and Bruce Frantzis asserts that ultimately, although it has real and tangible health

560

benefits, "qigong is only a preparatory practice for Taoist meditation techniques."

The ways in which qigong can be seen to be a way to help men and women to access the invisible aspect of their nature and the invisible side of the universe around (and within) us, and thus to be a form of "raising the Djed" column or of incorporating the "raising-up" imperative into daily life, should be fairly self-evident.

A powerful demonstration of the beneficial "raising-up" and ex-stasis nature of internal energy-work can be seen in this video from more than twenty years ago, in which Bruce Frantzis was invited to teach these techniques to prisoners, who testify in the video to the positive impact it has had on their lives:
https://www.youtube.com/watch?v=o5lNoYeEBa8

Towards the end of the clip, Master Frantzis makes a very profound and illuminating statement based on his work at the prison:

> A couple of things that I know are happening in here because I've gotten it from people: one is that they're getting a sense of family with the people they're working with, because they're doing something in common. Second of all, they're doing something that's really improving the insides of them -- they can't do anything about their external environment: they're told what to do from the morning, the minute they get up to the minute they sleep, but internally they can find a place where they're free inside. Now most people on the outside, most people in the world in general are not free: they don't have walls around them, they have a wall in their own mind, they have a wall in their own body.

A better definition of *ex-stasis* can hardly be desired.

The Djed column every day:
Tantra and Fong Zhong Shu

2015 April 28

If all the world's sacred scriptures and mythology actually consist of stories in which the motions of the celestial spheres take on the personalities of men, women, gods, goddesses, angels, demons, monsters, djinn, and other mystical creatures (and they most certainly do), then we are left with a very important question: *Why?*

I believe the answer certainly includes as a central feature the profound teaching embodied in the Great Cross of the Year, formed by the solstices and equinoxes, and associated with the concept symbolized in ancient Egypt by the "casting down" of the Djed column of Osiris and the subsequent "raising-up again" of the same: an esoteric concept which depicts the entire nature of human existence as a divine soul thrown down into incarnation, while voyaging through, reflecting and in some mysterious way embodying the infinite universe at the same time – a universe which is itself composed of both a visible realm and an even more important and subtle invisible realm.

Numerous recent posts and videos which have attempted to outline this critically-important central teaching (found, I believe, in virtually all of the world's sacred traditions in varying depictions and disguises).

Those discussions presented evidence that the concept of "raising the Djed" conveys a powerful message regarding the long process of our realization of the infinite divine sleeping within ourselves and indeed within every atom of the living universe around us, a process which takes place during the entire cycle of our earthly existence and perhaps over the course of many successive "existences" -- but it is also (we saw) a message which appears to urge upon us the practice of "raising the Djed" every single day, through the practice of blessing, through the recognition and elevation of the divine in ourselves and others, and through the

563

special form of spiritual elevation leading to the state of ecstasy or ecstatic trance, in which our perception actually transcends the physical body and makes contact with the invisible world.

And, while the entry into the ecstatic state is perhaps the most intense and most transcendent of the forms of recognizing and reconnecting with and calling forth and raising up the infinite divine spiritual realm which is always present, around us and within us, we have also seen evidence that in addition to incorporating techniques of ecstasy into our lives on a regular basis, we can also practice other forms of "raising the Djed" into our lives as well, even when we are not in the ecstatic state (since it is not possible to exist in a state of ecstasy at all times). It seems likely that consciously incorporating more than one of these into our lives is quite possible and probably beneficial -- and that they are not at all "mutually exclusive" (incorporating one does not require that we renounce all the others, although there is obviously a limit to how many we can choose to really pursue seriously).

In order to simply provide a very cursory pointer towards some practices which have been developed in different cultures from very ancient times, for those who may wish to learn more about them on their own, I started a short "mini-series" of posts discussing a few such practices which seem to fit into the general category of "raising the Djed." The first one we mentioned briefly was the practice commonly called qigong or chi gung, which clearly involves contact with "the invisible" in some way (the "invisible within," the "invisible without," or both), and which enables its practitioners to directly and tangibly experience the fact that we are made of more than just physical substance.

The goal of this little mini-series is not to try to teach these practices, or even to point to specific teachers or resources where people can learn more about these practices, but rather to simply make people aware of the existence of these many different disciplines which fit into the general category of "raising the Djed" and which some readers may find very beneficial if they

564

choose to pursue them. Many of these practices, while extremely ancient, are not well known in "the west" -- that is to say, in the parts of the world in which the ancient esoteric knowledge was largely replaced by a literalistic rather than esoteric understanding of the ancient sacred stories and myths.

Another discipline which clearly falls into this same category is the practice of techniques known in some cultures (especially India and Tibet) as *maithuna* and usually known in China and Taoism (or Daoism) as *fong zhong shu* or

房中術

The above calligraphy shows traditional characters, but in simplified characters the final character above is changed to **the** symbol found in the middle of the traditional version of that character) and so the same phrase would be rendered as

房中术

In either case, the three symbols stand for "bedroom - within - skill" (pronounced *fong zhong shu* in Mandarin and *fohng jung seuht* in Cantonese) and are usually rendered into English using the phrases "bedroom arts" or "art of the bedchamber" and corresponding very generally to what is often referred to in the west as "Tantra" (although apparently that word actually encompasses a much wider landscape of transformative disciplines involving meditation, mantras, mandalas, visualization, and other practices in addition to what most people in the west today envision when they think of Tantra).

In general, these related arts involve transformation through sexual ritual, a practice which can be seen to have been highly developed in ancient China, ancient Japan, ancient India, ancient Tibet, and many other cultures around the world, including some Native American cultures. There is some evidence that the spiritual potential of this aspect of human existence was also developed in "western" cultures in various forms prior to being largely rejected or suppressed with the advent of literalist Christianity.

Although still perhaps not so very widely known, excellent books on Taoist *fong zhong shu* have been available in English for many years, including the work of Daniel P. Reid and Mantak Chia, among others.

Additionally, some of the ancient Chinese texts that traditionally formed the foundation for the preservation and development of the knowledge of *fong zhong shu* have survived in varying degrees of completeness.

Of these, perhaps the most important, and almost certainly the most often-cited and well known is the *Su Nu Jing*, or

素女經

The title is often translated as "Classic of the Plain Girl," but the three characters actually stand for "natural-colored [often used to describe natural-colored or undyed silk]" - "woman" - "classic or canonical text" and because the first word can also mean "plain" as in "unspotted" or "without markings" or simply "white, pure, or undyed," the same title is also sometimes translated as the Classic of the "Immaculate Woman" or the "Pure Woman."

This figure appears in some aspects to be a goddess or divine figure, who is in some cases associated with grain and hence may connect to the celestial figure of Virgo (this would not be a surprise). Interestingly enough, this would also connect her to the Greek goddess Demeter, whom Plutarch uses as part of his powerful argument against the consumption of animals for food, and the same word and symbol sometimes translated "Plain" that is used to describe her in China can also be used to mean "vegetarian." She is sometimes depicted as giving instruction to the Yellow Emperor or Huangdi (sometimes spelled Huang Ti), whom Hertha von Dechend and Giorgio de Santillana identify as a Saturnian figure in *Hamlet's Mill*.

566

So, *Su Nu Jing* means "Pure-Undyed-Silk Woman Classic" in Mandarin, and would be pronounced *Seuh Neuih Ching* in Cantonese, and the last word in the title (Jing or Ching) is the same word found in the title of the Tao Te Ching. It is certainly at least as old as the Sui Dynasty (AD 590 - AD 618) and may be even older, perhaps originating in the Han Dynasty (221 BC - 207 BC) -- and the knowledge it contains may of course have come from an even earlier source.

As explained in *Sexual Life in Ancient China: A Preliminary Survey of Chinese Sex and Society circa 1500 BC till 1644 AD*, by R. H. Van Gulik (1961), no complete original text of the *Su Nu Jing* nor of several other ancient Taoist *fong zhong shu* texts has survived. However, much of the text of the Su Nu Jing was preserved in a different text that quotes large portions of it, which is called the Tung-hsuan-tzu and which may have been written by the scholar Li Tung Hsuan in the 7th century AD.

The text of the Tung-hsuan-tzu begins as follows (as translated in 1961, when conventions were slightly different than they are today -- the modern reader may wish to mentally substitute "humanity" for the general "man," which in previous decades was generally used to mean all of humanity and not specifically men to the exclusion of women):

> Master Tung-hsuan said: Of all the ten thousand things created by Heaven, man is the most precious. Of all the things that make man prosper none can be compared to sexual intercourse. It is modeled after Heaven and takes its pattern by Earth, it regulates Yin and rules Yang. Those who understand its significance can nurture their nature and prolong their years; those who miss its true meaning will harm themselves and die before their time. 135.

This introduction is extremely significant, and author R. H. Van Gulik notes that most of the more ancient Taoist sexual texts also begin with an expression of the cosmological aspect of human sexuality, which was seen to "model Heaven and [. . .] Earth."

567

Later, we reach a portion of the text in which the *Su Nu Jing* is quoted extensively. In the introductory chapter, entitled "The Supreme Significance of the Sexual Act," the Plain Girl declares that in sex:

> Woman is superior to man in the same respect as water is superior to fire. [. . .] The union of man and woman is like the mating of Heaven and Earth. It is because of their correct mating that Heaven and Earth last forever. Man, however, has lost this secret, therefore his age has gradually decreased. If a man could learn to stop this decline of his power and how to avoid ills by the art of Yin and Yang, he will attain immortality. 135 - 136.

Here we again see the explicit "macrocosm-microcosm" understanding that the motions of men and women on earth mirror the motions of the great cycles of the heavenly objects, and also mirror the motions of the earth which contribute to our interaction with the celestial mechanics in the heavens above. We are also introduced to one of the central concepts in Taoist *fong zhong shu* and related disciplines, which is the inherent superiority of the woman to the man, in that she is already capable of multiple, progressive, and basically unlimited orgasms (leading to the raising of chi, prana, or the kundalini, and ultimately to ecstasy), while the man must learn to achieve this capability and does not usually obtain it without the cultivation of *fong zhong shu*, primarily through the ability to separate orgasm and ejaculation and achieve multiple orgasms without ejaculation.

Without going any further into the specifics of that subject, which interested readers can pursue for themselves, it is worth noting that in this ancient text, the *Natural-Silk Woman* or *Immaculate Goddess* uses the expression "as water is superior to fire." This phrase is loaded with esoteric symbolism, the concept of fire plunging into water is an esoteric metaphor for the process of incarnation itself, by which the divine spark of spirit is plunged

568

into and submerged within the physical material realm and a physical material body.

Because of this understanding, we can then gain a better appreciation for the insistence in these ancient texts that human sexuality itself somehow "models Heaven and Earth" and becomes an esoteric symbol for our incarnation itself . . . and for our ability to be spiritually transformed and elevated by our experience in a physical body, an experience which ultimately leads to transcendence of the physical nature.

Rather than being extinguished by and completely subsumed within the material nature in which we find ourselves, our task is to hold on to the spiritual, call it forth from within this physical world, and ultimately to transform both matter and spirit together -- "raising the Djed." It can readily be perceived that the arts that are often referred to as Tantric are esoterically and experientially involved in just this very purpose as well.

Just as the myths themselves "bring the stars down to earth" by depicting the sun, moon, stars and planets as human beings and as gods and goddesses walking among humanity, rituals which we undertake that mirror and embody the motions of the heavens and the earth (as the Plain Silk Girl tells us that *fong zhong shu* most certainly does) connect us to the motions of the universe, and "bring the heavenly motions" down into the human realm, the microcosm reflecting and embodying the macrocosm of the infinite cosmos.

Finally, it is worth noting that here that, as in so many other places where the esoteric ancient wisdom has somehow been subverted, a practice and a body of knowledge which is clearly intended for the elevation and liberation and positive transformation of individual men and women has instead been turned too often into a negative force for degradation, dehumanization, oppression, and powerful feelings of shame, hurt, and alienation.

569

The fact that, as we are told by the Immaculate Woman, in the bedroom "the woman is superior to the man as water is superior to fire" can lead to tremendous insecurity and resentment on both sides, when these ancient practices are not known and understood -- but when they *are* understood and put into practice, they can lead to tremendous security and empowerment for everyone involved.

This subject provides yet another example of how vitally important it is to understand what the ancient texts and the ancient treasures which were entrusted to humanity are actually trying to tell us, and how we can learn to incorporate them into our lives on a very practical level -- and what a great tragedy it is that this ancient inheritance imparted to the human race has somehow so often been turned completely upside down.

The Djed column every day: Yoga

2015 April 30

The past several posts have examined the concept of "raising the Djed" (cultivating and evoking and amplifying the spiritual, invisible, divine spark in ourselves and in the universe we travel through), the possibility that we can in some way incorporate this concept into daily life, and some of the multitude of practices for doing so which cultures from around the world have preserved from ancient times.

Part of the reason for this short survey is to bring the discussion of what may seem at times to be a very esoteric and philosophical topic "down to earth" and suggest that it is actually an intensely practical topic and one which may be tremendously beneficial to our seemingly mundane day-to-day existence.

Another reason to look at some different methods from different cultures is to show that no one method should be considered a "monopoly" -- that there probably dozens or perhaps hundreds of different ways that human beings can choose to pursue in this important area of life, and that although they do share some important similarities they are different enough that they can appeal to different people of different backgrounds or needs.

A third reason might be to familiarize readers with some techniques which may be less well known, such as previous "Djed-raising" disciplines explored in the previous posts on qigong[555] and on Tantra and *fong zhong shu.*[563] If just one reader who has not previously heard of a certain practice decides to examine it further and it becomes a beneficial part of his or her life for years to come, that would seem to justify the entire "mini-series" right there.

The next daily discipline probably cannot be classified as one that any reader has not yet heard about, because it is a tradition that is so strong and so rich in teachers and followers and the level of ancient wisdom which continues to be passed along in its broad and powerful stream, but it is very clearly related to the concept that the ancient Egyptians symbolized by the raising of the Djed, and by the symbol of life carried by almost all of the gods and goddesses, the Ankh -- which may in fact be linguistically related to the name by which this discipline has been known for millennia.

We're talking, of course, about Yoga -- a subject that could withstand a lifetime of deep consideration without ever exhausting its possibilities.

While most of us upon hearing the word "Yoga" immediately think of the *asanas* ("postures") which are undoubtedly the most well-known aspect of Yogic practice, Yoga in fact is a very comprehensive discipline of transformation incorporating meditation, concentration, study of ancient texts and tradition, true conduct in daily life, nonviolence, freedom from anger, and other practices designed to reawaken and elevate the spiritual, and ultimately to lead to deep contact with the divine and the ultimate. Asanas are an important aspect of Yoga but only one of its many "limbs."

In *Light on Yoga*, first published in 1966, B.K.S. Iyengar explains:

> The word Yoga is derived from the Sanskrit root yuj meaning to bind, join, attach and yoke, to direct and concentrate one's attention on, to use and apply. It also means union or communion. [. . .]

> In Indian thought, everything is permeated by the Supreme Universal Spirit (Paramatma or God) of which the individual human spirit (jivatma) is a part. The system of yoga is so called because it teaches the means by which the jivatma can be united to, or be in communion with the Paramatma, and so secure liberation (moksa). 19

The letter "s" in the final word, *moksa*, has a diacritical "dot" underneath it, indicating that the "s" is pronounced more like a "sh," and you will sometimes see the same word spelled *moksha*.

The video linked at the end of this paragraph, entitled *Yoga Ruins Your Life*, by Richard Freeman of Yoga Workshop in Boulder, Colorado, may make you want to take up Yoga, even if you have never wanted to try it out before (that is, if the above passage from *Light on Yoga* has not already led you to stop reading and start a search for a Yoga shala in your area).
https://www.youtube.com/watch?v=1zwtlZwbUmo&feature=youtu.be

During the video, we hear the perspective offered by someone who has pursued the path of Yoga for many years and who has dedicated a great deal of energy to passing it on to others and helping others on their own Yoga journeys:

> So I've often said that Yoga ruins your life, and by that I mean it ruins your Samsaric life, because once you get a taste of Yoga, you kind of "lose interest" in all the things that are kind of dim reflections of that taste. [. . .] Yoga can also ruin your career, because you feel so nice when you do it that you're less aggressive, and you tend to *like* people more. And when you practice Yoga, you no longer take political extremes in your mind, and so . . . what are you going to fight about? Or, religious extremes either, because, you get to the -- kind of the *root* experience that all these different religions are looking for, but in a very generic and very natural, human way, so you don't have to clasp onto the fantastic or the otherworldly.

There are several important concepts in that short video worthy of careful consideration and further examination -- far more than can be pursued here in one sitting. We will explore just a few here.

One concept which is expressed in the opening sentence (and in the provocative title of the video) is the idea that Yoga "ruins" your Samsaric life, the life of attachment to the physical and the temporary into which we are "cast down" upon our incarnation, what the video's description section calls our "auto-pilot" life. It is a vehicle for transforming and transcending the illusions of the material world -- but doing so in part through the vehicle of our incarnation in this material world.

This idea of being on "auto-pilot" for a certain part of our life in this world, and then beginning to wake up more and more to a higher reality is expressed in the extended passage of a lecture by Alvin Boyd Kuhn quoted previously in "Easter: the Birth-Day of the Gods" (April 5, 2015) in which he traces the cycle of the soul which is "cast down" at the fall equinox (representative of being incarnated in the body) and continues to plunge downward even after that until it finally reaches a turning point at the winter solstice, the very point that creates the "vertical line" of the annual cross of the zodiac which represents the "raising back up" of the Djed column, and the point at which the inner divine is esoterically described as being "born in a manger." Kuhn says of the incarnate soul:

> It is born then as the soul of a human; but at first and for a long period it lies like a seed in the ground before germination, inert, unawakened, dormant, in the relative sense of the word, "dead." This is the young god lying in the manger, asleep in his cradle of the body, or as in the Jonah-fish allegory and the story of Jesus in the boat in the storm on the lake, asleep in the "hold" of the "ship" of life, with the tempest of the body's elemental passions raging all about him. He must be awakened, arise, exert himself and use his divine powers to still the storm, for the elements in the end will obey his mighty will.

This "sleeping semblance of life," which Kuhn also says is life "unawakened" and "inert," "dormant," and "dead," is the condition that the video above promises that Yoga can "ruin."

In case you're new to this theory, Kuhn is arguing that the story of Jesus asleep in the boat on Lake Galilee in the storm, or Jonah asleep in the hold of the ship of Joppa bound for Tarshish, are both allegorical or esoteric stories intended to describe a condition which each and every one of us experiences -- the condition of our own soul upon being "cast down" into this life, wedded to a human body like Prometheus nailed to a rock (to use yet another picture from a different set of allegorical myths), and temporarily "unawakened," "inert," "dormant," and "in the relative sense of the word, 'dead'." From this condition, Yoga promises a path that leads to liberation, or moksa -- but in doing so, it "ruins your life" of comfortable dozing in the hold of the ship.

Those stories, Kuhn tells us, are not literal and historical accounts -- and they were never intended to be taken that way (they are, he says elsewhere, "a thousand times more precious" as myths than as supposed histories). They are pointing to a profound truth that is in many ways even more mysterious than any fantastic or otherworldly story -- a truth that you can experience for yourself and a truth that "all these different religions are looking for," in the words of the Yoga video above.

In the words of someone who has walked the path of Yoga for decades, in pursuing that path you get to actually "get to the *root* experience" that these sacred myths are pointing to. In doing so, you lose the need to "clasp" onto someone else's story about it, because you experience it for yourself -- you *know* it. This is the concept that was anciently contained in the word *gnosis* -- first-hand experience *of* the ultimate, rather than second-hand faith *in* it.

The same idea was expressed by Gerald Massey (1828 - 1907) in a passage cited previously in "The Centrality of ecstasy, according to ancient wisdom,"[162] when he says:

> What do you think is the use of telling the adept [. . .] that he must live by faith, or be saved by belief? He will reply that he lives by knowledge, and walks by open sight; and that another life is thus demonstrated to him in this. As for death, the practical Gnostic will tell you, he sees through it,

and death itself is no more for him! Such have no doubt, because they know.

And yet, to make one more observation on the wonderful avenues of discussion that this subject opens up, those stories are not to be disdained on account of their being "fantastic" or "otherworldly" or simply "allegory" -- those powerful metaphors can help us to grasp the meaning of these spiritual concepts which deal with things that by their nature are invisible and which in fact are even beyond the ability of the mind to reason out using ordinary logic.

In fact, in attempting to convey the meaning of Yoga, B.K.S. Iyengar himself alludes to the "fantastic" or "otherworldly" story contained in the *Bhagavad Gita*, which is a portion of the ancient Hindu *Mahabharata*, in which Krishna expounds upon the meaning of Yoga to the disciple, Arjuna, and calls it a knowledge that the yogi (one who follows the path of Yoga) will experience that is "beyond the pale of the senses which his reason cannot grasp" (*Bhagavad Gita* 6.21, cited in *Light on Yoga* 19).

Interestingly enough, in a different part of the same *Bhagavad Gita* (a passage not, to my knowledge, cited by B.K.S. Iyengar, at least not in the book quoted above), Krishna tells Arjuna:

O Arjuna, now I shall describe different paths departing by which, during death, the yogis do or do not come back. Fire, light, daytime, the bright lunar fortnight, and the six months of the northern solstice of the sun; departing by the path of these gods the yogis, who know Brahman, attain nirvana. Smoke, night, the dark lunar fortnight, and the six months of southern solstice of the sun; departing by these paths, the righteous person attains lunar light and reincarnates. The path of light and the path of darkness are thought to be the world's two eternal paths. The former leads to nirvana and the latter leads to rebirth. Knowing these two paths, O Arjuna, a yogi is not bewildered at all. Therefore, O Arjuna, be steadfast in yoga at all times. *Bhagavad Gita* chapter 8, verses 23 - 27.

This is very noteworthy. Krishna has just revealed to us that the annual wheel, with its "upper half" consisting of the six months containing the summer solstice ("the northern solstice of the sun") and its "lower half" consisting of the six months containing the winter solstice ("the southern solstice of the sun," both of these expressions being geared towards an observer in the northern hemisphere) are esoteric allegories for two different paths through this life, one of which will lead to reincarnation (the cycle of Samsara) and one to liberation and nirvana.

This is the exact same cycle that we have seen formed the allegory of the "casting down" of the Djed column (into the lower half of the year) and the "raising up again" of the same (on the way back to the upper half of the year, and the summer solstice):

DAYS LONGER THAN NIGHTS:
Heaven, Promised Land, Greece, etc.

Horizontal Column:
The Djed cast down

Vertical Column:
The Djed raised up

NIGHTS LONGER THAN DAYS:
Hell, Egypt, Troy, etc.

Clearly, Yoga is a discipline designed to "raise the Djed column" (to use the terminology of ancient Egypt) and ultimately to transcend the cycle of being "cast down" into the lower half of the wheel.

Elsewhere in *Light on Yoga*, and in reference to concepts described in other sacred ancient texts, we see hints that this "transcending of the lower half" involves transcending the "shifting forms" or the "endless changes" that characterize the material half of our dual universe and a reconnection with the realm of pure potential. B.K.S. Iyengar says that the *Kathopanishad* tells us:

> When the senses are stilled, when the mind is at rest, when the intellect wavers not -- then, say the wise, is reached the highest stage. This steady control of the senses and mind has been defined as Yoga. He who attains it is free from delusion. 20.

Patanjali, Sri Iyengar notes, calls this condition *chitta vrtti nirodhah*, which means "the restraint (nirodhah) of mental (chitta) modifications (vrtti)," or the "suppression (nirodhah) of the fluctuations (vrtti) of consciousness (chitta)" (20).

And in the Bhagavad Gita, Krishna describes this concept to Arjuna thusly:

> When his mind, intellect and self (ahamkara) are under control, freed from restless desire, so that they rest in the spirit within, a man becomes a Yukta -- one in communion with God. A lamp does not flicker in a place where no winds blow; so it is with a yogi, who controls his mind, intellect and self, being absorbed in the spirit within him. When the restlessness of the mind, intellect and self is stilled through the practice of Yoga, the yogi by the grace of the Spirit within himself finds fulfillment. *Light on Yoga* 19, citing *Bhagavad Gita*, chapter 6 and verses 18 - 20.

This concept appears to be very closely aligned and perhaps even essentially identical to the practice that Peter Kingsley discusses in his 1999 text *In the Dark Places of Wisdom*, and which Dr. Kingsley believes was being practiced and passed down through a "master to disciple" method of transmission in certain groups of mystic philosophers prior to Socrates and Plato, and including Parmenides (or Parmeneides).

It is interesting that Yoga as well is traditionally passed down through just such a master-to-disciple relationship (the Guru, whose name literally means "light out of darkness," and the sisya, or disciple).

Fortunately, unlike so many other ancient traditions for the transmission of such profound transcendental gnosis, Yoga has survived into the present day, and can be followed as a means of daily transformation and "raising of the Djed."

Note, however, that B.K.S. Iyengar tells us that:

> All the important texts on Yoga lay great emphasis on sadhana or abhyasa (constant practice). Sadhana is not just a theoretical study of Yoga texts. It is a spiritual endeavour. Oil seeds must be pressed to yield oil. Wood must be heated to ignite it and bring out the hidden fire within. In the same way, the sadhaka must by constant practice light the divine flame within himself. 30.

But, he also quotes the following encouraging passage, from the *Hatha Yoga Pradipika*, chapter 1 and verses 64 - 66:

> The young, the old, the extremely aged, even the sick and infirm obtain perfection in Yoga by constant practice. Success will follow him who practices, not him who practices not. Success in Yoga is not obtained by the mere theoretical reading of sacred texts. Success is not obtained by wearing the dress of a yogi or a sanyasi (a recluse), nor by talking about it. Constant practice alone is the secret of success. Verily, there is no doubt of this. – Cited in *Light on Yoga*, 30.

So, that is encouraging, and argues that it is probably never to late to consider this ancient path.

Just beware that it may "ruin your life"!

Why divinities can appear in an instant

2015 July 03

Why do the deities in the Mahabharata often appear instantly, upon the recitation of a mantra, the singing of a hymn, or even simply upon being remembered?

I believe that this characteristic was included in the ancient scriptures in order to show us that we have access to the infinite at all times -- and indeed that in a very real sense we can and should avail ourselves of that access on a regular basis, in this life.

Many previous posts have explored the critically important assertion of Alvin Boyd Kuhn which is in many ways a key to our understanding of the ancient myths, scriptures and sacred stories of humanity, in which Kuhn (addressing the stories of the Bible in particular) declares:

> Bible stories are in no sense a record of what happened to a man or a people as historical occurrence. As such they would have little significance for mankind. They would be

the experience of a people not ourselves, and would not bear a relation to our life. But they are a record, under pictorial forms, of that which is ever occurring as a reality of the present in all lives. They mean nothing as outward events; but they mean everything as picturizations of that which is our living experience at all times. The actors are not old kings, priests and warriors; the one actor in every portrayal, in every scene, is the human soul. The Bible is the drama of our history here and now; and it is not apprehended in its full force and applicability until every reader discerns himself [or herself] to be the central figure in it! [For full quotation and source see "Winter Solstice, 2014"[331].

Now, what Kuhn asserts in the above paragraph is just as true for the world's other myths. Let's see how it applies to the specific aspect of the Mahabharata mentioned above (the ability to summon the gods and goddesses at a moment's notice).

If we apply this paragraph directly to the Mahabharata, we can paraphrase some of these assertions as follows:

The episodes in the Mahabharata in which men or women are depicted as summoning powerful deities through the recitation of a mantra, the singing of a hymn of praise, or even by simply thinking upon that deity and wishing for him or her to appear, are in no sense a record of what happened to a man or woman long ago in a more magical (or imaginary time and place). As such, while they might be tremendously entertaining, they would have little significance for our lives today. They would be the (miraculous and extraordinary) experience of a people not ourselves, and would not bear a relation to our life. But these events are actually recorded in these myths to provide us with a vivid picture of something that is in fact a verifiable reality of a situation that is present in your life and in mine – indeed, a reality in all lives. They mean nothing as outward events: the beautiful wives of Pandu, for instance, did not summon gods outwardly. Nor was Arjuna's invocation of the goddess Durga an outward event. These are picturizations of truths which are part of our living experience at all times. We indeed are in contact with those same mighty supernatural powers -- with Krishna and Durga and the heavenly Twins or Ashvins -- right at this present moment. The actors in these myths are not beautiful wives or powerful warriors: in every single episode, these actors are none other than the human soul possessed by each and every one of us. The Mahabharata (and all the other myths and scriptures and sacred stories) is a drama of our lives -- our lives right here, right now, in this modern life, in the city where you live, in the situations you experience -- and it is not apprehended in its full force and applicability until every reader discerns himself or herself to be the central figure, present in every single scene!

In a post from 06/29/15, we discussed some of the unusual marriage activity recorded in the Mahabharat, in which the two wives of Pandu take five different divine gods to be the fathers of the five powerful sons who collectively become the heroes of the story, the Pandavas (a name which means descendants of Pandu). The summoning of the five different gods is done through the recitation of a mantra: immediately upon its recitation, the desired god appears.

Elsewhere in the Mahabharata, as we saw, Arjuna (one of the Pandavas) recites a hymn of praise to the goddess Durga, at which the powerful goddess appears and blesses him, telling Arjuna that he will be victorious and that in fact it would be completely impossible for him to be defeated in the upcoming battle.

At other points in the epic poem, such as in Book I and section 3, the celestial Twins called the Ashvins are summoned by a disciple named Upamanyu, who has consumed some leaves of a tree that made him blind, causing him to stumble into a deep well, where he was trapped until he called upon the Ashvins for succor.

And there is also a powerful sage or rishi named Vyasa or Vyasadeva who is the mythical author of the Mahabharata itself and who also appears as a character who weaves in and out of the various scenes, appearing when he is needed before retreating again to his contemplation and disciplines in the remote mountains. Vyasa also has the characteristic of being able to appear whenever he is thought upon: at his birth (recounted in Book I and section 63) he tells his mother "As soon as thou remembers me when occasion comes, I shall appear unto thee."

What are we to make of these wondrous episodes in the Mahabharata, each one of which is surrounded by all kinds of memorable action and human drama? These depictions of the gods and goddesses (and, in the case of Vyasa, this epic poet and bringer of inspired verse) appearing at an instant when a human man or woman concentrates upon them are not to be understood *as outward events,* in Kuhn's argument, but rather as an inward reality, as a depiction of our experience in the here and now.

If Kuhn is right, then what (oh what) could these specific episodes be depicting?

I believe the answer is hinted at in yet another earlier post exploring the powerful teaching contained in the Mahabharata -- an examination of the Bhagavad Gita, which is a section within the Mahabharata itself. There, we saw compelling evidence that the conversation between the semi-divine bowman Arjuna and his companion and divine charioteer, the Lord Krishna, relate to the "metaphor of the chariot" found in other ancient Sanskrit scriptures.

In that metaphor, the chariot helps us understand aspects of our incarnate condition. The war-cart itself is our body, and the mighty horses which pull it are our senses and our desires (both of which can easily run completely out of control, and threaten to wreck the entire enterprise). The reins in the metaphor, we are told in another Sanskrit scripture, are our mind, through which the horses can be controlled.

But obviously, there must be someone or something else *behind and above* the reins in order to direct the chariot: behind and above "the mind" itself, that is. This concept of a someone or something else, standing apart from the mind and above it, was discussed in the first blog post of this series, entitled "Self, the senses, and the mind." This higher self is referred to by many names, among them the True Self, the Supreme Self, the Lord in the chariot, and (in the Sanskrit text cited for this metaphor) the Atman. In other cultures and other traditions there are many other names to refer to the same concept.

But in all cases we are dealing with a Higher Self who is in some sense and to some degree connected to the infinite and the ultimate. This is the infinite, the ultimate, the un-definable: the divine charioteer who is beyond the "chattering" and the "endless transforming" and the "labeling and defining and delineating" of the mind (and again, the mind is not a negative or bad tool, any more than the reins on the chariot are a bad tool -- it is an essential tool, but it is not the one who should be driving the chariot).

We get in contact with this infinite aspect by standing apart from our mind, our senses, and our desires (not by getting these to somehow "go away" or "stop" being what they are -- the horses on the chariot will not go away, nor will they turn into something other than horses -- but we can stand apart from and above them in order to see that we are not them and we do not have to go wherever they want to pull us, that in fact we can tell them where we want them to take us).

Practices we have at our disposal for getting into contact with the infinite include mantras, chanting or singing of hymns, prayer, meditation, yoga, rhythmic drumming, and more.

The gods and goddesses in the stories show up quite suddenly and instantly because they are, in a very real sense, already there. We are already connected with them. This does not mean that they are simply "our imagination" or "not real" (as if our "imagination" is not connected to the very same vital flow of

584

infinity that is completely unlimited in its potential and its power). As we see in Kuhn's quotation above, which is so valuable that we can and should return to it in analysis like this, just because the myths are depicting inner realities *as outward events* does not mean that they are not "real" if they do not take place in the outward space. These myths are dramatizing truths about *our living experience at all times*. You and I are in contact with Krishna and with Durga right now: if we do not realize it, that is only because we are allowing the chatter of our minds or the horses of our senses to keep us from connecting with the power of the unbounded, the undefined, and the infinite (unbounded aspects of which Krishna and Durga show themselves to be in the Mahabharata).

It is also noteworthy to point out that divinities who can appear at a moment's notice are also found in other esoteric mythologies and scriptures around the world. The Norse god Thor, for instance, was notable for being able to appear whenever his name was called by the other gods, in time of need (which they had to do on more than one occasion). The other gods usually had to call on him when they were being bested by a powerful jotun, and thus Thor usually appeared in a fighting rage (or, if he wasn't in a rage when he appeared, one glance at the menacing jotun usually caused Thor to go into battle mode).

Thor, der Donnergott.

But, it should be noted that Thor's ability to appear in an instant means that he, too, is somehow representative of that divine charioteer who is above mind and above even the physical world, and yet somehow available to us at all times, if we just learn how to direct our focus in the right direction.

It is also not inappropriate, I believe, to point out that the risen Christ in the stories of the New Testament also displays the ability to simply appear out of nowhere amongst the disciples, sometimes when they are least expecting him to do so.

In the preceding post, which looked at the two wives of Pandu who used a mantra to call upon divine gods to appear (06/29/15), we also saw that the pattern of five husbands in the Mahabharata appears to have an echo in the New Testament episode of the encounter of Jesus with the Samaritan woman at the well, who likewise is said to have had five husbands. In that encounter, the previous post points out that Jesus tells the woman that she can have everlasting water, living water, springing up unto everlasting life -- and that this living water is somehow "within."

I believe that this again is a "pictorial form" (in Kuhn's words) of something that is in fact a "present reality" in the life of each and every human soul. This "picture" is one of an *unbounded*, an *infinite*, and a *life-giving stream*, available for the asking because it is already "within" us. We already have access to this living water, but we need someone to tell us that it is something that we can actually get in touch with. That is what the ancient myths and scriptures are there to do.

By his demonstrated ability to simply appear out of nowhere and disappear again at will, the risen Christ in the gospels would also, under this interpretation, be pointing us towards connecting with the infinite *within ourselves*. And this, according to some analysts, is exactly what Paul in his epistles declares to his listeners, using the strongest language possible in some cases:

O foolish Galatians, who hath bewitched you [. . .]? Are ye so foolish? having begun in the spirit, are ye now made perfect by the flesh? (Galatians 3:1-3)

Gerald Massey (1828 - 1907) and others have argued that the writer who calls himself Paul is pointing his listeners to a spiritual truth, not an external flesh-and-blood individual. He is pointing them to what he elsewhere declares to be "Christ in you, the hope of glory" (Colossians 1:27).

This is not to say that Paul did not believe what he was talking about to be "real" or that he did not believe it to have life-altering power: on the contrary, the tenor of his letters indicates that he knew what he spoke of to be absolutely real, and absolutely earth-shaking in its ability to transform. Nowhere in the above discussion should anything be taken to indicate that the infinite, the ultimate, the un-limitable and truly un-bounded divine power -- which the Bhagavad Gita describes as the Lord Krishna and which the Hymn to Durga addresses as Kali, as Maha-Kali, as Uma, and as "Durga, who dwelleth in accessible regions," and as "identical with *Brahman*" -- is in any way not *real*.

But, as the quotation from Alvin Boyd Kuhn tells us, these are not stories about ancient events that happened to someone else: these are aspects of our life, right here and right now. They are telling us about a divine aspect to which we have access right here and right now, and with which we are already internally connected in some mysterious way.

As the verse in the Old Testament wisdom-book of Proverbs tells us, "There is a friend that sticketh closer than a brother" (Proverbs 18:24).

Even closer than a brother, because not external to us at all.

A meditation upon the Thirteen Postures Song
十三勢歌

十三勢歌

十三總勢莫輕視　命意源頭在腰際　變換虛實需留意　氣遍身軀不少滯
　　　　　　　　　因敵變化示神奇　勢勢存心揆用意　得來不覺費功夫
刻刻留心在腰間　腹內松淨氣騰然　尾閭中正神貫頂　滿身輕利頂頭懸
仔細留心向推求　屈伸開合聽自由　入門引路需口授　功夫無息法自修
若言體用何為准　意氣君來骨肉臣　想推用意終何在　益壽延年不老春
歌兮歌兮百四十　字字真切義無遺　若不向此推求去　枉費工夫貽歎息

2015 August 02

The Wudang Mountains of China are associated with the ancient internal arts: practices and disciplines designed to facilitate one's cultivation of and connection with chi.

Whether or not they actually originated there, the association between the internal arts and the Wudang region is justifiable, because the area has been a center for both ascetic and monastic pursuit of the Way of the Tao for at least 1200 years and possibly even longer than that. There are direct parallels between concepts conveyed in the Tao Te Ching and teachings associated with the specific "internal" martial arts and disciplines associated with Wudang.

According to legend, it was to the Wudang Mountains that the mysterious Zhang Sanfeng retired to live an ascetic life, leaving a promising career in the government ministries and giving away all his possessions. The traditions say that Zhang was already an accomplished martial artist who became more and more attracted to the development of internal kung fu, and whose prowess became greater and greater even as he became less and less interested in external displays of power, until he eventually made his way to the mountains . . .

There, he would in time master the internal arts, develop one (or more) of the most famous systems for cultivating internal power, and ultimately become a Taoist immortal or 仙 -- a word that is pronounced *Xian* in Mandarin and *Sin* in Cantonese, and which when used as a verb means "to ascend" or "to transcend," and which thus when used as a noun means by extension "a transcendent one" or "an ascended one" (the character itself is composed of the symbol on the left for person and the three-pronged symbol on the right which means "mountain").

Among the many texts sometimes attributed to this legendary personage or associated with the internal arts he imparted, one intriguing representative of their style and content is the *Song of the Thirteen Postures*, a short poem whose actual origin and date and author(s) are all unknown, but which is counted among the Tai Chi Classics: texts belonging to the art of Tai Chi Chuan, one of the three main Chinese martial arts associated most closely with Wudang and with the practice of attuning oneself to the flow of chi. The other two are Xing-Yi Chuan (形意拳 -- pronounced *Jing Ji Kyun* in Cantonese, and translating to something like "Form and Conscience Fighting Style [literally "fist"]") and Bagua Chuan (八卦掌 -- pronounced *Baat Gwaa Jeung* in Cantonese, and translating to something like "The Eight Divination-Trigrams Palm").

Zhang Sanfeng is traditionally credited with creating the original system of Tai Chi Chuan itself. An earlier name for Tai Chi Chuan was in fact "The Thirteen Postures" or 十三勢 -- the first two characters and syllables of which literally mean "Ten - Three" (which is the standard way of saying "thirteen") and the final character and word translating more literally as "powers" or "energies" or "forces" or "dynamics." Thus, "The Thirteen Dynamics" or "The Thirteen Forces" might be a more accurate translation of the sense of the original, although it is so commonly referred to in English as "The Thirteen Postures" that this is probably what we should use to refer to the poem in question.

It is also worth noting that the reason for the "Thirteen" in the title comes from the connection of the different "forces" or projections of energy used in the motions of Tai Chi were traditionally eight in number and connected to the eight angles or Eight Divinatory Trigrams of the BaGua, and to these were added five directions or ways of stepping or directing the body (going forward, going backwards, going left, going right, and holding at the center), to bring the total to thirteen.

The *Thirteen Postures Song* is reproduced in the Wudang Mountains image above, and is available in various English translations (some more literal than others) in a variety of places on the web, including:

- http://www.scheele.org/lee/classics.html
- http://qi-encyclopedia.com/?article=Song%20of%20the%20Thirteen%20Postures
- https://brennantranslation.wordpress.com/2013/05/25/the-taiji-classics/

Borrowing from these sources as well as from the literal meanings of the characters themselves (with apologies for any misinterpretations which I myself introduce in the process), a fairly literal translation might be:

Thirteen Collected Dynamics: Do Not Lightly Esteem ["do not take them lightly"].

[Their] Life-Heart and Head: [It] Issues from the Waist / Kidney Region.

The Transformations and Turnings of Empty and Solid: [You] Must Keep in Heart-Soul-Mind.

590

Chi Everywhere in the Body, the Human Body: Not Steered into an Obstacle [usually translated to mean "not hindered or obstructed"].

Stillness [in the] Center of Initiating-Action: Action Like Stillness.

Because of it, the way that you Adapt to the Opponent's Moves: Indeed Mysterious and Uncanny.

Each Posture [each "dynamic" or "force"] Learn by Heart: Come to Know its Usefulness and its Deepest Essence.

Acquire / Will Come all-Unconscious: Effortless Mastery or Advanced Skill [literally "kung fu"].

Deeply Engrave and Hold the Heart-Mind in the Place of the Waist / Kidney Region.

In the Abdomen area [be] Relaxed and Still: Chi Gallops, Flying-up – Yes!

Tailbone Centered and Straight: Divine Energy [from there up through] The Top of the Head (like a string through a thousand coins).

The Benefit of a Body Filled with Lightness and Agility: [it is achieved by] Hoisting or Suspending the Top of the Head (as if hanging from above).

Follow the Slender Thread [perhaps meaning "to the deepest, thinnest ends of the roots"]: Push Towards what you Seek.

Flexing and Opening and Closing: You will Hear it or Know it from Within Yourself.

The One who Begins this Path: Must necessarily have this teaching Transmitted from the Mouth [of a teacher].

Practice your Skill [literally "kung fu"] Without Stopping, Without Resting: the Way is by Your Own Study – your own Cultivation.

Regarding the Usefulness of this System: What Guideline or Standard shall we Make or Observe?

The Heart-soul and the Chi Arrive as the Sovereign: the Bones and the Flesh are the Monarch's Ministers and Officials.

Towards What Goal does all of this Push or Impel us?

The Benefit of Desired Long Life and Delay of Aging: a Never-Aging Springtime.

A Song -- Ah! A Song -- Oh! A Hundred and Forty.

These Written Characters -- Genuine, Clear-cut: Right in Conduct, Without any Suspicion.

If one does Not Toward this Direction Push, Seek, and Go . . .

In Vain all that is Spent on Achieving Skill [literally "kung fu"]: Sighing, Loss, and Regret.

This is a remarkable poem, filled with important teachings with far-reaching implications.

Foremost among them, perhaps: the connection of the cultivation of chi and the concept of stillness in the midst of action.

The poem imparts specific images to aid in attuning oneself to the invisible force of chi. Chi itself is written

and it is pronounced *hei* in Cantonese: both *chi* and *hei* mean "breath" and "spirit," which just as in English can refer to either literal breathing and also to the entire realm of spirit, the life-force, that which animates all beings (the in*spir*ation) and which also permeates all things in the cosmos.

While the actual date and authorship of this specific poem is unknown (and some scholars place its origin to within only the past few hundred years or so), texts which explicitly refer to the raising of chi exist from as early as 380 BC, as Professor Victor H. Mair (an accomplished scholar of Chinese culture, language and history who has taught at the University of Pennsylvania since 1979) notes in his valuable translation of the Ma Wang Deui "silk texts" containing an early arrangement of the Tao Te Ching (discovered in 1972). Describing an inscription on ten pieces of jade which once formed a small knob, he gives this translation:

592

In moving the vital breath (*hsing ch'i*) [through the body, hold it deep and] thereby accumulate it. Having accumulated it, let it extend (*shen*). When it extends, it goes downward. After it goes downward, it settles. Once it is settled, it becomes firm. Having become firm, it sprouts [compare Yogic *bija* ("seed" or "germ")]. After it sprouts, it grows. Once grown, then it withdraws. Having withdrawn, it becomes celestial [that is, yang]. The celestial potency presses upward, the terrestrial potency presses downward. [He who] follows along [with this natural propensity of the vital breath] lives; [he who] goes against it dies. [cited on page 159 of the paperback edition of 1990 of Victor H. Mair's translation of *Tao Te Ching: The Classic Book of Integrity and the Way*].

The harmonies in this jade inscription from 380 BC and the teachings contained in *The Song of the Thirteen Postures* should be self-evident.

Additionally, as Professor Mair references in a bracketed parenthetical comment upon one specific part of the above-quoted jade inscription, some clear connections can be perceived between the teachings in these ancient Chinese texts and the teachings preserved in the Yogic traditions and texts. Professor Mair addresses in some detail these conceptual connections in the Afterword and the Appendix of his translation of the Tao Te Ching -- not referring specifically to the Thirteen Postures Song but rather to the Tao Te Ching itself, which also contains numerous admonitions to have stillness or inaction even in the midst of action.

Professor Mair points out some of the passages in the Tao Te Ching concerning action-inaction, and connects their teachings directly to the direction given by the Lord Krishna to Arjuna in the Bhagavad Gita. For example, in the section of the Tao Te Ching traditionally numbered 38 (but in fact arranged as the very first section in the Ma Wang Deui texts), we read -- in part -- that:

The person of superior integrity takes no action, nor has he a purpose for acting.

The person of superior humaneness takes action, but has no purpose for acting.

The person of superior righteousness takes action, but has a purpose for acting. [From the passage found on page 3 of Professor Mair's 1990 translation].

All this taking action and taking no action, without a purpose for acting, may seem confusing, but when we examine (as Professor Mair does) the words of Lord Krishna in the Bhagavad Gita, we may begin to understand what is being advised.

And again, a few sections later in that part of the Tao Te Ching traditionally numbered 43 but arranged as section 6 in the older Ma Wang Deui texts, we read:

The softest thing under heaven
gallops triumphantly over
The hardest thing under heaven.
Nonbeing penetrates nonspace.
Hence,
I know the advantages of nonaction.

The doctrine without words,
The advantage of nonaction --
few under heaven can realize these! [page 11].

It is interesting to wonder, given the explicit description in the tenth line of the *Thirteen Postures Song* of chi as "galloping," whether the Tao Te Ching in this passage is not referring to the invisible spirit-force of chi when it describes the triumphant nature of "the softest thing under heaven."

For a fairly detailed examination of the importance of the teachings given to Arjuna by Lord Krishna in the Bhagavad Gita, see the post from 06/21/2015 (which also contains a video).

That post examines the fact that Krishna's direction to Arjuna, given in a variety of different ways using a variety of powerful

594

metaphors, can be summarized as "do what is right, without attachment to the results" and thus, without ulterior motive -- without concern for reward or even without concern for the outcome whatsoever. This can clearly be seen as throwing some light upon the Tao Te Ching's admonitions regarding "taking action" or "taking no action" but having "no reason" (no ulterior motive, no concern for or connection to the results) for it.

Professor Mair explains in his Afterword that the concept of *wu-wei* or nonaction is one of the most important concepts in the Tao Te Ching, which tells us that "through nonaction, no action is left undone" (see discussion on page 142). He explains that an understanding of the Bhagavad Gita helps us to realize that this teaching about nonaction may in fact mean *action* -- but action *as though* not acting (because totally nonattached to the action) [this is at least my interpretation of what Professor Mair is expounding on pages 142 and surrounding].

My earlier post and video discussing the Bhagavad Gita, which on a literal level is portrayed as Krishna's advice to Arjuna prior to entry into the great Battle of Kurukshetra, may in fact be seen as guidance given to the human soul prior to descending into incarnation itself, which is by its very nature a great battle or struggle or interplay between the "forces" of matter and spirit. If so, then the Bhagavad Gita teaches us that one of the most important principles in this life is to do what is right, but without attachment. And, the Bhagavad Gita shows that in order to do this, one must be connected to the divine charioteer -- portrayed in the Gita as the divine Lord Krishna, who in the text of the Gita itself reveals himself to be the Infinite, the Supreme, the Undefinable (beyond words or categorization).

In other words, in order to be able to act without acting (without attachment), we must cultivate connection with the Infinite: with the invisible force which pervades everything. And this is exactly what the Tao Te Ching teaches as well (note that it describes the Tao as beyond categorization, beyond labeling with words, beyond definition).

595

And it is exactly what the *Song of the Thirteen Postures* appears to be telling us also! In order to achieve action without action, we must attune ourselves to the invisible force which is inside us and which permeates the universe around us as well (which it explicitly calls 氣 or *chi*).

The line which most clearly deals with the concept of action while centered in complete stillness or lack of action (lack of attachment, lack of motion) is the fifth line of the *Song of the Thirteen Postures* (the line which is highlighted in red in the text shown above, superimposed on the photograph from Wudang), and which translated rather literally reads:

> *Stillness [in the] Center of Initiating-Action:*
> *Action Like Stillness.*

The poem could hardly be more clear and direct on this point.

Note also the important third line of the poem, which emphasizes guarding deep in our heart-soul-mind the endless interplay of *empty* and *solid*: this, I would argue, could well be the very same interplay or struggle allegorized in the Battle of Kurukshetra -- the endless interplay between the realm of Spirit and the realm of Matter (between "empty" and "solid").

Finally (although there is much more to discuss), we cannot end this brief examination of *The Song of the Thirteen Postures* without pointing out that fascinating fourth line from the end, which says:

> *A Song -- Ah! A Song -- Oh! One Hundred Forty.*

What is this supposed to be teaching us? Well, the very next line tells us the meaning of the "one hundred forty": it refers to the characters in the poem itself. So, the first half of that line which says "*A Song -- Ah! A Song -- Oh!*" must be talking about the *Song of the Thirteen Postures* itself.

It is advising, it would seem, a regular repetition of this song to oneself, as a way of calling to mind this important guidance for

596

our struggle in this life. It is telling us that this song is something we should sing to ourselves, perhaps daily – in much the same way that the sections of the Mahabharata which take place immediately prior to the Bhagavad Gita present us with a hymn to sing in order to summon the goddess Durga,[367] and then tell us that this is a song we should sing to summon the goddess every single morning!

Thus we see that the ancient texts were given as powerful helps for us in this life -- powerful tools to guide us towards the cultivation of our contact with the infinite (which is, in fact, already inside us and already all around us) and our cultivation of an effortless and unattached principle of action: doing what is right, without attachment to the outcome.

And, along with these ancient texts, there were given very specialized disciplines, including the practice of Yoga but also in China of martial arts which have a clear focus upon the cultivation of the internal power of chi.

These practices are for our daily use – and both the *Song of the Thirteen Postures* and the jade inscription from 380 BC advise us to pursue them diligently, because the benefits of practicing them are very great, but the penalty for neglecting them include sighing and loss and regret.

We are indeed fortunate that this ancient wisdom has survived and that we can avail ourselves of it, and that doing so does not necessarily entail a life of asceticism in the Wudang Mountains – although for some it might!

Meditation, detachment, and
working for the welfare of all living beings

2015 September 14

The Bhagavad Gita explicitly connects the state of "action without attachment to results" and the ability to bring the mind under the control of the Higher Self.

In order to achieve the state of acting without attachment -- which some verses also describe as "actually doing nothing at all" even while acting -- it says that the mind must be brought under control.

In the sixth verse of the sixth chapter of the Bhagavad Gita, Krishna tells Arjuna that, for the one who has learned how to control the mind, the mind is the best of friends -- but for the one who has not, the mind can resemble the worst of enemies.

The good news is that Krishna explains that we can, with practice, bring the mind under our control -- and that this practice is also the best way to come into contact with the Higher Self.

In chapter 6 verses 33 and 34, Arjuna expresses doubts about the ability to ever bring the mind under control, due to its flickering and unsteady nature.

"For the mind is restless, turbulent, obstinate, very strong, O Krishna," Arjuna says, "and to subdue it, I think, is more difficult than controlling the wind" (6.34).

Krishna acknowledges that it is "very difficult to curb" the mind, but also tells Arjuna that "it is possible by suitable practice and by detachment" (6.35).

He advises the regular practice of meditation in a secluded location, while alone, providing some very specific instructions which begin in verse 10 of chapter 6:

> 10 One perfecting the science of uniting the individual consciousness with the Ultimate Consciousness, consistently residing alone in a secluded place engaged in controlling the mind, desireless, free from proprietorship [translated here as "feelings of possessiveness"], should meditate on the inner self.
> 11 In a sacred and purified place after establishing a seat neither too high nor too low of kusa grass, deerskin or natural cloth;
> 12 thereupon sitting firmly on that seat controlling the mind and activities of the senses making the mind one-pointed; one in realization should meditate by the science of uniting the individual consciousness with the Ultimate Consciousness for purifying the mind.
> 13 Holding the physical body, head and neck straight, unmoving and stable, gazing upon the tip of the nose and not glancing in any direction, fixed in the vow of celibacy,
> 14 with an unagitated mind, fearless, completely subduing the mind; the renunciate should sit concentrating upon Me as the Ultimate Goal.

Thus, in verse 14, the Lord Krishna explains that the yogi should focus entirely upon Krishna -- while only a few verses later, in verse 19 of the same chapter, the Lord Krishna describes the mind thus concentrated as focused upon or resting steadily in the Higher Self:

> 19 As a lamp in a windless place does not waver, so the transcendentalist, whose mind is controlled, remains always steadying his meditation on the transcendent self.

From these passages, then, I believe we can make the case that the teaching described using the imagery of Arjuna and the divine Krishna is teaching very much the same thing as that which is being conveyed through the parallel imagery of Thomas and the Divine Twin found in the New Testament gospel of John and in texts found at Nag Hammadi such as the Gospel of Thomas and the Book of Thomas the Contender.

Or, stated the other way around, the metaphor of Thomas Didymus (the Twin) is trying to give us the same understanding that Krishna is here imparting to Arjuna in the Bhagavad Gita.

The mind is always unsteady, prone to rushing off in different directions -- as is Thomas in the New-Testament-era texts discussed in the posts linked above. It is full of doubts, as is Arjuna in the Bhagavad Gita, when Arjuna expresses to Krishna his doubt that the mind can ever be controlled at all.

And yet, Thomas has a divine counterpart with whom he is already inextricably linked -- with whom he is "twinned," one who is already "closer than a brother" to him, although he does not always act like it. This is the divine twin -- the transcendent self or the Higher Self described in the Bhagavad Gita using different language, and described in the letters of Paul as the "Christ within."

We can catch glimpses of this unwavering, transcendent divine twin below or beneath the endless flickering of our mind as we learn to stop letting the mind carry us wherever it is "blowing" from one moment to the next -- and when this happens our mind stops becoming our "worst enemy" and begins to become a tool that we control instead of one that controls us.

Beginning each day by following some of the specific recommendations given in the sixth chapter of the Gita (such as the recommendations of using a meditation cushion, one that is neither "too high nor too low," assuming good upright posture, not closing the eyes completely but rather gazing downward in the direction of the tip of the nose [or, as 5.27 and 28 say, "keeping the eyes and vision concentrated between the two eyebrows," in the region of the third eye], and focusing the mind on one point, while "suspending the inward and outward breaths within the nostrils [5.27-28]) can help us to begin to gain control over the mind, which Lord Krishna tells Arjuna in 6.36 is -- in his opinion -- the most practical and appropriate path towards uniting with the Higher Self.

Interestingly enough, in light of the assertion being made that the allegory of Thomas and the Divine Twin is intended to convey the same message found in the Bhagavad Gita chapters 5 and 6, Krishna tells Arjuna that the one who gains full consciousness of the Higher Self -- the one who attains awareness of Krishna -- will "attain peace" (5.29).

It is quite evident from a reading of the New Testament scriptures that the Christ regularly greets his followers with the word "peace" and that he very memorably promises them his peace, such as in John 14:27.

The Bhagavad Gita suggests that it is only by bringing the mind out of its "flickering, doubting" mode and under the control of the Higher Self that we are able to achieve this peace, which is characterized by complete detachment from either the fear of the consequences of right action or desire for benefits from its successful outcome. Instead, we can simply focus on doing what is right, without being beset by doubt. In 5.25, for example, Krishna tells Arjuna:

> Those who are beyond the dualities that arise from doubts, whose minds are engaged within, who are always busy working for the welfare of all living beings, and who are free from all sins [which 5.10 has already explained comes from freedom of attachment and surrender to the Supreme], achieve liberation in the Supreme.

Incredibly enough, the bringing of the mind beneath the control of the Lord, and the release from doubt through the surrender to the Supreme, appears to be *exactly* what is dramatized in the famous episode of "Doubting Thomas"[499] found in John, chapter 20.

I believe that the Gita tells us how to pursue this uniting of the "Thomas-mind" to the Supreme Lord *every single day*, through very practical direction and the promise that, while at first difficult, with discipline "it is possible," through suitable practice and detachment.

When the two "twins" are united, then our mind can begin to become our "best friend" rather than our "worst enemy."

Because, as we learn to become less attached to the outcome and more in touch with the Higher Self, we are less "blown about by the wind" (the exact same metaphor used in the Bhagavad Gita 6.19, 33 and 34 is also used in the New Testament, in James 1:6), less wracked by doubt, and more prepared to do what is right, without attachment to the outcome.

And, Krishna also appears to be telling us in the Gita, this detachment is essential in order to be "always working for the welfare of all living beings," which is our duty.

Note that by explaining what I see in these ancient texts, I am not trying to imply (at all) that I myself have achieved this state! It is one thing to know what the texts are telling us, and quite another to achieve what they are describing.

But, even though Arjuna expresses his doubts to Krishna, and says that all of this seems to be harder even than controlling the wind, Krishna promises him that the path can be successfully followed – and what is more, that eventually we are assured of success in this regard (although perhaps over the course of many incarnations).

The incredible importance of the INNER connection to the Infinite

2015 October 01

The previous post[137] discussing the significance of the Moon Festival explains (following the incisive and inspired analysis of Alvin Boyd Kuhn in *Lost Light*) that the moon is used in ancient myth and ancient wisdom the world over as the symbol of the body, which is illumined during the night by the light of the unseen sun, which is analogous in this particular metaphorical construct to the light of the divine, invisible spirit-fire.

That and other posts have explained that the crossing down to the "lower half" of the year was analogous to the incarnation of the soul within the human form. The lower half of the year is the half of the year in which night-time dominates over daylight --

when nights "cross over" to begin to be longer than the daylight hours, on the way down to the very lowest part of the year, the very "pit" of the winter solstice.

This toiling across the lower part of the year, from the "crossing down" point at fall equinox to the "crossing up" point at spring equinox, was used to convey subtle and powerful truths about our human existence. This incarnate life, in other words, was analogized by the lower half of the year during which *night* dominates.

During the night, we cannot see the sun. But, we *can* see the *reflected* rays of the invisible sun (the sun being invisible to us at night-time).

We can see the reflected rays of the invisible sun shining upon and reflecting out of the body of the moon. The moon actually has no light of its own, but it shines with a glory of light transmitted to it from the invisible sun.

In this way, the moon helps us to better understand our condition, for we are in a body that the ancient myths and metaphors associate with the moon, and yet we too are made alive by a light from the invisible and divine realm of spirit.

We are crossing through the lower-valley, the valley of night and of the shadow of death, where the reality of the spirit world is hidden from our plain sight. And yet we are not without access to the power and the glory of the unseen world: just as the moon at night is illuminated by the rays of the invisible sun, we and everyone we meet in this incarnate world have access to the light and warmth of the divine rays which come to us from a world which we cannot presently see (at least not with our physical eyes).

But, although like the moon in having access to the divine light of spirit, our situation is unlike the moon in that our access is to a divine light that is in fact within us and not completely external to us (unlike the case of the actual moon in the heavens, which does receive light from an external sun outside of itself). The moon metaphor is helpful in that the sun is invisible at night, as is the source of the invisible divine light to which we have access (we cannot physically see the divine world, just as we cannot see the sun at night).

But our connection to the Infinite is in fact *internal* and not completely external.

This is a *very* important point, and one that the ancient scriptures take great pains to explain and portray to us: the divine is *within*, and is not separate from you, not external to you. In the New Testament scriptures, for example, Jesus is portrayed saying, "For behold, the kingdom of God is within you" (Luke 17:21) and Paul's letters declare, "Know ye not your own selves, how that Jesus Christ is in you" (2 Corinthians 13:5).

As we have seen in many other discussions of this subject, this same message can be found being conveyed by various different sacred metaphors in myths the world over -- see for instance the specific discussion here, and a host of other related posts listed here.

For instance, in the image above depicting an important episode in the Mahabharata, Kunti is reciting a mantra to summon the divine Indra, who appears to her in an instant (and who is depicted here with a halo, which ties in to the metaphor of moon and sun discussed above). Kunti is depicted with her hands together in the Anjali mudra, the gesture of namaskaram: the acknowledgement of deity.

But note that this gesture is an acknowledgement of deity both in *oneself* and in another.

Thus the artwork above contains the message that Indra appears to Kunti instantly because Indra is *always* present with Kunti -- and what is true for her is true for each and every man and woman in this incarnate existence. We have an ever-present and unbreakable inner connection to the divine Infinite.

605

As Alvin Boyd Kuhn explains at great length in a very early passage in *Lost Light* (beginning around page 45), it is very important to understand what this means and what it does not mean. To acknowledge our own divine spark is *not* to deny the existence of an external divine Infinite. Alvin Boyd Kuhn explains:

> Ancient religion was suspected of having left the monotheistic God out of its picture. It did not leave it out, but it had the discretion to leave it alone! The sage theologists reverenced it by a becoming silence! [. . .] But the pagan world provided a contact with a god dwelling immediately within the human breast. [. . .] And this implied no spirit of vaunting humanism or affront to deity. It was just the recognition of deity at the point where it was accessible. The real heresy and apostasy, the gross heathenism, is to miss deity where it is to be had in the blind effort to seek it where it is not available. [. . .] The kingdom of heaven and the hope of glory are within. They lurk within the unfathomed depths of consciousness. Divinity lies buried under the heavier motions of the sensual nature and the incessant scurrying of the superficial mind. 45-46.

He goes on to argue that the *externalization* that took place through a literalistic interpretation of the ancient myths and scriptures moves all deity to someone or something outside of us, and thus "has left the rest of mortals unsanctified" by cutting us off from our own divinity (47).

This may seem like a philosophical, theological, or intellectual discussion -- but it has tremendous practical application and consequences.

Because it leads directly to the problem of the "Two Visions" which the Lakota holy man Black Elk described, between the vision of grasping and grabbing and chasing after the external (which leads to destruction of the earth around us, and ultimately to our own self destruction) and the knowledge that what we are really seeking is already available to us, all along (a vision of plenty, a vision of security, and a vision of peace and harmony).

Look again at the assertions that Alvin Boyd Kuhn articulates above. He says that externalizing the divine *cuts us off from our divinity* (the quotation is halfway down page 47). In the very next sentence he states that this externalization *robs humanity at large of its birthright.*

Cut off from accessing our inner connection to the infinite, we pursue it externally instead. And, I believe, under the vision of scarcity and lack that stems from the very concept we are here discussing, there is the urgency that we have to "race to get it" before someone else does -- leading to exactly the mindset of "little islands" amidst a "gnawing flood dirty with lies and greed" that Black Elk warns us against, the mindset of "everybody for himself," rushing down "a fearful road."

This is exactly the mindset that we can see running absolutely rampant to the point of literally threatening to tear the world apart around us right now -- the mindset that seeks to reduce humanity into little islands in the midst of a gnawing flood, to the point that they have no ability to band together to stand up for the dignity that is ultimately rooted in the acknowledgement that each and every man and woman you meet contains the spark of infinite divinity within and thus has infinite worth.

Examples are so numerous they hardly need to be listed here -- most readers can think of dozens without even trying. But for the sake of illustration, one might think of the blasting away of any attempts by communities to restrict the rush towards genetic alteration of the food supply: this is akin to the gnawing flood of lies and greed that seeks to isolate individuals into little islands, hanging on for dear life ("if you don't like it, then don't buy it, little individual island -- but you have no right to hold hands together with others and raise your voices together to make demands for your community or region regarding genetic modification of the crops, or the chemicals used on those crops and what they might be doing to bees or water tables or anything else").

If these crops or the chemicals that they are designed to have sprayed over them cause harm to men and women and children for the sake of "lower input costs" (and higher profit margins), then the vision which Black Elk warns us about is plainly taking place before our eyes: a rushing after getting more and more, regardless of the cost to others – everybody for himself, rushing down a fearful road (a "race to the bottom").

Another example are wars of aggression waged under false pretenses in which thousands of men, women and children are killed, injured, or have their homes destroyed. Men and women in the United States and western Europe need to examine the available evidence very closely and objectively in order to determine whether or not they have been supporting or participating in criminal wars of aggression for the past fourteen (or more) years.

I would submit that the willingness to excuse, overlook, or attempt to justify the unjustifiable killing of other men, women and children in wars of aggression can be directly connected back to the denial of the inherent divinity present in each and every human being and the infinite worth and sacred nature of each and every man or woman regardless of where they live.

And also to an artificially-induced insecurity that comes from artificially externalizing something to which we already have access, something so profoundly valuable that Alvin Boyd Kuhn describes it as the "birthright" of humanity at large: its own divinity. And by externalizing what is internal, this error sets humanity off on a tremendous chase, a running after external answers, which then manifests itself on just about every other layer of the human experience.

The cutting off of humanity from deity "at the one point where it was accessible," to use Kuhn's phrase above, is what starts the whole wheel spinning in the wrong direction -- and it seems to be spinning faster and faster as (to use the metaphor of Black Elk) the flood eliminates the last interconnecting land forms between the "little islands" and leaves everyone isolated in the midst of the dirty rushing tide.

But, as I have said in previous posts on this subject, the good news is that this disconnect from what we actually need the most -- this cutting off from our most precious birthright, our constant connection to our own divine inner spark -- is in fact an *artificial* disconnect.

We are never *actually* disconnected from our own inner connection to the Infinite. What can be artificially induced by slight-of-hand or erroneous hermeneutics can be "un-induced" and corrected, as if in a flash.

We can wake up in an instant to our own actual connection to divine wisdom and knowledge, and to our connection with every other human being and indeed with the earth and with all of nature around us.

This may, in fact, be what we are here in this incarnate life to do: to wake up to that inner connection to the Infinite, to raise the awareness of and reflection of the divine within ourselves and as much as possible in others around us.

Indeed, the ancient scriptures and myths of humanity seem to say that this is precisely what we are down here to do in the first place (or a very big part of what we are here to do).

"The Thirteen Postures Song" and
the Cultivation of the Way

2016 April 11

A previous post from last August[588] explored some of the many lessons contained in the "Thirteen Postures Song," one of the mysterious Tai Chi Classics.

Last month, while participating as the "Author of the Month" on Graham Hancock's site, I revisited the "Thirteen Postures Song"[610] to comment on a different aspect of that amazing poem, one which I find very pertinent in the discussion of the ancient myths of humanity. It is so important, in fact, that I believe it deserves comment again here, for the benefit of those who may have missed it previously.

The "Thirteen Postures Song" initially appears to be concerned with cultivating skill in martial arts for the purpose of defeating opponents in physical combat when necessary. It describes the attainment of such a level of *kung fu* that one achieves "action like stillness," and a "mysterious and uncanny" ability to adapt to the opponent's every move.

610

However, as the previous discussion linked above demonstrates, this achievement of "action like stillness" clearly has a spiritual dimension, as well as important parallels to the central message of other sacred texts from around the globe, including the central message of the Bhagavad Gita.

And indeed, as we continue reading the rest of the "Thirteen Postures Song," we see that it does not really seem to spend much additional thought on defeating opponents, but instead begins to discuss the cultivation of the invisible force of *qi* or *chi*, and the achievement of a state which it describes with the metaphor of a "never-ending springtime" or (literally) a "never-aging springtime."

The entire poem, along with my best effort at a fairly literal translation of the characters is reproduced again below for convenient reference:

十三勢行工歌訣

十三總勢莫輕識。命意源頭在腰隙。

變轉虛實須留意。氣遍身軀不稍癡。

靜中觸動動猶靜。因敵變化是神奇。

勢勢存心揆用意。得來不覺費工夫。

刻刻留心在腰間。腹內鬆靜氣騰然。

尾閭正中神貫頂。滿身輕利頂頭懸。

仔細留心向推求。屈伸開合聽自由。

入門引路須口授。工用無息法自休。

若言體用何為準。意氣君來骨肉臣。

詳推用意終何在。益壽延年不老春。

歌兮歌兮百四十。字字真切義無疑。

若不向此推求去。枉費工夫遺歎惜。

(those characters found online in various places – the above are found in this online version here: https://brennantranslation.wordpress.com/2013/05/25/the-taiji-classics/).

And my translation (with apologies in advance for any errors, although I believe this translation to be fairly accurate, if very literal):

Thirteen Collected Dynamics: Do Not Lightly Esteem ["do not take them lightly"].

[Their] Life-Heart and Head: [It] Issues from the Waist / Kidney Region.

The Transformations and Turnings of Empty and Solid: [You] Must Keep in Heart-Soul-Mind.

Chi Everywhere in the Body, the Human Body: Not Steered into an Obstacle [usually translated to mean "not hindered or obstructed"].

Stillness [in the] Center of Initiating-Action: Action Like Stillness.

Because of it, the way that you Adapt to the Opponent's Moves: Indeed Mysterious and Uncanny.

Each Posture [each "dynamic" or "force"] Learn by Heart: Come to Know its Usefulness and its Deepest Essence.

Acquire / Will Come all-Unconscious: Effortless Mastery or Advanced Skill [literally "kung fu"].

Deeply Engrave and Hold the Heart-Mind in the Place of the Waist / Kidney Region.

In the Abdomen area [be] Relaxed and Still: Chi Gallops, Flying-up – Yes!

Tailbone Centered and Straight: Divine Energy [from there up through] The Top of the Head (like a string through a thousand coins).

The Benefit of a Body Filled with Lightness and Agility: [it is achieved by] Hoisting or Suspending the Top of the Head (as if hanging from above).

612

Follow the Slender Thread [perhaps meaning "to the deepest, thinnest ends of the roots"]: Push Towards what you Seek.

Flexing and Opening and Closing: You will Hear it or Know it from Within Yourself.

The One who Begins this Path: Must necessarily have this teaching Transmitted from the Mouth [of a teacher].

Practice your Skill [literally "kung fu"] Without Stopping, Without Resting: the Way is by Your Own Study -- your own Cultivation.

Regarding the Usefulness of this System: What Guideline or Standard shall we Make or Observe?

The Heart-soul and the Chi Arrive as the Sovereign: the Bones and the Flesh are the Monarch's Ministers and Officials.

Towards What Goal does all of this Push or Impel us?

The Benefit of Desired Long Life and Delay of Aging: a Never-Aging Springtime.

A Song -- Ah! A Song -- Oh! A Hundred and Forty.

These Written Characters -- Genuine, Clear-cut: Right in Conduct, Without any Suspicion.

If one does Not Toward this Direction Push, Seek, and Go . . .

In Vain all that is Spent on Achieving Skill [literally "kung fu"]: Sighing, Loss, and Regret.

The metaphor of a "never-aging springtime" is extremely important, because (along with the focus on the cultivation of the "divine energy" of *chi*) it appears to point towards the connection with that realm which the ancient texts and myths of humanity point towards as the source for all life in this world -- the Invisible Realm, which is even sometimes explicitly referred to as the "seed realm."

If indeed the "Thirteen Postures Song" is about the cultivation of connection with the invisible world, the way in which it tells us

613

that we cultivate that connection is worth examining more closely.

I believe that there are two lines in the poem, found "back-to-back," which provide us with a very important insight into the path we take towards the goal that the poem says we should be pursuing.

Here are those two lines:

- *The One who Begins this Path: Must necessarily have this teaching Transmitted from the Mouth [of a teacher].*
- *Practice your Skill [literally "kung fu"] Without Stopping, Without Resting: the Way is by Your Own Study – your own Cultivation.*

These lines are extremely interesting and worthy of examination. They almost seem to be contradictory. The first line tells us that we must necessarily be shown the way by the mouth of a teacher, and the second tells us that the pursuit is somehow on our own.

What does it mean?

The first line selected above says that "the one who begins this path must necessarily have this teaching transmitted from the mouth." In other words, it is generally necessary for the meaning to be shown in some way. After Daniel-san waxed the cars and painted the fence, he "knew the moves" but he did not know what he knew. He needed Mr. Miyagi to actually enable him to see it.

It is very common to find ancient wisdom passed down through a master-to-disciple relationship. This is found in Yoga, in the martial arts, and in some of the texts describing ancient Greek philosophy, for instance (and see also the important text by Peter Kingsley, entitled *In the Dark Places of Wisdom*).

But then follows the next line, which adds an additional angle (and note that these lines are found right together in the poem itself, one following the other, just as presented above).

614

There we see the poem telling us: "Practice your Skill [literally "kung fu"] Without Stopping, Without Resting: the Way is by Your Own Study -- your own Cultivation."

So, there is the necessity of one to show us, but then the poem tells us that "the Way is by Your Own Study -- your own Cultivation." We must both "be shown" and then "Practice without stopping" in order to actually cultivate this "Way" that the "Thirteen Postures Song" is describing.

I believe that this pair of lines is very appropriate to the study of they sacred myths given to humanity as well. I cannot tell you what they mean -- and neither can anyone else. The act of receiving what they have to say and the knowledge they wish to impart to you "is by your own study -- your own cultivation."

However, as mentioned above, if Daniel-san is never shown the layers of meaning that are beyond the act of waxing the car or painting the fence, then he would never be able to then "practice and cultivate by his own study." This situation, I believe, describes much of what has happened in places where the esoteric understanding of the myths and scriptures has been suppressed or cut off (particularly in "the west" -- which is what led to the realization over the years, accelerating in the twentieth century, of men and women from western cultures looking for and searching out answers which could be found in places where this tradition had not been so thoroughly suppressed or cut off).

So both lines are very important to our understanding of the myths and what they have to tell us.

The "Thirteen Postures Song" itself is actually talking about the cultivation not only of prowess at martial arts or kung fu but also about the cultivation of *chi*, and of the connection with the Invisible Realm – as (I argue) are the ancient myths of the human race.

One additional aspect in the poem which points towards that Infinite Realm, which I had not noticed back in August 2015, is

615

the fact that the number of the "Thirteen Dynamics" themselves comes from the "five directions" and the Eight Angles (of the Bagua, which contains the Eight Divinatory Trigrams – divination being rather indisputably involved with connecting to something beyond the purely material realm).

That much I did mention in the blog post linked above – but note that five and eight (adding to thirteen) also create what we call a Fibonacci progression (5, 8, 13 . . . the next in the series would be 21, etc). This connects to the concept of the Golden Ratio and to Phi – and to the Infinite Realm as well (Phi being a numerical concept which, like Pi as well, "touches infinity" in that it goes on forever without any known repetition).

Much more could be said here, but the main point is this: the cultivation of the wisdom and the Way involves the Infinite Realm, it is usually conveyed to us through esoteric means, which requires some initial guidance "from the mouth" (traditionally, this has been in the form of a "master-to-disciple" relationship), but that ultimately each man or woman must "go in" themselves, gain the *gnosis* themselves -- no one else can do it for them.

I believe there is much to meditate upon in these words from the Tai Chi Classic text (and indeed, to meditate upon *daily*, as the "Thirteen Postures Song" in fact instructs us to do).

There is no member of mine devoid of a god

2017 May 25

In a passage I have cited many times before, from a 1936 lecture entitled *The Stable and the Manger*, Alvin Boyd Kuhn declares that the ancient myths and scriptures "mean nothing as outward events" (that is, as supposed narratives of literal events in terrestrial history) but rather that

> they mean everything as picturizations of that which is our living experience at all times. The actors are not old kings, priests and warriors; the one actor in every portrayal, in every scene, is the human soul.

All of the myths and sacred stories, according to Kuhn's analysis, deal with one theme: "The epic of the human soul in earthly embodiment" (*Lost Light*, 67). In all of the varied episodes, he maintains, whether wrestling with serpents or being tempted by Sirens or struggling to return home, "the history of the divine Ego in its progress from Earth back to the skies was allegorically portrayed" (67).

The profound ramifications of this perspective upon the ancient myths may be easy to overlook, unless we take the time to let the weighty import of these assertions sink in. If Kuhn is correct (and

I believe that he is), then it means that you and I and every other man or woman or child that we ever meet are much more than we have been conditioned to believe.

In his view, the world's ancient wisdom is declaring again and again that each and every human soul is in fact divinity "buried in clay," or "chopped up" like Osiris and scattered across the country, or entwined in "the coils of a serpent," or "shorn of power" like Samson – but that our experience here in this "lower passage" involves the recovery of that "buried" aspect of our nature, and the "re-integration" or "reclamation" of that divine aspect which has been temporarily lost or forgotten.

In support of his assertion, Kuhn points to a powerful passage from the ancient Egyptian *Ritual of the Coming Forth by Day* (also commonly referred to in modern times as the Egyptian Book of the Dead). There, in the 42nd chapter (illustrated in what is known as "Plate 32" or "Sheet 32" of the version inscribed upon the famous version known as the Papyrus of Ani, which is seventy-eight feet in length and beautifully illustrated), we read the following lines (as translated by E. A. Wallis Budge in 1901, with occasional clarification from the 1972 translation by Raymond O. Faulkner added in brackets):

> The hair of Osiris Ani, triumphant, is the hair of Nu [or Nun]
> The face of Osiris Ani, triumphant, is the face of Ra
> The eyes of Osiris Ani, triumphant, are the eyes of Hathor
> The ears of Osiris Ani, triumphant, are the ears of Ap-uat [or Wepwawet]
> The lips of Osiris Ani, triumphant, are the lips of Anpu [or Anubis]
> The teeth [or molars] of Osiris Ani, triumphant, are the teeth of Serqet [or Selket]
> The neck of Osiris Ani, triumphant, is the neck of Isis
> The hands of Osiris Ani, triumphant, are the hands of Ba-neb-Tattu
> The shoulder of Osiris Ani, triumphant, is the shoulder of Uatchet
> The throat of Osiris Ani, triumphant, is the throat of Mert

The forearms of Osiris Ani, triumphant, are the forearms of the Lady of Sais

The backbone of Osiris Ani, triumphant, is the backbone of Set

The chest of Osiris Ani, triumphant, is the chest of the lords of Kheraha

The flesh [or chest] of Osiris Ani, triumphant, is the flesh of the Mighty One of Terror [or He who is Greatly Majestic]

The reins and back [or belly and spine] of Osiris Ani, triumphant, are the reins and back of Skehet [or Sekhmet]

The buttocks of Osiris Ani, triumphant, are the buttocks of the Eye of Horus

The phallus of Osiris Ani, triumphant, is the phallus of Osiris

The legs [or thighs and calves] of Osiris Ani, triumphant, are the legs of Nut

The feet of Osiris Ani, triumphant, are the feet of Ptah

The fingers of Osiris Ani, triumphant, are the fingers of Orion

The leg-bones [or toes] of Osiris Ani, triumphant, are the leg-bones of the living uraei [the rearing cobras of the goddess Wadjet]

There is no member of my body which is not the member of some god ["There is no member of mine devoid of a god"]

The god Thoth shieldeth my body altogether ["And Thoth is the protection of all my flesh"]

What this profound passage appears to be teaching us is that we are in some way intended to incorporate the powers of the gods into our lives -- and that the gods who dwell in the Invisible Realm somehow have their home and exercise their power through the persons of men and women in this realm.

In *Lost Light*, on page 549, Alvin Boyd Kuhn interprets this very passage, saying:

> The god himself, fallen into carnal mire, buried and inert, had to be raised and restored to sound condition. As he awakened his faculties and sloughed off the imprisoning vesture of decay, it was as if every member of his body was resuscitated and made over. [. . .] This work is gradual and is accomplished piecemeal. The god finds glorification coming day by day, feature by feature; he is reconstituted limb by limb, member by member, until he says there is no part of him that remains mortal. He is given the hair of

Nu or heaven (solar rays); the eyes of Hathor; the ears of Apuat; the nose of Khenti-Kas; the lips of Anup; the teeth of Serkh; the neck of Isis; the hands of the mighty lord of Tattu; the shoulders of Neith; the back of Sut; the phallus of Osiris; the legs and thighs of Nut; the feet of Ptah; and the nails and bones of the living Uraei "until there is no limb of him that is without a god." "My leg-bones are the leg-bones of the living gods. There is no member of my body that is not the member of some god. I am Yesterday, and Seer of Millions of Years is my name." Here is notice to man that he must traverse every kingdom in order that he may absorb and embody in himself every aspect of nature's power, the efficacy of every god. Mighty truth is this. 549 - 550.

In this explication of the passage, Kuhn echoes Gerald Massey, who wrote in *Ancient Egypt: the Light of the World* (Volume One, published in 1907) that:

Before the mortal Manes could attain the ultimate state of spirit in the image of Horus the immortal, he must be put together part by part as was Osiris, the dismembered god. He is divinized in the likeness of various divinities, all of whom had been included as powers in the person of the one true god, Neb-er-ter, the lord entire. Every member and part of the Manes in Amenta has to be fashioned afresh in a new creation. The new heart is said to be shaped by certain gods in the nether world, according to the deeds done in the body whilst the person was living on earth. He assumes the glorified body that is formed feature by feature and limb after limb in the likeness of the gods until there is no part of the Manes that remains undivided. He is given the hair of Nu, or heaven, the eyes of Hathor, ears of Apuat, nose of Khenti-Kas, lips of Anup, teeth of Serk, neck of Isisi, hand of the mighty lord of Tattu, shoulders of Neith, back of Sut, phallus of Osiris, legs and thighs of Nut, feet of Ptah, with nails and bones of the living Uraei, until there is not a limb of him that is without a god. 198 - 199.

While Kuhn admired the work of Massey and praised him as one of the most discerning and insightful to have studied ancient Egypt, he disagreed that the passages have to do with the soul in the afterlife -- rather, Kuhn argues, the soul in this incarnate life (which is indeed "the Underworld," compared to the intangible Realm of Spirit) is indicated. Thus, we are not to understand that we must incorporate the attributes and powers of the various gods and goddesses *in the next life*, but rather in this one.

As Kuhn says, "Mighty truth is this" -- and a profound mystery.

But it is clearly and forcefully proclaimed in the ancient Egyptian texts -- and it opens an entirely new paradigm on our mission and purpose here in this incarnate life.

And it opens an entirely new paradigm upon the way we treat others and structure our society -- because it means that the gods in a very real sense can be understood to be present in those around us (just as they are also present in ourselves).

In what ways is our society -- which itself is a product of a civilization that has been violently cut off from the world's ancient wisdom for at least seventeen centuries -- dishonoring the gods or even *stealing from them* in the way that it treats individual men and women in whose members the gods manifest (according to the ancient texts)?

In what ways do we steal from the gods when we do not teach these truths to others, or allow them to be known -- and when we do not pursue the kind of integration *in ourselves* that is described in the ancient Book of the Coming Forth by Day?

I am quite convinced that the world's ancient wisdom was given to humanity for our benefit and blessing -- and that their ancient teaching belongs to each and every one of us.

Because, as the Papyrus of Ani illustrates and declares, the gods and goddesses are properly present in us: in you, and in me, and in everyone you ever meet.

Humanity's Forgotten History

Did mankind know the precise size and shape of our earth many thousands of years ago?

2012 June 06

Did mankind understand the size of the spherical earth to an astonishing degree of precision many thousands of years ago? Here is a link to a fascinating discussion of the analysis presented by Jim Alison at his website, entitled "The Prehistoric Alignment of World Wonders: A New Look at an Old Design." http://home.hiwaay.net/%7Ejalison/index.html

Mr. Alison presents evidence that ancient sites including the Giza pyramids, Ollantaytambo, Nazca, Easter Island, and many others are all located along a single great circle centered on a point in Alaska. This is a startling assertion, as these sites are clearly very widely distributed geographically, and the massive monuments located at each site, while sharing some remarkable similarities, are also thought to have been the products of cultures separated in time by many centuries or even millennia.

Even more remarkable are the relationships that he demonstrates between many of these sites, such as his finding that Ollantaytambo (in Peru) is 108° along the great circle from Giza (in Egypt). As he points out, 108 is an extremely significant number, and the location of these historic sites 108° apart is unlikely to be a coincidence (the idea that all the location of all these sites along a great circle of the globe could be a coincidence is likewise extremely unlikely -- when we find that two of the sites are separated by 108° along a global great circle, the possibility of coincidence becomes even more remote).

The number 108 is an important precessional number, as is the number 72, which Mr. Alison demonstrates to be operating between Giza and other sites. He notes in a different article on the same subject, this one published on Graham Hancock's website (http://www.grahamhancock.com/forum/AlisonJ1-p1.htm) that Giza and the site of Angkor Wat in Cambodia are very nearly 72° apart along the same great circle, and that Easter Island

624

(Rapa Nui) is located at a point on the circle that is very nearly equal angles of arc to Angkor going in one direction and Giza going in the other. Mr. Hancock also discusses the significant separation angles of various sites around the globe (including Rapa Nui, Angkor, and Giza) in his gorgeous book *Heaven's Mirror* (and movie of the same name).

Also extremely significant is the fact that Easter Island / Rapa Nui is located directly across this great circle from the Indus Valley. This relationship (as well as the one discussed above for Angkor, Rapa Nui, and Giza) can be easily seen in the diagram of the great circle on this page of Mr. Alison's essay (http://home.hiwaay.net/%7Ejalison/easter2.html -- bottom diagram on that page). The mysterious (and as-yet undeciphered) writing systems of both of these sites (Rapa Nui and the Indus Valley) are almost certainly related. The undeciphered script of the Indus Valley is generally known as the "Indus Valley Script" and the undeciphered writing of the tablets found on Easter Island (preserved, in fact, by the people living there when the first European ships arrived) is known as "Rongo Rongo" (and the tablets known as "Rongo Rongo Tablets"). The similarities between these mysterious writing systems -- separated by half the globe along this great circle discovered by Mr. Alison -- can be clearly seen on this web page comparing the two writing systems: http://www.ancient-wisdom.co.uk/easterislandindusvalley.htm.

The idea that these sites could be completely unrelated and just somehow happen to lie upon the same great circle of the globe -- at significant intervals on that circle, including angles of arc corresponding to clear precessional numbers – is ridiculous. When evidence such as the similarity of the Rongo Rongo Tablets to the Indus Valley Script is added to the discussion, the case for a connection becomes even stronger. Also note the very strong evidence at Ollantaytambo for a connection with other ancient sites across the oceans which was discussed in this previous post: http://mathisencorollary.blogspot.com/2012/04/face-at-ollantaytambo.html.

Not only does this analysis by Mr. Alison argue for an ancient connection of some sort between all these sites, but also that mankind knew the size and shape of the earth in great antiquity -- long before the conventional timeline of human history says that humans could have had such precise scientific knowledge. As he says on page 10 of the article on Graham Hancock's website:

> Many similarities between these sites have been well documented, including the use of perfectly cut and precisely placed monolithic stones, exact orientations to the cardinal points and astronomical orientations. The prevailing view of world history dismisses these similarities as coincidental developments of separate stone-age cultures. Unless it is also a coincidence that these sites are located at mathematically and geometrically significant points on a single line around the center of the Earth, it may be time to reconsider the idea that Europeans of the present era were the first to know the size and shape of the Earth.

Here is a different website (http://ancientcartography.net/geoAN.html) which refers to the work of Mr. Alison, presenting other evidence that an ancient culture or cultures knew the size and shape of the earth quite well, and deliberately positioned significant sites around the globe as a sort of world-wide web of coordinates or reference points.

Mr. Alison presents even more additional evidence to support the conclusion that the ancients knew the size and shape of the earth with great precision. He points to other great circles which can be drawn through other significant ancient sites, such as another great circle that connects the Great Pyramid and the Serpent Mound of Ohio and other ancient significant sites (Ross Hamilton presents extensive analysis using completely different evidence which appears to support a connection between the Great Pyramid and the Serpent Mound in his amazing book *The Mystery of the Serpent Mound: in Search of the Alphabet of the Gods*).

626

All of this analysis appears to be very strong supporting evidence for the conclusion that mankind in the very ancient past had a very sophisticated level of knowledge and scientific achievement (and that they could and did travel over the entire globe), an assertion put forward in the *Mathisen Corollary* book as well (supported by different evidence and analysis) as well as in many other previous posts on this blog. While the monuments currently located in some of those sites may have been erected much later, it is fairly clear that many of those points along the great circles found by Mr. Alison (including Giza, Malta, and the Indus Valley) are of extreme antiquity, and thus it is quite likely that this system of worldwide coordinates is very ancient indeed.

The discovery of the Nag Hammadi library

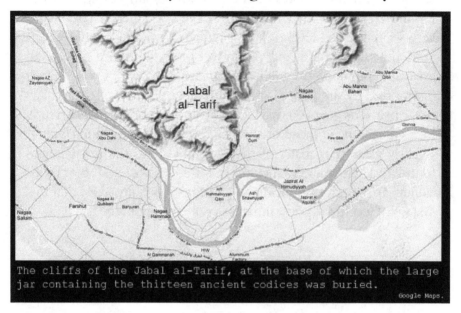

The cliffs of the Jabal al-Tarif, at the base of which the large
jar containing the thirteen ancient codices was buried.

Google Maps.

2012 December 31

It was sometime in December of 1945 when three brothers,
fellahin from the village of al-Qasr unearthed an ancient jar at the
foot of the cliffs of the Jabal al-Tarif west of the village of Hamrah
Dun. Although they could not know it at the time, it was a
discovery of enormous import.

The jar contained texts written on papyrus and bound into codex
form, and the contents of those codices turned out to be texts that
were most likely declared to be heretical in the fourth century AD
and were thus buried to escape detection and destruction --
destruction that was so thorough that the teachings in these texts
were almost completely eradicated and could only be pieced
together by inference from writings by those who were on the
side of those ordering their destruction and who were
denouncing the doctrines that were preserved in the codices
inside this ancient jar.

The discovery, then, opened a window onto a part of the ancient
world that had been sealed off for centuries, an imperfect window

628

to be sure, but since all the other windows onto that view had been deliberately smashed and bricked over (so to speak), it was an important window indeed.

These texts would come to be called the "Nag Hammadi library," and they would take a rather circuitous path to publication and translation, but they are now available to the public (since 1975) and can be found online in various places, such as here: http://gnosis.org/naghamm/nhl.html

The story of their discovery has been told by Professor James M. Robinson, who was responsible for tracking down the discoverers and determining the approximate date of their discovery (which had not been previously specified beyond a range of years), as well as for much of the analysis of the texts and their significance. He spoke in person with the field hand who discovered the texts, and in an essay entitled "Nag Hammadi: the First Fifty Years" published in the Proceedings of the 1995 Society of Biblical Literature Commemoration, explains that the three brothers, named Muhammad Ali (the eldest brother), Khalifa Ali, and Abu al-Magd, were digging for fertilizer in the talus at the foot of the Jabal al-Tarif. From his interviews with Muhammad Ali, Professor Robinson discovered:

> When the local sugarcane harvest was over and the land lay fallow during the brief winter, he regularly dug soft earth at the foot of the cliff that served as fertilizer for the fields. He had been digging fertilizer, he recalled, just a few weeks before the Coptic Christmas, which is January 7, when he made the discovery. [. . .]

> Muhammad Ali at first feared to open the jar (sealed with a bowl attached with bitumen to the mouth of the jar) lest it contain a jinn. But then it occurred to him it might contain gold. This gave him courage enough to break it with his mattock. Out flew, into the air, what he thought might be an airy golden jinn, but which I suspect was only papyrus fragments. He was very let down to find only worthless old books in the jar.

629

He tore some up to share with some of the other camel drivers who were present, which explains some of the damage and loss which does not fit the pattern of what one would expect from the gradual deterioration of the centuries. Since the other camel drivers, no doubt out of fear of Muhammad Ali, declined his insincere offer to share, he stacked it all back up together, unrolled his turban from around his head, put the codices in it, slung it over his shoulder, unhobbled his camel, drove back home, and dumped the junk in the enclosed patio in his house where the animals and their fodder were kept. His mother confirmed to me that she had in fact burnt some along with straw as kindling in the outdoor clay oven.

We will probably never know what was on those texts that helped light the outdoor clay oven. However, scholars have since determined that the jar contained thirteen codices (the twelfth was probably sacrificed as kindling, according to the analysis done by Professor Robinson, and only one text and the opening of another text from the thirteenth survive, having anciently been stuffed inside the cover material of the sixth codex), and that the surviving library of texts number forty-seven (not counting additional fragments and duplicates). They have been given the name "the Nag Hammadi texts" or the "Nag Hammadi library" after the largest village near the cliffs where they were found.

Nag Hammadi is located near the dramatic bend in the Nile just north of Luxor, which in ancient Egypt was called Thebes, the mighty ancient capitol of Upper Egypt (southern Egypt, the "upstream" and thus "upper" portion of Egypt, since the Nile flows south-to-north).

Getting back to the incredibly important texts found buried in that ancient jar at the base of the Jabal al-Tarif, they have been generally categorized as reflecting a "gnostic" understanding of the origin of mankind and our purpose here in this life, a perspective that is at odds with the understanding that would become the teaching of orthodox Christianity and which would therefore be violently declaimed against and apparently was stamped out. The texts are described by Marvin Meyer in his 2005 book *The Gnostic Discoveries: The Impact of the Nag Hammadi Library* as follows:

> Research on the Nag Hammadi library and the Berlin Gnostic Codex [also found in Egypt but in the late 1800s, which appears related in its contents to the Nag Hammadi texts and which contains four texts] has disclosed a broad spectrum of perspectives among the texts that may be identified as gnostic or gnosticizing, and the texts seem to fall roughly into five groups. These five groups may reflect, for several of the groups, gnostic schools of thought embraced by teachers and students in communities.

> The first group of gnostic texts in the Nag Hammadi library consists of the Thomas texts: the Gospel of Thomas, the Book of Thomas, and probably the Dialogue of the Savior. [. . .]

> The second group of texts derives from the Sethian school of thought. Sethian texts reflect traditions of Seth, son of Adam and Eve, as a paradigmatic human being. [. . .]

> The third group of gnostic texts represents the Valentinian school of thought. Valentinus was a second-century Egyptian who became a Christian gnostic teacher and preacher in Alexandria and Rome. [. . .]

> The fourth group of gnostic texts in the Nag Hammadi library comes from the Hermetic heritage. The Hermetic tradition has been known for a long time, and Hermetic texts, collected in the Corpus Hermeticum, have assumed a prominent place in discussions of mystical religion in antiquity and late antiquity. [. . .]

> [. . .] Within the texts of the Nag Hammadi library there are three Hermetic texts, two previously known, an excerpt

631

from the Perfect Discourse and the Prayer of Thanksgiving, and one new Hermetic text, the Discourse on the Eighth and Ninth.

The fifth group of gnostic texts in the Nag Hammadi library and the Berlin Gnostic Codex is hardly a definable group, but instead consists of those gnostic texts that defy classification. These texts seem to incorporate leading gnostic themes, as suggested above, and may show similarities to other gnostic texts and traditions, but they do not fit neatly into the other groups of gnostic texts. 48 - 52.

Such is the categorization suggested by Marvin Meyer and other scholars. Others may perhaps organize or categorize them differently. However they are categorized, their significance is profound on many levels and for many reasons. First, as Marvin Meyer explains elsewhere in the same book, "Prior to the discovery of the Nag Hammadi library, 'gnosticism' typically was considered to be an early and pernicious Christian heresy, and much of our knowledge of gnostic religion was gleaned from the writings of the Christian heresiologists, those authors who attempted to establish orthodoxy and expose heresy in the early church. [. . .] Since the discovery of the Nag Hammadi library and related texts, the study of gnostic religion and its impact upon ancient and modern religion has been fundamentally transformed" (1-6).

Second, the deliberate burial at the base of a cliff after sealing the texts into a jar suggests that those who valued these texts were hiding them from those who wanted to suppress or even destroy them, and this brings up the entire theme of the destruction of ancient knowledge which has appeared numerous previous posts. Marvin Meyer provides evidence that these texts may have been buried upon the publication of the Festal Letter of Athanasius, Archbishop of Alexandria, in AD 367 (also known as the 39th Festal Letter (Meyer 30-31). This letter lists the texts considered canonical and condemns as heretical those that are "an invention of heretics."

632

Finally, the texts have great importance to us on their own merit, for the light they may shed upon the meaning of human existence and the nature of human consciousness. For this, the reader is encouraged to examine them for himself or herself.

There is also the matter of that mysterious report from the discoverer of the long-lost jar containing the library, who said that when he broke open the jar, out flew what "might be an airy golden jinn," but which Professor Robinson says he suspects was "only papyrus fragments." I wonder . . .

Angkor Wat, Giza, Paracas and the World-Wide Grid

2014 February 10

In their outstanding book *Heaven's Mirror*, Graham Hancock and Santha Faiia point out the undeniable fact that Angkor Wat is located seventy-two degrees east of Giza in Egypt (page 254).

Seventy-two is an important precessional number. It is highly unlikely that the site of Angkor Wat would be located such an important number of degrees of longitude from the site of the Great Pyramid at Giza simply by accident, especially because both the Great Pyramid and the art and architecture of Angkor Wat deliberately employ precessional numbers and symbology in their construction and (in the case of Angkor Wat) in their symbology, as Graham Hancock discusses in *Heaven's Mirror* and other books (including *Fingerprints of the Gods*).

The fact that Angkor Wat bears a name which is made up of two sacred Egyptian words -- Ankh and Horus -- makes the connection between Giza and Angkor even more certain to be deliberate and not a coincidence. Further, as Joseph P. Farrell and Scott D. de Hart discuss in their book *Grid of the Gods*, both sites contain pyramid structures; the pyramids at Angkor (shown above in an image from Wikimedia commons) happen to be shaped differently than the pyramids of Giza (being taller than they are wide, unlike those at Giza), but that may well be because their purpose is different, as Drs. Farrell and de Hart discuss in their book.

The fact that these two sites are separated by seventy-two degrees of longitude is an important piece of evidence supporting the assertion made by Drs. Farrell and de Hart, and by Graham Hancock and others, that the ancient astronomically-aligned sites on our planet are part of a global "grid" laid out by an ancient civilization or civilizations possessed of deep wisdom and sophisticated scientific understanding.

Interestingly enough, *Heaven's Mirror* also reveals that one hundred eight degrees to the *west* of Giza lies another mysterious site: Paracas, on the Pacific coast of Peru. One hundred eight is one of the most important and widely used precessional numbers in the mythology and traditions of the world. It is, of course, 1.5 times 72. Multiples of 108 which are found in numerous ancient myths and legends include 216, 432, and 540.

Paracas is located in the Ica region of Peru, home to the mysterious elongated skulls which were recently in the news due to the results of DNA tests which are discussed in this article on the *Ancient Origins* website. Paracas is also the home of an ancient geoglyph known as the "Paracas Candelabra" (shown below in an image from Wikimedia commons), which may reflect the constellation of the Southern Cross according to Graham Hancock's analysis in *Heaven's Mirror*.

As Graham Hancock also points out in Fingerprints of the Gods, the ability to measure longitude accurately eluded "western civilization" until the 1700s, when John Harrison finally developed a chronometer accurate enough for precise longitudinal calculation (see discussion here regarding the "Longitude Prize"). The fact that Giza is seventy-two degrees west of Angkor Wat and one hundred eight degrees east of

Paracas suggests that the ancients had a way of accurately measuring the globe.

The fact of an ancient world-wide grid is not simply a piece of "gee-whiz" trivia. As Drs. Farrell and de Hart discuss in the book linked above, such a grid may have been constructed in order to harness the enormous power of the earth, and potentially the power created by the motions of other planets in our solar system as well. The fact that we today know little or nothing about such a grid -- and the fact that conventional academia reflexively suppresses such information and ridicules and marginalizes those who choose to investigate the evidence regarding this and other mysteries of mankind's ancient past -- suggests the possibility that vital information about our planet and our relationship to has been deliberately concealed for centuries.

The Smithsonian Cover-up

2014 April 27

John Wesley Powell (1834 - 1902, image above from Wikimedia commons) was made the first director of the Bureau of Ethnology in the United States in 1879, which was established that same year by an Act of Congress, a position he held until his death in 1902.

That bureau, which changed its name to the Bureau of American Ethnology in 1897, was directly connected to the Smithsonian Institution, which had been established in 1846 through the will of the British chemist James Smithson (1765 - 1829) and funded by his bequest of 105 sacks of about 1,000 gold sovereigns each, and pursued the mission of organizing all the anthropological research in the nation.

In his first year as head of the Bureau of Ethnology, Powell submitted the first of his Annual Reports of the Bureau of Ethnology to the Secretary of the Smithsonian Institute, dated July 1880 and covering the Bureau of Ethnology's efforts for 1879-1880. The entire report can be seen online in Google Books among other places.

Beginning on page 73 of that publication is a famous essay by Powell entitled "On Limitations to the Use of Some Anthropologic Data." In it, Powell sets forth the doctrine which would become the guiding principle of his Bureau of American Ethnology and of the Smithsonian at large all the way through the present day, a strictly isolationist doctrine which flatly declares that it is "illegitimate" to entertain any line of analysis which attempts to connect any artifacts found in the New World with any "peoples or so-called races of antiquity in other portions of the world."

A reproduction of the letter, with the passages emphasizing this isolationist doctrine highlighted in yellow, can be found online here as well:

http://www.scienceviews.com/lostcivilizations/powelldoctrine.html

The motivations behind this strict imposition of the isolationist paradigm and flat rejection of the examination of any possibility of diffusionist explanations (which propose the possibility that there was contact across the oceans prior to the arrival of Columbus) can and have been debated. Many biographers and vignettes emphasize the "tremendous respect" Powell had for the native tribes of North America and some have suggested that his support for an isolationist doctrine was based upon that respect for the Native Americans and the view that any theory proposing ancient pre-Columbian contact with "peoples or so-called races of antiquity in other portions of the world" must automatically be disrespectful to the native peoples here, or even based upon some kind of racist animus.

It is certainly likely from some of the episodes of Powell's life that he did in fact have tremendous respect for the Native Americans. However, it is undeniable that Powell's own 1879 essay displays some extremely paternalistic and disrespectful generalizations, including his assertions in the second part of the essay (entitled "Picture Writing," beginning on page 75 of the above-linked version of the 1879 report) that the "pictographs" found in North America are "simply the beginning of pictorial

638

art" and in almost all cases "simply mnemonic" -- possessed of no systematic or as Powell calls it, "conventional," structure by which ideas could be preserved using symbols that possessed a common meaning agreed upon by all who understood that system (i.e., whose meaning was agreed-upon by *convention* across a large number of people, thus constituting a writing-system).

In this astonishing denial of the existence of writing systems, Powell explicitly includes even the obvious writing-systems of the Maya and the Inca and other cultures of Central and South America, whose artifacts were by no means unknown to him and to the other employees of the Bureau of Ethnology (in fact, the 1879 report contains long sections dealing with "Central American Picture Writing," and many of the other annual reports discuss the artifacts and culture of the Maya and Inca and other civilizations in detail). Nevertheless, Powell asserts in his letter that:

> To some slight extent pictographs are found with characters more or less conventional, and the number of such is quite large in Mexico and Central America. Yet even these conventional characters are used with others less conventional in such a manner that perfect records were never made.

Such a statement is extremely paternalistic, and effectively denies the existence of any true systematic writing systems, even among the cultures of Mexico and Central America! Based upon this false assertion, Powell then declares: "Hence it will be seen that it is illegitimate to use any pictographic matter of a date anterior to the discovery of the continent by Columbus for historic purposes." By this declaration, Powell effectively discarded any and all artifacts containing writing from consideration of historic analysis, and in doing so protected his earlier declaration that any contact with peoples from "other portions of the world" is plainly "illegitimate."

Thus, none of the numerous inscriptions and artifacts which clearly attest to the possibility of ancient contact -- many of which have been discussed in previous posts on this blog and many more of which have been detailed in numerous published books -- could be considered as evidence which might challenge the isolationist dogma. Some of those artifacts containing evidence of writing which strongly supports the possibility of ancient contact are discussed in the following previous posts (not included in this collection, but available online):

- "Have you heard of this artifact? The Grave Creek Stone"
- "Chief Joseph, 1840 - 1904"
- "Beltane and ancient inscriptions in the New World"
- "Ogham inscriptions in Colorado"
- "The staggering implications of the ancient inscriptions at Hidden Mountain near Los Lunas, New Mexico"
- "The case of the Micmac hieroglyphs: a powerful blow to isolationist theories"

And there are hundreds of other examples which could be discussed in addition to the evidence discussed in those posts. To simply refuse to consider any such evidence at all is unscientific to the extreme, and yet it has been the implicit or explicit policy of the Smithsonian since the days of John Wesley Powell.

The fact that the Smithsonian has not changed their policy of refusing to consider any artifacts which might suggest the possibility of pre-Columbian trans-oceanic contact with the "New World" is evident from the controversy over the Bat Creek Stone found in 1889 in Tennessee, which the Smithsonian recently (early in 2014) called "an obvious fraud" in their response to Scott Wolter's discussion of the artifact on his *America Unearthed* program on the History Channel. Scott Wolter's response to the Smithsonian's dismissive belittling of his examination of the Bat Creek Stone, and their *ad hominem* attacks on Wolter himself as lacking in "qualifications and reputation as a researcher," can be seen here. His response also includes expressions of regret towards the Smithsonian's dismissal of the Bat Creek Stone from representatives of the Cherokee people, who did possess a system of writing and who told the Smithsonian that if they are so sure that the stone is a fraud, the Cherokee can take the stone back and rebury it where it was found out of respect to those who originally produced it.

Recently, a new aspect of the Smithsonian's policy of refusing to countenance any artifacts that might pose a challenge to Powell's "doctrine" of isolationism has received a lot of publicity in light of the publication of Richard Dewhurst's new book *Ancient Giants Who Ruled North America: the Missing Skeletons and the Great Smithsonian Cover-Up*. Richard Dewhurst used the capabilities of modern search engines to examine the archives of US newspapers going back to the early 1800s and found hundreds of published descriptions of giant skeletons being unearthed across the North American continent, many of them containing photographs.

He also found evidence that, while the Smithsonian in its early years was an enthusiastic documenter of such discoveries, the arrival of John Wesley Powell marked a dramatic change in the Smithsonian's attitude and policy towards such finds, to such an extent that Dewhurst was forced to conclude that: "What my research has revealed is that the Smithsonian has been at the center of a vast cover-up of America's true history since the 1880s" (3). He documents numerous cases in which representatives from the Smithsonian arrived on the scene of any reported discoveries of giant skeletons with remarkable rapidity (sometimes within one or two days, even in the late 1800s and even when the archaeological find was in remote regions of the American west) and in which skeletons reported as being turned over to the Smithsonian were never seen again.

Today, if one searches the internet for the terms "Smithsonian cover up," the predominant results will have to do with the cover-up of giant skeletons. Richard Dewhurst believes that the motives for what he calls the "Powell doctrine" of suppressing and denying any archaeological evidence that could indicate the presence of other ancient peoples in the Americas or contact with ancient cultures from across the oceans may have sprang from the fact that John Wesley Powell's father was a Methodist preacher in Palmyra, New York (Powell himself was obviously named after John Wesley, 1703 - 1791, the founder of Methodism), where Joseph Smith first published the Book of Mormon in 1830 and where the early enthusiasm of the people of the area for the new revelations caused Powell's father to lose his congregation (as Richard Dewhurst explains in a footnote on page 6 as a likely motive for Powell's animus towards any diffusionist theories).

Richard Dewhurst also believes that the distasteful US policy of "Manifest Destiny" and the efforts of the federal government following the Civil War to seize the territory to the west of the Mississippi and to suppress the Native Americans who lived there played a role in the Smithsonian's (and Powell's) desire to characterize the native peoples of the continent as primitive barbarians, incapable of producing anything more than "the most rudimentary picture making," (Dewhurst, 6). Dewhurst proposes that such a doctrine may have been deployed in order to help convince the population to support the aggressive plans to exploit the lands of the Native Americans.

If so, then the "Powell doctrine" probably did not originate with Powell himself, but would have likely been the determined policy of a number of other government officials. At the front of Powell's first annual report (containing his essay declaring as "illegitimate" any attempts to connect any artifacts found in the New World with cultures from anywhere else) is an introductory letter from Powell to Spencer F. Baird, the Secretary of the Smithsonian Institute, in which Powell says the following:

> Sir: I have the honor to transmit herewith the first annual report of the operations of the Bureau of Ethnology.
>
> By act of Congress, an appropriation was made to continue researches in North American anthropology, **the general**

direction of which was confided to yourself. As chief executive officer of the Smithsonian Institution, you entrusted to me the immediate control of the affairs of the Bureau. This report, with its appended papers, is designed to exhibit the methods and results of my administration of this trust.

If any measure of success has been attained, it is largely due to **general instructions received from yourself and the advice you have ever patiently given me on all matters of importance.**

I am indebted to my assistants, whose labors are delineated in the report, for their industry, hearty co-operation, and enthusiastic love of the science. Only through their zeal have **your plans** been executed.

Much assistance has been rendered the Bureau by a large body of scientific men engaged in the study of anthropology, some of whose names have been mentioned in the report and accompanying papers, and others will be put on record when the subject-matter of their writings is fully published.

I am, with respect, your obedient servant,
J.W. POWELL

(**Bold emphasis** added by me).

While this introductory and dedicatory letter may simply be an example of "polite formalities" or conventional platitudes within a government bureaucracy, in keeping with the style and traditions of the period, it is also possible in light of the topic being discussed that it contains evidence that Powell's doctrine did not originate with Powell himself, but was part of a policy transmitted by the Secretary of the Smithsonian Institution whose office was in Washington, DC, and of other men in Washington as well. The bold-print areas (all boldface "highlighting" is my own and is not found in the original document) seem to support such a possibility, with Powell referencing a "general direction" which "was confided" to the Secretary of the Smithsonian Spencer Baird by some unnamed parties (presumably parties connected with Congress, whose authorizing act for the creation of the Bureau of Ethnology was mentioned immediately prior to this mysterious assertion), and "the advice you have ever patiently given me on all matters of importance," and his declaration that "your plans [have] been executed."

The likelihood that what Dewhurst calls "the Powell doctrine" has roots far deeper than Powell himself (or even Powell's animus towards diffusionist theories due to the loss of his father's congregation) is evident from the fact that the Smithsonian's policy of refusing to entertain any possibility of ancient contact across the oceans and its haste to declare any artifacts containing inscriptions which might employ the known writing systems of ancient Mediterranean cultures as frauds or hoaxes has continued long after the death of John Wesley Powell, and continues to this day.

This continuing refusal to examine artifacts containing inscriptions such as those mentioned in the list of previous posts above and reflexive labeling of such artifacts as either fraudulent or the products of post-Columbian contact cannot be explained by the Powell family's personal experiences in Palmyra, New York. Nor, it seems, can the benighted and repulsive nineteenth-century belief in "Manifest Destiny" be the reason that the Smithsonian continued to enforce the "Powell doctrine" throughout the twentieth century, long after the United States had seized all of the lands of the Native Americans between the Mississippi and the Pacific Ocean, and most of the citizens of the country had forgotten that their land had once belonged to someone else. Is it possible that there is some other motive which lies behind the Smithsonian's ongoing policy of anti-diffusionism?

Personally, I am not an expert on the "giant skeletons" controversy. While it certainly seems, based upon the prodigious volume of reports and descriptions and even photographs, that such skeletons have been found throughout the Americas in some numbers, and that the absence of any such skeletons on display at the Smithsonian National Museum is suspicious, I also believe it is a mistake to focus entirely on giant skeletons when talking about a "Smithsonian cover-up."

The easiest way for defenders of the Powell doctrine to deflect such cover-up arguments is to argue that such "giant skeletons" were simply the remains of some isolated individuals exhibiting traits of giantism, to point out that enthusiasm over giants and the possibility of ancient trans-oceanic contact was rife in the nineteenth century (much of it fueled, it must be noted, by religious agendas and a desire to support literalist interpretations of the Bible or by the newly published Book of Mormon), and to argue that whatever skeletons may have been uncovered in those early decades were lost or crumbled to dust and were not maliciously squirreled-away in the bowels of the Smithsonian's warehouses.

I certainly do not agree that these counter-arguments settle the case, and believe that Richard Dewhurst's analysis of the evidence of giant remains (and other such analysis by other researchers, such as the analysis in this essay found in several places on the web) is extremely valuable and worthy of careful consideration. I also believe that all dogmatic declarations that the facts of the matter are settled and that no further analysis is legitimate (*whatever* the subject) should be treated with great suspicion. Nevertheless, I also believe that the "giant skeleton" aspect of the "Smithsonian cover-up" question could become a huge red herring which falsely divides the debate in the eyes of the general public into two camps, those who believe America was once home to a race of giants, and those who generally side with the Powell doctrine.

The Powell doctrine excludes a whole lot more evidence than giant skeletons, as the recent Bat Creek Stone controversy demonstrates. There is abundant evidence that there was ancient contact across the oceans, most of it involving human beings of what we might call "normal" (or at least non-gigantic) stature. As far as I know, no one is maintaining that the giants whose skeletons have been found throughout the Americas were the authors of inscriptions using known "Old World" writing systems including Hebrew, Egyptian (both hieroglyphic and hieratic), Phoenician/Punic, Ogham, cuneiform, runic, Iberian, Libyan, and Roman, but many of these have been found in the Americas and conventional scholars either ignore them, declare them to be frauds or hoaxes, or explain them away as artifacts which were brought to the Americas by Europeans after Columbus and either lost or given to Native Americans (this is the explanation for the small cuneiform tablet which Chief Joseph had in his possession when he surrendered to the US Army, described in the previous post mentioned above). Many other forms of evidence for ancient trans-oceanic contact have been found, such as the amphorae at the bottom of Guanabara Bay in Brazil, and the mummies and other evidence listed in a post from 09/17/11, not included in this collection but available online, describing the "Calixtlahuaca head" (which is itself another artifact attesting to ancient trans-oceanic contact).

To the extent that the Powell doctrine and the ongoing policy of the Smithsonian and the rest of conventional academia ignores or devalues these artifacts, and discourages their honest appraisal by professional scholars, the search for the truth is greatly inhibited. What professional scholar wants to risk ridicule and marginalization by publishing an examination of any of these pieces of evidence, at least one that reaches conclusions which contradict the oppressive official policy of the Powell doctrine?

Clearly, the so-called Powell doctrine did not originate with John Wesley Powell alone, and its ongoing enforcement throughout academia (and at the Smithsonian) is evidence that its roots go far deeper than John Wesley Powell himself. Its continuing effect of suppressing open-minded examination of the evidence cannot simply be explained by Powell's personal views of the Native American peoples, or the personal impact his family may have experienced due to the "lost tribes" enthusiasms of the nineteenth century in general and the beginnings of the Mormon religion in particular. Nor can its continuing impact be attributable to the nineteenth-century doctrine of "Manifest Destiny" (although perhaps related to the latest incarnation of that vicious doctrine).

I believe that there is a bigger reason why powerful forces believe that evidence of ancient trans-oceanic contact must be suppressed, one that involves the spreading of illusions about history which powerful interests find extremely valuable for the public to accept. The *control of history* can certainly be a form of very powerful mind control -- and the single-mindedness evident in the efforts of John Wesley Powell (and of the Smithsonian Institute since 1879) demonstrates just how important this control of history must be to someone's agenda.

Piercing the fog of deception
that hides the contours of history

2014 May 06

I believe that we are living at a crucial juncture in human history, but that in order to see why, it helps to have an accurate "map" of history, and of the contours and terrain which led up to this particular point in time.

Unfortunately, I also believe that the control of history has proven to be a powerful tool for those who want to control the thinking of others, and to condition their acceptance of certain actions and depredations and violations of human liberty and of natural universal law. False historical narratives can be used to lend a "veil of legitimacy" to actions which are anything but legitimate. These have acted like a blanket of fog to cloak the true outlines of history under a cloud of deception.

False historical narratives can act like a well-crafted movie, into which audiences immerse themselves and -- through the "suspension of disbelief" -- which huge numbers of people come to see as real, imbuing them with a kind of reality that is a

648

function of their desire to believe that the narrative is true (the *Star Wars* movies might be a good example of fictional "fantasy" which large numbers of people imbue with enough reality that they actually take on a sort of life of their own, and are treated as if they are real events with real people inhabiting real places, even though they are clearly a work of fiction created in movie studios using cameras and special effects).

There is a huge amount of evidence which suggests that the conventional historical narratives which have been institutionalized in many "western" countries over the course of the past few centuries -- beginning during the "Enlightenment" and refined and reinforced and strengthened in each succeeding century -- are severely flawed, particularly with regards to ancient history but also regarding the history of "the west" since the so-called "fall" of the Roman Empire.

Using such an intentionally false historical "map" to try to determine where we are in history will almost certainly lead to wildly incorrect conclusions. This is why the control of the historical narrative is often a very central component of mind control and the control of populations not primarily through the threat of physical force but rather through propaganda, misinformation, and the creation of "fantasy worlds" which they buy into and imbue with a sort of artificial life.

The Undying Stars presents abundant evidence which suggests that the real narrative of history is far different -- and far more bizarre -- than the conventional fantasy narrative which is force-fed to the population (primarily through the school system from kindergarten through college, but also through various media outlets and historical programs). Many aspects of this evidence have been discussed to some degree on the pages of this blog over the course of the past four years; below is a simplified list of some of the assertions explored in *The Undying Stars*, along with links to blog posts from the past which touch on the various assertions in the list. The evidence examined strongly supports the following conclusions:

- The scriptures of both the Old and New Testaments are founded upon celestial allegories, ingeniously incorporating allegories relating to the human body at the same time, likely designed to impart profound esoteric teachings regarding the nature of the universe and the nature of human existence.
- The esoteric teachings and the system of celestial and human-body allegories indicate that the scriptures of the Old and New Testaments are in fact close kin to other ancient sacred traditions found the world over, designed to impart the same ancient wisdom to humanity (contrary to the conventional view that they somehow stand apart and are of a completely different character and convey a completely different message than that found in the "pagan" mythologies).
- These ancient traditions seem to have included an understanding of the universe that is what we today might call "holographic," and that it included the knowledge of the possibility and even of the necessity of various forms of shamanic travel or ecstasy, including contact with or travel to the "spirit realm," also described as the "hidden realm" or (in modern terms) the "implicate realm."
- This ancient understanding of the universe may also have included advanced technologies very different from today's technology, and may help explain some of the amazing accomplishments of whatever civilization or civilizations preceded the known civilizations of history, accomplishments that the conventional historical paradigm absolutely cannot explain, and which include the construction of what appears to be a "world-wide grid" which demonstrates an understanding of our planet which appears to go beyond the full grasp of even our most modern science.
- While the full glory of this extremely ancient knowledge appears to have vanished before the arrival of the first historically-known ancient civilizations, some strong

650

remnants of the ancient knowledge clearly survived into ancient historical times. The end of the ancient understanding in "the west" -- and its attendant "shamanic-holographic" rituals and techniques, appears to closely coincide with the advent of the literalist-historicist interpretation of the ancient scriptures, especially those which we call today the Old and New Testaments.

- The rise of literalism corresponded with a deliberate and sometimes violent suppression of the esoteric and the gnostic interpretation, and of those who were teachers of such an interpretation. Evidence of the suppression of texts that were difficult or impossible to paper-over with a literalist interpretation include the texts that were found buried at Nag Hammadi. However, the scriptures that did survive into the Old and New Testaments, while given a literalist spin, still testify clearly to their original esoteric origin and intent.

- On the European continent, the new literalist religion (wedded to the power of the Roman Empire) waged long and bloody but ultimately successful campaigns to absorb the Germanic and Celtic cultures and others in the broader region, and replace their original sacred traditions with the literalist religion.

- The suppression of those who understood the shamanic-holographic vision and who opposed the literalist revolution may have led to the escape of at least some of the non-literalist contingent westward across the oceans -- to the lands we call the "New World" (which had been known to the ancients for many centuries, even prior to the literalist revolution we are discussing here). There is evidence that they interacted with the Native American people and cultures they encountered there.

- Later, when the literalists gained enough power and the technology to do so, they also crossed the ocean, and treated the people they found there with ferocious violence and barbarity -- possibly because they were still incensed at the escape of many non-literalists to that continent, centuries before. They also deliberately destroyed as much of their literature as they could get their hands on, perhaps as part of the cover-up for the literalist revolution they had perpetrated and the legitimacy of which they still wished to maintain.

- The suppression of the shamanic-holographic and of the esoteric appears to continue to this day. With it, of course, comes a suppression (and oppression) of men and women, a suppression of freedom, and a suppression of the pursuit of consciousness.

I hope that you will take the time to examine the evidence and lines of argument presented in *The Undying Stars*. I believe that if _replacing the truth about the history of mankind with a fabricated cover-story_ can act as a component of mind control, then _seeing through this cover-story and beginning to perceive the outlines of the real contours of human history_ can be an important step on the pathway to freedom, consciousness, and the escape from mind control.

Common symbology between Mithraic temples and the Knights Templar, and what it might mean

2014 June 23

Previous posts have explained that in the ancient system of metaphor found in the world's ancient mythologies, the "summer half" of the year (in which days are longer than nights) was variously allegorized as a heavenly mountain, a high hill, a gleaming city, or the land of Paradise or Heaven.

The "winter half" of the year (in which nights are longer than days) was variously allegorized as a deep pit, a land of bondage or toil or slavery, Tartaros, Hades, Sheol, the Underworld, Amenta, or Hell.

In between these two halves of the year were two "crossing points," where the fiery path of the sun (the ecliptic path) crosses the celestial equator each year -- the two equinoxes (one in the spring and one in the fall).

Previous posts have demonstrated that the ancient systems of metaphor often depicted sacrifices at these "crossing" points (including, appropriately enough, the crucifixion of Christ, which is replete with both autumnal and vernal imagery). For more detailed examination of some of the equinox sacrifice metaphors, and the celestial clues which indicate that these sacrifices align with the equinox in the ancient esoteric system of astronomical allegory, please see the first three chapters of *The Undying Stars*, chapters which are available to read online here (in particular, you'll want to read the third chapter, which begins on page 26 of the book, using the page numbers as they appear on the book pages themselves).

However, the ancient system did not always depict these equinox crossing points with sacrifice myths: sometimes they involved passage through a narrow and dangerous doorway, gateway, or channel between two rocks (such as the Symplegades encountered by Jason and the Argonauts of Greek myth), and sometimes they involved other metaphors (see for instance the series of three examinations of Virgo myths through various ancient cultures, including those of Scandinavia, the Americas, and Japan).

Another way that the equinox "crossings" have been allegorized in ancient symbology is the use of figures with their legs distinctively crossed, in the symbolism employed in the cult of Sol Invictus Mithras. The cult of Mithras was an exclusive secret society, which met in underground grottos called mithraea, or in buildings designed to feel as if they were underground. As David Ulansey explains in his important 1989 publication, *Origins of the Mithraic Mysteries: Cosmology and Salvation in the Ancient World*, the Mithraic mysteries were so secret that virtually nothing of their inner workings was ever written down – or if it was, none of it has been known to survive into the present. He writes that they:

> centered around a secret which was revealed only to those who were initiated into the cult. As a result of this secrecy, the teachings of the cult were, as far as we know, never written down. Modern scholars attempting to understand

654

the nature of Mithraism, therefore, have been left with practically no literary evidence relating to the cult which could help them reconstruct its esoteric doctrines. 3.

However, he explains, the remains of the mithraea which have been discovered scattered throughout the lands of the former Roman Empire do provide important material for modern analysts to examine, in particular, the symbols found in the scenes which are found upon the walls of these ancient meeting-places. Ulansey writes:

> But the Mithraists did leave to posterity a key for unlocking the inner mysteries of their religion. For although the iconography of the cult varies a great deal from temple to temple, there is one element of the cult's iconography which was present in essentially the same form in every mithraeum and which, moreover, was clearly of the utmost importance to the cult's ideology: namely, the so-called tauroctony, or bull-slaying scene, in which the god Mithras, accompanied by a series of other figures, is depicted in the act of killing a bull. This scene was always located in the central cult-niche of the mithraeum. 6.

Professor Ulansey's 1989 book is important in that in it, Ulansey challenges the conventional theories that had been accepted up until that time regarding the origin of the symbols (which held that they must have come from Persia and ancient Persian myth, since most scholars accepted the idea that Mithraism somehow came into the Roman Empire from Persia, an idea which Ulansey shows to have been almost entirely championed in modern times by a single nineteenth-century and early twentieth-century scholar, Franz Cumont).

Ulansey's text labors to advance an alternative thesis, that the symbolism of the tauroctony is almost entirely celestial and primarily zodiacal, and that its central scene of slaying the bull has clear ties to the precession of the equinoxes. Towards this end Ulansey musters overwhelming evidence, and it is safe to say that

655

on this point his arguments are decisive in favor of the fact that the imagery present relates to the zodiac signs and neighboring constellations, and the ages-long motion of precession.

One of the extremely interesting parts of Ulansey's argument concerns his interpretation of two mysterious figures who appear in many (but not all) of the tauroctonies, two torchbearers known as Cautes and Cautopates (we know their names from dedicatory inscriptions, as Ulansey explains on page 62). These figures often (but not always) have crossed legs, and in most (but not all) of the tauroctony scenes in which they appear, one of them (Cautes) has his torch pointing upwards, and the other (Cautopates) has his torch pointing downwards.

Ulansey presents cogent arguments for identifying these figures, with their *crossed* legs and *torches*, as indicative of the crossing of the fiery arc of the sun's path *down* into the lower or wintery half of the year (at the fall equinox, indicated by Cautopates with his lowered torch) and *up* into the upper or summery half (at the spring equinox, indicated by Cautes). I discussed Ulansey's arguments, along with supporting arguments from *Hamlet's Mill* (which show that fire-imagery is very common at the points of the equinoxes in many of the world's ancient sacred mythologies) in my first book, *The Mathisen Corollary* (in chapter 10). Some scholars have challenged Ulansey's identification of Cautes and Cautopates with the vernal and autumnal equinoxes, but he presents counters to their attacks in his book.

You may be able to spot Cautes and Cautopates in the tauroctony scene above, which is from an ancient mithraeum and which is currently on display in the Kunsthistorisches museum in Vienna.

In that particular tauroctony, Cautes appears on the right side of the bull-slaying scene as we look at it, with his torch's flame pointing upwards, and Cautopates appears on the left as we look at it, with his torch's flame distinctly pointed downwards. Below is another image of the same scene, this time with Cautes and

656

Cautopates outlined with red rectangles, and the direction of their torches indicated by red arrows (the point of the arrow going towards the respective flames of the torches):

Below is another tauroctony scene from a different ancient mithraeum, which also features Cautes and Cautopates. Can you spot them and their crossed legs and torches (one pointing down for Cautopates and one pointing up for Cautes)?

Once again, they should be relatively easy to spot. Their crossed legs are very clear in the above image, due to the way they happen to stand out in the photograph. Below is the same image, with boxes drawn around Cautes and Cautopates to indicate their location, and arrows on each torch pointing in the direction of the flame, which is pointing down in the case of Cautopates on the left, and up in the case of Cautes on the right:

Based on what we have discussed in many previous posts at this point, Ulansey's argument that these two figures represent the two equinoxes is almost certainly correct. Below is the now-familiar image of the zodiac wheel of the year, which has a large "X" at each equinox to indicate a "crossing."

DAYS LONGER THAN NIGHTS:
Heaven, Promised Land, Greece, etc.

NIGHTS LONGER THAN DAYS:
Hell, Egypt, Troy, etc.

Mithraism (the cult of Sol Invictus Mithras) may play a far more important role in world history than most people (or even most conventional scholars of Mithraism or of ancient history in general) realize at this time. Historian and author Flavio Barbiero has published a book entitled *The Secret Society of Moses: The Mosaic Bloodline and a Conspiracy Spanning Three Millennia* (2010) in which he presents evidence that the cult of Sol Invictus Mithras was used as an underground "nerve center" for certain former priestly families of Judea whom Barbiero argues were brought to Rome after the conquest of Judea and the fall of the Temple in Jerusalem in AD 70, at the hands of the general (and future emperor) Titus, who was prosecuting the military campaign in conjunction with his father, Vespasian.

Admiral Barbiero argues that Mithraism basically functioned as an extremely effective secret society, and one which spread through certain strategically-chosen institutions in the Roman Empire, including the Praetorian Guard, the Roman army at large, the centers of trade and commerce (in particular the ports and customs-facilities) and the organs of the political bureaucracy. It took some time (almost two hundred years), but this "nerve center" eventually gained so much power that it was

659

able to install and remove emperors at will. The extensive evidence to support this amazing claim is discussed at length in Barbiero's book, and it is also discussed in *The Undying Stars* in conjunction with that book's examination of the question of "what happened to the ancient wisdom?" Interested readers can also get an overview of the theory in this article which Flavio Barbiero published in 2010 on the Graham Hancock website: http://www.grahamhancock.com/forum/BarbieroF2.php

According to this theory, the nobility which controlled Europe during the Middle Ages (as well as the leaders of the western church) almost certainly descended directly from the same lines of priestly families who came to Rome with Vespasian and Titus after the fall of Jerusalem. Interestingly enough, Barbiero finds evidence for this theory (in addition to the bigger pieces of evidence which are discussed in that linked article and which make up the bigger part of his argument) in the fraternal orders which formed among the European nobility during the Crusades – including the most famous of these, the Knights Templar.

Those familiar with the history of the Knights Templar may have already been struck by the distinctive "legs crossed" symbology in the foregoing discussion, from the temples of Sol Invictus Mithras. Barbiero argues that it was probably within "some associations of nobles in which the most authentic spirit of the original institution of Sol Invictus could have survived" (333). He makes note of the connection between the fact that the funeral monuments of Templar knights represent their effigies with their legs crossed, and says "we cannot imagine that it is a simple coincidence that in all the mithraea there are always two characters with their legs crossed in the same way" (337).

That the members of the noble families who were descended from those original priestly lines who defended the land of Judea and the Temple of Jerusalem against the invading Roman armies under Vespasian and Titus would form dedicated military orders which had secret rituals and shared symbology with the ancient cult of Sol Invictus perfectly accords with Barbiero's thesis. In

660

fact, he writes of knights who made up the top rank of these military orders (such as the Templars):

> To all effects, they were professional warriors dedicated to war, which always appeared to be a striking anomaly in the Catholic religion, in flagrant contrast to the pacifism preached by Christ. In reality, this was no anomaly, but was instead a perfect continuation of the traditions of the priestly family. Josephus Flavius was a priest but also a warrior and a military leader. The followers of Sol Invictus had taken control of the Roman army and were, first of all, military men. 335.

Below are some images of the tombs of various Templar knights. In the first one, for example, you can see that the two knights on the right-hand side of the photograph (as we look at it) have their legs crossed:

Below is a drawing of another effigy of a knight from his tomb. You can clearly see that his legs are crossed:

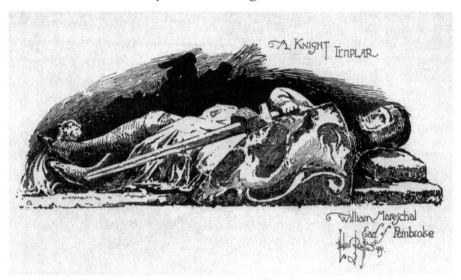

And finally, one more drawing of effigies of knights from medieval tombs. The tomb closer to the viewer clearly depicts an effigy with the legs crossed in the same distinctive manner:

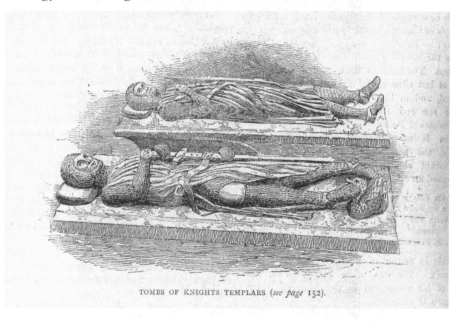

TOMBS OF KNIGHTS TEMPLARS (*see page* 152).

662

There is much more to this subject, but this connection alone constitutes yet another piece of evidence supporting Flavio Barbiero's theory as presented in his book, a theory which is very important for the question of "what happened" to the common system of celestial allegory which underlies virtually all of the world's ancient sacred texts (including those which found their way into what today we refer to as the Bible), and to the knowledge that this ancient wisdom should actually unite mankind, instead of dividing us.

Ten reasons to suspect a close connection between ancient Roman Mithraism and ancient Roman Christianity

2014 June 24

Conventional scholars continue to debate the origin of the Roman cult of Sol Invictus Mithras, which (based upon the archaeological evidence of the mithraea) arose circa AD 100 and ends in AD 396. Although scholars today are more circumspect in their pronouncements regarding the origins of this institution than they have been in previous decades (prior to the 1970s), it is still common for well-regarded Mithraic scholars to assert that Mithraism and Christianity were bitter rivals.

For instance, this essay published in a collection in 1994 tells us that: "Between the second and fourth centuries C.E. Mithraism may have vied with Christianity for domination of the Roman world."
[http://books.google.com/books?id=QRfhSBLmAK8C&pg=PA147#v=one page&q&f=false]

664

The author continues:

> The Christians' view of this rival religion is extremely
> negative, because they regarded it as a demonic mockery of
> their own faith. One also learns of Mithraism from brief
> statements in classical Greek and Roman authors.

While it is certainly true that Christian polemicists, including
Tertullian, attacked Mithraism on these grounds, this does not
necessarily indicate that the two systems were indeed at cross-
purposes. Author Flavio Barbiero, whose work is discussed in
The Undying Stars and in this previous post,[63] has put forward a
theory which argues that the cult of Sol Invictus Mithras was
actually the secret society through which decisions were made
and strategy enacted to gain control of the "command-and-
control" centers of the Roman Empire, and that this exclusive
institution, whose proceedings were kept entirely secret, operated
in the background, using literalist Christianity as a public and
nonexclusive shield – one that it controlled, and one that would
take the brunt of those who wanted to stand against the
underground campaign.

Flavio Barbiero offers a host of evidence to support this view of
events, and the conclusion that this campaign was ultimately
tremendously successful – successful to the point that it shaped
European history and then world history for the following
seventeen centuries, and continues to do so to this day. The
following points are taken from his 2010 publication *The Secret
Society of Moses: The Mosaic Bloodline and a Conspiracy
Spanning Three Millennia*. Many of these pieces of evidence are
also discussed in his 2010 article entitled "Mithras and Jesus:
Two sides of the same coin" on the website of Graham Hancock.

Note that the following points are not intended to be aimed at
any particular branch of Christianity as it has existed since the
fourth century, but rather to shed light upon the possible origins
of *all* of literalist Christianity, which deliberately chose to take a
very different approach to the interpretation of the Biblical

665

scriptures, and one which intentionally cut itself off from all the "pagan" traditions of the world as well as from the esoteric, gnostic, Sethian, Valentinian and Hermetic forms of Christianity which existed prior to this juncture in history.

- Mithraism was neither a "religion" nor a "mystery cult" – unlike other ancient religions, it was extremely exclusive and met in special mithraea which were so small that, "At most, forty people could be seated in each of them" (158). The majority of mithraea could not hold more than twenty.
- Numerous mithraea have been found underneath ancient Christian basilica or churches, indicating that there may have been some kind of symbiotic connection between the leadership of the cult of Sol Invictus Mithras and that of the Christian church. While it is possible to explain this fact away by saying that the Christian church triumphantly took over the sites of its old rival and built churches on top of their sacred sites (as it later did around the world), there is evidence that this explanation was not the case for Mithraism and Christianity. Specifically, Barbiero notes that the Basilica of St. Peter on Vatican Hill was built above the Phrygianum, the most central mithraeum in Roman Mithraism, where the "Father of Fathers" (head of the entire order of Sol Invictus Mithras) held sway. Most significantly, the Christian Basilica of St. Peter was built by the emperor Constantine in AD 322, but the last "Father of Fathers" of Mithraism did not die until AD 384, and he continued to use the mithraeum in the Phrygianum for all those years! It would be remarkable if these two supposedly "rival religions" coexisted for even two years with their "headquarters" co-located, but the dates indicate that this coexistance lasted for *sixty*-two years. Barbiero writes:

> In this light, we are forced to conclude that Sol Invictus Mithras and Christianity were not two religions in competition, as we often read, but were

two institutions of a different nature that were closely connected. Rather than being a simple hypothesis, this is practically a certainty. It is unthinkable that the Roman church continued to extend hospitality to the head of a rival pagan religion for more than half a century and at the heart of its most exclusive property, the basilica dedicated to the prince of apostles. The Mithraic *pater patrum* and the bishop of Rome must necessarily have been closely linked. 163-164.

- As the passage just cited indicates, the title of the supreme head of the Mithraic organization was *pater patrum*, or "Father of Fathers." The Mithraic system had a hierarchy of seven Mithraic grades, with the highest being the *Pater* or "Father" (the head of any particular mithraeum). The head of the entire system, of all the Mithraic "lodges," was the "Father of Fathers," or *pater patrum* (pa-pa, for short). It is most significant that, after the death of the last Mithraic *pater patrum*, in AD 384, the bishop of Rome adopted this same title, which is still used to this day (and which is rendered in English "the Pope," but in Italian and Spanish is still *papa*). This evidence is discussed in Barbiero, 163 and elsewhere.

- As part of the same discussion, Flavio Barbiero notes that specific aspects of Mithraic ritual and attire were adopted into the rituals of Christianity, including the distinctive headgear of Christian bishops, which is still called a *mitre*, a word with linguistic connections to *Mithras* or *Mitra*.

- There is powerful evidence of early prominent Christian leaders who were also members of the Sol Invictus Mithras organization, right up to the point that they declared themselves Christians, or took holy orders to become high-ranking leaders of Christianity. The most prominent of these whom Barbiero notes is the emperor Constantine himself (Barbiero, 166-167). Others include St. Ambrose, whom Barbiero notes "passes directly from being a pagan to being bishop of one of the most important sees of the period" (166). St. Ambrose was the son of a father who was a member of Sol Invictus Mithras, as was the Christian apologist and polemicist Tertullian (AD 160 - AD 225), as well as church fathers St. Jerome and St. Augustine (Barbiero 167-168). This fact is highly significant and indicates that these early Christian "Fathers" were descended from the same family lines that Barbiero discusses in his thesis.

- Constantine continued minting coins with clear Sol Invictus symbology and imagery, even after his vision of the heavenly "Chi-Rho" sign in some cases minted coins containing both sets of symbology, Christian and Mithraic. This is a clear indication that the two systems were not actually seen as antagonistic, at least during the early stages of establishing Christianity as official to the empire (later, Mithraism would be dismantled and the family lines would use Christianity as their open system of control, the "underground" mechanism of Sol Invictus Mithras having served its purpose). This use of Sol Invictus symbology on his coins is discussed in Barbiero page 165, and is also attested to in the notes to a translation of the works of the Christian polemicist and apologist Eusebius (c. AD 260 - c. AD 340). On page 207 of this edition of the works of Eusebius, we read a note from the editor to Eusebius' mention of a chi-rho coin which informs us that Constantine claimed to have seen the Christian chi-rho sign in the sky "resting over the sun," and that thereafter Constantine "continued to commemorate [the sun] on his coins as Sol Invictus (see Bruun, 'Sol'), whether out of numismatic conservatism (Barnes) or as a sign of solar monotheism."
- There is evidence that early Christian leaders saw reverence to the sun as not at all incompatible with Christianity, with Pope Leo in a famous passage in his Christmas sermon of AD 460 declaring that: "This religion of the Sun is so highly respected that some

668

Christians, before entering the basilica of St. Peter the apostle, dedicated to the one true living God, after climbing the steps that lead to the upper entrance hall, turn towards the Sun and bow their heads in honor of the bright star" (cited in Barbiero, 161). Tertullian also writes that "it is a well-known fact that we pray turning towards the rising sun" (*Ad Nationes* 1.13, cited in Barbiero "Two sides of the same coin," page 3). This connection between the sun and the "one true living God" described in the sermon by Pope Leo is in keeping with Constantine's use of both Sol Invictus imagery and Christian "chi-rho" symbology on his coins (Constantine evidently did not see anything contradictory or conflicted about the use of both).

- In AD 386, a decree by the emperor Aurelian changed the name of the Christian day of worship from "the day of the sun" (Sunday being the first day of the week, in a significant change from the seventh-day Sabbath of antiquity) to "the day of the Lord" (Barbiero, 237).

- The spread of the mithraea throughout the western empire (particularly in the vicinity of army barracks and organs of the government bureaucracy) parallels the spread of Christianity. Barbiero writes, "Wherever the representatives of Mithras arrived, there a Christian community immediately sprang up" ("Two sides of the same coin," page 9). Early bishop's sees were located in Britannia, Gaul, Spain, and North Africa -- the same places that legions were located and which are the sites of mithraea (*Ibid*).

- Barbiero traces the progress through which the new Roman class of *equites* or "equestrians," to which the descendants of the family lines who had come to Rome with Titus and Vespasian after the fall of Judea belonged, gained access to the Senate and then progressively grew more and more powerful in the Senate. Dedicatory inscriptions reveal that as this process took place, more and more senators were members of Sol Invictus Mithras. However, upon the death of the last *pater patrum* of Sol Invictus Mithras, Flavio Barbiero notes that the entire Senate, that "stronghold of the cult of Mithras, discovered that it was totally Christian" (163, see also 241). In other words, the transition was remarkably smooth and bloodless -- indicating that Mithraism and Christianity were not at all the bitter rivals that the conventional narrative often paints them as being. They were, as Barbiero says, "two sides of the same coin."

These are by no means all the pieces of historical evidence which Flavio Barbiero musters to support his assertion that the institutions of Sol Invictus Mithras and literalist Christianity actually worked "hand in glove." Further, while this is a central part of his overall theory, there is much more to the theory, and that "much more" is itself supported by *still further* extensive evidence from other aspects of history.

In short, there is so much evidence to support this thesis that it simply cannot be ignored, and deserves careful consideration by everyone who wishes to explore the possible reasons for the suppression of the ancient celestial system of allegory which (I believe) was meant to preserve and to convey a sophisticated shamanic-holographic cosmology that was once widespread around the globe and which flourished in "the west" right up until the fourth century AD. The loss of this ancient wisdom, an inheritance belonging to all of humanity, is an absolutely watershed event in human history, and one which continues to impact our lives right up to the present day.

Commodus and Marcus Aurelius

2014 July 25

The movie *Gladiator* (2000), starring Russell Crowe and Joaquin Phoenix, presents the transition from the rule of the Roman emperor Marcus Aurelius to the rule of his son Commodus as a crucial turning point in the history of the empire.

In the film, Marcus Aurelius recognizes the pathological twist in his son's character and decides he will not appoint Commodus as his successor, instead desiring to return Rome to a republic, and appointing the virtuous Maximus to act as "protector" during the transition. As fans of the movie know, Commodus was none too pleased with this arrangement and took matters into his own hands, eliminating both his father and eventually Maximus as well, and ascending to the throne to become one of Rome's most megalomaniacal rulers.

While the above plot takes considerable historical license and inserts an entire series of fictional characters and events surrounding the memorable but entirely imaginary general-turned-gladiator, Maximus, the transition between Marcus Aurelius and Commodus was in fact an enormous turning point in world history, and one that is worthy of careful study and consideration.

According to the theory put forward by Flavio Barbiero in his 2010 book, *The Secret Society of Moses*, the transition from Marcus Aurelius to Commodus was critical in that Commodus was the first emperor who was an initiate into the secret society of Sol Invictus Mithras. As explained in my previous post entitled "Ten reasons to suspect a close connection between ancient Roman Mithraism and ancient Roman Christianity,"[664] and articulated at greater length by Flavio Barbiero in an online article entitled "Mithras and Jesus: Two sides of the same coin," there is evidence to support the thesis that the secret society of Sol Invictus Mithras was the primary vehicle through which the priestly families from Judea took over the levers of control of the entire Roman Empire.

Judea and Jerusalem fell to the Roman legions led by Vespasian and his son Titus in AD 70. Vespasian and Titus brought back certain members of the leading priestly families from Judea to Rome -- including the crucially important historical personage of Josephus. Once in Rome, according to the thesis expounded by Flavio Barbiero and backed up by extensive historical evidence (starting with the writings of Josephus himself), these priestly families began a secret campaign to gain control of the levers of power, beginning with the Praetorian Guard and then extending steadily to the centers of commerce, the bureaucracy of the empire, and of course the Roman army itself (especially the officer corps).

In order to accomplish this takeover, these families used the twin vehicles of Mithraism and literalist, hierarchical, ecclesiastical Christianity (which they created, and which slowly took over

672

from and supplanted the earlier gnostic and esoteric forms of Christianity that had existed prior to the campaign of Vespasian and Titus and their destruction of the Temple at Jerusalem).

If that seems difficult to believe, remember that these families were extremely experienced at running a system which we could call a system of reality creation. Previous posts have explored the likelihood that the ancient esoteric wisdom which forms the foundation of *all* the world's ancient sacred traditions articulated a vision of our universe as one that is shaped at least in part by *human consciousness*, and taught that through consciousness we can actually *create realities*. As a post from 07/23/14 articulates, I believe there is evidence that this wisdom was intended for (and anciently used for) benevolent purposes, but it can also be used for purposes of control, domination, and the general suppression of human consciousness in others.

So, if the families that came to Rome after the fall of Jerusalem were experts in "reality creation," were they more disposed to use that knowledge for the more benevolent purposes of enhancing human consciousness and freedom, or for the more oppressive purposes of control and domination?

Well, there is clear evidence which demonstrates that most repositories of the ancient wisdom were destroyed *after* the arrival of these families in Rome, and in fact after the time that Sol Invictus Mithras began appointing emperors and thus demonstrating that it had gained control of the levers of power of the Roman Empire. Examples of this destruction of the esoteric ancient wisdom include the destruction and suppression of gnostic and esoteric texts within the Roman Empire itself (see discussions here and here and here), as well as the burning or seizing of ancient texts stored at the library of Alexandria. It also includes the shuttering of the sites that carried on the various mystery cults within the borders of the Roman Empire, which (as I explain in *The Undying Stars*) also appear to have preserved aspects of the ancient knowledge that the new order set about to suppress or eliminate (for a discussion of one of the most important of these ancient mystery cults, see this previous post exploring the Eleusinian Mysteries).[173]

Based on the fact that these suppressions all took place within the Roman Empire *after* the time that Sol Invictus began appointing emperors, and especially after Constantine made literalist Christianity the official religion of the empire, it is safe to say that those expert practitioners of reality creation who took over the Roman Empire were generally more interested in the "control and domination" side of the art.

In his book, Flavio Barbiero points out that the use of the two-pronged strategy which included both the public-facing literalist-Christian vehicle and the private, exclusive, and extremely secret society of Sol Invictus Mithras was critical to the success of the takeover. The old Roman families, especially those who controlled the Senate of Rome until the Senate was slowly infiltrated by equestrian-class newcomers, never actually realized that the leaders at the top of Sol Invictus were the ones calling the shots. The representatives of the old Roman families generally saw Christianity as the threat, and tried to attack *it* instead -- thus the spread of Christianity served as the perfect distraction or decoy to misdirect their attention and enable the secret society of Sol Invictus to move its pieces across the chessboard until it was able to emplace emperors at will.

At first, the leaders of Sol Invictus used emperors who were from the old Roman families but had been initiated into the Sol Invictus cult (not knowing that they were only shown the "lower-level" activities of the secret society, and were not invited to the high-level inner-circle meetings where the real strategy was enacted). However, at some point, Sol Invictus had enough power (backed up by their control of the Praetorian Guard) to appoint descendants of their own priestly families to the office of emperor.

According to Flavio Barbiero's research, the first emperor to be a member of Sol Invictus was none other than Commodus, who took the throne in AD 177, just over one hundred years after the fall of Jerusalem and the arrival of Josephus and the other members of the priestly families in Rome.

The fact that Commodus was closely associated with Sol Invictus is clear from several historical details. For one, he took the name "Invictus," and when he renamed the months of the year after his own several names and appellations (an example of his egomania which caused tremendous resentment among the traditional Roman families), he chose to name one of the months "Invictus." Another piece of evidence can be seen in the coin below, which has the image of Commodus on one side and the image of a solar god or figure on the obverse side, with legs crossed and leaning against a pillar:

We have already examined at some length in previous posts the fact that crossed legs is a form of metaphorical solar symbology which is extremely characteristic of the iconography found in Mithraic meeting-places (mithraea). The work of Mithraic scholar David Ulansey clearly establishes that the crossed legs refers specifically to the sun's crossing of the celestial equator, which takes place twice a year at equinox. I have also argued in previous discussions that there is extensive evidence to conclude that the pillar refers to the line running from the winter solstice to the summer solstice.

In addition to this, there are other ancient sources which indicate that Commodus was affiliated with Sol Invictus Mithras. Marcus Aurelius, on the other hand, seems to have sensed the rising threat to the ancient traditions and belief systems and to have attempted to stem the tide, thus placing his reign and that of his son on two different sides of the crucial power struggle over the future of the western world.

Flavio Barbiero points out that there is some evidence that Marcus Aurelius actively persecuted Christianity to some degree (185). Other scholars argue that it is not entirely certain to what degree Marcus Aurelius actively encouraged the persecution of Christians that took place under his reign (although it is hard to explain how that could have gone on against his will or without his knowledge). Thus, it is quite possible that Marcus Aurelius, who was himself a Stoic and an important philosopher in his own right, perceived that dangerous forces were at work to supplant the old ways, and incorrectly believed that Christianity was the primary threat and targeted its adherents, not perceiving that Sol Invictus was the more important nerve center that was directing the long-term campaign.

With the passing of Marcus Aurelius and the accession to the throne by Commodus, there was a definitive shift in power. Flavio Barbiero says that "the imperial office from Commodus onward" was "conferred almost exclusively on members of the Sol Invictus organization, independent of the rank they held in the organization and whether or not they belonged to whatever branch of the priestly family" (209).

Marcus Aurelius is often included in the category of the "five good emperors" (a category which also included Nerva, Trajan, Hadrian, and Antoninus Pius). In his massive 1,500,000-word magnum opus *The Decline and Fall of the Roman Empire*, Edward Gibbon (1737 - 1794) says of the events that took place after the reign of Antoninus Pius (the end of whose reign he saw as the beginning of the long decline and ultimate fall of the empire), it was "a revolution which will ever be remembered, and is still felt by the nations of the earth" (page 11 of this version).

How right he was, especially on that second point about it being "still felt by the nations of the earth."

Of the reign of Commodus, Flavio Barbiero writes:

> On the death of Marcus Aurelius, probably the most zealous and efficient persecutor of Christianity of the emperors who succeeded Nero, the Empire passed to his son Commodus, who was initiated into the Mithraic organization. For more efficient influence over Commodus, he was given a Christian concubine, Marcia, who, for the entire duration of his reign (AD 180 - 192), had the prerogatives and powers of an empress. Commodus has gone down in history as one of the most ferocious and extravagant of the Roman emperors; he sent thousands of people to their deaths for the sheer pleasure of it. Among these people, however, there was not a single Christian, because he put an end to the persecutions of his father and showed favor to Christianity in every way.
>
> Commodus was certainly not of priestly lineage, and his unpredictability made him difficult to maneuver for the Mithraic organization, which eventually decided to eliminate him. [. . .] Marcia was the instrument of his elimination, and was helped by Quintus Aemilius Laetus, prefect of the Praetorian Guard, which, by then, was completely under the control of Sol Invictus. 186-187.

Note that Laetus appears in the 2000 movie *Gladiator*, and plays a rather important role throughout the film (he is clearly no fan of Commodus).

The fact that Commodus, a member of Sol Invictus, had as his favorite concubine (who was given the powers and prerogatives of an empress) a woman who was a known Christian should be seen as yet another piece of evidence (in addition to those listed here[664] and the many other pieces of evidence listed in the works of Flavio Barbiero) that Christianity and Mithraism were not arch-rival religions the way they are often portrayed in conventional scholarship.

The fact that the Praetorian Guard was perfectly capable of removing emperors by assassination is also demonstrated from the above passage, as it is also demonstrated by the numerous emperors after Commodus who reigned for only a few weeks or months before being assassinated themselves.

The graphic below shows the majority of the emperors from Vespasian to Constantine, with important milestones indicated in highlighted-yellow type. Emperors who reigned for very short periods of time are depicted in smaller images than those who reigned for longer. Dates are indicated in red lettering and all of them are AD. The images are for the most part those found on this Wikipedia page. Not all dates are listed, but enough are listed to give a general idea of the timeline. Years listed are for the year the emperor began to reign:

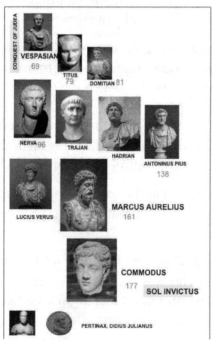

678

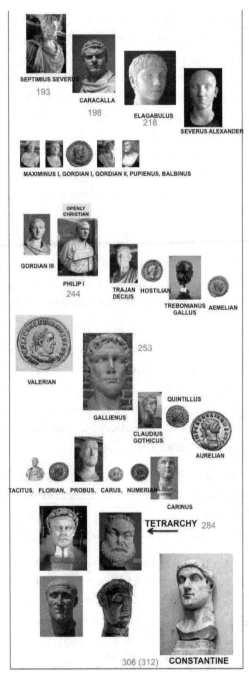

From the above chart, we can see that (if the analysis of Flavio Barbiero is correct, and I believe that it is) the priestly families worked towards the ability to get an initiate of Sol Invictus into the imperial office from their arrival in Rome around AD 70, and finally succeeded with the accession to the throne by Commodus in AD 177. After his reign (and later assassination), there were two emperors removed in quick succession, followed by several more emperors closely affiliated with Sol Invictus starting with Septimius Severus in AD 193: this indicates that Sol Invictus continued to decide who would become emperor (and whether that emperor would stay the emperor) from Commodus onwards (there was only one big setback to their plans, during the period of the tetrarchy, discussed briefly below).

This excellent web page from ancient coin collector Bill Welch shows Roman coins with clear Sol Invictus imagery beginning most especially with coins minted during the reign of Septimius Severus: https://www.forumancientcoins.com/moonmoth/reverse_sol.html

Flavio Barbiero discusses the evidence that one of the primary missions assigned to the emperor Septimius Severus and his immediate successors was the drastic reduction of the power of the old senatorial families of Italic stock, and the gradual infiltration into the Senate of newly-wealthy equestrians (who generally came from the bureaucratic offices of the empire, and from the military)(187 and following).

Also notable among the emperors in the group that begins with Septimius Severus is the emperor Elagabulus, of whom Flavio Barbiero finds evidence suggesting that he may have been the first emperor to be descended from one of the priestly family lines.

Following another internecine period of rapidly-assassinated and replaced emperors, Gordian III emerged to reign from AD 238 - AD 244, followed in AD 244 by Philip I (also known as "Philip the Arab"), the first openly Christian emperor. It should be noted that emperors could be both Christian and members of Sol Invictus all the way up until after the time of Constantine.

Also noteworthy, especially in light of the huge number of emperors who only lasted for a period of weeks or months, is this quotation from Flavio Barbiero: "It is quite likely that the heads of the branches of the priestly family, who monopolized the higher levels of the Mithraic organization, were reluctant to take on the office themselves and preferred to govern through expendable pawns affiliated at the first levels -- and were ready to eliminate them as soon as they deviated from their instructions or disappointed expectations" (197).

Finally, Flavio Barbiero explains that Diocletian almost certainly devised the unwieldy mechanism of the "tetrarchy" in order to protect himself from the threat of rapid assassination by the Sol Invictus powerbrokers who had allowed him to become emperor (something that happened to many emperors who stepped out of line, including many of those who immediately preceded Diocletian). As a result of Diocletian's move, according to Barbiero, the leaders of Sol Invictus decided to make Christianity the official religion of Rome and to move their nerve center from the hidden organization of Mithraism to the more-open organization of the Christian church, which took place during the reign of Constantine (who reigned from AD 313, when he gained undisputed control over the empire after the Battle of the Milvian Bridge, until his death in AD 337).

In order to facilitate that control, they also had Constantine move the seat of the imperial office out of Rome and over to the new city of Constantinople.

Clearly, there is historical evidence which strongly supports Barbiero's thesis. If he is correct, the transition of power between Marcus Aurelius and his son Commodus, was one of the turning points in human history. It is also clear that Commodus went off in quite a different direction than that of his father -- Marcus Aurelius working as best he saw how to try to stave off the threat that was working to take over the empire from within, while Commodus actually signed up for the program that was maneuvering that takeover.

I wonder if the makers of the film *Gladiator* (which was a joint British and American production; Flavio Barbiero provides evidence that the British Isles became an early and important stronghold of both the priestly families discussed above and the Sol Invictus Mithras organization) knew all of this when they chose to make a film about that critical transition that took place in AD 177 . . .

Ross Hamilton's Star Mounds

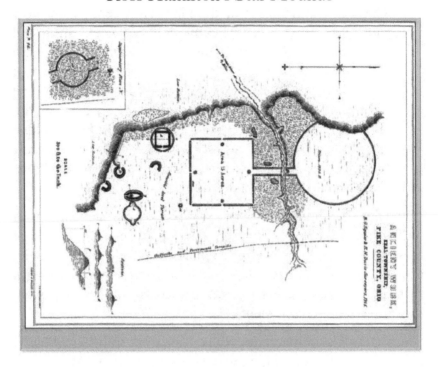

2014 September 10

The following is the text of a review I just posted of Ross Hamilton's *Star Mounds: Legacy of a Native American Mystery* (2012):

> Ross Hamilton has done and continues to do a great service to our generation and future generations in his thorough (and ongoing) exploration of the mysterious "Star Mounds" of the Ohio Valley, and his analysis and attempts to unlock their ancient messages, alignments and meanings. His deep knowledge of geometry, sacred geometry, and astronomy, as well as his profound respect for the Native American traditions and legends and teachings, shine through in this beautiful book. The geometric analysis was astonishing, and the connections to the constellations equally significant: these earthworks are truly an often-overlooked wonder of

the world (many of the earthworks surveyed by Squier & Davis in the first half of the nineteenth century are now sadly lost), and Ross Hamilton's careful and insightful analysis should increase our awareness of and desire to learn more about these treasures from the distant past, and about those who envisioned them and caused them to take shape upon the terrain of our amazing and wonder-filled world. Through his patient and careful examination and his unique set of skills and abilities, Ross brings to light the truly breathtaking hidden patterns and connections designed into these mounds.

In the first line, I might also have said "and to past generations as well" -- because through his prodigious study and examination of these ancient earthworks Ross Hamilton has helped to discover, express, and preserve at least some of the multi-layered messages that the ancient builders of these incredible monuments were conveying in their monumental design.

The earthworks of the greater Ohio Valley are truly incredible in their size, scope, durability, and layers of mystery. Most well-known among them, perhaps, is the Great Serpent Mound, which Ross shows to function as the central "key" uniting all the far-flung structures, and to which he devoted an entire equally-essential earlier book, *The Mystery of the Serpent Mound: In Search of the Alphabet of the Gods* (1993), which I discuss in earlier blog posts available online.

In that earlier book, Ross Hamilton presents convincing arguments and evidence to demonstrate that the Great Serpent Mound is a terrestrial model and reflection of the celestial serpent found in the constellation Draco; in Star Mounds the author provides additional evidence to demonstrate that the entire network of earthworks in the greater Ohio Valley region (covering -- at least -- an area claimed by four different modern states in the US) represents a vast mirror of the wider heavens, including the constellations of the zodiac band, the circumpolar stars of the north celestial pole, and even the Southern Cross.

684

Using diagrams of the earthworks created in past centuries by professional surveyors (before some of these precious monuments were obliterated or damaged almost beyond recognition), as well as modern LiDAR imagery, Ross illustrates the celestial connections -- and ties the constellations into Native American myth and legend (the *Star Mounds* are tied in to *Star Myths!* incredible!). Some of the most satisfying and difficult-to-dispute connections to celestial formations are those that Ross illustrates between the Newark Complex and the constellation Pisces (especially in his diagram at the top of page 169) and between the works known as the Cross and the constellation Cygnus (page 131), but there are many others, and there are so many correspondences that the resonance cannot be dismissed as being either imagination or mere "coincidence."

This analysis is of tremendous importance, for it demonstrates that the designers of the "Star Mounds" of the Ohio Valley region were involved in the very same kind of monumental terraforming to make the terrestrial landscape reflect the celestial that can be shown to have also taken place in South America, the British Isles, and the Nile River valley of ancient Egypt: the entire "Orion's Belt" thesis of Graham Hancock and Robert Bauval was an important recognition and discussion of this ancient world-wide endeavor, as were earlier works such as John Michell's *View Over Atlantis (1969)* and *New View Over Atlantis* (1983) and Kathryn Maltwood's even earlier *Guide to Glastonbury's Temple of the Stars* (1929).

These undeniable connections to similar projects around our planet make Ross Hamilton's demonstration of the celestial patterns in the North American complex of the Ohio Valley region extremely significant not just to North American history, but to the mysterious history of the entire globe, and to the ancient dictum (which can be shown to operate in all the world's mythologies) of "as above, so below," the microcosm and the macrocosm.

This accomplishment alone would make *Star Mounds* an essential reference to place in your library next to the other works just mentioned dealing with other regions, but that's not all for the hidden messages that Ross Hamilton has found in these ancient monuments -- not by a long ways. With his formidable knowledge of geometry and proportion, Ross Hamilton discerns undeniable evidence of advanced sacred geometry at work in many of the ancient earthworks -- some of them simply breathtaking.

As just a small example, take for instance his examination of the "works" located in Seal Township, in Pike County, Ohio, and drawn by Squier & Davis in the illustration shown at top (the entire text of Squier & Davis's *Ancient Monuments of the Mississippi Valley*, published in 1848, can be viewed online here).

Ross Hamilton notes that the Seal Work monument itself consists of a square and a circle, connected by a "neck" feature that is aligned (along with the sides of the square) to true north with near-perfect precision. He points out that this "neck" feature in the Seal Work is the longest known among the Star Mounds. He then proceeds to demonstrate why the length of that neck is so important, and how the distance between the square and the circle was surely no accident or simple random decision by the ancient designers.

Taking the outline of the circle, which is incomplete due to its interruption by a steep cliff, he completes the circle -- shown below in green (these are my illustrations of the explanations that Ross Hamilton shows on page 41 of his book, which the interested reader should consult -- any mistakes in the discussion that follows or the diagrams that I've made to try to illustrate it are my own and should in no way reflect badly upon Ross Hamilton's book):

686

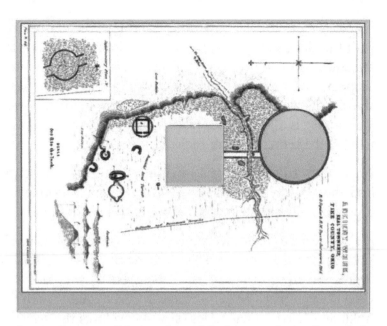

In the above diagram, I have also added a purple outline to the circular feature, because what happens next is that we slide that purple outline towards the square feature, until the closest edge of the square forms a tangent with the perimeter of the circle, as shown below:

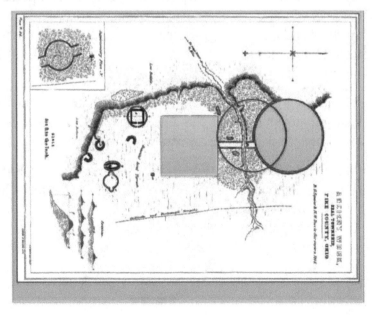

As Ross Hamilton discovered, performing the above operation with the circle of the Seal Works and sliding its outline towards the square forms "a true vesica piscis" (the extremely important shape made by the overlapping outlines of the two circles shown above – the words *vesica piscis* are Latin for "bladder of the fish"). The outline is a "true vesica piscis" if the widest point of the vesica touches the center-point of each of the two circles.

This geometric figure was obviously incorporated deliberately into the Seal Works, and it provides tangible evidence of the sophistication and geometric knowledge of the designers.

It also speaks to their incorporation of sacred spiritual symbology in their Star Mound architecture: for the vesica piscis symbol relates directly to the message conveyed by the Ankh and the Djed discussed in a series of recent posts beginning with "Scarab, Ankh and Djed." That series of posts explores the symbology of the cross (including the Ankh, but also the Djed, which was depicted horizontally "cast down" and then vertically "raised up"), and presents evidence that this common and potent ancient symbol typifies our human condition in our incarnate material form: a horizontal "cast down" or "animal" component (our material, physical body of earth and water) and a vertical "spiritual and immortal" component (the Christ within, the divine spark that comes down from the realm of air and fire to inhabit this body of earth and water).

In *The Jesus Mysteries* (1999), Timothy Freke and Peter Gandy make the argument that the ancient symbol of the vesica piscis was known to convey *the exact same message!*

On page 40 of that text, Freke and Gandy provide a vesica piscis illustration, showing how it is the origin of the familiar Christian "ichthys" symbol shown below:

In the caption beneath their illustration of the two circles forming a vesica piscis with superimposed ichthys, they write:

> The sign of the fish is widely used today as a symbol of Christianity, but originated in Pagan sacred geometry. Two circles, symbolic of spirit and matter, are brought together in a sacred marriage. When the circumference of one touches the center of the other they combine to produce the fish shape known as the vesica piscis. The ratio of height to length of this shape is 153:265, a formula known to Archimedes in the third century BCE as the "measure of the fish." It is a powerful mathematical tool, being the nearest whole number approximation of the square root of three and the controlling ratio of the equilateral triangle. 40.

The mathematical approximation of the square root of three that they are referring to is the number 1.732, which is also the result (or quotient) that we get if we divide 265 by 153 (265/153 = 1.732).

Is it not extremely noteworthy that the vesica piscis was anciently seen to be emblematic of the unification of spirit and matter, just as we have seen the crosses of antiquity to have been as well?

Freke and Gandy point out the fact that in the 21st chapter of the New Testament gospel of John, the risen Jesus showed himself to the disciples, who have been fishing all night but without success. From the shore, he directs them to cast their nets on the right side of the ship, and upon doing so they catch so many fish that "they were not able to draw it for the multitude of fishes" (John 21:6). After they get the catch to shore (with the help of another boat), we are told the exact number of the fishes that were caught, in verse 11: "Simon Peter went up, and drew the net to land full of great fishes, an hundred and fifty and three: and for all there were so many, yet was not the net broken" (John 21:11). Freke and Gandy draw the direct connection to the anciently-revered sign of the vesica piscis and its known ratio of 153:265, and point out that the ancient Pythagoreans "were renowned for their knowledge of mathematics and regarded 153 as a sacred number" (39).

689

But this subtle incorporation of the sacred vesica piscis is not the only sophisticated mathematical message that Ross Hamilton finds hidden within the Seal Works of Pike County, Ohio. In the illustration below, the area of the square that is connected to the circle is designated as one square unit (for a square with sides of one unit), and the area is placed adjacent to the square, as shown below. A circle is then traced out which touches upon the four corners of the "doubled square" as shown in the illustration:

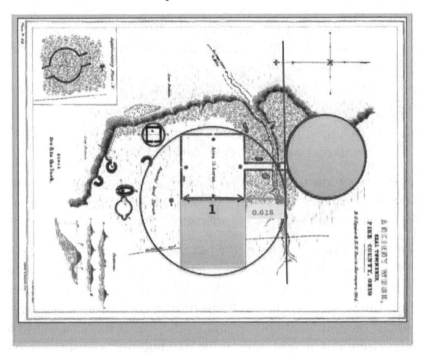

Here again we see that the distance between the circle and the square at the Seal Works was carefully thought out. Because the new larger circle thus indicated happens to reach precisely to a line drawn tangent to the edge of the original circle, and parallel to the edge of the original square (see diagram). In other words, we can see right away that the size of the square was no accident, if doubling it leads us to be able to draw a circle that precisely reaches to the nearest edge of the earthwork's physical circular space.

690

But that is not all, because the size of the square, and the distance between the square and the circle, incorporate an even more amazing connection than the one described in the preceding paragraph (as obviously deliberate as that preceding connection must be). Because Ross Hamilton demonstrates that the distance to the outer edge of the circle would be 0.618 units, if the edge of the Seal Works square is 1 unit. In other words, there is a golden ratio indicated by the ratio of the distance between the square and the new larger circle just described. Specifically, this aspect of the golden ratio (the 0.618) is called the *sacred cut* or the *golden cut* (for some discussion see here and also look at the subsection entitled "Golden Ratio Conjugate").

It would be very difficult to argue that these mathematically demonstrable aspects of the Seal Works were not intentional. Think for just a little while about the beauty and subtlety of the design of this ancient earthwork, and how the size of the circle feature is just right to create a vesica piscis with the distance along the "neck" to the square, and then how the size of the square feature was likewise made to be just right to create a double-square that yields a circle that indicates the golden ratio when placed at that exact same distance (along the "neck" again) back to the original circle!

And this is just one of the amazing earthworks in the Ohio Valley region and the subtle geometrical knowledge that it contains and that it preserved through the centuries (in addition to its celestial mirroring aspects!). There are many more which are equally if not more astonishing, and which Ross Hamilton demonstrates and illustrates in his remarkable book. I should point out that I myself would never have seen these subtle geometrical messages within the Seal Works if Ross Hamilton had not taken the time to discover and present them. He has truly done humanity an important service.

The book also illustrates that most if not all of the works also have a feature which can function as a sort of "hub" around which the outline of the entire earthwork can be (mentally) rotated in a

691

full circle, and that the number of "complete outlines" which fit around such a circle is different in most cases but has some significance to the message of that particular earthwork. Often the fit is a perfect whole number, indicating still further levels of astonishing sophistication and subtlety in the design of these amazing treasures in the landscape.

Knowing what we now know about the incredible ratios built into the Seal Works shape discussed above, the reader will no doubt be as sickened as I was to learn that much of this incredible monument has now been destroyed in order to *dig for gravel*. The short-sightedness of that fact is difficult to express in words, and it seems distressingly symbolic of much of our modern disregard for the sacred landscape and for the incredible efforts and achievements of those who occupied long before us.

These incredible Star Mounds of North America should be known to all humanity, and should be revered -- not ransacked in order to dig gravel pits.

Ross Hamilton's encyclopedic examination of these earthworks and their stories should go a long way towards bringing this vast ancient heritage site the recognition and reverence that it deserves. We, and future generations, owe him a debt of gratitude for his work.

Ambrose and Theodosius

2014 September 15

At the death of the Roman Emperor Theodosius I (AD 347 - 394), the formal panegyric was given by Ambrose, the Archbishop of Milan (AD 340 - 397), and amidst all the eulogy's praise of the departed emperor, Ambrose makes reference to the penitence of Theodosius, weaving this incident quite effortlessly and eloquently into a very beautiful metaphor within a larger theme of humanity's need for mercy and therefore the need to be merciful and forgiving to one another.

The reference itself refers to an incident that took place in AD 390, in which citizens of the region of Thessalonika revolted, apparently in anger at the presence of Gothic soldiers in the service of the empire stationed in their midst. It is worth pointing out that the stationing of military forces among the citizenry is one of the hallmarks of tyrannical states, and the use of foreign-born troops to do it is another pattern in history, as they are less likely to feel an affinity with or sympathy for the local populace.

Note that both of these specific grievances were part of those listed by the authors and signatories of the Declaration of Independence on July 4, 1776 against the King of Great Britain to support their argument that he showed "a history of repeated injuries and usurpations" with the "direct object the establishment of an absolute Tyranny over these States":

> He has kept among us, in times of peace, Standing Armies without the Consent of our legislatures.
>
> He has affected to render the Military independent of and superior to the Civil power.
>
> He has combined with others to subject us to a jurisdiction foreign to our constitution, and unacknowledged by our laws; giving his Assent to their Acts of pretended Legislation:
>
> For Quartering large bodies of armed troops among us:
>
> For protecting them, by a mock Trial, from punishment for any Murders which they should commit on the Inhabitants of these States [...]
>
> He is at this time transporting large Armies of foreign Mercenaries to compleat the works of death, desolation and tyranny, already begun with circumstances of Cruelty & perfidy scarcely paralleled in the most barbarous ages, and totally unworthy the Head of a civilized nation.

We don't have a similar statement from the Thessalonikans who revolted, but we can image that they were similarly outraged by the behavior of the foreign "mercenaries" stationed among them by the Empire (these happened to be Goths), and the impunity with which those mercenaries were allowed to behave and the violations of natural law which they perpetrated – hence the revolt.

Contemporary historians of the time tell us that Theodosius reacted to their revolt by authorizing the Goth commander to slaughter a stadium full of the Thessalonikans, cutting down innocent and guilty alike, as if they were stalks of wheat at harvest time.

Ambrose apparently criticized Theodosius for this ruthless slaughter, barring the emperor from entering church or taking communion for several months, and ordering him to do penance for several months before he could enter again and receive the host (the painting above, from around 1620 or 1621, depicts Ambrose on the right as we look at it, wearing a gold mitre on his head and gold-and-blue robes, barring the entrance to the Milan Cathedral from the hopeful but disappointed Theodosius, who is on the left as we look at the painting, wearing the royal purple, which looks more like what we would probably call crimson today).

Ambrose makes reference to this penance of the emperor in the official panegyric, which can be read in an English translation online at https://archive.org/details/fathersofthechur01281812mbp (beginning on page 307 of that 1953 text, which is actually page 335 of the "e-text" linked, since the e-text includes some front matter in its page count). There, on page 319 in the original book's pagination, or page 347 in the e-text reader linked above, Ambrose says of Theodosius:

> And so because Theodosius, the emperor, showed himself humble and, when sin had stolen upon him, asked for pardon, his soul has turned to its rest, as Scripture has it, saying 'Turn my soul unto thy rest, for the Lord hath been bountiful unto thee.'

The scriptural reference is to Psalm 116:7. The paragraph itself is numbered 28 in the text of Ambrose's speech.

Let's just pause to note that this is actually a fairly astonishing situation. The absolute ruler of the entire Roman Empire, Theodosius I, who is basically the supreme authority and seemingly answers to no one, is apparently being refused entrance to the Mass by the Archbishop of Milan (it is important to know that Milan, located in northern Italy, was then the western seat of the empire, after Constantine earlier moved the center of political power east to Constantinople, a fact which plays a part in the theory discussed below). Not only that, but the emperor is being ordered to repent, humble himself, and do penance by the Archbishop, and the emperor does so for several months before being reinstated to the privilege of taking communion.

The fact that this incident is mentioned in the official eulogy of the emperor by Ambrose is a pretty good indication that it actually happened: if it did not, there would have been plenty of people who could have said so at the time. And so, we can see here an indication that the emperor himself was answerable to the most powerful bishops in some matters, who were obviously seen as representatives of an even higher power.

We might also note that the relatives of those several thousands who were slaughtered in Thessalonika were probably not particularly satisfied at this evidence of the accountability of Theodosius for his war crime -- a few months of being barred from taking communion, and all was forgiven.

In fact, this incident -- and the larger significance of the reign of Theodosius and his actions as emperor -- along with other important pieces of evidence preserved in that eulogy written by Ambrose, provides remarkable support for the revolutionary theory presented by Flavio Barbiero in *The Secret Society of Moses: The Mosaic Bloodline and a Conspiracy Spanning Three Millennia* (2010). In the analysis of Flavio Barbiero, the hierarchical Christianity that Ambrose represented was part of an

696

incredible conspiracy to take over the Roman Empire from the inside, launched centuries earlier by survivors of the destruction of the Temple of Jerusalem and the Roman campaign to suppress revolt in Judea, led by the generals Vespasian and his son Titus in AD 67 - 70 (the main years of the First Jewish-Roman War).

According to Barbiero's analysis, and backed up by a compelling chain of evidence, the culminating actions of that three-hundred-year-long conspiracy took place with the installation of Constantine as emperor in AD 312, and the finishing touches on the victory were overseen by the ruthless Theodosius I, who became emperor in AD 392. For more details on this theory of ancient history, which I believe to be very convincing, see previous posts such as this one[671] and this one,[664] as well as this online explanation of some important aspects of the theory that Flavio Barbiero wrote for the Graham Hancock website (http://www.grahamhancock.com/forum/BarbieroF2.php), and of course see his book itself, which is filled with historical detail and contains a complete blow-by-blow description of the entire takeover and its aftermath.

You may also wish to see *The Undying Stars*, in which I discuss this theory in the context of the evidence that the ancient history of the human race is far different from what we have been taught -- that there is evidence of advanced scientific, technological, spiritual, and shamanic knowledge in humanity's ancient past, at least some of which endured in "the West" right up until the takeover that Barbiero talks about, but which was deliberately and systematically stamped out in the West after this literalistic hierarchical Christian takeover -- a suppression which may have continued in the following centuries and even right up to this day! (Note that my incorporation of his theory into my own analysis should *not* be interpreted as an indication that Flavio Barbiero supports any of my analysis in any way).

The actions taken by Theodosius during his reign can be seen as powerful confirmation of the theory of Flavio Barbiero, many of which are discussed in Barbiero's book. The relationship between Ambrose and Theodosius can also be seen as confirmation of the larger pattern Barbiero describes.

Some of the metaphors and anecdotes used by Ambrose in the panegyric at the death of the emperor, I believe, can additionally be seen as startling confirmation of my placement of the revolutionary theory of Flavio Barbiero within the larger context of the deliberate subversion of the ancient esoteric system that connects the scriptures that became the Old and New Testaments with the ancient wisdom of the rest of the world's cultures -- with an especially close tie to the expression of the ancient wisdom in Egypt. These additional metaphors and anecdotes from the eulogy delivered by Ambrose are not part of Flavio Barbiero's analysis.

According to the theory of Flavio Barbiero, the reason that the generals who put down the rebellion in Judea in AD 67 to 70 were able to become emperors and found the Flavian dynasty (first Vespasian and then his son Titus upon the death of Vespasian) was the financial assistance they received from a vanquished leader of the rebels, who gave them access to the vast hidden treasures of the Temple of Jerusalem itself. This leader and those he selected to come with him were spared from the summary execution that Vespasian and Titus meted out to most of the rebel leaders, and brought back to Rome to enjoy privileged status for the rest of their lives.

Barbiero finds evidence that these leaders, who possessed deep experience running a religious system, decided to set about building a "spiritual Temple" to replace the one that had been burned down by the Romans, and to use it to advance their fortunes in their new setting. They succeeded to a degree that is absolutely astonishing.

Using the twin devices of literalist Christianity (the public and open religious system which they co-opted upon or shortly following their arrival in Rome) and the secret society of Sol Invictus Mithras (the secret and exclusive underground society

698

which they created and operated behind the scenes), they created a mechanism for passing on their vision and accomplishing their goals many generations after the original group which had been brought from Judea to Rome passed from the scene.

The secret society of Mithraism became influential in the Praetorian Guard, then among the imperial bureaucracy, the important imperial checkpoints over commerce such as the customs service, and finally and most importantly over the officers of the Roman Army stationed in various provinces and along the far-flung frontiers of the empire.

All the while, the leaders of the Mithraic system operated in utmost secrecy, protected by strict oaths of secrecy sworn to by all those invited into the society. The old aristocratic families of the empire perceived that something was going on that was a major threat to their control, but they never identified the real nerve center of that threat -- instead, they concentrated on the public face of the two-pronged attack, which was the fledgling literalist Christian religion. Safe from persecution, the leaders of the Mithraic conspiracy could maneuver their pieces on the political chessboard with steady and unwavering focus over the years.

One of their first major victories was the accession to the throne of the murderous Commodus, who reigned from AD 177 to 192, and whom Flavio Barbiero believes to have been the first emperor to have actually been an initiate into the Mithraic society (more discussion of the significance of Commodus, and some of the evidence to support this conclusion, is found in this previous post).[671]

There were many significant setbacks after the reign of Commodus, but in general the interruptions (usually by generals who stormed into power with the support of their legions, only in most cases to be quickly assassinated by members of the Praetorian Guard or other agents of the secret society) were fairly brief (albeit violent and tumultuous), and a new emperor whom they controlled would eventually be maneuvered into the throne by the society of Sol Invictus. Each time, the network's grip over the levers of power increased, and it looked as though their careful campaign was succeeding famously.

This pattern was seriously disrupted in AD 284, upon the accession to the throne of Diocletian (who was born in AD 244 and died in AD 311), who was in fact an initiate of the cult of Mithras. However, Diocletian instituted a cunning (if unwieldy) strategy to ensure that he did not end up being eliminated by the secret society behind the throne if he crossed their will (as had happened to so many of his predecessors). He created what has come to be referred to as "the Tetrarchy," splitting up power with an ally, Marcus Aurelius Maximian, and determining that each of the two co-emperors (Diocletian and Maximian) would name a "caesar" who would be the successor to each when they decided to retire, and who would have certain authority and powers even before that time.

It was a terrible way to run things (just imagine a major corporation with four CEOs at the same time, or even two CEOs plus two "sub-CEOs" with nearly as much power as the CEOs), but by dividing the empire up this way and having it run by an alliance of four, it helped to ensure that the secret society of Sol Invictus Mithras could not bring in a new power-hungry general from another part of the empire to overthrow the existing emperor when he did something they didn't like, as had happened so many times before.

According to Flavio Barbiero, this was the turning point which determined a change in the strategy by the "power behind the throne" -- from now on, they would rule through the *public* mechanism that they had created, literalist Christianity, instead of the *secret society* of Mithras.

And that is why, after a violent power struggle that ensued after Diocletian and Maximian left office, the emperor who took control decided to proclaim openly that he was now a Christian, and that Christianity would henceforth be officially tolerated in the empire. That emperor was Constantine I, who ruled from AD 312 until his death in AD 337.

One of the important moves which Constantine made, and one which adds credence to this theory of Flavio Barbiero, was his decision to move the seat of the emperor to Constantinople and out of Rome. The "power behind the throne" (which had been hidden in the society of Mithras, and which was now identical with the upper reaches of the hierarchy of the Christian church) could thus operate out of Rome without interference from the emperor or from pesky usurpers marching in from other parts of the empire to try to seize the throne.

However, Constantine did not declare Christianity to be the sole religion of the Roman Empire: far from it. His reign was only the first decisive move in the "endgame" of this centuries-long chess-match for control of the Roman world. The "checkmate" would come during the reign of Theodosius I, who was born about ten years after the end of Constantine's reign and who came to power in AD 392. It was Theodosius who administered what Flavio Barbiero calls "the fatal blow to paganism and what little still remained of the ancient Roman senatorial aristocracy" of the Roman Empire (216).

Theodosius ordered in AD 380 that all Christians must profess their faith in the bishop of Rome, thus outlawing alternative dogmas besides the one promulgated by the hierarchical structure controlled by the descendants of those long-ago transplants from Judea.

He outlawed paganism outright in AD 392, decreeing the death penalty for anyone practicing augury or some of the other practices of the traditional Roman pagan rites.

He closed the ancient Oracle at Delphi in AD 390, and ended the Eleusinian Mysteries in AD 392, as well as (according to some scholars) the Olympic games after that same year.

Prior to his death, Theodosius issued decrees that the rule would be split between his two sons (who were quite young at the time of his death), one ruling in the east and the other in the west. It was a decision that would eventually lead to the breakup of the empire: Theodosius I was the last to rule a united empire, and eventually the trappings of the western empire would evaporate completely, and western Europe would be run by a variety of different kings and nobles and -- exercising tremendous control over their actions -- the hierarchy of the Christian religion centered in Rome.

In light of this theory, we can see that Ambrose and Theodosius were actually close allies in the final execution of a long-reaching plan. We can also see that Ambrose, as a high-ranking member of the hierarchy that represented the "power behind the throne" was actually in some ways the superior of the emperor himself -- and his ability to impose sanctions on Theodosius may be an indication of exactly that.

All of the actions above from the life of Theodosius certainly can be interpreted as strong supporting evidence for Flavio Barbiero's theory (among many other pieces of supporting evidence from the preceding centuries, which are recounted in detail in Flavio Barbiero's book).

But there is another incredible piece of evidence hidden in the speech of Ambrose given at the death of Theodosius which also tends to confirm this revolutionary alternative view of Roman history, and to tie it into the larger theory that I expound (and which, as I said, is not part of Flavio Barbiero's theory and should not be taken as an assertion that he supports this wider theory).

In *The Undying Stars*, and in many previous blog posts, I argue that virtually *all* of the world's sacred traditions and scriptures appear to be based upon a common esoteric system of sophisticated celestial metaphor.

I believe that these esoteric ancient myths are actually designed to convey a worldview which is best described as shamanic (and in fact, as what I would call "shamanic-holographic"[167]). This worldview included the belief in an unseen world which actually contains the "source code" for this ordinary, material world which is in fact only a *projection* of the "real world that is behind this one."

It also included techniques for *making contact with* and *actually traveling to* that unseen realm, in order to gain information or make changes to the "source code" there, which could have tremendous impact on events back here in "the ordinary world."

If my theory is correct, then the high-ranking priests who escaped the destruction of the Temple of Jerusalem and the wars of Judea during the time of Vespasian and Titus may well have understood that worldview, and may even have known how to use that knowledge to help them in their plans. They created a literalist, hierarchical religion which suppressed this shamanic ancient knowledge, but they themselves may have known the esoteric secrets and continued to pass them down within their inner circle.

In the eloquent speech of the Archbishop Ambrose given at the death of his fellow-worker Theodosius the emperor, I believe we can see amazing confirmation of this possibility.

In paragraph 40 of his speech (just one paragraph after Ambrose has invoked a literalist vision of heaven and declared that Theodosius and his predecessor Gratian are both enjoying everlasting light and the company of all the saints, as well as a literalist vision of hell and declared that Theodosius' enemies "Maximus and Eugenius are in hell" to "teach by their wicked example how wicked it is for men to take up arms against their princes," thus showing how useful these literalist interpretations are for supporting the divine right of the ruler), Ambrose takes up the example of Constantine's mother Helena, whom he introduces as "great Helena of holy memory, who was inspired by the Spirit of God" (found on page 325 of the original pagination of the 1953 text linked above).

703

Specifically, Ambrose at this part of his panegyric, tells us that once Constantine had killed the last of his enemies and become the sole emperor (Ambrose puts it more tactfully, saying that she was "solicitous for her son to whom the sovereignty of the Roman world had fallen," as if Constantine was just innocently eating lunch one day and the rule of the entire empire happened to fall into his lap), Helena "hastened to Jerusalem and explored the scene of the Lord's Passion" to see if she could find any relics from the Crucifixion with which to aid her son in his new job (paragraph 41, on page 325 of the original pagination of the 1953 text linked above).

As Ambrose explains, it turns out that she did find "three fork-shaped gibbets thrown together, covered by debris and hidden by the Enemy" [that is to say, by the Devil, whom she addresses rhetorically in the preceding paragraph, declaring that she will find proof of the resurrection in spite of the Devil's attempts to conceal it] (paragraph 45, on page 327 of the original pagination of the 1953 text at https://archive.org/details/fathersofthechur012812mbp).

One of these "fork-shaped gibbets," Ambrose tells us, was "the Cross of salvation," which Helena was able to recognize by the fact that "on the middle gibbet a title had been displayed, 'Jesus of Nazareth, King of the Jews'" (paragraph 45, on page 327 of the original pagination). In paragraph 47 Ambrose tells us in addition that:

> She sought the nails with which the Lord was crucified, and found them. From one nail she ordered a bridle to be made, from the other she wove a diadem (page 328 of the original text linked above).

Ambrose, who displays the allegorical virtuosity for which he is noted by historians, then expounds further upon this decision to make one nail into a crown and the other into a bridle:

> On the head, a crown; in the hands, reins. A crown made from the Cross, that faith might shine forth; reins likewise from the Cross, that authority might govern, and that there might be just rule, not unjust legislation. May the princes

704

also consider that this has been granted to them by Christ's generosity, that in imitation of the Lord it may be said of the Roman emperor: 'Thou has set on his head a crown of precious stones.'

[...]

But I ask: Why was the holy relic upon the bridle if not to curb the insolence of emperors, to check the wantonness of tyrants, who as horses neigh after lust that they may be allowed to commit adultery unpunished? What infamy do we not find in the Neros, the Caligulas, and the rest, for whom there was nothing holy upon the bridle?

What else, then, did Helena accomplish by her desire to guide the reins than to seem to say to all emperors through the Holy Spirit: 'Do not become like the horse and mule,' and with the bridle and bit to restrain the jaws of those who did not realize that they were kings to rule those subject to them? [Paragraphs 48 - 51, pages 328 - 330 in the original pagination of the 1953 text].

Now, this is truly remarkable, to anyone who understands the ancient system of celestial allegory – as Ambrose here indicates that he thoroughly and masterfully did.

We have discussed in a series of previous posts, beginning with "Scarab, Ankh, and Djed,"[42] that the Cross of the New Testament clearly parallels the "Djed-column raised up," which also closely parallels the sacrifice of Odin upon the World-Tree, and the Vajra-Thunderbolt of the Vedas, and many other important images around the world.

All of those posts discuss the fact that this sacred symbol of profound significance is also closely connected to the "vertical pillar" of the zodiac wheel, which runs from the winter solstice at the "bottom of the year" straight up to the summer solstice at the "top of the year."

Ambrose has just told us that, upon finding the True Cross, Helena the mother of Constantine took two of the original nails, and made one into a crown and the other into a bridle. The choice to incorporate one nail into a crown is pretty obvious for the "top of the column," which represents both the dome of heaven and also the "dome of heaven" at the top of each human being, the head (microcosm and macrocosm). But how could the other nail's incorporation into a horse-bridle have anything to do with the bottom of the zodiac-Djed-column?

Have a looking at the zodiac wheel below and see if there are any zodiac signs at the bottom which could help explain this choice:

If you said Sagittarius (who at the bottom of the wheel just before winter solstice, peeking out below and partially obscured by the yellow label that says "The Djed raised up"), then I would agree. Sagittarius is an archer, but he is also a horseman (often a centaur). He is indicated in many myths within the system of celestial metaphor by "horse" imagery. It is almost a certainty that Ambrose is indicating an understanding of this ancient system, with this story about Helena and the discovery of the Djed-column / True Cross, and the fashioning of one nail into a crown for the head, and of the other into a bridle for a horse.

But we don't actually have to guess about whether or not Ambrose understood this esoteric system: he provides breathtaking confirmation of the fact in paragraph 46, when he declares:

> She discovered, then, the title. She adored the King, not the wood, indeed, because this is an error of the Gentiles and a vanity of the wicked. But she adored Him who hung on the tree, whose name was inscribed in the title; Him, I say, who, as a scarabaeus, cried out to His Father to forgive the sins of His persecutors [327 - 328 of the original pagination].

What was that metaphor? It is certainly not one which, I would venture to say, most modern Christians are accustomed to hearing their preachers use regarding Christ upon the Cross. But it is one which, the footnote tells us, Ambrose used quite a lot. It is the metaphor of Christ on the Cross as *a scarab*!

And here we see that Ambrose undoubtedly understood the ancient system of celestial metaphor, and what is more that he understood it to the degree that he could employ metaphors common to ancient Egypt when referring to the top of the Djed-pillar! We have seen, in more than one of the previous posts linked above, that the scarab beetle with its upraised arms was connected directly to the sign of Cancer the Crab (at the top of the zodiac wheel, beginning at the point of summer solstice).

The scarab was also esoterically connected to the crown of the skull, as shown in the image below:

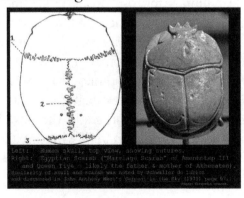

Left: Human skull, top view, showing sutures.
Right: Egyptian Scarab ("Marriage Scarab" of Amenhotep III and Queen Tiye - likely the father & mother of Akhenaten). Similarity of skull and scarab was noted by Schwaller de Lubicz and discussed in John Anthony West's Serpent in the Sky (1979) page 67.

This should suffice to explain why the first nail was made into a crown, while the other went into a bridle -- one is connected to the sign of Cancer and the other to the sign of Sagittarius, two signs at either end of the Djed-column or the "solstice pillar," of which the True Cross is yet another manifestation (and which, we might add, is a central image in many shamanic rituals around the globe).

This deep familiarity with these symbols by Ambrose should also suffice to convince most readers that Ambrose knew full well that there was no physical Cross or nails to be found -- any more than one could "find" the vertical pillar that connects the winter and summer solstices and fashion some kind of crown or bridle out of it (he almost certainly would also have known that both "heaven" and "hell" were zodiac metaphors and not literal eternal destinations for the departed soul). But it should demonstrate conclusively that he was a master of the ancient esoteric system of celestial metaphor -- and that we can assume this knowledge had been passed down to him from his predecessors stretching back through the centuries, no doubt to generations long before the destruction of the Temple of Jerusalem.

Together, Ambrose and Theodosius offer powerful evidence which supports the revolutionary alternative theory published by Flavio Barbiero.

Additionally, pieces of evidence such as the silencing of the Oracle at Delphi and the Mysteries of Eleusis by Theodosius, as well as the clear hints Ambrose drops indicating his masterful understanding of the esoteric system, provide powerful confirmation of the wider theory that this ancient successful conspiracy to take over the Roman Empire was also part of something far bigger: a conspiracy to smash the ancient shamanic-holographic wisdom bequeathed to humanity, and to keep it within a small group who could then use it to "rule the school" (in the "Cobra Kai metaphor") while giving everyone else a literalistic interpretation which they themselves knew to be false.

708

The Therapeutae

2015 February 12

The ancient philosopher Philo of Alexandria (c. 25 BC or 20 BC - c. AD 50), devotes the bulk of the text in one of his most well-known surviving works, *De Vita Contemplativa* ("On the Contemplative Life"), to discussing the important group of followers of the ancient Hebrew Scriptures who were known as the Therapeutae.

Some of the aspects of the Therapeutae as described by Philo include the following:

- They lived in ascetic communities which were open to both men and women, although living most of the time separated by sexes and coming together for special meals and celebrations in which all participated.
- Philo tells us that while such communities could be found in many countries, they were most prevalent in Egypt.
- They gave away their possessions and left the bonds of society and of family, not (Philo explains) out of any misanthropy, but rather out of desire to benefit others by

709

giving away their wealth and to be free of "undue care for money and wealth" and to devote their time to the pursuit of holy mysteries.

- They typically sought out desert places in order to retreat from the crowded life of cities and pursue a spiritual path with a balance between solitary contemplation and communal activity.
- They made their dwelling places far enough apart from one another to give themselves plenty of room for solitude and contemplation, but close enough together to be able to defend each other in the case of attack by robbers.
- They held the ancient scriptures in extremely high regard and devoted much of their time to their study.
- They spent much of their time in meditation and prayer, with prayer specifically mentioned as being offered at the time of the rising of the sun and the setting of the same.
- They favored very simple clothing and food, nothing that was expensive or ostentatious.
- They followed a vegetarian diet, bringing nothing to their table (Philo tells us) that has blood.
- They did not drink wine but rather water.
- They fasted regularly, and in fact seem to have fasted throughout the daylight hours each day according to Philo, saving food and drink for after sunset, as well as at times fasting for longer periods, such as three days or even six days.
- They did not use slaves at a time when slavery was commonly accepted, but instead "look[ed] upon the possession of servants or slaves to be a thing absolutely and wholly contrary to nature, for nature has created all men free" and regarded slavery as a product of injustice, covetousness, and evil.
- They had a high regard for singing and sang sacred songs, psalms, or chants, and that they did so with a dignified rhythm and sometimes with men and women all together, forming two choruses which at times sing different parts

and at times all sing the same, and at times break into stately forms of dance and choreographic expression to accompany their singing.

Translations of Philo's text are easily found on the web, where those interested can consult his descriptions for themselves – one such site can be found here.

Readers who are familiar with some of the texts that have come to be known as the New Testament will recognize some of the characteristics attributed to these Therapeutae in some of the admonitions and recommendations in certain New Testament passages, including the singing of hymns, psalms and sacred psalms (urged in Ephesians 5 and Colossians 3), and the passage in Luke in which Jesus says his disciples must "hate" father and mother and wife and children and brethren (Luke 14:26), which is very similar to Philo's statement that those who left society to join these spiritual communities "desert[ed] their brethren, their children, their wives, their parents, their numerous families, their affectionate bands of companions . . ."

Indeed, the later author and polemecist Eusebius (c. AD 260 or 265 - AD 339 or 340), who was a bishop in the hierarchical and literalist Christian church, recognizes so much in the descriptions given by Philo that Eusebius states very plainly that these ascetic communities described by Philo represented the "multitude of believers" converted by the gospel author Mark when he traveled to Egypt: see chapter 16 of Book II of the *Ecclesiastical History* written by Eusebius (links to all the Books of the work are available online here, and the link to Book II is here). Eusebius further declares in Chapter 17 of Book II (which contains numbered paragraphs -- the paragraph numbers are preserved below in the quotation):

> 3. In the work to which he gave the title *On a Contemplative Life* or on *Suppliants,* after affirming in the first place that he will add to those things which he is about to relate nothing contrary to truth or of his own invention,

he says that these men were called Therapeutae and the women that were with them Terapeutrides. He then adds the reasons for such a name, explaining it from the fact that they applied remedies and healed the souls of those who came to them, by relieving them like physicians, of evil passions, or from the fact that they served and worshipped the Deity in purity and sincerity.

4. Whether Philo himself gave them this name, employing an epithet well suited to their mode of life, or whether the first of them really called themselves so in the beginning, since the name of Christians was not yet everywhere known, we need not discuss here.

Eusebius is here plainly declaring that the Therapeutae and Therapeutrides were the first Christians, going by that name prior to the common use of the term "Christian" itself!

This, Gerald Massey points out (whose arguments regarding the suppression of the original Gnostic nature of the Biblical scriptures by the later literalists was discussed in the preceding post,[239] among other previous posts), is a "fatal admission" on the part of Eusebius, because in arguing that the description given by Philo indicates that the Therapeutae must have been early Christians, and in arguing (as he later does in paragraph 12 of Book II, Chapter 17) that the texts the Therapeutae esteemed so highly were very probably "the Gospels and the writings of the apostles, and probably some expositions of the ancient prophets contained in the Epistle to the Hebrews, and in many others of Paul's Epistles," Eusebius is either completely overestimating the speed with which all those "New Testament" writings were produced (since Philo's description of the Therapeutae was most likely published in AD 40), or else he is inadvertently revealing the truth that all those writings listed were in existence much earlier than AD 40, or were based upon texts that were in existence much earlier than AD 40 (see Massey's *Gnostic and Historic Christianity*, paragraph 34).

712

Now, let's examine this argument a little bit. It seems at first to be fairly flimsy: Massey seems to be placing too much weight on the writings of a literalist bishop who was writing sometime around the first two decades of the fourth century (probably completing it prior to AD 323), long after Philo wrote his *De Vita Contemplativa*. Of course Eusebius could have been making a mistake (or being deliberately disingenuous), so what's the big deal?

And, based on the timeframe of Philo's publication (not later than about AD 50, when Philo died), it would seem that Eusebius was "too hasty" in claiming the Therapeutae as early Christians, and in assuming that the texts they revered and meditated upon must have been early copies of the Gospels and the Epistles. From our perspective in history, it seems very unlikely that the "multitudes" of Therapeutae described by Philo could have possibly had time to spring up and develop the rather rigorous patterns and traditions of ascetic living and worship that Philo describes, and extremely unlikely to the point of impossibility that they could have been doing all that rigorous textual study and exegesis described by Philo upon New Testament texts like the Gospels and Epistles, since virtually no scholar today believes that all of those Christian texts were even written down by the time Philo penned his treatise. Certainly we can ascribe the remarks of Eusebius as simply overly-optimistic or over-zealous, and move on -- right?

And yet Massey, whose analysis often proves to be extremely penetrating, even if there are areas of his analysis with which I strongly disagree, sees in these assertions by Eusebius a "fatal admission" (meaning that Massey believes this admission is "fatal" to the historicist or literalist position which Eusebius held which treats the characters in the scriptures as literal historic persons, and which attacks "pagans," "Platonists," and those who do not share this literalist and historicist version of Christian faith).

Massey does not explain very much further to help us see why this position from Eusebius is so damaging to the historicist approach. He only states by way of explanation that:

> it is impossible to claim the Essenic Scriptures [Massey presents arguments to support his conclusion that the Therapeutae and the Essenes were closely related or indeed the same general school] as being identical with the Canonical records, without, at the same time, admitting their pre-historic existence, their non-historical nature, and their anti-historical testimony. They could only be the same in the time of Eusebius by the non-historical having been falsely converted into the historical.

Again, it would seem that the rebuttal that "Eusebius just made an error" would defeat Massey's argument here . . . except for the fact that Eusebius himself identifies the actual *actions and practices* of the Therapeutae as obviously reflecting the teachings found in the Gospels and Epistles!

In other words, even if the Therapeutae described by Philo did not have the *texts* Eusebius says that they had (and there is no way that they could have, unless those texts were more ancient than the time period during which the Christ of the historicists was said to have lived, which is the possibility that Massey believes is the correct solution), the very fact that these Therapeutae were described by Philo doing things that would *later be incorporated* in the Gospels and Epistles (a couple examples of which were mentioned above) is a strong indication that the New Testament concepts and teachings pre-dated the historical period during which the literalist Christ is said to have lived. This is especially true because Philo, who probably wrote this treatise by AD 40 and certainly by AD 50, is describing these practices as though they are already long traditions.

This is why Massey believes that the descriptions in Philo's text are so damaging to the literalist position. Massey believes that the literalist approach was a later invention, in fact a subterfuge,

714

through which a group of men converted a "non-historical" (that is to say, "allegorical" or "metaphorical" or "esoteric" or "Gnostic") set of spiritual teachings into a "historical" (that is to say, "literalistic, describing events that literally took place in history") faith.

And, in fact, we can find some *additional* extremely interesting aspects of Philo's description of the Therapeutae which appear to add further powerful support to the argument Massey is making regarding the later appropriation by historicists such as Eusebius of teachings or practices that were essentially anti-historical or esoteric and Gnostic.

Interestingly enough, they are the same two characteristics that were argued in the preceding post which declared that the scriptures of the Old and New Testaments of the Bible are essentially shamanic! That is to say, the two features which that post argues generally *go together*: an understanding of the techniques of what can be termed ecstatic trance or shamanic out-of-body travel, and an understanding that the ancient scriptures of the world (to include those texts found in the Bible) are allegorical in nature and that their allegorical nature is intended to point to this shamanic understanding.

In Philo's description of the Therapeutae, he distinctly says more than once that their long study of the sacred texts, and their group exposition of the meanings of these ancient texts, involved an *allegorical* approach, and a search for the hidden (or esoteric) meanings in those texts. For example, in his description of their reading and interpretation of sacred writings, Philo says that the Therapeutae would finish their communal meals and then wait in great anticipation and an even deeper and more reverential silence than that with which their conduct is ordinarily marked as they waited for some one of their number to rise and carefully, patiently, and without any attempts at showy eloquence or cleverness, explain the deeper aspects of some passage of their sacred scriptures. The words with which Philo describes their approach to scripture exposition are significant:

the writings are delivered by mystic expressions in allegories, for the whole of the law appears to these men to resemble a living animal, and its express commandments seem to be the body, and the invisible meaning concealed under and lying beneath the plain words resembles the soul [. . .]

The approach to the scriptures as primarily containing *mystic expressions in allegories*, and the statement that their *invisible meaning* is *concealed under and lies beneath the plain words*, could not be more clear in indicating that the Therapeutae understood their sacred texts to be esoteric in nature.

This, all by itself, appears to demolish the attempts by Eusebius at co-opting the Therapeutae described by Eusebius into the literalistic faith that Eusebius and his colleagues were enforcing during the reign of Constantine. The approach as described is the opposite of the historicist approach. It would also seem to be highly unlikely to have developed to the degree described by Philo in such a short time after the publication of early New Testament texts, even if anyone still believed the Therapeutae could have gotten access to those texts at such an early date. The presence of the type of austere communities devoted to perceiving the esoteric meanings behind and beneath the plain words of the texts speaks to the fact that these texts were undoubtedly of great age themselves.

It is also significant that the Therapeutae appear to have contrasted the "plain words" (what is also called the "exoteric" sense of the passage) as perceived on the surface with the "spirit" that is invisible, and to compare the exoteric sense of the words to "a living *animal.*" The metaphor Philo uses (and which he may well have repeated from the Therapeutae themselves) is most telling. Previous posts have noted the penetrating arguments of Alvin Boyd Kuhn, who maintained that the ancient system used the symbol of the Cross in exactly the same way: with the horizontal component of the Cross symbolizing the "animal" nature of our material existence, when we are "cast down" into

716

this physical world, with that horizontal bar running parallel to the ground in the same way that an animal does, and the vertical component of the Cross represents the spirit which is hidden inside each one of us and in fact within all of creation, and which -- while invisible -- is no less real and which is in fact the truly important aspect of our existence which must be remembered, recognized, and "raised back up," so to speak.

And, in a pattern found throughout the world, where allegorical myths can also be shown to be essentially shamanic in nature, these Therapeutae who valued the ability to seek out the invisible meaning of their sacred texts also appear to have valued and practiced the techniques of traveling to what has been called "non-ordinary reality" or by a host of other names, including the Invisible Realm, the Spirit Realm, and the Dreamtime, and brining back communications from that non-ordinary reality.

Philo tells us that among these communities:

> Therefore they always retain an imperishable recollection of God, so that not even in their dreams is any other object ever presented to their eyes except the beauty of the divine virtues and of the divine powers. Therefore many persons speak in their sleep, divulging and publishing the celebrated doctrines of the sacred philosophy.

Philo does not go further than this, and at first glance it is easy to simply skip over it as a rhetorical exaggeration on Philo's part, going over-the-top in his idealized description of the Therapeutae to the point of saying that they even *dream* of only virtuous and spiritual matters (no impure or even simply mundane dreams among this community).

But, while we might write these lines off as a clumsy and unbelievable embellishment by Philo, he doesn't merely state that they only dream of spiritual and virtuous matters: he states quite clearly that many persons *speak* in their sleep, and when they do so they divulge sacred matters which might otherwise have remained hidden.

717

When he adds that detail, it changes the tone of what Philo is saying altogether. He is not simply saying that the Therapeutae are so single-minded that they even dream about spiritual things: he appears to be indicating that many members of their communities regularly enter into a state in which they speak messages divulging hidden teachings. This mode of communication is strongly suggestive of the messages brought from the Invisible World by other practitioners of sacred ecstasy or trance, such as the Pythia of Delphi.

Philo also states during his descriptions of their communal songs and chants and even dances that the participants seem to enter a state of "intoxication" at times (especially when they are continued all night until sunrise).

Both of these features -- an esoteric approach to sacred scripture, and a regular use of the techniques of ecstatic trance -- have been strongly condemned by the literalistic and historicist Christianity that polemicists such as Eusebius advanced (some might counter that church fathers including Eusebius did not deny the allegorical aspects of scripture, but no one can argue that they would have strongly condemned any suggestion that the scriptures were primarily or even *exclusively* allegorical, and that they were not intended to be understood literally and historically).

And this evidence appears to be powerful support for Massey's general argument, which is that the historicist bishops and polemicists, such as Eusebius, successfully stamped out a much older approach and co-opted many aspects of its teachings and many of its scriptures and turned them to their own ends.

In fact, Massey provides substantial evidence that the ancient wisdom that was historicized and co-opted by the literalists stretched back into much greater antiquity -- and that it can be clearly seen in some of the most ancient texts and teachings of Egypt in forms which suggest that the outlines of the doctrines of the Therapeutae, and the outlines of the texts that the literalists later appropriated, existed for millennia before showing up in the writings of Eusebius or Philo.

718

Indeed, it can hardly be denied that many of the features of the Therapeutae lifestyle shown in the list above have not characterized most of what we would recognize as "Christian teaching" through the centuries.

Christianity is not generally associated, for instance, with vegetarianism.

Christianity is not widely associated with an emphasis on communal living and the renunciation of possessions and property (with some notable exceptions from time to time).

Christianity is not historically associated with the rejection of the idea of having slaves or even servants, and the teaching that to do so is evil and contrary to nature (again, with some important exceptions).

While there are notable historical exceptions, which could be profitably examined and discussed, it cannot be denied that historic, literalistic Christianity has generally taught quite emphatically that the killing of animals for food, the amassing of property, and even the keeping of slaves are all *explicitly condoned* by the sacred scriptures (not condemned: *condoned*).

However, there are some other traditions around the world where the above teachings were widely taught, and practiced, and where they influenced entire cultures and civilizations -- in some places (especially those which were not conquered by the Roman Empire, which by the time of Constantine was increasingly dominated by literalist Christianity) aspects of some of these teachings continue right down to the modern era.

Clearly, the descriptions of the Therapeutae by ancient authors (as well as the possibly-related sect of the Essenes, of whom more at a later time) constitute an extremely profitable line of study, and one which appears to contain powerful evidence to support the theory that a literalist re-interpretation was mistakenly -- or, as other evidence seems to suggest, deliberately and deceptively -- substituted for a far more ancient esoteric approach, and that this switch took place during the first four or five centuries AD within the Roman Empire.

Examining some further aspects of this line of investigation may well turn up some additional surprises, which will be the subject of future posts to follow!

Fragments of a lost whole

2015 October 16

In a justifiably well-known quotation from the introduction to the ground-breaking 1969 book *Hamlet's Mill: An essay on myth and the frame of time*, Giorgio de Santillana quotes from an earlier essay of his which he wrote in 1959 (and which, although I may be incorrect on this, I believe was published in 1962 or 1963 as an article in a literary journal, under the title "In the High and Far-off Times").

Often quoted for its most memorable lines, the full quotation is worth considering carefully (from pages 4 and 5 of the edition linked above):

> The dust of centuries had settled upon the remains of this great world-wide archaic construction when the Greeks came upon the scene. Yet something of it survived in traditional rites, in myths and fairy tales no longer understood. Taken verbally, it matured the bloody cults

intended to procure fertility, based on the belief in a dark universal force of an ambivalent nature, which seems now to monopolize our interest. Yet its original themes could flash out again, preserved almost intact, in the later thought of the Pythagoreans and of Plato.

But they are tantalizing fragments of a lost whole. They make one think of those "mist landscapes" of which Chinese painters are masters, which show here a rock, here a gable, there the tip of a tree, and leave the rest to imagination. Even when the code shall have yielded, when the techniques shall be known, we cannot expect to gauge the thought of those remote ancestors of ours, wrapped as it is in its symbols.

> Their words are no more heard again
> Through lapse of many ages . . .

Think for a moment of the implications of what is being asserted in the above statements. The author is saying that all of the world's surviving myths, rites, and even fairy tales are "tantalizing fragments of a lost whole" -- in other words, that they are all *connected*, or were at one time.

But, he says, like the "mist landscapes" of Chinese art, vast portions of this "great world-wide archaic construction" are now hidden from our sight, and all the little pieces or bits of ground which we can see today *seem* at first glance to be disconnected.

Now, with the great ancient structure now in ruins, we can only pick out "here a rock . . . there the tip of a tree," or perhaps the corner of a gable on a roof, or even some mysterious artifact whose original purpose is now unknown.

Modern "isolationist" theory, and most conventional understanding of ancient sacred scriptures, treats this landscape as if each part *is not* and *never was* part of a single unified whole, that they are not descended from some single vastly ancient unified design.

It is as if someone were to stand on one little "island in the mist" in a Chinese painting and declare that they are part of their own separate landscape, and that they have no connection to what is going on over on the other side of the mist-covered parts of the same painting.

Or, to use the metaphor from the first line of the above quotation, it is as if a vast ancient ruin of great antiquity can be seen poking up in various places around our entire planet, sometimes scattered in the sands of deserts, or peeking out of the vines of jungles, or rising up out of lonely swamps or marshes or even windswept tundras . . . and all of it apparently once functioning together as a single overarching and interconnected structure.

After making this remarkable assertion, the authors of *Hamlet's Mill*, both university professors, then proceed to fill hundreds of pages with evidence to back up their reconstruction of the now-vanished lines of connection, connections which are now obscured by thick mists, or which must be lying deep beneath the shifting sands and the jungle overgrowth between the few fragments of ancient ruins that we can still see above the surface. In support of their argument, they provide hundreds of examples from mythology, images of the stars and constellations and of artifacts and art from ancient times, and extensive quotations and footnotes from scholars during the twentieth and previous centuries in the fields of mythology, religion, anthropology, history, linguistics, and literary criticism.

The "great world-wide archaic construction" whose now-fragmented pieces they are examining appears to have encompassed aspects of measure (of distance and of time), of music, of art and proportion, of architecture, of history, of astronomy, of cosmology, of physics, of what the authors call a "kind of *Naturphilosophie*" (p70), of consciousness, of the gods and their realm . . . and perhaps of much, much more.

And of course, central to this structure in some way appears to have been the myths -- like pillars holding up the vast over-arching design, not one single pillar perhaps but numerous pillars located in every single part of the globe, all different in some way and yet all connected and all mutually-supporting.

Carrying further the work that von Dechend and de Santillana have done in tracing the interconnectedness of these myths and myth-systems and cosmologies, we can see that it becomes harder and harder to deny that all the world's sacred traditions appear to share a common system, an esoteric system, a system founded upon the heavenly cycles and on celestial metaphor.

Since first encountering *Hamlet's Mill,* I have over the course of many years of investigating and pondering and even dreaming about various ancient myths and stories, begun to see the outlines of this system -- particularly in regard to the myths of the world -- more and more clearly, and now believe that its outlines and connections can be seen even more extensively than even was visible at the time that de Santillana and von Dechend were writing. In addition to the "rock here" and "tip of a tree there," additional features have occasionally emerged out of the flowing mist, to the point that the existence of this vast ancient system cannot be denied.

The preceding post, explaining in fairly extensive detail all the celestial connections and clues preserved for our understanding in the ancient Hebrew scroll of Numbers (part of the Pentateuch) and specifically in the story of Balaam and the Ass and their encounter with the Angel on the way, is just one more example of an analysis that could be repeated again and again and again, using other sacred stories from around the world, including from Africa, Scandinavia, ancient Greece, ancient India, ancient Egypt, the Americas, China, Japan, and many more.

Significantly, the same celestial analysis can also be repeated again and again in the stories of the Old Testament and New Testament (see the partial list here -- and dozens more could be provided).

724

The fact that the *very same system of celestial metaphor* can be seen forming the foundation for the sacred stories of the Bible that forms the foundation for virtually all the other myths, scriptures, and sacred traditions, from every part of the planet, shows that -- far from being somehow set apart from the rest of humanity, as some literalist interpretations have maintained -- the ancient wisdom preserved in these particular scriptures, and the stories themselves, are almost certainly also fragments of the same ancient structure.

To maintain that they are somehow disconnected and independent and self-contained is akin to some isolated group dwelling amidst some small part of the remains of this vast world-wide construction, ignorant of its original ancient purpose, standing on top of their own local clump of ruined blocks, and declaring that their piece of the whatever it used to be is superior to all the others, and that their portion cannot possibly be connected to the rest of the ruins, because obviously there are now great sections of wasteland in between the various places, and they could never have once functioned all together (and besides, some of them even say, all those other ruins are fakes, or at best copies of this one section over here in our territory).

The question, then, of whether the myths are all fragments of an ancient whole thus becomes one of incredible importance, and the evidence (overwhelming in its abundance once the system begins to emerge) that they are all connected by a common system of celestial metaphor thus becomes evidence which argues that the myths of humanity actually unite us, rather than divide us.

Taken *literally*, the myths and stories tend strongly towards *dividing* us from one another. Their literal interpretation has been very frequently used in the past (and indeed right up through the present) to divide instead of to unite different groups and branches of the human family.

In part, literal interpretations tend strongly towards division because, taken literally, the myths are understood to be about *external*, literal, historical individuals and groups -- individuals who are the ancestors of *some* of us but not of all of us.

For example, the Old Testament story of Shem, Ham and Japheth has been used in previous centuries to divide all people on earth into the supposed descendants of one or another of these three sons of Noah -- and to justify all kinds of oppression, denigration and mistreatment of one or another supposed set of descendants based on a literal interpretation of what I believe can be shown quite conclusively to be based upon a celestial metaphor.

But when the story is seen to be, like virtually every other sacred myth or tradition from around the globe, a celestial allegory, then it can no longer be used to claim that some are descended from one or another of the figures (if the figures are constellations in the heavens). Once this allegorical aspect is perceived, then the esoteric meaning of the myths can begin to impress itself upon our understanding: they cannot be about human progenitors of various people-groups, and so they must be about something else -- and that "something else" that they are about, I believe, is our human condition as simultaneously spiritual and material beings, inhabiting a cosmos which is also simultaneously spiritual and material . . . and all the incredible ramifications of those twin aspects of our incarnate existence.

The Star Myths of humanity are not about someone else: they are about each and every one of us, and the motions of the stars and the other heavenly cycles were selected because they perfectly allegorize our own spiritual experience of descending from the spiritual "realms above" into this apparently-physical material existence, and the necessity of our remembering the spiritual realm from whence we originally came, and of reconnecting with it and elevating it both within our own lives and in everyone and everything else around us, while we are "down here" in this "valley below."

And if they are about each and every one of us, then again it is clear that the myths *unite us.* If their message is primarily esoteric, and applicable to each and every human soul, then they are about you, and for your benefit, and they are about me and for my benefit as well.

When taken literally, however, the opposite can tend to happen -- they are seen as and taught as being *about and for* one group, and not for anyone else. That group can be defined by supposed physical descent from this or that historical person (as in the case of Shem, Ham and Japheth cited above), or it can be based upon acceptance of one set of literal interpretations and assertions, in which case those who believe and accept those assertions separate themselves from everyone else who does not.

All of these literal misinterpretations can be seen as a form of "living in one part of the mist landscape, and denying its connection to the rest of the picture," of falsely dividing and isolating and, if you will, "getting lost in the fog."

They are also a form of "physicalizing" teachings that, properly understood, are spiritual and esoteric in nature. De Santillana strongly hints at this mistake when he says, in the original quotation cited above, that "Taken verbally . . . " [and by this I believe he means the same thing I am saying when I say 'taken literally' or 'literalistically'] these ancient myths were mainly incorporated into "bloody cults intended to procure fertility" and to bend the ambivalent forces of nature for one's own interests.

In other words, they were basically "physicalized" and turned towards material ends, almost towards "animal ends," rather than pointing us towards the invisible and spiritual truths and the elevation of the divine spark (although occasionally, as in the teachings of Pythagoras and Plato, their "original themes could flash out again" and light up the general darkness).

Today, thanks to the tireless work of many, many researchers in the decades since de Santillana and von Dechend wrote *Hamlet's*

Mill (many of those researchers inspired in their own work very directly by the "tantalizing fragments" which they encountered in *Hamlet's Mill*), we can see even more clearly than when that book was published that we do indeed stand within and among the mighty ruins of a great, world-wide archaic construction.

Indeed, new and astonishing aspects of the *physical* remains of that ancient construct continue to come to light -- such as Nabta Playa and Göbekli Tepe, neither of which were known in 1969 when *Hamlet's Mill* came out.

Much of this "mist landscape" still remains shrouded with mist, to be sure.

But the fact that these fragments are part of what was once some ancient and very sophisticated unified system of understanding is now almost impossible to deny.

The ramifications of this fact are extraordinary.

But we can be encouraged in the knowledge that at least one of the important ramifications of the assertions made in that quotation, so long ago, and the evidence which continues to be found in its support, is the fact that we are all connected, that all of us as human beings share an incredible (if still mysterious) past history, and that we are all inheritors of the precious treasure of the ancient wisdom that was preserved in the world's myths for our benefit -- an inheritance that belongs to us all, and not just to some.

Geology, mythology, cataclysms, and the world's ancient wisdom

2016 October 13

In the late 1970s, the pioneering researcher, author, lecturer, guide and "rogue Egyptologist" John Anthony West published *Serpent in the Sky: The High Wisdom of Ancient Egypt.* There, he set forth a revolutionary proclamation: *overwhelming* evidence points to the conclusion that "Egypt did not 'develop' her civilization, but inherited it."

In the final chapter of that book, he set forth evidence that the weathering on the Sphinx indicates a date of original construction far older than the conventional dating for its construction or conception between 2500 BC and 2600 BC (conventional history dates the construction of the Sphinx to the reign of Khafra aka Chepren, c. 2570 BC).

John Anthony West attributes the germ of this revolutionary idea to a comment included in a book published in the early 1960s in France (and not yet published in English when Serpent in the Sky was first written) by R. A. Schwaller de Lubicz, entitled *Sacred Science: the King of Pharaonic Theocracy* (published in France in 1961 as *Le Roi de la theocratie Pharaonique,* first published in English in 1982).

729

There, on page 96, Schwaller writes:

> A great civilization must have preceded the vast movements of water that passed over Egypt, which leads us to assume that the Sphinx already existed, sculptured in the rock of the west cliff at Gizeh, that Sphinx whose leonine body, except for the head, shows indisputable signs of aquatic erosion.

At this point, Schwaller includes a footnote, where he continues:

> It is maintained that this erosion was wrought by desert sands, but the entire body of the Sphinx is protected from all desert winds coming from the West, the only winds that could effect erosion. Only the head protrudes from this hollow, and it shows no signs of erosion. [footnote 28 on page 96].

Schwaller then goes on to cite some ancient sources saying that the Sphinx was actually buried in the sand even in ancient times, and that pharaohs (specifically Thothmes IV, an 18th dynasty king who reigned until his death which may have been in 1391 BC, and who is thought to have been the father of Amenhotep III, the father of Akhenaten, thus making Thothmes or Thutmose IV the grandfather of Akhenaten) were visited in dreams and advised to clear away this sand buildup, a task that Schwaller suggests probably took place more than once over the centuries during dynastic Egypt.

All of this, John West perceived, meant that the Sphinx must be much older than 2500 or 2600 BC (to have sustained such weathering, despite spending long centuries buried up to the neck in sand – and, what is more, to have sustained weathering that strongly suggests *water* erosion rather than wind erosion).

John Anthony West's realization of the *significance* of this remark by Schwaller de Lubicz was absolutely essential to the research which was to follow, regarding the evidence pointing to the existence of advanced ancient civilizations centuries or even millennia before dynastic Egypt.

730

Equally significant was the move that John West made next, after publishing the first edition of *Serpent in the Sky*: he contacted a friend who was an English professor at Boston University, to see if that professor knew any geology professors who might be interested in looking at the evidence in order to help corroborate this revolutionary hypothesis.

Fortunately for those who value the quest to uncover the true story of our planet's past and humanity's ancient history, that English professor decided to put geologist Dr. Robert Schoch in touch with John Anthony West – a Yale-trained professor who was open-minded enough to travel to Egypt to see the Sphinx for himself in 1990 . . . and the rest is history.

Dr. Schoch confirmed that the body of the Sphinx as well as the walls of the Sphinx "enclosure" showed unmistakable signs of water erosion -- and not just a little water erosion, but what may have been many centuries of water erosion. Because the available evidence and the accepted climate models posit that the Giza plateau has been extremely dry and arid since at least 5,000 BC, this undeniable evidence of massive water erosion on the Sphinx suggests that it was originally conceived and created no later than 5,000 BC, and by the extent of the weathering probably several centuries or even millennia earlier than that.

Along with his lifelong curiosity about ancient history and his extensive training and expertise in the field of geology, Dr. Schoch has the ability to see important connections between different sites and different evidence which relates to this critical subject, and in the decades since that first visit to Egypt he has published many books detailing this evidence, and its revolutionary implications for our view of humanity's ancient past.

These include *Voices of the Rocks* (1999) and *Forgotten Civilization* (2012), and he will be releasing a new book written with co-author Robert Bauval and detailing additional developments which support the overall thesis in March of 2017, entitled *Origins of the Sphinx: Celestial Guardian of Pre-Pharaonic Civilization.*

When Dr. Schoch first lent his expertise in geology to the observations of John Anthony West (who caught the idea from Schwaller de Lubicz's 1961 articulation of the idea) beginning in 1990, the idea that a culture capable of creating the Sphinx could have existed prior to 5,000 BC was so revolutionary that it was attacked mercilessly by conventional Egyptologists as well as conventional historians, archaeologists, and defenders of the prevailing paradigm that argues humans were primitive hunter-gatherers or nomadic shepherds and herders prior to the first agricultural civilizations in the third or fourth millennia BC. As the pioneering geologist supporting that idea, Dr. Schoch's work and professional reputation were also viciously attacked by the same defenders of the conventional paradigm.

However, the evidence for the existence of extremely sophisticated civilizations prior to 5,000 BC received incredible confirmation in the late 1990s and early 2000s with the excavation of the complex at the site known today as Gobekli Tepe, which dates to at least 10,000 BC, and which was deliberately buried for reasons unknown by massive amounts of transported dirt, rubble and debris not later than about 8,000 BC.

This fact of the burial of the site means that whatever civilization conceived of and executed the breathtaking artwork and massive but exquisitely engineered stones in the Gobekli Tepe complex must have done so well before 8,000 BC.

This discovery vindicates the arguments of Robert Schoch and John Anthony West (and Schwaller de Lubicz) about the evident antiquity of the Sphinx, and the undeniable evidence pointing to the existence of an incredibly sophisticated ancient culture long before conventional historical timelines will admit to such a possibility.

I had the opportunity to listen to both Robert Schoch and John Anthony West deliver terrific presentations at the 2016 *Conference on Precession and Ancient Knowledge* about two weeks ago, and it is no exaggeration to say that the magnitude of the importance of the paradigm shift which they and subsequent researchers have ushered in since the first publication of *Serpent in the Sky* in 1979 is difficult to quantify.

All those who realize the importance of better understanding the true story of the ancient history of our planet and of humanity should be extremely grateful for the insight – and the courage -- that both of them have displayed in tackling this vital question, and the graciousness with which they both continue to share their knowledge with the world. The discovery at Gobekli Tepe was not even *dreamed about* when their study of the Sphinx began -- but fortunately their earlier work (and earlier "trial by fire" of defending their observations and insights) enabled them to immediately grasp its significance and see how it (and other evidence around the world) may fit into the big picture, and their work also set the stage for subsequent researchers and enabled new insights from others who have brought other new evidence to light.

It should at this point be widely accepted that some sort of very advanced and very sophisticated civilization existed on this planet *thousands of years* prior to the time that the conventional paradigm has been telling us that human civilization first appeared.

In addition to the overwhelming *geological* evidence and overwhelming *archaeological* evidence which points to such a conclusion, I would like to suggest that there is also undeniable *mythological* evidence which supports the very same conclusion of an advanced but now forgotten ancient civilization.

As I have detailed with several hundred examples in this blog and in my multi-volume series *Star Myths of the World, and how to interpret them* (Volumes One, Two and Three so far, totaling

1,966 pages) as well as *The Undying Stars* (444 more pages of evidence), the ancient myths, scriptures and sacred stories found in *virtually every culture on our planet*, on every inhabited continent and island, can be shown to share a common foundation which employs a very specific way of looking at the constellations and celestial cycles, and a very specific way of encoding those constellations and cycles into myth.

This system can be shown to be operating in myths and sacred stories from Africa, Australia, ancient China, ancient India, ancient Japan, ancient Mesopotamia, ancient Egypt, ancient Greece, northern Europe, other parts of Asia, the Americas, and across the length and breadth of the Pacific.

Elements of myth-systems founded upon this metaphorical methodology can be seen in the Pyramid Texts of ancient Egypt, dated to the period between 2300 BC and 2400 BC -- and it is fairly evident that the system was already extremely mature and well understood at that point in time (it was not being introduced for the first time in the Pyramid Texts: of that we can be fairly certain).

Clear examples of this same worldwide system can be seen to be in full operation throughout the Gilgamesh series of myths found in the cuneiform tablets of ancient Sumer and Babylon, some of them thought to have been etched in the same third millennium BC that saw the inscription of the Pyramid Texts in Egypt.

Star Myths of the World, Volume One, gives some examples and explication of the celestial system as it is seen in these ancient texts from Egypt and Mesopotamia.

But Volume One also provides conclusive evidence that this *very same system* of celestial metaphor can be seen to be at work in the sacred stories of the Aboriginal peoples of Australia -- thought to be one of the oldest continuous cultures to be found anywhere on the planet, with some scholars attributing their arrival in Australia to dates of unbelievable antiquity, perhaps 45,000 years ago or even earlier according to some theories. It is thought that

734

their culture may easily go back 10,000 years in fairly unbroken succession, making the Aboriginal people of Australia contenders for the title of very oldest continuous and surviving culture to be found anywhere on Earth.

If they somehow possess this same system, then this seems to be strong evidence arguing for an origin in extreme antiquity, perhaps among the now-forgotten civilizations or cultures responsible for the Sphinx and for the ruins at what we know today as Gobekli Tepe. It is possible to argue that the Aborigines of Australia or their predecessors originated the system, and somehow spread it to all the other cultures of the world in the subsequent millennia, but given the general consensus regarding the relative isolation of the Aboriginal peoples of Australia from ongoing contact with peoples on other continents, this hypothesis seems to be less likely at this time.

It seems to be much more likely that some very sophisticated ancient culture (both *technologically* sophisticated -- judging by the incredible workmanship at Gobekli Tepe and Giza, as well as their apparent understanding of phenomena such as precession and other advanced astronomical and mathematical concepts -- and also *spiritually* sophisticated, judging by the beauty and power and subtlety of the worldwide esoteric system of myth and celestial metaphor) was nearly wiped out by some sort of massive catastrophe that may have operated on a global scale, at some time in the distant past (probably earlier than 9000 BC).

Robert Schoch puts forth reasoned arguments backed up by ice-core data and scientific papers published in academic journals which indicate the possibility of a Coronal Mass Ejection from our Sun at a point in time he believes would correspond to 9700 BC. For details of his argument, please see his book *Forgotten Civilization*, and also his online article here.

Other researchers have proposed comet strikes or other catastrophes. Longtime readers of this blog and of my first book will know that I also believe there is overwhelming evidence on

735

our planet which points to one or more cataclysms of almost unimaginable violence in Earth's distant past.

Intriguingly enough, the sacred site known as Painted Rock (located in the same county in California where I live), appears to provide important additional evidence supporting the revolutionary paradigm of ancient human history described above. I have written two extensive blog posts about this incredibly significant ancient sacred site, available online).

Even conventional archaeologists agree that Painted Rock, located in a long narrow valley known today as the Carrizo Plain, and the surrounding country was a site of extensive human occupation and activity stretching back to at least 10,000 years before the present (to at least 8,000 BC).

That date in itself is obviously significant in regards to the paradigm of ancient human history -- but even more so because the Painted Rock formation itself is home to a class of unusual and distinctive rock art which (even in its tragically damaged and vandalized condition) bears the unmistakable evidence of the very same system of constellational outlining that informs the myths, scriptures and sacred stories found everywhere else in the world, from Australia to Africa to ancient China to ancient India, ancient Japan, ancient Greece, ancient Mesopotamia, across the Pacific Islands, and even in just about every story in the Old and New Testaments of the Bible!

Even in its now criminally-defaced condition, the sacred enclosure can be seen (to this very day) to display outlines that I am convinced correspond directly to the constellations Scorpio, Ophiucus, and Aquarius (some recent photographs of these are shown in the previously linked posts). Additionally, photographs taken prior to the deliberate destruction of some of the artwork appear to indicate undeniable correspondences to the constellations Pisces, Aquila, and Gemini (and probably to others as well).

736

Although some scholars believe that the rock art itself may not date all the way back to 8,000 BC, that is not firmly established, and because there are datable markers of human habitation that may go back all the way to 8,000 BC, the presence of sacred rock art with clear use of recognizable constellational (and indeed *zodiacal*) outlines in the most distinctive rock formation in the landscape of the Carrizo Plain appears to be a strong argument for the use of these constellations in a sacred capacity in North America at dates approaching or even surpassing the antiquity of the earliest surviving Pyramid Texts of Egypt or cuneiform tablets of Mesopotamia.

Perhaps equally important and intriguing is the fact that this ancient sacred stone enclosure, which preserves the rock art described above, is itself evidence of potentially catastrophic or cataclysmic geological processes in Earth's distant past.

The massive stone ring known today as Painted Rock is a sandstone extrusion located towards one edge of an elongated basin or canyon, which even now contains an alkali dry lakebed consisting of a layer of mineral salts, similar to the salt flats of Utah.

Below is a screenshot from Google Maps showing the location of Painted Rock in the Carrizo Plain:

Note in the above image the white patch in the center of the plain which is the satellite image of Soda Lake, a dry lakebed composed of alkali salts in powdery form. On either side of this dry lakebed we can see ranges of mountains extending in a kind of "V-shaped" pattern that comes together in the southeast (bottom right of the screen). The upper chain of uplifted ridges or mountains is the Monte Diablo range, and it follows and is undoubtedly a product of the long, straight line of the San Andreas fault, which can be seen running alongside the southern edge of the ridgeline.

In the image, I have labeled Soda Lake and pointed towards it with a yellow arrow. I have labeled the Monte Diablo Range in a cream or tan color. And I have labeled the San Andreas Fault in bright green, with a series of small green arrows pointing to the fault line itself.

Painted Rock itself is located due south from the southernmost point of Soda Lake.

Here is a closer view, zoomed in to the horseshoe-shaped natural stone enclosure of Painted Rock:

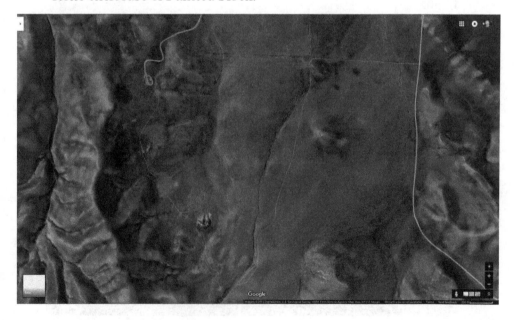

738

And here is an aerial photograph of Painted Rock, showing the same outline. Note that in the above image, we can clearly see that the opening of Painted Rock is to the north (even though Painted Rock itself looks very small in the satellite photo above). The image below, from Wikipedia, is taken from an angle that positions the opening to the lower-right corner of the screen (that way is generally north in the photo below):

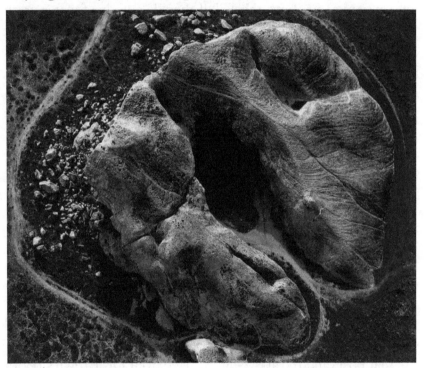

Note that the upper arm of the formation as we face the image shown above is smooth and contains countless parallel striations over its surface. Most geological descriptions of this formation describe it as sandstone.

I believe that Painted Rock shows clear signs of being a *liquefaction mound*, as defined and described by hydroplate theory originator Dr. Walt Brown, in a chapter of his book devoted entirely to the important geological phenomenon of liquefaction: http://www.creationscience.com/onlinebook/Liquefaction.html

Liquefaction is a geological phenomenon that can be described as generally "catastrophic" in nature, or as a phenomenon that is primarily found in conjunction with catastrophic processes -- specifically earthquakes and flooding. It is a phenomenon in which the water already present in soil is forced upwards by the sinking and settling motion of the soil particles (usually caused by the vibration of earthquake forces, or by the compressing action of waves moving above the soil, if there is a body of water or an ocean above the soil in question). This upward-moving water can cause the soil particles to suddenly lose their cohesion and solidity and turn into a liquid slurry.

Here is the Wikipedia article describing liquefaction:
https://en.wikipedia.org/wiki/Soil_liquefaction

Here is a page in Walt Brown's online book describing liquefaction, its causes and results (in much more detail than the Wikipedia article):
http://www.creationscience.com/onlinebook/Liquefaction7.html

And here is a blog post I published almost five years ago, discussing the evidence that the mighty terrain feature known as Uluru in Australia (also known as Ayers Rock) is a *liquefaction mound*, based on the discussion of this geological phenomenon presented by Dr. Brown:
http://mathisencorollary.blogspot.com/2011/11/geography-of-uluru-and-kata-tjuta.html

As that post (and the above-linked page from Dr. Brown's book) explain, a liquefaction mound is created when the upwelling slurry in liquefacted soil erupts out of the ground and spills out in a kind of slumped-bowl shape. You can see examples of such eruptions in photographs and videos from recent major earthquakes of the past five years, such as in Christchurch, New Zealand or Tokyo, Japan. Here is a photo of some such eruptions from Christchurch:

As Dr. Brown explains (in a quotation I cited in my discussion of Ayers Rock / Uluru), these slushy "volcanoes" will usually wash away -- but if they happen to erupt into a basin of warm water that remains relatively still for hundreds of years and cools, the sandy mound can harden into sandstone -- and then it may survive, even after the water in the natural basin eventually breaches and drains away, or evaporates over hundreds or thousands of years.

In fact, as I point out in that article discussing Uluru, liquefaction mounds (and their cousins, liquefaction plumes) are often found in ancient basins that once contained trapped inland lakes or seas.

And this description matches exactly what we find in the formation of Painted Rock in the Carrizo Plain (even though I was completely unaware of the *existence* of Painted Rock and the Carrizo Plain back in 2011 when I wrote that post about Uluru in Australia).

Clearly, Painted Rock is situated in an ancient basin – the Carrizo Plain.

Equally clearly, that ancient basin once held a trapped inland sea -- the dry lakebed known today as Soda Lake is unmistakable evidence of this fact (and the fact that the water in the basin probably had no outlet, leading eventually to the formation of a *salt flat* once the shallow inland sea dried up).

Additionally, the Carrizo Plain is traversed along its long axis by the line of the San Andreas fault – which indicates the possibility of earthquakes in the distant past. Earthquakes are often associated with the eruption of liquefacted soil, or the creation of liquefaction mounds.

Below are some recent photographs I took of the awe-inspiring terrain feature of Painted Rock, showing that it possesses all of the attributes described by Dr. Brown in his discussion of liquefaction mounds:

The above image is looking towards the southeast, and approaching from the northwest. The opening is to the left of the photo as we face it (the leftmost portion of the formation is pointing almost due north – our left). You can see the first-quarter waxing moon of October 08, 2016 directly above the highest portion of the liquefaction mound that forms Painted Rock. The moon in this photo is moving across the sky from the east but has yet to reach its zenith.

Deep striation can be seen along the back of the mound's shoulder, as we face the formation, towards the right side as we view it.

It may not be easy to view from this distance, but the distinctive "undercut edge" which is seen in many liquefaction mounds (including some on the page from Dr. Brown's book, linked above) can be observed on the far right lower edge of the formation of Painted Rock. This undercut edge will be more visible in subsequent photos.

Below we see another aspect which is very characteristic of liquefaction mounds: "water vents" along the *sides* of the feature. These vent-holes are caused by water escaping the liquefacted soil when the mound is still erupting. They are not caused by later weathering -- and thus are not found along the top surfaces of liquefaction mounds, as we might expect if they were caused by rainfall over the centuries, for example.

The vent-holes in the side of Painted Rock as we approach it from the northwest resemble a face giving a fierce warning to all who approach (certainly appropriate, given the disrespectful vandalism that took place to the sacred site's artwork during the first part of the twentieth century):

Once again, we can clearly see the smooth striation which curves over the "shoulder" of this ancient liquefaction mound. Similar striation can be observed on Uluru / Ayers Rock in Australia.

Below is another view of the smooth striation of Painted Rock:

The very existence of this terrain feature argues that the entire Carrizo Plain was once under water, and for a long enough period of time that the erupted sand from below the surface was able to harden to the point that it would not be swept away when the water in the basin either drained out or evaporated.

Note also that the presence of massive salt deposits in the Soda Lake formation argues that the water that was trapped in the Carrizo Plain geographic feature was *not* "freshwater," if the mineral salt remaining in the dry lakebed was once *in the water* that later evaporated away and left the salts behind. Below is an image looking east towards Soda Lake (visible as a thin white stripe), with the line of the Monte Diablo range in the background:

I personally am not prepared to make definitive pronouncements regarding the time period in which this flooding and liquefaction may have taken place. The event (or events) that produced Ayers Rock and Painted Rock and the other liquefaction features around the globe (and especially in the American southwest) may have happened long before the catastrophes that researchers such

745

as Dr. Schoch are proposing for the event that may have disrupted the cultures that created the Sphinx or Göbekli Tepi.

However, the fact that Ayers Rock (which is far larger than Painted Rock) also appears to have been formed by similar processes (but in a basin of water that must have been much deeper and more extensive) does appear to argue for the possibility of a massive flood at some point in our planet's distant past.

If these features in California and Australia were formed under the water remaining from the same flood, that would argue for a massive flood indeed (and note that the entire American southwest is home to numerous liquefaction features and other evidence of massive trapped ancient bodies of water).

What I *am* prepared to say is that the above discussion should demonstrate very clearly that the subjects of geology and of the possible ancient catastrophes which may have wracked our planet is vitally connected to the question of a now-forgotten ancient civilization. The evidence on our planet for both (for both an advanced ancient culture, and for some sort of cataclysm or cataclysms which may explain the disappearance of that culture) appears to be virtually undeniable.

Finally, it is very noteworthy (I believe) to observe that both Ayers Rock and Painted Rock are sacred sites to the indigenous cultures of the continents on which these two ancient terrain features are found. We have evidence that these places were (or are) places of contact with the Spirit Realm, and possibly with beings from the Other World.

Is it possible that there are aspects in their formation which cause these natural formations to have certain qualities that are conducive to such contact? The above discussion touched on some of the aspects of the conditions under which liquefaction mounds tend to form, usually involving the deep wave action of earthquakes or of powerful ocean waves, setting up a settling motion in the soil particles and an upwelling motion in the saturating water molecules, creating a lifting motion and sometimes an eruption of soil that has temporarily turned to liquid.

Is it possible that the liquefaction mounds themselves retain some of the structural alignments caused by these ancient low-frequency vibrations (from the earthquake or ocean wave-pounding that set them off in the first place, so long ago)? Perhaps in some way, they function as physical conductors of wave energy, even to this day.

Note also that the sacred natural rock temple of Painted Rock is located in a long, very linear valley, with the distinctive line of the San Andreas Fault itself clearly visible for many miles: a possible natural line conducting Earth energy (see previous post on "dragon lines" and related subjects, not included in this collection: http://mathisencorollary.blogspot.com/2013/10/john-michell-on-dragon-current-or-lung.html).

Clearly, ancient traditional cultures understood that these specific places on our Earth's surface could be used for making contact with or even traveling to the Invisible Realm.

Perhaps, in the year 8,000 BC (when we know men and women were conducting activity in and around Painted Rock on the Carrizo Plain), these generations who were not so far removed from the cataclysm that may have struck around the year we call 9,700 BC (or BCE) still retained much of the sophisticated understanding that informed the now-forgotten culture or cultures responsible for giving us some of the mysterious monuments we still see at significant geological locations around the globe -- and responsible as well, perhaps, for the incredible worldwide system of celestial metaphor which informs the myths of all the cultures on earth.

The other pressing question we might ask ourselves would be -- how is it that so many today have forgotten it, and how was this ancient understanding lost (or even deliberately stamped out) in so many parts of the human family at some point?

Two Visions

Black Elk and the sacred circle

2014 November 07

The sayings of Black Elk, who walked on this earth between the years that we call 1863 and 1950, preserved in the book *Black Elk Speaks: Being the Life Story of a Holy Man of the Oglala Sioux* as told through John G. Neihardt, first published in 1932, have many profound messages of great importance to individuals and society today.

Black Elk's message and his words should be pondered deeply.

This particular essay will examine only one small aspect of his message. Many, many more things are there to return to on another day. Other aspects of his message have been explored in previous discussions.

One very important thread running through the words of Black Elk as recorded by John Neihardt concerns the sacred numbers

750

of his people, which often appear in his visions, and which he often explains at the appropriate point in his narrative. Related to this concept is the concept of the sacred shapes which he discusses, and their meanings, especially the shapes of the circle and the square.

As an example of the importance of sacred number, Black Elk often discusses the number six as it relates to all the directions of the world: the four directions in horizontal space (which we might say relate to an X-axis and a Y-axis, as well as to the four cardinal directions of East, South, West and North), plus the two vertical directions of above and below, or up and down (and which we might also describe as including the Z-axis, and thus providing all the necessary referents for any point in space).

In addition to the number six, he also describes the importance of the number four, and relates it to the four stations of the sacred hoop which sustains all life on earth, and which he associates at some points in his narrative with the earth. The concept of the sacred hoop of the Lakotas and other American Indian nations was often expressed as a circle containing the equal lines of a cross, and it can be seen to be closely related to the outline of the zodiac wheel with its horizontal and vertical cross-lines which is discussed in many previous posts, including here[42] and here.[186]

Here is a passage in which Black Elk explains these sacred numbers, which appear often in his great vision:

> I sent a pipe to Running Elk, who was Standing Bear's uncle and a good and wise old man. He came and was willing to help me. We set up a sacred tepee at the center as before. I had to use six elks and four virgins. The elks are of the south, but the power that they represented in my vision is nourished by the four quarters and from the sky and the earth; so there were six of them. The four virgins represented the life of the nation's hoop, which has four quarters; so there were four virgins. Running Elk chose two of the elks, and I, who stood between the Power of the

World and the nation's hoop, chose the four others, for my duty was to the life of the hoop on earth. The six elk men wore complete elk hides on their backs and over their heads. 209.

In another passage, a very poignant passage, Black Elk discusses the difference between the circle and the square, two sacred shapes which each played a very important role in his culture and in his visions. As he explains in this passage, even though the number four is related to the earth's four directions, and has an important role to play in the world, it is the circle which manifests the vitality and power of life. Note, for instance, that in the passage above, he actually associates the number four with the four quarters of the sacred hoop, preserved by the figures of the four virgins, who are connected with our life here on earth. In other words, he connects the number four to a quartered circle, rather than to a square, in the enactment of that particular vision.

The importance of the circle takes on added meaning as Black Elk explains how it is at the heart of everything in his culture, and how he feels that his people's connection to the circle has been taken away, to their terrible detriment:

After the heyoka ceremony, I came to live here where I am now between Wounded Knee Creek and Grass Creek. Others came too, and we made these little gray houses of logs that you see, and they are square. It is a bad way to live, for there can be no power in a square.

You have noticed that everything an Indian does is in a circle, and that is because the power of the World always works in circles, and everything tries to be round. In the old days when we were a strong and happy people, all our power came to us from the sacred hoop of the nation, and so long as the hoop was unbroken, the people flourished. The flowering tree was the living center of the hoop, and the circle of the four quarters nourished it. The east gave peace and light, and south gave warmth, the west gave rain, and the north with its cold and mighty wind gave strength and endurance. This knowledge came to us from the outer world with our religion. Everything the Power of the World does is in a circle. The sky is round, and I have heard that the earth is round like a ball, and so are all the stars. The wind, in its greatest power, whirls. Birds make their nests in circles, for theirs is the same religion as ours. The sun comes forth and goes down again in a circle. The moon does the same, and both are round. Even the seasons form a great circle in their changing, and always come back again to where they were. The life of a man is a circle from childhood to childhood, and so it is in everything where power moves. Our tepees were round like the nests of birds, and these were always set in a circle, the nation's hoop, a nest of many nests, where the Great Spirit meant for us to hatch our children.

But the Wasichus have put us in these square boxes. Our power is gone and we are dying, for the power is not in us any more. [. . .].

We are prisoners of war while we are waiting here. But there is another world. 195 - 196.

This passage is very profound: it is like a deep pool into which we could dive and go down for a long time without ever reaching the bottom of it. One thing it tells us is that the heavenly realms are associated with the circle, for the sky itself appears round, and the heavenly beings such as the sun and moon are round themselves and also follow cycles that make endless circles: "the sun comes forth and goes down again in a circle. The moon does the same, and both are round."

Black Elk explains that the round shape of the tepee, and the arrangement of the tepees in a circle, is connected to the power of the circle, to the cycles of life, and the connection to nature (expressed in the reference to the round nests of birds). But now his people have been forced to live in squares instead of in circles.

There is much here to consider. One connection which I believe might be of value to draw out is the insights expressed in a recent interview by Marty Leeds, who has thought long and deeply on the significance of numbers and shapes and who arrives at many of the same conclusions about the numbers and shapes expressed above that Black Elk and his people seem to have known.

In an interview with Greg Carlwood of *The Higherside Chats* published in August of 2014, Marty explains beginning at about 0:41:35 into the interview:

> Heaven is known as a circle: it's a three; and earth is known as a four: it's a square. OK, let's break this down: the circle encapsulates the most amount of space. [. . .] So, really what you have there with the inherent or innate symbolism of the circle is you have the All or the Entirety or the Wholeness, that its you know encapsulating.
>
> Now, what's important about the circle is that we can never truly find its area! Why? Because we have to use Pi. We have to use Pi. And Pi, at one point, we have to approximate, as we were talking about before. So we can never, ever truly find the area of a circle. Why this is important is: why is itrelated to heaven? Because we can never truly measure the heavens! It's the infinite! Right? That makes sense – that makes sense that it would be attributed to a circle.
>
> Because, you know, like I said, Pi is this infinite number, we can't see it's "tail," we look out into the heavens [. . .] where is the edge of the universe? I don't know! You know? But now, look at a square. A square is earth. Well we can measure the earth. [. . .] The square: we can always find the area of a square. Always! Because all we have to do is square something. So if the length of the side of a square is two, well two times two or two squared is four, so therefore we know the area of a square.

754

So this is why heaven is known as a circle, and the infinite, and earth is known as a square.

Marty then goes on to relate the square and the circle to the *stupa* of Buddhism, which can be seen as bringing the sacred circle of the heavens down to earth, in much the same way that the tepee and the ring of tepees connected to the sacred circles of the sun and moon and cyclical motions in the passage spoken by Black Elk, above.

What I find very interesting in the Marty Leeds passage cited here is the symbolic association of the circle with the "unmeasurable," which Marty finds due to his focus on the transcendental number of Pi. He explains that Pi is a number which goes on forever, and can never be completely known, but only approximated by us, since we must at some point "cut off its tail" and thus use an approximation of Pi rather than Pi itself. Thus the circle in some sense moves beyond this material world and into the other world – the realm which cannot be actually perceived by the five-senses, or measured with the tools used to measure things in the material realm. Pi becomes a symbol for the transcendental and the realm beyond the material.

And this connection to the realm of spirit – and the Power that comes from that other world -- is exactly what Black Elk is lamenting when he sees that his people now live in square houses, and are cut off from their connection to the circle. The circle is a symbol of the unseen world, the spirit world: Black Elk often refers to it as the "outer world," as he does in the passage cited above.

Elsewhere he makes very clear that, although that outer world may be invisible to our normal senses, it is very real – and it is the source of power for those in this world.

His message speaks to the critical importance of the spirit world, and of maintaining a connection to the spirit world. It can even be seen as a message of the importance of connecting to the other world in daily life, even in details which might at first seem mundane or unimportant (such as the shape of the dwelling in which we choose to live).

His message also speaks to the tragedy of being cut off from the power of the other world.

This is something to think about deeply.

Literalism, colonization, and conquest

2014 December 05

A previous post explored some of the important issues raised by Mark Plotkin's recent TED talk entitled "What the people of the Amazon know that you don't" (the post from December 3, 2014).

Specifically, it explored the contrasts offered in Dr. Plotkin's talk between those who are acting as part of the world-encompassing western system and those who have thus far managed to avoid being absorbed by that system and whose people have called the Amazon rainforest their home for centuries or for millennia.

It suggested that contrast may spring from the fact that one group is characterized by harmony with the natural world and the spirit world (and indeed, it could be said that this group sees no hard-and-fast distinction between the visible, material, natural world and the invisible, immaterial, spirit world) and that the other group is characterized by a disconnection with the world of nature (if not an antagonism towards it) and an almost total disregard for the spirit world (if not an antagonism towards the very idea of a spirit world, as understood in shamanic cultures).

It further noted that this antagonism in earlier centuries stemmed primarily from literalist Christian dogma and in later centuries

757

has stemmed from the "ideology of materialism" which has in some important western circles become a replacement religion for literalist Christianity).

This divide can be seen as central to the very different approaches highlighted in the TED talk between "western medicine" and shamanic healing, between living in harmony with the rainforest and clearing it out to create grazing land for a few skinny cows, between pursuing the old ways while avoiding western contact and pursuing uncontacted groups in order to take pictures with them, enslave them, or try to convert them to literalist Christianity.

Regrettably, there is a very real and ongoing doctrine among literalist Christians that they are under divine commission to reach every people group on the planet in order to attempt to replace the indigenous or traditional belief with literal Christianity. For an example of the seriousness of this ongoing belief, and the numerous groups that have been organized to pursue this "mission" or "great commission" of converting some members of every culture on earth to literalist Christianity, simply type the words "*reaching the unreached*" into a decent search engine and visit some of the links that come up as results.

This doctrine of a "great commission" to convert everyone is regrettable because, as it turns out, there is substantial evidence that the Biblical scriptures were never intended to be understood literally, being built upon a foundation of celestial metaphor. Ironically, I believe that there is extensive evidence to suggest that *this exact same system of celestial metaphor* can also be shown to be the foundation of the sacred traditions of nearly every culture on the planet, including those in the so-called "New World."

For this reason alone (along with many others which have to do with not trying to conquer other men and women), I believe that the idea of aggressively working to teach "unreached" people to reject their traditional sacred knowledge and replace it with

758

literalist interpretations of the Biblical scriptures is profoundly misguided.

Among some literalist Christians, this mission is also joined to an apocalyptic vision regarding the end of the world, the end of the age, and the prophesied return of the literal and historical Christ. This connection is generally based specifically on words attributed to Jesus in Matthew 24:14, which declare: "And this gospel of the kingdom shall be preached in all the world for a witness unto all nations; and then shall the end come."

In fact, it can be demonstrated that Christopher Columbus wrote quite extensively on his own belief that the scriptures teach that end of the world and the return of Christ require the conversion of the people of the new continent to the Christian faith, as well as the physical rebuilding of the Temple in Jerusalem -- and he believed that his voyages to the Americas were instrumental in both of those requirements (the second he felt would be aided by the opening of a new westward route to the Holy Land for the western European monarchs, bypassing some of the obstacles of the eastward route from western Europe, and aided as well by the gold which could now be brought back from the Americas and put to good use in facilitating the rebuilding of the Temple).

In his *Libro de las Profecias* ("Book of the Prophecies"), which Columbus wrote in the years 1501-1502 in Spain, in conjunction with a monk named Gaspar Garricio, he explains his belief that the Americas serve this important apocalyptic purpose in fulfillment of Biblical prophecy, and cites extensive scriptural references from the Old and New Testament to back up his claims, often commenting on them to tie them to his thesis. Strangely enough for a work of such apparent historic importance, it has only very rarely been translated into English, and even those translations can be difficult to obtain (it's not as though Columbus is some kind of marginal figure of minimal historical importance, so the scarcity of this work in easy-to-access online English translations is somewhat puzzling and perhaps worthy of comment -- especially in light of the fact that one of the most important sources Columbus cites in his work, the medieval Joachim of Fiore, is also rather difficult if not impossible to find in English translation as well).

But, the link above will take you to an online transcription of the original text as it was written in Spanish, albeit with frequent archaic spelling conventions (for instance, places in which the letter "i" would be used in the Spanish spelling of a word today often use the letter "y" instead, and some words which today would be spelled using the letter "v" contain the letter "b" where we would expect to see a "v," which is consistent with the pronunciation but not the modern spelling -- and that in some places we would expect a "b" we find a "v" instead). However, it is fairly readable for those who can read modern Spanish. There, you will find that Columbus declares that:

> El abad Johachin, calabres, diso que habia de salir de Espana quien havia de redificar la Casa del monte Sion (see Folio 6, "B").
> This translates roughly to: "The abbott Joachim [of Fiore], of Calabria, said that he has to come from Spain the one who is going to re-build the House of the Mount of Sion [or Zion]."

Earlier, at the end of the first side of Folio 5, Columbus states of the prophet Isaiah (according to the interpretations of San Geronimo and Saint Augustine) says that, "Este puso toda su diligencia a escrevir lo venidero y llamar toda la gente a nuestra santa fee catolica" which I translate roughly to mean "This one exerted all his diligence to write of what is coming and to call *all the people* to our holy catholic faith."

Columbus then begins to cite extensive passages from the scriptures on the subject of the end of the world, as well as passages from the writings of Augustine and others. When he gets to the important passage from Matthew 24:14 quoted above (regarding the requirement for the gospel of the kingdom to be preached "in all the world" and then "shall the end come"),

760

Columbus comments:

> <<En todo el mundo>>: es evidente que antes de la destruccion de la ciudad [Jerusalen] por Tito y Vespansiano, el evangelo fue predicado en las tres partes del mundo, es decir, en Asia, Africa y Europa, pues viviendo todavia Pedro, la fe fue predicada en Italia &c. Hay que inquirir [estas cosas], si le place a uno.

My rough translation of this passage might be as follows:

> "In all the world": it is evident that before the destruction of the city by Titus and Vespasian, the gospel was preached in the three parts of the world, that is to say, in Asia, Africa and Europe: even more, within the life of Peter, the faith was preached in Italy etc. It needs to be examined, if it pleases him to [have it preached] in one more.

While it is undeniable that the historical context of the writing of this *Book of Prophecies* by Columbus included his desire for the rulers of Spain to send him back on another mission to the Americas, no one who reads it can come away unconvinced that Columbus was deeply versed in the scriptures and that he possessed a thoroughly-developed framework of eschatology, predicated upon the rebuilding of the Temple at Jerusalem and the conversion of all the unreached nations of the globe to the literalist Christian faith -- and that he could back up his vision with an interlocking lattice of verses from both the Old and New Testaments. It is difficult to argue that this vision was not dominant (or at least extremely important) in his desire to undertake voyages across the Atlantic from the outset.

In his most-recent book, *Thrice Great Hermetica and the Janus Age*, the insightful and extremely thorough researcher Joseph Farrell makes this very argument regarding the purpose of the voyage of Columbus: that it was part of a carefully-planned vision for bringing about the fulfillment of prophecy by powerful groups at the top of the power structure of western Europe (see pages 156-157 in particular). Of the reference to Joachim of Fiore, Dr.

761

Farrell says:

> Joachim, in other words, more than anyone else, is responsible for viewing prophecy as a code to be decrypted, and once decrypted, as a playbook or agenda to be followed by the power elite of his day. [. . .] Thus, in terms of the hidden "prophetic" agenda driving Columbus and his backers, his voyage of 1492 was not a chance discovery, but a planned revelation whose every last detail was coordinated, including especially those details meant to exhibit "the fulfillment of prophecy." 156 - 157.

That there remain to this day those who continue to believe some version of this "playbook or agenda" and who see both the Americas and Mount Zion as important to that prophecy's ultimate fulfillment is hardly possible to doubt. Some of those who continue to hold to these beliefs may also tie the "reaching" of every last "unreached" culture into their vision of the fulfillment of such "end times" prophecies.

Again, I believe that there is extensive evidence from within the Biblical scriptures themselves to support the conclusion that they were not intended to be interpreted as literally and historically as they are often interpreted. For example, both Joachim and Columbus published specific predictions for the year in which the Apocalyptic events predicted in the scriptures would take place on earth -- and yet I believe that the scriptures in general and the Apocalypse of John in particular (often called the Revelation today) are celestial in nature and were intended to convey esoteric teaching and not historical or literal predictions.

Some discussion of the celestial foundations for the events described in the Revelation of John (particularly in chapter 9, where the celestial connections are very clear) can be found in a post from July 6, 2012, as well as in the three chapters of my book *The Undying Stars*, which can be read online (see pages 9 through 13 of the book, which are part of the selection that is posted online).

Ultimately, I believe that the above discussion points to some of the very substantial evidence which suggests that literalist Christianity itself can be seen to encourage a kind of "colonizing mindset," in that literal misinterpretations of its content can lead to the regrettable conclusion that it should be "forced upon" others, either by persuasive or even aggressive arguments or -- in some extreme but by no means isolated instances -- by physical force or violence (see the record of Charlemagne in Europe, for example, as well as many other cases in later centuries). The connection between this mindset and the other forms of imposing the western world-system on others who might be more disposed to live without it or outside of it should be clear.

Further, I believe there is strong evidence to support the theory that literalist Christianity was deliberately designed as a vehicle for *taking over the Roman Empire from the inside,* and that it turned out to be a very effective vehicle for doing so. If this theory is in fact correct, then we should hardly be surprised that it continued to be an effective tool for colonizing and taking over other cultures around the planet in subsequent centuries, and that it continues to do so today.

Some may object at this point by saying that there have been plenty of non-Christian examples of conquest at the point of the sword, and colonization and cultural takeover of one people by another throughout history, and of this there is no doubt. But it is also extremely notable that western Europe, where the literalists who took over the Roman Empire had the most power and influence for the longest period according to the theory mentioned above, has proven to be the most aggressive and most "successful" (if taking over the culture of others can be measured as a success) colonizing entity the world has ever known (at least, as far as history is known to this point).

It might also be pointed out that, unlike sheer physical conquest by the force of arms, if Christianity was designed to take over a culture from the inside primarily by tactics other than physical force, it can be said to have a powerful "built-in" propensity for what might be called "mental colonization" or "mental conquest" -- or, to use a term which has been defined more precisely in other posts: "mind control."

Thus, I believe that it is no small item that Mark Plotkin mentioned the efforts of Christian missionaries alongside the other deleterious impacts of the western world-system upon the human and natural ecosystems of the Amazon. In many ways, it can be said that literalist Christianity is at the heart of this entire pattern, and has been for many centuries -- stretching back to Columbus, and perhaps even for centuries before that.

"Vision A" or "Vision B"

2015 January 13

When Black Elk, a holy man of the Lakota people, expressed the difference between the life before the arrival of the European invasion and after, he said:

> Once we were happy in our own country and we were seldom hungry, for then the two-leggeds and the four-leggeds lived together like relatives, and there was plenty for them and for us. But the Wasichus came, and they have made little islands for us and other little islands for the four-leggeds, and always these islands are becoming smaller, for around them surges the gnawing flood of the Wasichus, and it is dirty with lies and greed. [*Black Elk Speaks*, 8].

There is a lot to notice in these two sentences. Black Elk chooses to characterize the difference between the two cultures by saying that his culture saw humanity as part of nature: they lived *together* with the earth's other creatures *like relatives.* In contrast, the bringers of the new culture clearly saw themselves as divided from nature, and created what Black Elk describes as "little islands" to physically separate people from the earth's other creatures.

765

This short passage also implies that directly related to these two opposite views of humanity's place in nature are two opposite views of nature itself: in the first, a vision of abundance, that "there is plenty for them and for us," and in the second, a vision of scarcity and a "gnawing flood [. . .] dirty with lies and greed."

I would argue that in these two sentences, Black Elk has pinpointed the most important negative consequence of the literalist twist that was imposed upon ancient scriptures in Europe (in the time of the Roman Empire) that actually changed their teaching from a message that is closer to the first position Black Elk articulates (we could call this "vision A" for ease of discussion) to the horrific vision of the "gnawing flood" and the ever-shrinking "little islands" described in the second half (we could call this "vision B").

In other words, the ancient scriptures actually articulate "vision A," but at a certain point in history they were twisted into "vision B."

For example, previous posts [as well as later books] discuss the Genesis account of Adam and Eve, as well as the Genesis account of Noah's three sons Shem, Ham and Japheth, and argue that if they are understood allegorically, or esoterically, they clearly convey a message that applies to *all* men and women equally, and a message that our physical form is only a "coat of skin" and that our common spiritual origin unites us all. Understood this way, they also convey a message that unites humanity with all of nature, including the infinite starry heavens -- often expressed in the teaching "as above, so below."

However, the same stories when interpreted as describing literal and historic men and women named Adam and Eve, or Noah and his sons Shem, Ham and Japheth, have historically led to all kinds of racist ideologies, and have been used to divide men and women, to elevate one group and devalue another, and even to divide humanity from the other creatures (based on literalistic

766

misinterpretations of the enmity with the serpent, the teachings that man has "dominion" over all the earth and its animals, and the teachings given to Noah about domesticating animals, for example).

In other words, the scriptures that became central to western European culture, and which should be seen as teaching "vision A," were given a literalistic twist at a specific point in history, which led to a culture that was largely guided by "vision B."

In fact, the Biblical scriptures when understood esoterically can be seen as teaching a vision of the natural world, and humanity's place in it, which can be accurately characterized as shamanic. This is because they actually can be shown to be clearly built upon the same foundation as the sacred stories and traditions found around the world, all of which contain clear shamanic elements and teachings.

The literalist takeover of these scriptures, and the campaign to deliberately eliminate texts and teachers who taught an approach which challenged this literalistic "vision B" view of the scriptures, can be demonstrated to have taken place during the years that western historians call the first four centuries AD.

It is very important to understand that, whatever good things western European civilization and culture produced in the centuries that followed (and it cannot be denied that it did produce many good things), this fundamental "vision B" understanding guided much of its development, and that it in fact continues to inform "western civilization" in very powerful and sometimes very destructive ways.

Because, as Black Elk so incisively explains in just two sentences, the vision that shaped "western" thought contains a powerful tendency towards self-imposed division of humanity from nature, as well as antagonistic division between humanity itself. Connected to this division, in Black Elk's view, is a vision of scarcity rather than plenty.

Perhaps nothing illustrates the ongoing influence of this "vision B" attitude better than the rush to create and release genetically-engineered plants and animals into nature. Previous posts have cited ancient philosophers, who wrote prior to the triumph of the literalist takeover, admonishing those whose vision of scarcity led them to horrible treatment of animals and mistrust of nature's bounty -- see for example the arguments of Plutarch and Ovid, both of whom articulate a vision of humanity as related to the animals and to the rest of nature.

Since those posts were written, a new and even more horrific example of what we might call a "runaway vision B" has emerged, with the deliberate creation of genetically-engineered mosquitos, which have already been released *en masse* in at least two parts of the globe, and which are slated for release in Florida in either January or February of this year (no word yet on whether that has actually taken place already, or if it is set to occur within the next couple weeks).

If there is a better symbol of the terribly misguided decisions that the self-imposed division from nature that "vision B" produces than the decision to genetically alter an insect that regularly feeds on human blood, I don't know what it is -- unless it is the decision to start releasing clouds of them into the wild in an act that can never be un-done.

But just wait a few months and there will probably be a new example even more ominous and un-natural than this one.

It should be starting to become clear to even the most unthinking adherent of the "vision B mindset" that something has gone terribly wrong. Black Elk saw the problem with crystal clarity more than a hundred years ago.

But, the good news is that "vision A" is actually the vision that is at the heart of the shared ancient heritage of all of humanity. It was treacherously supplanted by and replaced with "vision B" in a certain part of the world, in a single culture, many centuries ago -- and the results have been catastrophic for many other cultures around the world in the intervening centuries since that takeover. But if "vision B" could replace "vision A," then that means that there is hope that the process could be reversed -- perhaps even more rapidly than the original switch.

People can and do change their entire outlook on the world, without violence and sometimes quite rapidly. I know this personally, as I have changed my own vision quite radically within the course of my own life.

The division from nature and from one another described above and in the quotation from Black Elk is clearly a self-imposed separation -- which means that it can also be "self-un-imposed."

We can still listen to the vision that Black Elk shared with the world -- before the gnawing dirty flood of lies and greed covers over the shrinking little islands altogether.

Two Visions

2015 May 22

In Peter Kingsley's remarkable book entitled *In the Dark Places of Wisdom*, he describes the general condition of "the hollowness we feel inside" and for which "the world fills us with substitute after substitute and tries to convince us that nothing is missing" (33-34).

"But nothing has the power to fill the hollowness," he says.

> Even religion and spirituality and humanity's higher aspirations become wonderful substitutes. And that's what happened to philosophy. What used to be ways to freedom for our ancestors become prisons and cages for us. We create schemes and structures, and climb up and down inside them. But these are just monkey tricks and parlour games to console us and distract us from the longing in our hearts. 35.

Dr. Kingsley states directly that this problem introduces a very negative aspect into the very heart of Western culture specifically. He says:

> Western culture is a past master at the art of substitution. It offers and never delivers because it can't. It has lost the power even to know what needs to be delivered. 35.

770

And yet, *In the Dark Places of Wisdom* explores evidence that at least some of the ancients – even in what was later to become "the West" -- knew what needed to be delivered, and what's more they knew exactly how to deliver it (or, perhaps more precisely, they knew "how it is delivered").

The paradox is that what we are looking for in all of the external substitutes and external systems cannot be found in that endless train of substitutes -- but that we actually "already have everything we need to know, in the darkness inside ourselves" (67). The ancient wisdom keepers, including the pre-Socratic philosopher Parmenides, Dr. Kingsley asserts, understood that the answer was "all a matter of finding their own link to the divine," and that the link was within us all along (64). It was "just" a matter of going within and turning ourselves "inside out until we find the sun and the moon and the stars inside" (67).

I believe that this profound message is very much at the heart of what the ancient knowledge in all its different manifestations was trying to convey to us, or to help us to discover. It is expressed in metaphor after metaphor -- the one I have used as a kind of "shorthand" for all of them is the concept of the "raising of the Djed," but it is also found in many other forms, such as the contrast between the good red road in the vision of Black Elk and the fearful black road of troubles.

It refers to the re-discovery of the divine within, an awakening to the fact that we are actually already connected to the entire universe outside, and the practice of going into the darkness and stillness of the invisible world and awakening to our connection to it on a regular basis -- "to find out," Peter Kingsley says, "how you're related to the world of the divine, know how you belong, how you're at home there just as much as here. It was to become adopted, a child of the gods" (64).

A society that has somehow lost or destroyed or buried this knowledge can be expected to be characterized by the kind of desperate pursuit of "substitutes" that Peter Kingsley describes in the quotations cited above.

But what a different attitude and approach to life is offered in the understanding that we already have the entire universe inside of us, that we are already inseparably connected to the invisible realm, described as "the realm of the gods," that in fact we are just as much "at home there" as we are in the material realm, the ordinary realm. That we are each somehow "a child of the gods" -- bearers of a divine nature lost and almost forgotten within our physical and animal nature.

What is taking shape in this discussion is a framework of two very different visions of the world, two very different "paths." They can be seen to be very closely related to the two "paths" or "visions" articulated in all the world's myths and sacred stories and scriptures which express this contrast using (among other things) the great cycles of the heavens, including the cycle of the year and the "cross" formed by the "horizontal line" between the equinoxes and the "vertical line" between the solstices, expressed in the mythology of ancient Egypt as the "Djed column cast down" and the "Djed column raised up."

The "great cross of the year" is itself a metaphor for the crossing of the *material* and the *spiritual, invisible, or divine* which together express the dual nature of every human being -- as well as the dual nature of the world around us. To the extent that our vision is operating along the "horizontal" or material line, we can be expected to exhibit the frantic pursuit of external substitutes described in Dr. Kingsley's quotations above.

But if we can awaken to the truth that we are already connected to the invisible realm, the divine realm -- that we actually "belong" in the invisible realm just as much as in the material realm with which we are more familiar -- the entire paradigm shifts (to use a phrase that has been ruined by over-use, unfortunately, but which I choose to deliberately employ here to describe the situation, because it expresses the complete transformation of the entire framework or *way of seeing*).

772

It should be evident that the first situation (frantic and endless substitution) would logically tend towards an attitude of scarcity, of always needing more (because desperately needing new substitutes when the old ones turn out to be as unfulfilling as all those that went before them).

It should be equally evident that the second situation, in which it is known that what we seek is already in our possession -- that in fact we contain the "entire universe inside," that we are in some way an "adopted child of the gods" -- points towards a vision of plenty (we don't have to worry that we won't get what we need, if it is already and always securely within ourselves, and impossible to be separated from ourselves).

The second situation, it can be seen, also leads towards a sense of connection with all other beings, if we ourselves are always in deep connection with the invisible realm, if we ourselves already reflect and embody the entire universe, which they (all other living beings) are also inseparably connected to and which they also contain.

But, if you are still in the mode of desperately cycling through "substitute after substitute" because you don't realize that you already have access to exactly what you seek, it might lead to profound division and competition and conflict between different men and women, and in fact it has.

I believe these two contrasting visions, and the attitudes of "plenty" versus "scarcity," and the understanding of "connectedness" versus "division and competition" can also be seen to be very close to the powerful message shared with the world by the Lakota holy man Black Elk and recorded in *Black Elk Speaks*, a message with tremendous importance for all people today.

Black Elk offers a very similar contrast between two approaches to life and the world, expressed in his vision of the two roads: the good red road which runs between the north and the south (which would correspond to the "vertical column" between the

773

solstices on the great cross of the year, and to the Djed column "raised up") and the black road which runs between the east and the west, "a fearful road, a road of troubles" (corresponding to the "horizontal column" between the equinoxes on the great cross of the year, and to the Djed column "cast down"):

In his explanation of his vision recorded in *Black Elk Speaks*, Black Elk contrasts the very different way of walking found on the two roads: traveling the black road, the fearful road, the road of troubles, he saw "everybody for himself and with little rules of his own" (215), while along the good red road he sees a fleeting vision of "the circled villages of people and every living thing with roots or legs or wings, and all were happy" (22), along with symbols of life blossoming forth in the world and in the heart of the people.

These characteristics are important: the attitude of "everybody for himself" is contrasted with the attitude of harmony and cooperation, and they relate directly to the contrast between what a previous post labeled "Vision A" and "Vision B,"[765] in which people and all other living beings lived together "like relatives" and in plenty (Vision A) versus everyone making "little islands" that separate people from nature and from one another, and the

774

islands are always becoming smaller and smaller as a "gnawing flood [. . .] dirty with lies and greed" seethes around them (8).

Upon further thought, the terms "Vision A" and "Vision B" don't really convey these concepts very well. It might be better to call them the "spirit road" and the "fearful road," or some other terms. But, we can lay out some of the characteristics inherent to each of these two opposing visions in a table below, each of which could easily become the topic for much more discussion and examination in the future:

- *Unity and community, "circled villages" vs. "Little Islands" and "Everybody for himself"*
- *Vision of plenty vs. Vision of scarcity*
- *Attitude of confidence vs. Attitude of fear and resentment*
- *Connection to nature and to other creatures vs. division from and hostility towards nature and other creatures, and a desire to subjugate them (all part of the same endless pursuit of something that can never be attained)*
- *Awareness of the dignity in each man or woman vs. racism, endless categorizing and divisions of humanity into "my allies" and "everyone else"*

And there are many more contrasts that could be added to the list above.

Some might object that the "spirit road" vision is nice, but not practical "here in the real world." This is a broad objection, but I would propose that it is at least possible that this objection basically stems from an attitude of fearfulness, rather than an attitude of confidence -- and from a vision of scarcity rather than a vision of plenty. The important ancient philosopher Plutarch addressed these kinds of arguments from his opponents, when he laid out his treatises against the eating of flesh.

Just think of the profound changes we might see in our lives if we were to suddenly realize, on a very deep level, that we already have access to that which we have spent so much of our lives chasing after. How it might change the way we speak, or drive in traffic, or go about our daily lives. And how it might change the way we think about some of the bigger issues that have impacts far beyond our daily lives.

I believe this message is very central to the message contained in all the world's ancient wisdom, bequeathed as a precious inheritance to all humanity, in the esoteric Star Myths found in the scriptures and sacred stories of virtually every culture on earth.

They are telling us that the connection to the entire universe is already right inside of each one of us, all the time. And every culture has (at some point in time) possessed knowledge of some of the different techniques for accessing that connection, and entering the realm of the gods, a realm we belong to just as much as we belong to this one.

In places where that knowledge has been lost (or deliberately destroyed), it is imperative that we find it again, for our own sanity and health -- and for that of the rest of the world.

The good news is that the answer is still there, in the stars over our heads -- which means that it is also right inside of us.

If we know about this, we owe it to others to tell them about it, because the substitutes "never deliver."

Blessings, food, economics, and libations to the gods

2016 May 12

Above is a striking image from an ancient artist depicting the god Apollo in the act of pouring a libation of wine, while a raven observes (evoking the spirit world, even as Apollo of course inhabits the realm of the gods, being a god himself).

In addition to evoking the spirit world, I believe that the raven may be included in this ancient artwork because Apollo in this image has certain elements corresponding to the posture typically associated with the constellation Virgo, including a seated position with one arm dramatically extended. The constellation of Corvus the Crow is closely associated with Virgo, and even though it is located on the opposite side of the constellation Virgo from the side on which the raven is depicted in this image of Apollo pouring a libation, this pairing definitely evokes the celestial figures of Virgo and Corvus, and thus the Infinite Realm or Spirit Realm.

The ancient myths use the stars and heavenly figures to convey powerful truths to us about the importance of, and our dependence upon, the Invisible Realm, the Infinite Realm, even in the most mundane aspects of our daily life.

In fact, the act of pouring a libation, which is described frequently in both the Iliad and the Odyssey, was done at the beginning of a meal, as a way of acknowledging that the food and drink which we enjoy, and upon which we depend for the very sustenance of our lives, is itself a gift from the gods, and that it flows down to us from the spirit world in a very real sense.

The invisible spark of life which causes the grain to grow from a seed into the wheat from which we make our bread, or which causes the vines to grow which produce the grapes which will later be turned into wine, must come from the Invisible World. It is not a product of the material realm -- and we are dependent upon the continued infusion of this purely physical world with the invisible, mysterious and divine outpouring which overflows from the spiritual world for the growth of the plants and vines and trees which sustain all life on earth.

The pouring out of libations described in the texts which have survived from ancient Greece seem to express the recognition that all food and wine (or *spirits*) are gifts from the divine realm.

Here are two examples from the Odyssey -- first, when Telemachus (along with the goddess Athena, disguised as his childhood guardian Mentor) arrives in the land of Nestor and his family, in sandy Pylos:

> There sat Nestor with his sons; and around were his companions preparing the feast, roasting flesh, and fixing

other things on spits. They then, when they saw the strangers, all came together, and in salutation took them by the hand, and bade them sit down. Pisistratus, son of Nestor, first coming near, took the hand of both, and placed the near the banquet, on soft fleeces, on the sand of the sea, near his brother Thrasymedes and his father. And then he gave them parts of the entrails, and poured wine into a golden cup; and stretching out the right hand, called upon Pallas Minerva, the daughter of Aegis-bearing Jove:

"Pray now, O stranger, to king Neptune; for at his feast have ye met coming here. But when ye have made libations and have prayed, as is the custom, then to him also give the cup of sweet wine to make a libation; since I think that he also prays to the immortals; for all me have need of the gods . . ." [3. 36 - 55, translation by Theodore Alois Buckley].

And again when Odysseus arrives among the Phaeacians:

[. . .] but much-enduring divine Ulysses drank and eat. And then the mighty Alcinous addressed the herald:

"O Pontonous, having mixed a cup, distribute wine to all throughout the palace, that we may make libations to thunder-rejoicing Jove, and who attends upon venerable suppliants."

Thus he spoke; and Pontonous mixed the grateful wine; and distributed to all, having first begun with the cups. But when they had made libations, and drunk as much as their mind wished, Alcinous harangued and addressed them . . . [7. 210 - 220, translation by Theodore Alois Buckley].

It is also noteworthy that, when Athena first arrives at the house of Odysseus, where Telemachus and Penelope are basically held hostage by the violent, destructive suitors, there is a feast going on, but no libations to the gods are described at all.

The practice of acknowledging, prior to a meal, our dependence upon the divine for every item of food that we eat and of outwardly proclaiming our gratitude for these gifts is traditionally known in English-speaking countries as "saying grace" or "giving the blessing." These terms are appropriate, in that the word "*blessing*" (I believe) encompasses the idea of acknowledging the importance of the divine realm, and of elevating that which is spiritual in ourselves and in others and in the wider world around us -- while the word "*grace*" evokes that which is seen as a gift from the other realm, something which cannot be forced or demanded or created by our own efforts but which comes basically as a gift from the gods, or as a gift from God.

Intriguingly enough, veteran economist Michael Hudson (who is also a scholar of ancient history, as well as having been an advisor to the United Nations Institute for Training and Development, a university professor, a manager at an early sovereign bond mutual fund, and a host of other high-level roles in academia, finance, and government) explains that "the great fight of the classical free market economists" was to oppose those who wanted to claim private ownership over "what was a gift of nature -- the sun that the Physiocrats cited as the source of agriculture's productive powers, inherent soil fertility according to Ricardo, or simply the rent of location as urbanization increased the value of residential and commercial sites" (*Killing the Host*, 30).

He has much more to say about this subject in that book and in his other writings. It would seem that the behavior of the suitors in the Odyssey could be used as an example of the kind of damaging, violent, host-destroying behavior that Michael Hudson -- and the classical liberal philosophers he invokes -- opposed. It is also noteworthy that when Telemachus publicly confronts the suitors about their destructive freeloading behavior, their leaders make speeches which essentially seek to exonerate themselves and blame Telemachus and his mother, just as the "suitors" that Michael Hudson describes in his book want to portray themselves as virtuous while blaming *their victims* for the horrendous damage that they themselves (the "suitors," or as the classical economists called them, the "rentiers") are causing.

The powerful insights which the ancient myths and sacred stories of the human race can offer and which can be applied to the very practical, and often very pressing, problems and situations we face today, are literally endless. A greater appreciation of the importance of the seemingly simple act of acknowledging our dependence on the gifts of the Invisible World, the gifts of the divine realm, before each meal, for every item of food or drink that we taste and upon which our lives depend here in this material realm can lead to a better understanding of the damage that arises from the reckless disregard of the divine source of these gifts, damage that actually acts as the very opposite of *blessing*.

More lessons from King Midas, who discovered he could not eat gold

2016 October 25

The previous post[415] explored some aspects of the extraordinarily bad (but all-too-human) judgment displayed by the famous King Midas of Phrygia, from ancient myth.

One of the signature aspects of the king's poor judgment involves his failure to recognize the primacy of the realm of spirit -- the realm of the gods -- over the realm of matter, and over all things material.

This failure to acknowledge the primacy of the realm of the gods is most evident in the story of the Judgment of Midas, in which the king fails to recognize the supremacy of the musical skill of the god Apollo -- the very god of music and thus the source of all musical talent, and the mythical leader of the Muses themselves.

For this failure, Midas is awarded the ears of an ass or donkey, a consummate symbol of his preference for the lower, the animal, the bestial and all that is proper to the material or bestial or lower realm, rather than what should be his proper recognition of things higher, things divine, and things spiritual.

782

The same pattern of failure to strive for what is higher, to recognize and pursue that which could elevate himself and help others, is also clearly evident in the more famous episode in the Midas story -- his calamitous request when granted one wish by the god Dionysus to have the ability to turn whatever he touches into gold.

In this case, the king is granted one boon from the divine realm, the very realm of the gods -- and he squanders that opportunity by making a wish that, far from elevating or enlivening himself and those around him, will instead turn everything he touches into a lifeless lump of metal. The horrifying consequences of this bad judgment soon become evident, when Midas realizes he has with his wish cut himself off from the very source of life itself, when even his food and drink turns to gold between his teeth or in his throat -- and then, in some versions of the story, the ultimate illustration of his disastrous wish become manifest when he turns his own daughter from a living, breathing, loving young woman into a lifeless golden statue.

As we saw, the gods were benevolent enough to undue the rash request of Midas, as well as all the horrifying transformations it had wrought, when Midas threw himself upon their mercy, and followed the command to immerse himself in the river Pactolus, at its source.

The failure to acknowledge the realm of spirit -- from which all life in this material realm flows, and indeed within which (the ancient wisdom tells us) all things in the material realm actually find their fount and source -- ultimately debases and objectifies and turns that which should be a blessing into a curse.

Although this lesson may seem to be so obvious, and the folly of Midas may seem to be so self-evident that none could ever be foolish enough to make the mistakes that he makes, in reality we find that our society even in our supposedly enlightened modern age has in fact rejected many of the profound lessons illustrated in the Midas myth, and suffers from the very same horrendous consequences depicted in the ancient story of the king who failed to honor and recognize the primacy of the realm of the gods, focusing only on the gold.

One example of this pattern is seen in the refusal to acknowledge the gifts of the gods which provide blessing and the very source of life to all men and women (which we could also call the "gifts of nature"), such as the abundance of the soil or the abundance of mineral wealth beneath the soil, by some who would argue that these gifts of nature or the gods "belong" to some and not to others, and can and should be "privatized" so that their abundance redounds only to a few and not to society at large.

In an article published in April of 1891, Simon Patten -- the very first economics professor at the Wharton School of Business at the University of Pennsylvania (the very first business school in any of the united states of America) -- addressed this exact problem.

The article was entitled "Another View of the Ethics of Land Tenure," and it was published in the *International Journal of Ethics* (pages 354 - 370). Note the word *ethics* in both the title of Simon Patten's article and in the title of the journal in which the article was published: Patten alleged that the failure he illustrates in this article, the failure to recognize and acknowledge and account for the gifts of nature (which the ancients might have also called the gifts of the gods) is an *ethical* failure. We might note that it is an ethical failure in exactly the same way that the failures Midas illustrates in the Midas myth are also ethical failures -- and the consequences can be equally disastrous.

In his article, Patten provides the example of fields under cultivation to provide food crops. He explicitly states that this example is only one of many others that could be used, but that it is a fairly easy way to illustrate the point he is trying to make, which also applies in other parts of the economy.

He asks the reader to consider the fact that different soils will naturally yield different harvests, depending on a variety of factors

784

including the quality of the land, the amount of rain received at different times during the growing season, and so forth. He then notes that as society grows in number, more land will have to be brought under cultivation in order to provide more food, with the best land usually being brought under cultivation first, and later as the needs of society increase, land with poorer prospects may have to be planted with crops, even though at first that land was not deemed to be as worthy of the effort.

At the same time, however, he notes that the market for wages paid to the workers on the various plots of land will tend to fall towards the price that makes economic sense for the *least-productive* plots of land. A farmer hiring laborers for a low-yielding piece of land will not be able to *afford* to pay those laborers prices that are higher than what the relatively low output of the field can support. This will tend to benefit the farmers hiring laborers for their higher-yielding pieces of land, because they can then pay wages that are about as low, or just as low, as that paid to the workers hired on the lower-yielding land.

If this seems confusing, just imagine the situation at an empty parking lot, where large numbers of migrant workers are sitting around in the early morning in their sweatshirts and work boots, blowing on their hands in order to try to keep their fingertips warm, and hoping that farmers will drive up and hire some of them for the day. When various farmers show up to hire laborers, the farmer owning a field that can produce 400 bushels per acre per year will not offer a wage that is higher than the wage that is offered by the farmer owning a field that can only produce 325 bushels per acre per year. If the farmer from the less-productive field can afford to offer X dollars per hour, then the farmer from the more-productive field can also offer X dollars per hour.

The workers waiting in the parking lot do not ask, "How productive is your field?" when they are offered a job, and they have no way of knowing whether they are being taken to a more-productive or less-productive field when they agree to work for the day and drive off to the respective locations.

785

Note that this means that everyone with more productive fields will then be getting labor at a cost that gives them a greater margin of profit, basically as a benefit from the extra yield that comes not from better labor but rather from nature (or, as we might say, from the blessings bestowed by the gods -- in the case of crops, from the goddess Demeter and from the blessings of the sun and the rain and in fact the blessing of more men and women having babies in the society).

This extra bounty from nature (or from the realm of the gods) will go directly to the land owner, and not to the laborers or to society at large (even though it was the growth of society that necessitated bringing more marginal land under cultivation, creating the lower wages that the owner of the better land can then bank as an extra profit margin). The benefit will remain private, even though it was obtained not by work but as a blessing from the divine realm.

As Professor Patten, that first economics professor in the first business school in the united states of America, explained in his article in the journal of ethics:

> If all wealth was produced by labor alone, then the value of a workman to society would be a just measure of the claim that each workman has upon the wealth that society has to distribute. But nature helps in the production of wealth as well as man, and at the end of each productive period society has to distribute the wealth produced by man, plus the wealth produced by nature. To illustrate, in the case of land, the poorest land means the land where nature does the least to aid man to produce food. The measure of the differences in soil is the difference win the aid of nature in production. If on the poorest land a man can raise 325 bushels of wheat, while on the best he can raise 400 bushels, the aid of nature on the best land is greater than that given to the poorest land by the equivalent of 75 bushels. [. . .] The difference between the better coal and

iron mines, water-powers, and other natural resources, and the poorest of these in use, is due to nature [. . .].

With every increase of the number of workmen, some of them work under conditions where they get less aid from nature, and if the value of each man is fixed by what society would lose if he ceased to work, then the value of all laborers is equal to what they could produce, if all of them worked on as poor land or with as poor instruments of production as the few laborers use that are at the margin of cultivation [that is to say, they will be getting paid a wage that is equal to the X dollars per hour mentioned above, a price that is set by the affordability of labor on the lowest-producing land that is being forced into production by the needs of society]. All the increase of wealth due to fertile fields or productive mines would be taken gradually from workmen with the growth of population, and given to more favored persons whose shares are not reduced by the use of poorer land. These privileged classes would then enjoy all the advantages due to better natural resources or to more productive instruments of other kinds. When it is said that the workingman under these conditions gets all he is worth to society, the term "society," if analyzed, means only the more favored classes who are contrasted with the workmen. They pay each laborer only the utility of the last laborer to them, and get the whole produce of the nation minus this amount. 365 - 366.

This is the very pattern, if we think about it for a bit, that characterized the folly of Midas, the king who failed to acknowledge and recognize and honor the very god of music when judging a contest of musical skill. The problem that Simon Patten was explaining – a problem he explicitly identifies as an ethical problem – is the structuring of society in such a way as to privatize the gifts of nature, distributing them more and more to the owners of the fields and mines at the expense of those working in the fields and the mines.

787

Patten actually argues that the profit should indeed go to the owners of the land, but that a claim should be assessed on such profits (in the form of taxes) in order to return part of that value to the laborer (the implication being that the tax burden should be upon passive income and in particular land tenancy of cropland or mining land which receive the benefits of nature, rather than placing the tax burden upon labor). He writes:

> If a laborer loses twenty dollars a year by a social change [such those just illustrated in his example], he is restored to his former condition, if the state pays twenty dollars of his school bills, or if it improves his sanitary condition so that he pays less doctor bills to that amount. He would also be put on his former footing, if the streets were improved so that he could live in places with lower rent, or if the cost of transportation was reduced so that he could get his food and fuel more cheaply. 367.

An astute reader who disagrees with Patten might argue that the owner of the better land, who is benefiting from the cheaper labor, probably had to pay more per acre for that better land than did the owner of the poorer-quality land, and this will often be the case. But, the fact that in this country the mortgage debt used to purchase such real estate can be deducted from taxes enables borrowing greater amounts than would otherwise be borrowed to make such purchases, and means that the additional productivity provided over and above the labor (by the gifts of nature, or the gifts of the gods, depending on your choice of terminology) will again be captured by the land owner, as well as the bank to which the interest on the loan is paid, rather than to society, leaving the laborer out of the equation and thus privatizing what is actually a gift from nature, or the gods.

This argument from Simon Patten is included as part of the larger argument made by Professor Michael Hudson, in his 2015 book *Killing the Host: How Financial Parasites and Debt Destroy the Global Economy* (see, for example, pages 83 - 86).

In that book, Professor Hudson provides a plethora of counter-arguments to objections such as the one just mentioned, and many others as well. He explains that many classical economists sought to place the tax burden on areas where natural monopolies enable the collection of economic rents, beginning for example with the French Physiocrats in the 1750s, who proposed taxing land rent (and only land rent) in order to, as Dr. Hudson puts it, "collect what nature provided freely (sunlight and land) and hence what should belong to the public sector as the tax base" (48).

As it stands today, however, the bounty that ancient society would have seen as flowing from the realm of the gods (the sun and rain and soil and waterways, and the treasures of gems and minerals and chemicals deep beneath the surface, each of which is presided over by multiple gods and goddesses, nymphs and naiads and other beings proper to the spirit realm) is often seen or described as wasted or squandered if it resides in the public sphere, by those who believe all these gifts should be privatized for the benefit of individual owners and corporations rather than to society at large.

Even more hideously, countries or leaders who have resisted the demands that the wealth of that country should also be privatized (usually by people and corporations from the outside) have found themselves threatened with regime change, or have had their countries invaded and in many cases the leaders of those countries met with a violent death.

This pattern has played out in Iran in the 1950s, in Afghanistan during the 1970s and 1980s, in Libya in 2011, and in numerous countries in Central and South America over the past several decades (to name just a few of many other examples of this very same pattern). The option of seeing the riches of minerals or oil under the surface as a gift from nature (or from the gods of mining and metallurgy and underground riches, such as Hephaestus or Plouton) to benefit the public as a whole was not

789

allowed: such riches could only be privatized, and wars would be (and continue to be) started in order to enforce privatization of nature's bounty to a given country as the only allowable option.

But it is not the only option -- it is not even the most life-affirming option. As Midas found out when he was offered one wish from the gods, and then focused only upon gaining more wealth, the refusal to observe the proper order, and the refusal to acknowledge the gifts that flow from the gods, only ends up making us bestial, as well as objectifying and deadening everything around us, and ourselves as well.

Professor Hudson in his book even compares the situation to the myth of Midas, saying:

> Ancient mythology asked how King Midas could survive with nothing to eat but his gold. This threatens to be a metaphor for today's finance capitalism -- a dream that one can live purely off money, without means of production and living labor. To avoid this fate, the remedy must add financial reform to the 19th century's unfinished revolution to sweep away the surviving inequities of post-feudal land grabbing, seizure of the Commons and creation of monopoly privileges. These are the vestiges of the past appropriations and insider dealing that underlie rent seeking and endowed a financial system that remains grounded in neofeudal practice instead of investing in industry and human well being. 398.

Even when his folly had reached its peak, and threatened to destroy everything he loved (and his own life as well), Midas did not find himself beyond mercy or beyond remedy for all he had done. He was permitted to go find the source of the stream, and dunk his head in it three times.

I believe that stream was a celestial stream, and that it thus represents a spiritual solution -- because the problem at hand is in fact a spiritual problem, and a moral and ethical one.

In his 1891 essay, Professor Patten of the Wharton School asks, "What ethical principals should we accept to bring our actions into harmony with the moral law?" (363).

This is actually an extremely well-framed question. By "the moral law," I would argue that Professor Patten means the same thing that the ancients would have recognized as *the law of the gods* -- the very powers inhabiting the very realm from which flows all that we have here in the material plane, including sun, rain, food, crops, babies, and life itself.

As Midas discovered, failing to answer that question properly can have disastrous consequences.

But as the ancient myth also teaches us, the consequences can be repaired, if we turn back towards the source, and immerse ourselves in it.

Siphoning-off the gifts of the gods
behind a cloak of Orwellian mind control

2017 April 02

Here is a recent interview between economist Michael Hudson and Adam Simpson of the *Next System Project*, published on March 23, 2017 at:
https://soundcloud.com/future-left/special-nsp-episode-discussing-junk-economics-with-michael-hudson.

An edited transcript is available here on Michael Hudson's website:
http://michael-hudson.com/2017/03/the-democracy-collaborative/

This may be the first podcast on economic matters to mention Gobekli Tepe, the incredibly important archaeological site which provides evidence for the existence of sophisticated civilization predating ancient Egypt and Mesopotamia by many thousands of years (as many thousands of years as those civilizations precede the modern day) and that the commonly-taught paradigm of ancient human history is deeply and fatally flawed.

As it turns out, the ancient texts and myths of the world which appear to preserve knowledge going back to that time long before the civilizations of Egypt and Mesopotamia and ancient India and China (because the earliest texts of those ancient civilizations already use the same world-wide system) have a lot to say about economic and political matters, including the concept of credit and debt -- which Professor Hudson sees as the most critical issue facing economies today.

The concepts of credit and debt are actually crucial to economic activity, but the ancients clearly recognized that these financial instruments also have the capability of destroying society, and that like many other powerful tools they can be used for the benefit of others but they can also be used to beat down, oppress, and enslave others.

Professor Hudson describes a model of economic activity in terms of "circular flow" of production and consumption, but notes that while financial activities such as credit and debt can contribute to production, they can also fall into the category of economic rents which act to divert or siphon-off the means of living out of the circle of the "real economy" through activity which basically amounts to the collecting of tolls on the activity of others. Certain forms of interest payments on debt could fall into this category.

Professor Hudson explains that the most ancient evidence (as well as many ancient scriptures) indicates that the "host-destroying" potential these diversions from circular flow were well understood in ancient times, and that there were systems in place to mitigate this danger which can be seen in the periodic debt-cancellation practiced in many ancient cultures, such as we find in ancient Babylon under the code of Hammurabi (for instance), or in the Jubilee system described in the ancient Hebrew scriptures.

These debt cancellations did not apply to business debts, which theoretically contribute to the growth of the circular flow (and if they don't, the business which borrowed the money will generally disappear and the lender will lose the money loaned) but rather to personal debts, which have the potential to reduce individuals to slavery.

In the interview, Professor Hudson describes the rise of Rome as a military power was a watershed event which overthrew the more ancient system -- and that Rome was "the first society not to cancel personal debts." This situation eventually led to the feudalism of the Middle Ages, during which a majority of the population was reduced to serfdom while a small percentage of the population enjoyed unearned income in the form of economic rents.

The classical economists of the 1700s and 1800s sought to find ways to undo the conditions that led to serfdom -- and they did so primarily by focusing on the *rentier* activities that siphoned-off the wealth from the circular flow, including land rent but also financial activity including non-productive credit and debt. But, as Professor Hudson explains, by the 1890s the rentier interests launched an effective counter-attack, which included an aggressive campaign to depict rent-seeking activities as *beneficial* rather than detrimental to society and the economy.

This campaign continues today -- to the point that (as he explains in the interview), many people have no idea what the classical economists really taught, because their teachings are diligently kept out of economics departments in major universities, and to the point that (as he explains in a different interview) economics is deliberately mystified in order to make it more difficult for people to figure out what is going on.

In the preface to his most-recent book, Michael Hudson says that "today's vocabulary of Orwellian Doublethink and Newspeak dominates the mainstream media, the teaching of economics and even the statistical representation of how the economy works -- as if there is no exploitation, barely any economic rent (unearned income), and no quantification of capital gains derived farm asset price inflation [. . .]" and that the promoted economic models "exclude the political, environmental and legal ramifications of debt in today's *rentier* economies" (11). He wryly notes that the motto seems to be, "If thine eye offends thee, pluck it out" (found in Matthew 18: 9 and Mark 9: 47).

Recognizing that the actual language of economics has been hijacked in order to *invert* the teachings of the classical economists who wanted to free societies from economic rents, unearned income and parasitic wealth, Professor Hudson's new book *J is for Junk Economics* goes through the alphabet to explain the terms that promoters of this inverted system use "to increase their time-honored 'free lunch' at society's expense" (5) -- and to provide an introduction to the classical economic thought "from Francois Quesnay and Adam Smith to E. Peshine Smith, John Stuart Mill and Karl Marx to Michael Flurscheim, Simon

794

Patten and Thorstein Veblen" and their recommendations for reducing the diversions of economic rent and thereby creating a more just and less oppressive society (12).

That this campaign of inversion has been incredibly effective is best evidenced by the popularity of the so-called "libertarian" model and the Austrian economists they revere -- including among those who recognize the deceptive, oppressive, and world-destroying aspects of the prevailing neoliberal model and who are therefore desperately seeking an alternative to the neoliberal world order.

As the entry for "libertarianism" in *J is for Junk Economics* notes (in part):

> The aim of libertarian planning is privatization, leading to economic polarization, oligarchy, debt peonage and neofeudalism. What the libertarian (that is, financialization) argument leaves out of account is that taxing land rent and other unearned rentier income requires a strong enough government to rein in the vested interests. Opposing government has the effect of blocking such public power. Libertarianism thus serves as a handmaiden to oligarchy as opposed to democracy. 142.

The interview linked above also goes into this issue and the subject of the Austrian economists (note that based on some of his other interviews and writings it appears to me that Professor Hudson is more positive on the economic thought of Joseph Schumpeter as opposed to the two other primary Austrian economists, Friedrich Hayek and Ludwig von Mises).

The issue of privatization is a crucial aspect of the economic model of circular flow and diversions from the circular flow described in the interview and in the discussion above. The classical economists (and the activities of ancient societies) argue that the gifts of nature – the land itself and the sunshine and the air and the mineral resources and the favorable ports and the fertility of the soil – provide blessings that basically come from the gods, and that at least some of the surplus value they produce (over and above the labor and other costs required to grow the crops on the land or run the ports at the sea's edge) should be yielded back to the gods and ultimately back to the population in the form of infrastructure which acts to benefit the population at large.

Note that in my understanding of the world's ancient myths, scriptures and sacred stories, the gods in some sense actually *live in* men and women as well as in the realm of the gods -- which is very helpful for understanding why the surplus of the land (which is a gift of the gods) should be used to benefit the population of that land as a whole and not to enrich just a few privileged individuals or families.

Infrastructure that is provided by governments out of such taxation actually reduces the siphoning-off from the circular flow of production and consumption of goods and services. In the United States, for example, the interstate system of highways has traditionally not been covered with tollbooths every few miles (although there are some parts of the country where it has been covered with tollbooths, and more are popping up in new parts of the country all the time); the ability to use these roads without paying tolls enables the production of goods and services (whether your business involves baking bread or playing in a rock band) without bleeding out additional costs to pay tolls every time you get on or off the highways.

Extreme libertarians (and privatizers in general) would like to sell off the roads to those who would then have the right to charge whatever the market will bear to anyone who wants to drive on them. The classical economists argued that by building up infrastructure in general (including the roads, the electric grid, the water system, the sewer system, and even things like education and healthcare) and providing it without added toll-booth fees and charges, the society's ability to increase the circular flow of goods and services is enhanced. The more costs and toll-booths one erects in front of necessities such as education, healthcare, access to electricity or water, etc., the more will be diverted out of the positive cycle of flow.

796

Privatization essentially takes the gifts of nature (which the ancient scriptures and myths would call the gifts of the gods) and gives connected individuals, families, or corporations a license to squat in front of them with a toll booth for their own enrichment, creating a diversion from the circular flow to the detriment of the society in general. Privatization takes what comes from the gods and should belong to the people as a whole and gives it to a few connected individuals, families or corporations (who then will usually erect toll-booths on it and further detract from society at large). Little wonder, then, that the fortunes of many of the wealthiest families in the world can be traced to the privatization of their country's telephone systems, oil and gas resources, mineral treasures, and especially real estate (and in particular, real estate in locations particularly blessed with the gifts of nature, or made more desirable by the infrastructure spending of the government in the forms of roads or railroads or ports).

Professor Hudson's criticism of the Austrian economists (especially Hayek and Mises) always prominently mentions their views on the origin of money and credit itself -- because, as it turns out, money and credit can also be seen as a form of infrastructure, which can and should be provided at the lowest possible cost rather than giving certain individuals, families or corporations a license to provide it at inflated and even exorbitant costs -- thus providing a cloak of invisibility to the most lucrative privatization in the world today. In the interview above, Professor Hudson explains that public banking rather than private banking (or at least a public option alongside of private options) would be an enormous step forward for circular flow.

One of the most important aspects of Professor Hudson's work and of his arguments is his perception of the existence of deep, widespread, and institutionalized deception and obfuscation in order to defraud, oppress, and enslave -- and his perception of the use of violence as the final backstop when the "softer" weapons of deception and obfuscation fail.

In other words, he is not afraid to declare that the real problem is *not* that we don't know the right answers and that well-meaning people ignorantly enact bad policies -- but rather, that principles that have been known for literally thousands of years have been deliberately obscured, suppressed, and written out of the history books *and* the economics books in order to facilitate widespread exploitation and oppression.

Ultimately, the issue comes down to one of mind control -- and to the control of language, with which thoughts and ideas are formed and communicated. Because George Orwell was the most incisive thinker and writer to express the power of language control and of mind control, it is fitting that Professor Hudson's book opens with a quotation from George Orwell, on page 13. But, as noted above, Professor Hudson also does not shy away from the reality that those who want to privatize resources do in fact resort to brutal violence when "Orwellian" mind control efforts fall short (actually, this violence is usually accompanied by, and "excused" or "justified" by, even more intense Orwellian reality distortion efforts). He mentions a few recent and ongoing examples in the interview above, including Central and South America, Libya, and Syria -- and literally hundreds more could be listed from the past seventy years alone.

Obviously, the various gifts from nature (gifts from the gods) described above should belong to the people of various countries, to be developed by them in order to benefit their own infrastructures and thus enhance the circular flow in their own country (which will ultimately feed into the circular flow of neighboring countries and theoretically the rest of the world as well) -- and with a mind to the long-term impacts on the country and the planet itself. An especially valuable part of the interview above is the fact that privatization of large-scale infrastructure projects tends to "externalize" damage to the environment in general, because private corporations are forced by their very nature to have a short-term perspective -- whereas a government which is truly representative of the people would only be acting in

798

accordance with its mandate if it takes a long-term view that involves *not* poisoning and destroying the environment and the planet itself.

I would personally recommend listening to the above-referenced interview more than one time – and then going to Michael Hudson's website to check out his other interviews, articles and books.

I would also submit that the patterns discussed in this interview and in Michael Hudson's work -- including the deliberate suppression of history and scholarship (including in the universities), the inversion of the principles taught in the ancient scriptures and traditions and practiced in the ancient world, a major break with the more-ancient ways during the rise of Rome, and the use of what can only be described as mind control and language control of the very kind that Orwell warned us about -- should be extremely familiar to readers of this blog and those familiar with the work of other researchers who have found evidence that humanity's ancient history has been deliberately obfuscated and suppressed.

I would even suggest that the inversion of the teachings found in the world's ancient myths and scriptures has been -- and continues to be -- used to assist and enable the stealthy inversion of economic philosophy that Professor Hudson observes taking place in history, going back to ancient times.

Earth Day, 2017: the choices of Midas and Solomon

2017 April 22

The egregiously bad judgment exhibited by King Midas[415] is legendary.

Offered any gift he wanted by the god Dionysus, Midas chose riches. Specifically, he chose – famously and foolishly -- to have anything he touched turn to gold. The results were, of course, devastating and life-destroying.

He soon regretted his awful choice, and the gods were merciful to him and provided a way for his foolishness to be undone.

It would seem that no one could possibly be as foolish as Midas. His tale is practically laughable. Midas is a rather unsympathetic character, because we all smugly assume that we could never make choices that would be as stupid as the choices of King Midas.

However, looking soberly at the world in which we find ourselves on Earth Day 2017, we might want to think again before we complacently congratulate ourselves that our judgment, at least, is not so tragically foolish as that of the mythical king.

And, we should be very clear that the conditions we see in the world today are indeed a function of *choices*. There is a very well-known line of argument which declares that there simply is no possible alternative -- that things are the way they are today because the way we have structured the world is the only possible way that it will work.

This line of argument, of course, is primarily advanced by those who benefit from the structures in place and who don't want to see them changed -- therefore, they argue that no change is even possible, and any alternative would either fail entirely or else be drastically worse than the current state of affairs.

But, as the ancient myths tell us, King Midas had a choice. He chose stupidly, and his choice if left un-changed would have led to his own death by starvation or thirst (as everything he consumed turned to metal as it crossed his lips and entered his throat) and to the destruction of the next generations (as he famously turned his daughter into a lifeless golden statue). But he did have a choice.

Other similar myths involving choices did not turn out so badly. For example, Solomon was similarly offered the granting of a single request, and chose wisdom -- specifically, wisdom in order to help the people, if you look closely at the actual text in the book of 1 Kings chapter 3. When he made that request, the text tells us that God was pleased, and specifically contrasted Solomon's choice with other possible choices, including riches or power over his enemies. Solomon in that ancient text chose rightly, in contrast to the bad judgment of King Midas.

There is an alternative -- but the world we have today has been shaped by choices of Midas-like bad judgment.

Professor Claudia von Werlhof, of the University of Innsbruck, gave a presentation in 2005 which was later turned into an essay and published in 2008, and given the title in English: "Globalization and Neoliberal Policies: Are there Alternatives to Plundering the Earth, Making War and Destroying the Planet?"

It was recently re-published on *Global Research* at the link given here:

http://www.globalresearch.ca/the-consequences-of-globalization-and-neoliberal-policies-what-are-the-alternatives/7973

In that article, Professor von Werlhof explains that neoliberalism -- a system with which we are all at least unconsciously familiar, since we are living in it, but about which we should all become much more familiar if we want to avoid the fate of King Midas -- was consciously implemented on a widespread scale in leading economies such as the United States, Great Britain, and (later) the European Union beginning in the 1980s, but that it had been carefully planned-for in advance and tested out in South American countries beginning with the violent US-backed coup in Chile in 1973.

It's not that there was no alternative or no possible "other choice" -- but rather that neoliberalism was deliberately and systematically selected and implemented as a conscious choice by those hoping to benefit from its implementation.

Other professors have noted that the roots of what is known today as neoliberalism go back even further -- especially to economic thought that arose during the 1930s, as explained by Professor Melinda Cooper from the University of Sydney in a recent interview on *This is Hell!* radio (an example of the kind of independent media that has always been marginalized to some degree but that is now coming under increasing pressure, and that you may want to consider supporting if possible):

https://thisishell.com/interviews/940-melinda-cooper.

In Professor von Werlhof's essay, she explains that neoliberalism is based upon a deliberate decision to choose values such as:

> self-interest and individualism; segregation of ethical principals and economic affairs, in other words: a process of 'de-bedding' economy from society; economic rationality as a mere cost-benefit calculation and profit maximization; competition as the essential driving force for growth and

progress; specialization and the replacement of a subsistence economy with profit-oriented trade ('comparative cost advantage'); and the proscription of public (state) interference with market forces. [quoting her colleague Maria Mies -- see the extended list of *works cited* at the bottom of Professor von Werlhof's essay].

In his new book *J is for Junk Economics* (discussed in this previous post[792]), Professor Michael Hudson adds some additional insights to the definition of neoliberalism, defining it (in part) as:

> An ideology to absolve banks, landlords and monopolists from accusations of predatory behavior.
>
> [. . .]
>
> Turning the tables on classical political economy, *rentier* interests act as plaintiffs *against* public regulation and taxation of their economic rents in contrast to Adam Smith and other classical liberals, today's neoliberals want to deregulate monopoly income and free markets for rent seeking, as well as replacing progressive income taxation and taxes on land and banking with a value-added tax (VAT) on consumers.
>
> Endorsing an oligarchic role of government to protect property and financial fortunes, neoliberalism loads the economy with an exponential growth of debt while depicting it in a way that avoids recognizing the rising *rentier* overhead (rent, interest and insurance) paid to the FIRE sector. Neoliberals want to privatize public infrastructure. They defend this granitization by depicting public ownership and regulation and less efficient than control by financial managers, despite their notorious short-termism. 167 - 168.

The result is a world of which King Midas (before his change of heart) might have been proud.

As Professor von Werlhof explains, the results of neoliberalism turn everyone and everything into commodities -- which is exactly what Midas (blinded by his lust for riches) was in the process of doing. She writes:

> Today, everything on earth is turned into commodities, i.e. everything becomes an object of "trade" and commercialization (which truly means "liquidation": the transformation of all into liquid money). In its neoliberal stage it is not enough for capitalism to globally pursue less cost-intensive and preferably "wageless" commodity production. The objective is to transform everyone and everything into commodities (Wallerstein 1979), including life itself. We are racing blindly towards the violent and absolute conclusion of this "mode of production," namely total capitalization / liquidation by "monetization" (Genth 2006).

She cites numerous examples. Perhaps the most visually-powerful involve the privatization of water. "In Nicaragua," she notes, "there exist water privatization plans that include fines of up to ten months' salary if one was to hand a bucket of water to a thirsty neighbor who cannot afford her own water connection (Sudwind 2003)." And, equally awful to contemplate:

> In India, whole rivers have been sold. Stories tell of women who came to the river banks with buffalos, children and their laundry, as they had done for generations, only to be called "water thieves" and chased away by the police. There are even plans to sell the "holy mother Ganges" (Shiva 2003).

This story is extremely telling -- because it shows how the use of force is inextricably connected to the implementation of neoliberalism (just as it was during its first big modern "test run" in Chile in 1973). Neoliberalism and the perpetual wars that are being waged by the most economically-developed countries on the planet (against people in the least-developed) are closely related, as Professor von Werlhof explains.

The reason violence is required for its implementation, beyond the obvious fact that it involves the taking of public resources for a smaller private group of beneficiaries, is that neoliberalism is inherently contrary to nature -- both to human nature and to Nature in general. In another visceral description, Professor von Werlhof describes its ultimate end, if left unchecked:

> One thing remains generally overlooked: The abstract wealth created for accumulation implies the destruction of nature as concrete wealth. The result is a "hole in the ground" (Galtung), and next to it a garbage dump with used commodities, outdated machinery, and money without value.

Once again, however, we should remind ourselves that this outcome is not a *necessary* outcome. The path of Midas is a choice, and one that the ancient wisdom of the world tells us is a terrible choice and a choice to be avoided. The ancient myths provide an example of a different choice, in the choice of Solomon, who did not choose riches but rather wisdom in order to judge rightly and help the people. Midas did not judge rightly. He chose gold over life itself.

To undo his choice, Midas turned to the gods for mercy, and was granted the ability to un-do his decision. Professor von Werlhof explains that neoliberalism also involves making the wrong choices on very much the same *moral* level (choosing the wrong gods, so to speak). She writes that,

> We are not only witnessing perpetual praise of the market -- we are witnessing what can be described as "market fundamentalism." People believe in the market as if it was a god.

And, it is true that massive amounts of propaganda-like reinforcement are employed in developed countries such as the US to inculcate just such a quasi-religious "market fundamentalism" which declares that any alternative to neoliberalism as defined above is not only mistaken but actually morally pernicious.

Clearly, this is not merely an "economic" issue but in fact a spiritual one.

Later, Professor von Werlhof expands on the spiritual aspect of this question, saying:

> We have to establish a new economy and a new technology; a new relationship with nature; a new relationship between men and women that will finally be defined by mutual respect; a new relationship between the generations that reaches even further than to the "seventh"; and a new political understanding based on egalitarianism and the acknowledgment of the dignity of each individual. But even once we have achieved all this, we will still need to establish an appropriate "spirituality" with regard to the earth (Werlhof 2007 c). The dominant religions cannot help us here. They have failed miserably.

I would argue that she is absolutely correct -- but I'd also argue that the *ancient wisdom* of the world as given in the myths and sacred traditions found in virtually every single culture on the planet, including the ancient cultures of Europe and the Mediterranean, did not fail miserably: but they have in some cases been hijacked and turned on their heads.

We need to think very carefully about the choice of Midas, because not only have we demonstrated that we are not "above" making the same kind of foolish choices that he displays in the ancient myths -- and in fact, as Professor von Werlhof so eloquently demonstrates in her essay, we have made those very same choices and are rapidly in the process of turning the world into lifeless gold (or perhaps plastic).

The ancient myths, however, are not about fantastic actions made by kings or heroes in the distant past (as Alvin Boyd Kuhn explains, in a lecture cited many times[331] on this blog). Solomon was not some external figure who was gifted with wisdom that we can never hope to access ourselves, and Midas was not some external figure who was filled with foolishness beyond any other human being. We ourselves are always capable of accessing the wisdom of the Infinite (like Solomon) or of foolishly ignoring the goodness of the gods (like Midas).

And we are also capable, like Midas, of turning to the divine realm and saying we have made a very foolish choice, and asking for help in un-doing it.

Before the world ends up as one giant hole in the ground, and next to it a garbage dump.

Collaborators against the gods

2017 July 12

The city of Athens was famously the scene of contention in ancient Greek myth between the god of the sea and the goddess of wisdom and art, to see which could prove the greater benefactor.

The wonderful 1962 edition of *Ingri and Edgar Parin D'Aulaire's Book of Greek Myths*, which was a big influence on me when I was growing up, describes the contest this way:

> Athena was very fond of a certain city in Greece, and so was her uncle, Poseidon. Both of them claimed the city, and after a long quarrel they decided that the one who could give it the finest gift should have it.
>
> Leading a procession of citizens, the two gods mounted the Acropolis, the flat-topped rock that crowned the city. Poseidon struck the cliff with his trident, and a spring welled up. The people marveled, but the water was salty as

the sea that Poseidon ruled, and not very useful. Then Athena gave the city her gift. She planted an olive tree in a crevice on the rock. It was the first olive tree the people had ever seen. Athena's gift was judged the better of the two, for it gave food, oil, and wood, and the city was hers. From her beautiful temple on top of the Acropolis, Athena watched over Athens, her city, with the wise owl, her bird, on her shoulder, and under her leadership the Athenians grew famous for their arts and crafts. 36 - 37.

You can see various ancient versions of this account, including references in the works of Plato and of Ovid.

The gifts of the olive tree -- and of access to the mighty sea itself and all of its blessings -- were rightfully seen as gifts from the gods in the ancient myths, the ancient wisdom, given to the people of Greece in remotest antiquity. And the same understanding, that the blessings of sunshine and rain and fertile soil and harvest and wood-bearing forests and mighty rolling oceans and even the treasures of gold and silver and other mineral wealth hidden deep in the earth were all gifts from the gods, can be found in the other ancient sacred traditions -- which Peter Kingsley calls the "original instructions" -- which were given to every other culture on our green Earth in the earliest times.

And indeed these ancient sacred myths or "original instructions" can all be shown to be closely related and based upon a common, worldwide system of celestial metaphor. Indeed, some of the likely celestial patterns upon which the famous contest between Athena and Poseidon for the patronage of the city of Athens are discussed in *Star Myths of the World, Volume Two*, which focuses primarily on the myths of ancient Greece. And the evidence that this system is extremely ancient -- perhaps preceding the earliest admitted civilizations to conventional history, such as ancient Egypt and ancient Mesopotamia and the ancient Indus-Saraswati region, by *at least* as many millennia as those ancient civilizations precede our own day -- is touched upon in my most recent book, *Astrotheology for Life*.

So we can say with a fairly high degree of certitude that the ancient pattern given to humanity and preserved from unbelievably ancient times includes the conviction that the blessings of the earth and the sea and the air and the sunshine, which modern economics generally categorize as natural resources, are blessings bestowed by the gods for our benefit. That these gifts are to be treated with respect, and not with disdain or ingratitude or with greed, is also attested to in many ancient myths and texts, including in the writings of the ancient philosopher Plutarch, who himself appears to have been an initiate in certain ancient *mysteria,* and who declares in the surviving fragments of his text "On the Eating of Flesh" that acting with greed to have more than the gods choose to give is in essence "slandering the earth" and acting shamefully towards the gods, in this case towards grain-giving Demeter and wine-giving Dionysus.

In light of this discussion, we should pause to consider the current situation in the modern country of Greece, where natural resources -- which under the world's ancient wisdom are seen as *blessings given by the gods to the people* -- are being sold off or "privatized" in order to benefit a few people at the expense of the people as a whole, to whom they were given by the gods. This is a pattern that can be seen to be taking place all over the world, and at an accelerated pace in recent decades, but it is especially visible in Greece, and especially poignant, given the fact that it is taking place in the very places spoken of by the myths of ancient Greece.

It is also especially significant, given that ancient Greece constitutes one of the cultures that gave birth to the concept of "the West," and whose ancient traditions and mythology and artistic achievements are preserved in abundance -- allowing us to see even more starkly the contrast between the ancient vision in which the fields and the mountains and the sunlight and the water and the rolling waves belonged to the gods and were given to humanity by the gods' good graces, and the vision that has

810

characterized much of "Western history" since becoming cut off from the ancient pattern handed down to humanity, an alternate or inverted vision in which those gifts of the gods can be taken away from the people to whom they were given, and taken over by a few for their own enrichment.

Last year, the government of Greece sold a majority ownership (67% ownership) of the port of Piraeus to a corporation (in this case, a global shipping company based in China, called Cosco), as described in this article from the *Financial Times*: http://uk.reuters.com/article/uk-eurozone-greece-privatisation-china-c-idUKKCN0X50XD

The Piraeus, of course, is the famous port of Athens, and can be seen near the left edge of the aerial image above of the Greek capitol city (where there is a long diagonal "cut" or "corridor" of water protruding into the shore and fanning out into a wide, three-part "cup-de-sac" at the end, creating a curving peninsula just to the right of it). During ancient times, as you may recall from history, the Piraeus port was connected to the city of Athens by the "Long Walls," which created a protected corridor down from the polis to the port itself. The primary harbors of Piraeus were fortified by Themistocles in the 470s BC, after the Athenian fleet had played a significant role in the repulsion of the Persian invasions, and the Long Walls were completed under Pericles in the 460s or 450s BC.

Other ports and natural resources are also being sold off by the Greek government to corporations or large consortiums of private investors, including the port of Thessaloniki, Greece's fourteen regional airports, the Greek train company TRAINOSE, the water system, and many other resources over which corporations and private investors salivate due to their ability to collect what amount to tolls or recurring revenues through the control of vital infrastructure necessary to the conduct of business or the sustenance of human life.

The government of Greece is being pressured to sell these assets off to the country's creditors, while at the same time cutting social programs including pensions, education spending, healthcare, and other parts of the budget, also to pay their creditors (in the familiar pattern dubbed "austerity"). But, as economist Michael Hudson explains in several different essays, books and interviews (including this one, for example) when lenders deliberately extend loans that they know are beyond the ability of the borrower to repay, with the ulterior motive of seizing the assets of the borrower which the lenders or their friends covet for themselves, this behavior is a form of predation or indeed warfare by financial means -- and in any case, predatory or not, "no country should be obliged to impose poverty on its population, and sell off the public domain in order to pay its foreign creditors."

In his excellent recent book *J is for Junk Economics*, Professor Hudson defines "public domain" as:

> The commons, consisting of land and natural resources, infrastructure and government enterprises. Natural monopolies such as canals, railroads, airlines, water and power, radio and television frequencies, telephone systems, roads, forests, airports and naval ports, schools and other public assets were long kept out of private hands. Their privatization since 1980 has turned them into rent-extracting opportunities for hitherto public services.
>
> Financing their purchase on credit (often at giveaway prices paid to debt-strapped or corrupt neoliberal governments) enables these monopolies to include interest, dividends, and high managerial salaries in their cost structure. The most rapidly rising consumer prices in the United States since 2008, for instance, are for health insurance [. . .], education and cable service. Privatization and economic polarization thus go together. 186.

Clearly, there is a connection between what the ancient wisdom describes as the "gifts of the gods" and what classical economists describe as the "public domain" or "the commons." In fact, some of the earliest classical economists, the French "Physiocrats" led

812

by Francois Quesnay, specifically argued that the economic wealth that came from the land and which was demanded in taxes by the "nobility" was actually "not produced by the nobility's labor or enterprise (contra John Locke) but by nature, ultimately from the sun's energy" -- that is to say, from the gods (or, if you prefer, from Nature), as Professor Hudson explains on page 177 of *J is for Junk Economics*.

Under his definition for the "commons," which are synonymous with the public domain, Professor Hudson provides another specific list which includes "land, water, mineral rights, airwaves and other public infrastructure" and argues that these create "natural monopolies" which are best administered in society's long-term interest via government or a community, not monopolized by *rentiers* as the ultimate takeover objective of financial capital" (60).

But because they are natural monopolies, they are coveted by those who want to privatize them and erect tollbooths around them -- depriving the people of the gifts that are given by the gods (or, in the terminology of the Physiocrats, given by the land and ultimately by the energy of the sun). To cite a portion of just one more definition from *J is for Junk Economics*, Professor Hudson's definition of "privatization," we read that:

> Since 1980 the main lever of privatization has been financial. Debt-strapped governments are forced to sell off the public domain as a conditionality imposed by the IMF in exchange for credit to avoid defaulting on bank debts or forcing debts (see Washington Consensus). The prime assets being privatized are natural monopolies able to extract economic rent by raising prices for hitherto public services. These rents tend to be paid out as tax-deductible interest to affiliates in offshore banking centers in order to deprive host economies of a public return on their land and natural resource patrimony or their immense capital investment in infrastructure -- much of which was financed by foreign debts for which governments remain liable.

Such privatization de-socializes public infrastructure, usually by rent extractors in partnership with government insiders. Access charges may be raised as high as users ("the market") will pay. Junk economics pretends that this will be more efficient than public investment to provide basic services at low prices. The reality is that countries that fail to invest in minimizing the cost of basic services (by avoiding tollbooths for financialized rent extraction) have a higher cost of living and doing business, making them less competitive in global markets. 181.

Note the clear resonance of what Professor Hudson is here arguing with the ancient idea that the bounties of nature (whether sheltered ports or mineral wealth below the earth's surface or even the electromagnetic spectrum) are actually each proper to a god or a goddess (the riches beneath the earth's surface, for instance, being proper to the god Hades or Plouton in ancient Greece, and the electromagnetic spectrum might be argued to belong to the god Zeus himself, wielder of the thunderbolt and ruler of cloudy air and the heights of Olympus) -- and that these resources are thus a sort of "patrimony" or inheritance given to all the people, and are not supposed to be fenced-off in order to enrich a few corporations or individuals who set themselves up in the place of the gods.

I would even go so far as to argue that the world's ancient wisdom teaches us that the gods in fact *have their home in* and *express themselves through* living men and women. We can see abundant evidence of this ancient understanding in passages found in the ancient Egyptian Book of Going Forth by Day as discussed in a blog post whose title is taken from a passage from that ancient Egyptian text: "There is no member of mine devoid of a god."[617] We can also see evidence of this ancient understanding preserved in sacred traditions from other cultures, such as ancient India, where the gods appear in an instant[580] when they are called -- indicating that they are always present and indeed that every man and woman has an inner connection to the

814

Infinite realm, the realm of the gods. That same post notes that the god Thor, known in northern Europe, would also appear in an instant when called by name.

Further, we could look at the people who are born in any given land as being given by the gods themselves as well. As I discuss in the aforementioned *Star Myths of the World, Volume Two*, the goddess Artemis was understood in ancient Greek myth to preside over each and every mother in childbirth -- and similar teachings can be found in the sacred traditions of virtually every other culture on our planet.

If you think about this line of reasoning, you will then see that it dovetails very nicely with Professor Hudson's assertion that privatization *deprives* the public, the people, of their "patrimony" or their inheritance, by taking away what belongs to the gods and which is given by the gods to all the people, and fencing it off for the benefit of a few (who thus by their actions attempt to usurp the place of the gods – and who also by their arguments usually attempt to convince others that their usurpation is natural and right and in the proper order of things, an argument that the people in general almost always see right through, either immediately or else eventually).

It should be abundantly evident to anyone who is even casually familiar with the myths of ancient Greece (as well as with any of the other myths given in antiquity to the families of humanity around the globe) that the attempt by greedy or vainglorious or prideful mortals to usurp the place of the gods, or to fail to properly acknowledge that their gifts came from the gods, always ends in complete disaster.

When the vain and prideful Arachne boasted that her skill at weaving was superior to that of the goddess Athena, for example (thus failing to acknowledge that Arachne's very skill at weaving was a gift from the goddess herself), the goddess famously turned Arachne into a spider. According to some ancient sources, the beautiful maiden Medusa did not think it excessive to have others compare her beauty to that of the same goddess – and was for her punishment turned into a monster. Many other myths involve maidens who declare that their beauty is as great as or even greater than that of one of the goddesses, always resulting in disaster.

It should not surprise us that societies characterized by the privatization by the few (usually by corporations or by well-connected individuals or families, and always in collusion with government insiders, who consent to giving away what is not really theirs to give away at all) of the gifts of the gods in the form of the natural resources of the land and the air and the sun and the waves will face similar negative consequences until the natural order is restored and the gods are again acknowledged and respected. As Professor Hudson explains in the "privatization" definition above (and as he demonstrates with more extensive proofs in many of his full-length books), when that which should belong to the people is instead fenced off with "tollbooths for financialized rent extraction," the cost of doing business and producing goods goes up (as does the price of hiring labor, since workers must necessarily spend more themselves in order just to survive, and thus will have to be paid higher wages as a result), and the country becomes less competitive, and is ultimately hollowed out economically if the process goes on long enough.

I would argue that this deprivation will also ultimately end up in the degradation of men and women, who are deprived of the blessings intended by the gods and properly belonging to the people themselves (the people that the gods cause, or allow, to be born in that land -- *all* the people, not just some of the people, but all of the people: who themselves come from the gods and through whom the gods work and in whom they have their home, in one sense, according to the ancient texts and traditions). When the people are not given the dignity they merit as men and women, when they are not allowed to manifest all the gifts of the gods, then the entire society suffers.

Ironically, or fittingly, it should be noted that the port of Piraeus was very close to the location that the participants in the ancient mysteria of Eleusis[173] -- the famed Eleusinian Mysteries which are believed to have gone on for two thousand years in a row before they were shut down during the reign of the Roman emperor Theodosius,[693] a literalist Christian, in AD 392 -- would go down to the sea and bathe while carrying a live piglet during the opening days of the annual ritual.

These sacred mysteries were open to all participants, male or female, free or slave, from whatever land of origin, just so long that the participant could speak Greek and had never committed the crime of murder. The rites of Eleusis were designed to elicit a connection to the realm of the gods in each participant, and can thus be said to have been a consummate representation of the ancient vision given to the cultures of the world that there is in fact a realm of the gods, and that we are supposed to recognize our dependence upon their gifts -- indeed, the dependence of everything in the material realm upon the invisible or infinite realm, which is the source or fountain of everything in this material realm, and of all good gifts.

Cut off from this vision, corporate entities, collaborating government insiders, and wealthy investors seeking to erect tollbooths upon the natural resources of the commons in order to enrich a few "within the fence" at the expense of everyone else "outside of the it" are thus setting themselves up against the gods, inverting the vision described in the world's ancient wisdom, and inviting disaster upon themselves and their entire societies.

I believe that the people in general know these truths, in deep way that comes from the fact that this wisdom was given so many thousands of years ago and has informed so many, many successive generations -- and that attempts to deny them or tell men and women that the privatization and degradation that they see accelerating around them is somehow natural, right, and just will be instinctually or intuitively rejected, even if they are not able to articulate the historical and economic connections the way that

Professor Hudson is able to do in his books and articles and interviews.

But we have been deliberately and systematically cut off from the ancient wisdom by those who want to collaborate against the gods, in a campaign that started at least seventeen centuries ago and which continues to this day.

The good news is that the wisdom contained in the world's ancient myths, scriptures and sacred traditions is still available to us. And what actually belongs to the gods and is given to all humanity by the gods cannot in reality be given away or legally transferred to the ownership of those who think it belongs only to themselves.

Choose Wisdom

Illustrations
listed by page number on which they appear

6. Temple of Apollo at Delphi. Wikimedia commons.
https://commons.wikimedia.org/wiki/File:Delphi-temple-to-appol01.jpg

26. Courses of the stars. Illustration by the author. Based on diagrams found on the website of the Polynesian Voyaging Society hokulea.com, and at the specific subpage entitled "The Celestial Sphere" found at:
http://www.hokulea.com/education-at-sea/polynesian-navigation/polynesian-non-instrument-wayfinding/the-celestial-sphere/

28. Polar stars. USAF Training Manual 64-3, 15 August 1969. Page 2-101.

42. Scarab with Ankhs, Djeds and Uraeus serpents from tomb of Tutankhamun. Wikimedia commons.
https://commons.wikimedia.org/wiki/File:Bijou_de_la_tombe_de_Toutânkh amon_(musée_du_Caire_Egypte)_(1815591264).jpg

45. Zodiac wheel of John Mylius, 1618. Wikimedia commons.
http://commons.wikimedia.org/wiki/File:Astrological_signs_by_J._D._Myliu s.jpg

47. Human Skull and Scarab Comparison. Wikimedia commons.
Left: https://commons.wikimedia.org/wiki/File:Kort-lang-skalle.gif
Right:
https://commons.wikimedia.org/wiki/File:WLA_brooklynmuseum_Marriage _Scarab_of_Amunhotep_III_3.jpg

48. Upraised arms in Yoga. Wikimedia commons.
https://commons.wikimedia.org/wiki/File:Yoga4Love_Freedom_Gratitude.jpg

49. Ancient deities carrying Ankh symbols. Wikimedia commons. Left to right,
Set: https://commons.wikimedia.org/wiki/File:Egypt.KV34.06.jpg
Isis - Hathor: https://commons.wikimedia.org/wiki/File:Flickr_-_schmuela_-_IMG_6672.jpg
Horus: https://commons.wikimedia.org/wiki/File:Edfu_Tempel_09.jpg

56. Angkor. Wikimedia commons.

https://commons.wikimedia.org/wiki/File:Angkor_Wat,_Camboya,_2013-08-15,_DD_032.JPG

57. Shrine with Entwined Serpents. Wikimedia commons.
https://commons.wikimedia.org/wiki/File:Serpent_deity_relief_at_Pogallapalli_in_Khammam_district.jpg

59. Pyramid of Kukulcan with Serpent. Wikimedia commons.
https://commons.wikimedia.org/wiki/File:Chichen_Itza_Temple_of_Kukulcan_Serpent.JPG

69. Composite image of the Temple of Isis at Saïs superimposed on background image of the Veil Nebula. Wikimedia commons.
Foreground: https://commons.wikimedia.org/wiki/File:Agilkia_Isis-Tempel_03.JPG
Background: https://commons.wikimedia.org/wiki/File:Veil_Nebula_-_NGC6960.jpg

76 - 80. Courses of the stars. Illustrations by the author, based on diagrams seen on the website of the Polynesian Voyaging Society: hokulea.com.

85 - 89. Sun-earth-ecliptic diagrams. Illustrations by the author.

91. Disney replica of *Columbia Rediviva*. Wikimedia commons.
https://commons.wikimedia.org/wiki/File:The_Sailing_Ship_Columbia.jpg

99. Hall of Judgment from Papyrus of Ani. Wikimedia commons.
https://commons.wikimedia.org/wiki/File:Papyrus_of_Ani.jpg

104. Sanctuary at Abu Simbel. Wikimedia commons.
http://commons.wikimedia.org/wiki/File:Flickr_-_archer10_(Dennis)_-_Egypt-10C-024.jpg

110. Chichen Itza Pyramid of Kukulcan. Wikimedia commons (colors have been altered by author of this book).
http://commons.wikimedia.org/wiki/File:ChichenItzaEquinox.jpg

115 and 124. Detail from 1485 painting of Lantern Festival. Wikimedia commons.
http://commons.wikimedia.org/wiki/File:Ming_Emperor_Xianzong_Enjoying_the_Lantern_Festival_(Ming_Dynasty).jpg

127. Isis and Nephthys from Papyrus of Ani (c. 1250 BC) , superimposed on background zodiac wheel from Mylius (see note for illustration on page 45).
https://commons.wikimedia.org/wiki/File:BD_Ankh,_Djed,_and_Sun.jpg

137. Chang Er flying to the Moon. Wikimedia commons.
https://en.wikipedia.org/wiki/File:Chang%27e_Flying_to_the_Moon_(Ren_S huai_Ying).jpg

138. Mooncake. Wikimedia commons.
https://commons.wikimedia.org/wiki/File:Mooncake.jpg

150. Stonehenge. Wikimedia commons.
https://commons.wikimedia.org/wiki/File:Stonehenge_(sun).jpg

156. Earth, Sun and Zodiac. Wikimedia commons.
https://commons.wikimedia.org/wiki/File:Ecliptic_path.jpg

162. Aegeus consults the Pythia. Wikimedia commons.
http://commons.wikimedia.org/wiki/File:Themis_Aigeus_Antikensammlung _Berlin_F2538_n2.jpg

167. Shamanic poles beside Lake Baikal. Wikimedia commons.
http://commons.wikimedia.org/wiki/File:Shamanic_poles.jpg

169. Hologram diagram. WIkimedia commons.
https://commons.wikimedia.org/wiki/File:Hologram-record1.jpg

173. Ancient Eleusis. Wikimedia commons.
http://commons.wikimedia.org/wiki/File:General_view_of_sanctuary_of_De meter_and_Kore_and_the_Telesterion_(Initiation_Hall),_center_for_the_El eusinian_Mysteries,_Eleusis_(8191841684).jpg

175 - 176. Maps of Athens and Greece. Google maps.

179. The votive-plaque of Ninnion. Wikimedia commons.
https://commons.wikimedia.org/wiki/File:NAMA_Mystères_d%27Eleusis.jpg

191. From Abydos: Isis receiving Djed. Wikimedia commons.
http://commons.wikimedia.org/wiki/File:Abydos_Tempelrelief_Sethos_I._20 .JPG

193. The Pietà of Michelangelo. Wikimedia commons.
http://commons.wikimedia.org/wiki/File:Michelangelo%27s_Pieta_5450_crop ncleaned_edit.jpg

198. Raising the Divine Spark. Wikimedia commons.
http://commons.wikimedia.org/wiki/File:Практика_туммо.jpg

199. Altaic shaman, c. 1911 - 1914. Wikimedia commons.
https://commons.wikimedia.org/wiki/File:SB_-
_Altay_shaman_with_drum.jpg

208. Stele of Lanza of the Andean Chavin. Wikimedia commons.
https://commons.wikimedia.org/wiki/File:Chavin_lanzon_stela2_cyark.jpg

217. Great Blue Heron. Wikimedia commons.
https://commons.wikimedia.org/wiki/File:Great_blue_heron_-
_natures_pics.jpg

222. Remains of the Pyramid of Unas. Wikimedia commons.
https://commons.wikimedia.org/wiki/File:02_unas_pyramid.jpg

224. Sarcophagus chamber Unas pyramid. Wikimedia commons.
https://commons.wikimedia.org/wiki/File:Ounas-chambre2.jpg

231. Remington painting. Wikimedia commons.
https://commons.wikimedia.org/wiki/File:Frederic_Remington_-
_Indian_Fire_God_(The_Going_of_the_Medicine-Horse)_-
_Google_Art_Project.jpg

239. Man of Tibet face down, c. 1938. Wikimedia commons.
https://commons.wikimedia.org/wiki/File:Bundesarchiv_Bild_135-S-13-18-
38,_Tibetexpedition,_Lhasa,_Pilger.jpg

248. Poussin's Ecstasy of St. Paul. Wikimedia commons.
https://commons.wikimedia.org/wiki/File:Poussin,_Nicolas_-
_Ecstasy_of_Saint_Paul_-_1643.jpg

265. Stele of Qadesh. Wikimedia commons.
https://commons.wikimedia.org/wiki/File:Stele_of_Qadesh_8976.jpg

266. Rhea on a Lion, Pergamon Altar. Wikimedia commons.
https://upload.wikimedia.org/wikipedia/commons/thumb/3/38/Pergamonmus
eum_-_Antikensammlung_-_Pergamonaltar_37.JPG/1280px-
Pergamonmuseum_-_Antikensammlung_-_Pergamonaltar_37.JPG

267. Runes from *The Hobbit*. Illustration by the author, based on
illustrations by Tolkien.

270. Odin and Gunnlod. Wikimedia commons.
https://commons.wikimedia.org/wiki/File:Odin_and_Gunnlöð_by_George_Wright.jpg

280. Suttung chasing Odin, form of eagles. Wikimedia commons.
https://commons.wikimedia.org/wiki/File:Processed_SAM_mjodr.jpg

282. Freya and the necklace Brisingamen.
Illustration by the author, after the illustrations in D'Aulaire's *Norse Gods and Giants*.

291. Yggdrasil. Wikimedia commons.
https://commons.wikimedia.org/wiki/File:DesenhodabandaYggDrasil.jpg

297. Jephthah and Daughter. Wikimedia commons.
https://commons.wikimedia.org/wiki/File:Giovanni_Antonio_Pellegrini_001.jpg

305. Durga Puja display from Burdwan. Wikimedia commons.
https://commons.wikimedia.org/wiki/File:Durga,_Burdwan,_2011.JPG

307. Durga slays Mahishasura. Wikimedia commons.
Outline of Virgo added by the author.
https://commons.wikimedia.org/wiki/File:Durga_Slays_Mahisasura.jpg

310. Illustration from *The Arabian Nights*, 1802. Wikimedia commons.
https://commons.wikimedia.org/wiki/File:Houghton_Lowell_4211.30_-_Smirke,_Arabian_Nights.jpg

320. Illustration from *The Arabian Nights*, 1802. Wikimedia commons.
https://commons.wikimedia.org/wiki/File:John_Tenniel_-_The_sleeping_genie_and_the_lady.jpg

331. Bethlehem by Night. Wikimedia commons.
https://commons.wikimedia.org/wiki/File:Josef_Langl_Blick_auf_Bethlehem_bei_Nacht.jpg

335. Isis with child Horus. Wikimedia commons.
https://commons.wikimedia.org/wiki/File:Egyptian_-_Isis_with_Horus_the_Child_-_Walters_54416_-_Three_Quarter_Right.jpg

338. Anandabodhi, sacred fig tree. Wikimedia commons.
https://commons.wikimedia.org/wiki/File:Anandabodhi.jpg

339. Meditating under the bodhi tree. Wikimedia commons.
https://commons.wikimedia.org/wiki/File:WLA_brooklynmuseum_Buddha_
Meditating_Under_the_Bodhi_Tree.jpg

340. Stylized bodhi leaf. Wikimedia commons.
https://commons.wikimedia.org/wiki/File:National_Museum_Vietnamese_H
istory_1_(cropped).jpg

343. Jonah under a gourd. Wikimedia commons.
https://commons.wikimedia.org/wiki/File:Hologram-record1.jpg

352. Daikoku, Otufuku, and Hotei. Wikimedia commons.
https://commons.wikimedia.org/wiki/File:Gyosai_Daikoku.jpg

359. Sacred tomb of the Baganda. Wikimedia commons.
https://commons.wikimedia.org/wiki/File:Kampala_Kasubi_Tombs.jpg

362. Buddha with Heracles (Vajrapani). Wikimedia commons.
https://en.wikipedia.org/wiki/File:Buddha-Vajrapani-Herakles.JPG

367. Durga -- Ahihole, India. Wikimedia commons.
https://commons.wikimedia.org/wiki/File:IMG_1697_Durga_Temple.jpg

371. Balaam and the Angel. Wikimedia commons.
https://commons.wikimedia.org/wiki/File:Gustav_Jaeger_Bileam_Engel.jpg

378. Examples of *equus asinus*. Wikimedia commons.
top left:
https://commons.wikimedia.org/wiki/File:Bonaire%27s_Critically-
Endangered_Nubian_Wild_Ass.jpg
top right:
https://commons.wikimedia.org/wiki/File:A_ANE_DE_SOMALIE.JPG
bottom left:
https://commons.wikimedia.org/wiki/File:Cap_d%27un_ase_a_Onda,_Plana
_Baixa.JPG
bottom right:
https://commons.wikimedia.org/wiki/File:Female_Donkey.jpg

383. Abraham and Isaac on Moriah. Wikimedia commons.
https://commons.wikimedia.org/wiki/File:Schnorr_von_Carolsfeld_Bibel_in
_Bildern_1860_028.png

393. Map of Turkish Empire, 1680. Wikimedia commons.
https://commons.wikimedia.org/wiki/File:1680_map_Turcicum_imperium_b
y_Frederik_de_Wit_BPL_15917_detail.png

405. Salmacis and Hermaphroditus. Wikimedia commons.
https://commons.wikimedia.org/wiki/File:Hermaphroditos_and_Salmacis_by
_Bartholomäus_Spranger.jpg

414. Salmacis and Hermaphroditus. Wikimedia commons.
https://commons.wikimedia.org/wiki/File:Magdalena_van_de_Passe_-
_Salmacis_and_Hermaphroditus_(after_J._Pynas).jpg

415. Midas turns his daughter to gold. Wikimedia commons.
https://commons.wikimedia.org/wiki/File:Midas_gold2.jpg

424. Contest between Apollo and Pan. Wikimedia commons.
https://commons.wikimedia.org/wiki/File:Clerck,_Hendrick_de_-_Midas_-
_c._1620.jpg

425. Judgment of Midas. Wikimedia commons.
https://commons.wikimedia.org/wiki/File:Lévy-Jugement_de_Midas.JPG

426. Midas and Silenos, c. 510 BC. Wikimedia commons.
https://commons.wikimedia.org/wiki/File:Silenos_Midas_Met_49.11.1.jpg

428. Phyrgian cap from Aphaia pediment. Wikimedia commons.
https://commons.wikimedia.org/wiki/File:Aphaia_pediment_Paris_W-
XI_Glyptothek_Munich_81.jpg

431. Shango and stars. Wikimedia commons.
foreground:
https://commons.wikimedia.org/wiki/File:Representação_de_Xangô_MN_01
.jpg
background:
https://commons.wikimedia.org/wiki/File:Large_Magellanic_Cloud.jpg

438. Hartebeest. Wikimedia commons.
https://commons.wikimedia.org/wiki/File:Serengeti_Kongoni2.jpg

441. Hirola antelope. Wikimedia commons.
https://commons.wikimedia.org/wiki/File:PZSL1889Plate42.png

443. Egyptian Book of the Dead. Wikimedia commons.
https://commons.wikimedia.org/wiki/File:The_Book_of_the_Dead.jpeg

450. David slaying Goliath. Wikimedia commons.
https://commons.wikimedia.org/wiki/File:David_Slaying_Goliath_by_Peter_
Paul_Rubens.jpg

462. Echo and Narcissus. Wikimedia commons.
https://commons.wikimedia.org/wiki/File:Echo_and_Narcissus_by_John_Wi
lliam_Waterhouse.jpg

466. Shankaracharya. Wikimedia commons.
https://commons.wikimedia.org/wiki/File:Shankaracharya_kutajadri.jpeg

475. Adi Shankar Shrine. Wikimedia commons.
https://commons.wikimedia.org/wiki/File:Adi_Shankara_Kodachadri_Kollur
.JPG

485. Hildebrand well-spring. Wikimedia commons.
https://commons.wikimedia.org/wiki/File:Hildebrand-
Quelle_Gottesbichl_02.jpg

492. Zodiac wheel with superimposed images of Isis, Nephthys,
Osiris supine and Horus raised-up. Background image is same as
that for page 45. Superimposed images:
top:
https://commons.wikimedia.org/wiki/File:Horus_as_depicted_in_Illustrated
_List_of_the_principal_Egyptian_Divinities_(1888)_-_TIMEA.jpg
bottom:
https://commons.wikimedia.org/wiki/File:Osiris_tombe_arbre.gif

498. Khepren with Horus. Wikimedia commons.
https://commons.wikimedia.org/wiki/File:Khephren%2BHorus.jpg

499. Incredulity of Thomas. Wikimedia commons.
https://commons.wikimedia.org/wiki/File:De_ongelovige_thomas.jpg

500. Incredulity of St. Thomas. Wikimedia commons.
https://commons.wikimedia.org/wiki/File:L%27incredulita_di_San_Tommaso.jpg

519. Incredulity of St. Thomas. Wikimedia commons.
https://commons.wikimedia.org/wiki/File:Brooklyn_Museum_-
_The_Disbelief_of_Saint_Thomas_(Incredulité_de_Saint_Thomas)_-
_James_Tissot.jpg

523. Eros and Psyche. Wikimedia commons.
https://commons.wikimedia.org/wiki/File:Annales_du_musée_et_de_l%27éco
le_moderne_des_beaux-arts_-
_recueil_de_gravures_au_trait,_d%27après_les_principaux_ouvrages_de_pein
ture,_sculpture,_ou_projets_d%27architecture,_qui,_chaque_année,_ont_rem
porté_(14762994904).jpg

525. Bhagavad Gita scene. Wikimedia commons. Labels added by the author and based upon the Katha Upanishad.
https://commons.wikimedia.org/wiki/File:Bhagavatgeeta.jpg

535. Hopi man, c. 1900. Wikimedia commons.
https://commons.wikimedia.org/wiki/File:Curtis_-_Snake_priest.JPG

536. Chakra diagram, 1899. Wikimedia commons.
https://commons.wikimedia.org/wiki/File:Sapta_Chakra,_1899.jpg

538. Sleeping Beauty. Wikimedia commons.
https://commons.wikimedia.org/wiki/File:Brewtnall_-_Sleeping_Beauty.jpg

544. Man in Trichy Srirangam temple. Wikimedia commons.
https://commons.wikimedia.org/wiki/File:Begging_in_Trichy.jpg

555. Qigong diagram, AD 1513. Wikimedia commons.
https://commons.wikimedia.org/wiki/File:Qigong_exercise_to_treat_distensi on_and_suffocation_Wellcome_L0038892.jpg

562. Dorje Chang, Tantric Art. Wikimedia commons.
https://commons.wikimedia.org/wiki/File:Vajradhara_L2012.81_01.jpg

580. Vision of Dhruva. Wikimedia commons.
https://commons.wikimedia.org/wiki/File:Raja_Ravi_Varma,_Dhruv_Narayan.jpg

585. Thor, the Thundergod. Wikimedia commons.
https://commons.wikimedia.org/wiki/File:Johannes_Gehrts_-_Thor,_der_Donnergott.jpg

588. Wudang Mountain. Wikimedia commons. Image cropped and text of the "Thirteen Postures Song" superimposed.
https://commons.wikimedia.org/wiki/File:Wudangshan_2003_10.jpg

598. Ellora Cave. Wikimedia commons.
https://commons.wikimedia.org/wiki/File:Ellora_Cave_12_sio243.jpg

603. Boon of Indra to Kunti. Wikimedia commons.
https://commons.wikimedia.org/wiki/File:Boon_of_Indra_to_Kunti.jpg

610. Wudang Mountains. Wikimedia commons.
https://commons.wikimedia.org/wiki/File:Wudangshan_pic_7.jpg

617. Papyrus of Ani. Wikimedia commons.
https://commons.wikimedia.org/wiki/File:Ani_LDM_18.jpg

628. Region of Nag Hammadi. Google Maps, 12/31/2012.

634. Angkor Wat with monks in foreground. Wikimedia commons.
https://commons.wikimedia.org/wiki/File:Buddhist_monks_in_front_of_the_Angkor_Wat.jpg

636. The Candelabra, Paracas. Wikimedia commons.
https://commons.wikimedia.org/wiki/File:Candelabro_de_Paracas.jpg

637. John Wesley Powell. Wikimedia commons.
https://commons.wikimedia.org/wiki/File:John_Wesley_Powell_USGS.jpg

648. Pyramids at Giza. Wikimedia commons.
https://commons.wikimedia.org/wiki/File:Pyramids_(5078723552).jpg

653. Tauroctony from Aquileia. Wikimedia commons.
https://commons.wikimedia.org/wiki/File:Kunsthistorisches_Museum_Mithras-Relief.jpg

657. Tauroctony from Nova Apulensis. Wikimedia commons.
https://commons.wikimedia.org/wiki/File:Mytras_Ceremony_SIBIU_Hystory_Museum.JPG

661. Knight effigy tombs, London. Wikimedia commons.
https://commons.wikimedia.org/wiki/File:TempleChurch-Effigies.jpg

662. Illustrations of effigy tombs. Wikimedia commons.
top (from 1891):
https://commons.wikimedia.org/wiki/File:Herbert_Railton_-_A_Knight_Templar-William_Mareschal,_Earl_of_Pembroke.jpg
bottom (from 1875):
https://commons.wikimedia.org/wiki/File:ONL_(1887)_1.156_-_Tombs_of_Knights_Templars.jpg

664. Mithraeum, Rome. Wikimedia commons.
https://commons.wikimedia.org/wiki/File:Mithraeum_San_Clemente_Rom.JPG

671. Bust of Marcus Aurelius. Wikimedia commons.
https://commons.wikimedia.org/wiki/File:L%27Image_et_le_Pouvoir_-_Buste_cuirassé_de_Marc_Aurèle_agé_-_3.jpg

675. Roman coin, Commodus. Wikimedia commons.
https://commons.wikimedia.org/wiki/File:Commodus_AR_Denarius_190_75
2187.jpg

678 - 679. Roman emperor timeline.
individual images: Wikimedia commons, Wikipedia.
http://en.wikipedia.org/wiki/List_of_Roman_emperors

683 through 690. Diagrams of the Seal Works, Ohio. From
Squier and Davis: *Ancient Monuments of the Mississippi Valley*,
1848. https://www.wdl.org/en/item/4301/

688. Ichthys symbol. Wikimedia commons.
https://commons.wikimedia.org/wiki/File:Ichthus2.svg

693. Ambrose barring Theodosius. Wikimedia commons.
https://commons.wikimedia.org/wiki/File:Anthonis_van_Dyck_005.jpg

708. Skull and Scarab. Wikimedia commons.
(Same images as page 47).

709. Qumran ruins. Wikimedia commons.
https://commons.wikimedia.org/wiki/File:Kumran-1-84.jpg

721. Landscape with Buddhist Temples. Wikimedia commons.
https://commons.wikimedia.org/wiki/File:%27Landscape_with_Buddhist_Te
mples%27_anonymous,_early_16th_century,_Honolulu_Museum_of_Art,_385
2.1.jpg

729. Giza photograph, 1859. Wikimedia commons.
https://commons.wikimedia.org/wiki/File:Frith-Sphinx.png

737 - 738. Carrizo Plain. Google Maps. 10/13/2016.

739. Painted Rock, aerial view. Wikimedia commons.
https://commons.wikimedia.org/wiki/File:Aerial-OverPaintedRock.jpg

741. Liquefaction, Christchurch. Wikimedia commons.
https://commons.wikimedia.org/wiki/File:Christchurch_quake,_2011-02-22.jpg

742. Painted Rock from northwest looking southeast.
Photograph by the author, 08 October 2016.

744. Painted Rock, two views
Top: Northwest portion of Painted Rock.
Bottom: North side of Painted Rock, just west of opening.
Both photographs by the author, 08 October 2016.

745. Carrizo Plain, looking east.
Photograph by the author, 08 October 2016.

750. *The Blue Tepee*, 1906. Wikimedia commons.
https://commons.wikimedia.org/wiki/File:Joseph_Henry_Sharp_-
_The_Blue_Tepee_(c.1906).jpg

757. Disembarkation of Columbus. Wikimedia commons.
https://commons.wikimedia.org/wiki/File:Desembarco_de_Colón_de_Diósc
oro_Puebla.jpg

765. Flooding on road. Wikimedia commons.
https://commons.wikimedia.org/wiki/File:Flooding_on_the_road_to_Delicat
e_Arch_Overlook_(8083724580).jpg

769. Stillaguamish River in flood. Wikimedia commons.
https://commons.wikimedia.org/wiki/File:Flooding_on_the_road_to_Delicat
e_Arch_Overlook_(8083724580).jpg

770. Holy man in meditation. Wikimedia commons.
https://commons.wikimedia.org/wiki/File:A_Holy_Man_in_Meditation.JPG

776. Holy man in meditation, with stars from Large Magellanic
Cloud superimposed. Wikimedia commons.
https://commons.wikimedia.org/wiki/File:Two_very_different_glowing_gas_
clouds_in_the_Large_Magellanic_Cloud.tiff

777. Apollo pours a libation, with Raven. Wikimedia commons.
https://commons.wikimedia.org/wiki/File:Apollo_black_bird_AM_Delphi_81
40.jpg

782. Field of Barley with Demeter superimposed. Wikimedia
commons.
Background:
https://upload.wikimedia.org/wikipedia/commons/thumb/d/db/Barley_field-
2007-02-22%28large%29.jpg/1280px-Barley_field-2007-02-22%28large%29.jpg
Goddess:
https://commons.wikimedia.org/wiki/File:Demeter_Altemps_Inv8596.jpg

800. Uranium mine, Namibia. Wikimedia commons.
https://commons.wikimedia.org/wiki/File:Rossing_Uranium_Mine_(0181046
7)_(12221017074).jpg

807. Mountain of trash, Nicaragua. Wikimedia commons.
https://commons.wikimedia.org/wiki/File:Montaña_de_basura_junto_al_lag
o_Managua.jpg

808. Athens, Greece. Wikimedia commons.
https://commons.wikimedia.org/wiki/File:Athens_(Greece).jpg

819. Athena, Goddess of Wisdom. Wikimedia commons.
https://commons.wikimedia.org/wiki/File:Athena-athena-polias.jpg
background: Large Magellanic Cloud. Wikimedia commons.
https://commons.wikimedia.org/wiki/File:LHA_120-
N11_in_the_Large_Magellanic_Cloud.jpg

Bibliography

Ambrose, Stephen F. *Crazy Horse and Custer: The Parallel Lives of Two American Warriors*. New York: Doubleday, 1975. First Anchor Books Edition, 1996.

Ancient Egyptian Pyramid Texts. James P. Allen, trans. Peter der Manuelian, ed. Atlanta: Society of Biblical Literature, 2005.

Apuleius. *Golden Ass*. Jack Lindsay, trans. Bloomington: Indiana UP, 1960.

Arabian Nights: Tales from a Thousand and One Nights. Richard Burton, trans. First Modern Library Edition. NY: Random House, 2009.

Bascom, William. *Sixteen Cowries: Yoruba Divination from Africa to the New World*. 2nd corrected ed. Bloomington: Indiana UP, 1980.

Barbiero, Flavio. *The Secret Society of Moses: The Mosaic Bloodline and a Conspiracy Spanning Three Millennia*. Rochester: Inner Traditions, 2010.

Black Elk and John G. Neihardt. *Black Elk Speaks: Being the Life Story of a Holy Man of the Oglala Sioux*. Omaha: University of Nebraska Press, 1988.

Book of the Dead. Samuel Birch, trans. and titled by him *The Funereal Ritual*. 1867. Accessed online at http://www.masseiana.org/ritual.htm.

D'Aulaire, Ingri and Edgar Parin. *Ingri and Edgar Parin D'Aulaire's Book of Greek Myths*. Garden City: Doubleday, 1962.

D'Aulaire, Ingri and Edgar Parin. *Norse Gods and Giants*. Garden City: Doubleday, 1967.

De Santillana, Giorgio, and Hertha von Dechend. *Hamlet's Mill: An Essay on Myth and the Frame of Time.* Boston: Godine, 1977.

Dewhurst, Richard. *Ancient Giants Who Ruled North America: the Missing Skeletons and the Great Smithsonian Cover-Up.* Rochester: Bear & Co, 2014.

Diodorus. *The Library of History.* Online version of the Loeb Classsical Library available at http://penelope.uchicago.edu/Thayer/e/roman/texts/diodorus_si culus/home.html.

Drake, Michael. *Shamanic Drum: A Guide to Sacred Drumming.* Salem: Talking Drum, 2002.

Egyptian Book of the Dead: The Book of Going Forth by Day. Raymond O. Faulkner, trans. Second Revised Edition. San Francisco: Chronicle Books, 1998.

Eliade, Mircea. *Shamanism: Archaic Techniques of Ecstasy.* Originally published in French under the title *Le Chamanisme et les techniques archaïques de l'extase* by Librarie Payot: Paris, 1951. English translation by Willard R. Trask. Princeton, NJ: Princeton UP, 1964. First Princeton / Bollingen paperback edition, 1972.

Enter the Dragon. Dir. Robert Clause. Golden Harvest, 1973. Film.

Epstein, Ronald. "The Shurangama-Sutra (T. 945): A Reappraisal of its Authenticity." Presented at the annual meeting of the American Oriental Society, March 16 - 18, 1976, in Philadelphia, Pennsylvania.
http://online.sfsu.edu/rone/Buddhism/authenticity.htm

Eusebius. *Ecclesiastical History.* Christian Frederick Cruse, trans. 1850.
https://archive.org/stream/ecclesiasticalhiooeuse/ecclesiasticalhi ooeuse_djvu.txt

Farrell, Joseph P. and Scott D. de Hart. *Grid of the Gods: The Aftermath of the Cosmic War and the Physics of the Pyramid Peoples.* Kempton: Adventures Unlimited, 2011.

Freeman, Gordon. *Hidden Stonehenge: Ancient Temple in North America Reveals the Key to Ancient Wonders.* London: Watkins, 2012

Freke, Timothy and Peter Gandy. *Jesus Mysteries: Was the "Original Jesus" a Pagan God?* NY: Three Rivers Press, 1999.

Gibbon, Edward. *Decline and Fall of the Roman Empire.* Hans-Friedrich Mueller, ed. Modern Library Paperback Edition. NY: Random House, 2003.

Gladiator. Dir. Ridley Scott. Universal, 2000. Film

Hamilton, Ross. *Mystery of the Serpent Mound: in Search of the Alphabet of the Gods.* Berkeley: Frog Books, 2001.

Hamilton, Ross. *Star Mounds: Legacy of a Native American Mystery.* Berkeley: North Atlantic Books, 2012.

Hancock, Graham and Santha Faiia. *Heaven's Mirror: Quest for the Lost Civilization.* New York: Three Rivers Press, 1998.

Homer. Odyssey. Theodore Alois Buckley, trans. Philadelphia: David McKay, 1896.

Hudson, Michael. *Killing the Host: How Financial Parasites and Debt Bondage Destroy the Global Economy.* Islet-Verlag, 2015.

Hudson, Michael. *J is for Junk Economics: A Guide to Reality in an Age of Deception.* Islet-Verlag, 2017.

Ingerman, Sandra. *Shamanic Journeying: A Beginner's Guide.* Boulder: Sounds True, 2008.

Ingerman, Sandra and Hank Wesselman. *Awakening to the spirit world: the shamanic path of direct revelation.* Boulder: Sounds True, 2010.

Iyengar, B. K. S. *Light on Yoga: Yoga Dipika.* Revised Edition. NY: Schocken Books, 1977.

Journey Through the Afterlife: Ancient Egyptian Book of the Dead. John H. Taylor, ed. Cambridge: Harvard UP, 2013.

Karate Kid. Dir. John G. Avildsen. Columbia, 1984.

Kingsley, Peter. *In the Dark Places of Wisdom.* Point Reyes: Golden Sufi Center, 1999.

Kuhn, Alvin Boyd. *Easter: The Birthday of the Gods.* 2nd edition. Theosophical Press: Wheaton, 1966.

Kuhn, Alvin Boyd. *Lost Light: An Interpretation of Ancient Scriptures.* Elizabeth, NJ: Academy Press, 1940.

Kuhn, Alvin Boyd. *Spiritual Symbolism of the Sun and the Moon.* nd.
https://archive.org/stream/SpiritualSymbolismOfTheSunAndMoon/AlvinBoydKuhn-SpiritualSymbolismOfTheSunAndMoon#page/n0/mode/2up

Kuhn, Alvin Boyd. *The Stable and the Manger.* 1936.
https://archive.org/stream/TheStableAndTheManager/AlvinBoydKuhn-TheStableAndTheManager#page/n0/mode/2up

Kuhn, Alvin Boyd. *Who is this King of Glory? A Critical Study of the Christos-Messiah Tradition.* Elizabeth, NJ: Academy Press, 1944.

Kvilhaug, Maria. *Maiden with the Mead: A Goddess of Initiation Rituals in Old Norse Mythology?* University of Oslo, 2004.

Lao Tzu. *Tao Te Ching: The Classic Book of Integrity and the Way.* Victor H. Mair, trans. NY: Bantam, 1990.

Liu Wu-chi. *An Introduction to Chinese Literature.* Bloomington: Indiana UP, 1966.

Mahabharata of Krishna-Dwaipayana Vyasa. Kisari Mohan Ganguli, trans. Calcutta: Pratap Chandra Roy, 1896. http://www.sacred-texts.com/hin/maha/index.htm.

Massey, Gerald. *Ancient Egypt, the Light of the World.* http://www.masseiana.org [site apparently no longer avaialable].

Massey, Gerald. *Gnostic and Historic Christianity.* http://www.masseiana.org

Massey, Gerald. "Luniolatry, Ancient and Modern." http://www.masseiana.org

Massey, Gerald. "Mankind in search of his soul during fifty thousand years, and how he found it!" http://www.masseiana.org

Massey, Gerald. "Paul the Gnostic Opponent of Peter, not an Apostle of Historic Christianity." http://www.masseiana.org

Mathisen, David Warner. *Mathisen Corollary: Connecting a Global Flood with the Mystery of Mankind's Ancient Past.* Paso Robles: Beowulf Books, 2011.

Mathisen, David Warner. *Star Myths of the World, and how to interpret them: Volume One.* Paso Robles: Beowulf Books, 2015.

Mathisen, David Warner. *Star Myths of the World, and how to interpret them: Volume Two.* Paso Robles: Beowulf Books, 2016.

Mathisen, David Warner. *Star Myths of the World, and how to interpret them: Volume Three (Star Myths of the Bible).* Paso Robles: Beowulf Books, 2016.

Mathisen, David Warner. *Undying Stars: The truth that unites the world's ancient wisdom and the conspiracy to keep it from you.* Paso Robles: Beowulf Books, 2014.

Meyer, Marvin. *Gnostic Discoveries: The Impact of the Nag Hammadi Library*. San Francisco: Harper, 2005.

Michell, John. New View Over Altlantis. First US edition. San Francisco: Harper & Row, 1983.

Nag Hammadi Scriptures: The Revised and Updated Translation of Sacred Gnostic Texts Complete in One Volume. Marvin W. Meyer, ed. Elaine H. Pagels, intro. New York: HarperCollins, 2007. First Paperback Edition, 2008.

Ollestad, Norman. *Crazy for the Storm: A Memoir of Survival*. First ed. New York: Ecco, 2009.

Ovid. *Metamorphoses*. Charles Martin, trans. NY: Norton, 2004.

Ovid. *Metamorphoses*. Roscoe Mongan, trans.

Quispel, Gilles. "Judaism, Judaic Christianity, and Gnosis." in Plotinus: The Enneads. Stephen MacKenna, trans. Burdett, NY: Larson, 1992.

Palmer, David A. *Qigong Fever: Body, Science, and Utopia in China*. NY: Columbia UP, 2007.

Parrinder, Geoffrey. *African Mythology*. London: Hamlyn, 1967.

Philo. *De Vita Contemplativa*. Frank William Tilden, trans. Indiana University Studies, number 52, volume IX. Bloomington: Indiana UP, 1922.

Plato. *Phaedrus*. R. Hackforth, trans. In *The Collected Dialogues of Plato, Including the Letters*. Edith Hamilton and Huntington Cairns, eds. Princeton: Princeton UP, 1961. 475 - 525.

Plotinus. Enneads. Stephen MacKenna and B. S. Page, trans. http://classics.met.edu/Plotinus/enneads.html

Plutarch. *Of Isis and Osiris*. In *Moralia* volume V. Frank Cole Babbitt, trans. Loeb Classical Library.

http://penelope.uchicago.edu/Thayer/e/roman/texts/plutarch/mo
ralia/isis)and)osiris*a.html

Plutarch. *On the 'E' at Delphi*. A. O. Prickard, trans. Oxford:
Clarendon Press, 1918.
http://penelope.uchicago.edu/misctracts/plutarche.html

Plutarch. *On the eating of flesh*. In *Moralia* volume XII. Harold
Cherniss and W. C. Helmbold, trans. Loeb Classical Library.
http://penelope.uchicago.edu/Thayer/E/Roman/Texts/Plutarch/
Moralia/De_esu_carnium*/1.html

Rey, H. A. *The Stars: A New Way to See Them*. Boston:
Houghton Mifflin, 1952. Enlarged World-Wide Edition, 1988.

Schoch, Robert M. *Voices of the Rocks: A Scientist Looks at
Catastrophes and Ancient Civilizations*. NY: Harmony, 1999.

Schoch, Robert M. *Forgotten Civilization: The Role of Solar
Outbursts in our Past and Future*. Rochester: Inner Traditions,
2012.

Schwaller de Lubicz, R. A. *Esoterism & Symbol*. Originally
published in French under the title *Propos sur Esotérisme et
Symbole* by La Colombe, Editions du Vieux Colombier, in 1960.
English translation by André and Goldian VandenBroeck, 1985.
Rochester: Inner Traditions, 1985.

Schwaller de Lubicz, R. A. *Sacred Science: the King of
Pharaonic Theocracy*. Originally published in French under the
title *Le Roi de la théocractie Pharaonique* by Flammarion, in
1961. English translation by André and Goldian VandenBroeck,
1985. Rochester: Inner Traditions, 1982.

Sellers, Jane B. *Death of Gods in Ancient Egypt: A Study of the
Threshold of Myth and the Frame of Time*. Lexington,
Kentucky: Lulu Books, 1992.

Taylor, Robert. *The Astronomico-theological Lectures of the
Rev. Robert Taylor, B. A*. NY: Calvin Blanchard, 1857.

840

Taylor, Robert. *Devil's Pulpit: or Astro-Theological Sermons by the Rev. Robert Taylor, B. A.* NY: Calvin Blanchard, 1857.

Tolkien, J. R. R. *The Hobbit: or, There and Back Again.* NY: Ballantine, 1967.

Ulansey, David. *Origins of the Mithraic Mysteries: Cosmology and Salvation in the Ancient World.* NY: Oxford UP, 1989.

Van Gulik. *Sexual Life in Ancient China: A Preliminary Survey of Chinese Sex and Society circa 1500 BC till 1644 AD.* Sinica Leidensia, volume 57. London: Brill, 2002.

Wallis Budge, E. A. *Book of the Dead: The Papyrus of Ani in the British Museum.* E. A. Wallis Budge, trans. London: Harrison and Sons, 1895.

Waters, Frank and Oswald White Bear Fredericks. *Book of the Hopi.* NY: Viking Penguin, 1963. Penguin Books edition, 1977.

Weil, Simone. "The Iliad, or the Poem of Force." First published in France in Cahiers du Sud, December 1940 and January 1941. Chicago Review edition published in 1965, Mary McCarthy, trans.
http://biblio3.url.edu.gt/SinParedes/08/Weil-Poem-LM.pdf

West, John Anthony. *Serpent in the Sky: The High Wisdom of Ancient Egypt.* NY: Julian Press, 1979. 1987 edition.

Wyatt, Lucy. *Approaching Chaos: Can an Ancient Archetype Save 21st Century Civilization.* Winchester: O Books, 2010.

List of all essays included in this book

with the page number on which they begin

845

Index